D0843361

FUZZY SYSTEMS
Modeling and Control

THE KLUWER HANDBOOK SERIES ON FUZZY SETS

Series Editors: Didier Dubois and Henri Prade

FUZZY SETS IN DECISION ANALYSIS, OPERATIONS RESEARCH AND STATISTICS, edited by Roman Słowiński
ISBN: 0-7923-8112-2

FUZZY SYSTEMS
Modeling and Control

edited by

Hung T. Nguyen
New Mexico State University

and

Michio Sugeno
Tokyo Institute of Technology

KLUWER ACADEMIC PUBLISHERS
Boston/London/Dordrecht

Distributors for North, Central and South America:
Kluwer Academic Publishers
101 Philip Drive
Assinippi Park
Norwell, Massachusetts 02061 USA

Distributors for all other countries:
Kluwer Academic Publishers
Distribution Centre
Post Office Box 322
3300 AH Dordrecht, THE NETHERLANDS

Library of Congress Cataloging-in-Publication Data

A C.I.P. Catalogue record for this book is available
from the Library of Congress.

Printed on acid-free paper.

Printed in the United States of America

Contents

Series Foreword

Fuzzy sets were introduced in 1965 by Lotfi Zadeh with a view to reconcile mathematical modeling and human knowledge in the engineering sciences. Since then, a considerable body of literature has blossomed around the concept of fuzzy sets in an incredibly wide range of areas, from mathematics and logics to traditional and advanced engineering methodologies (from civil engineering to computational intelligence). Applications are found in many contexts, from medicine to finance, from human factors to consumer products, from vehicle control to computational linguistics, and so on ... Fuzzy logic is now currently used in the industrial practice of advanced information technology.

As a consequence of this trend, the number of conferences and publications on fuzzy logic has grown exponentially, and it becomes very difficult for students, newcomers, and even scientists already familiar with some aspects of fuzzy sets, to find their way in the maze of fuzzy papers. Notwithstanding circumstancial edited volumes, numerous fuzzy books have appeared, but, if we except very few comprehensive balanced textbooks, they are either very specialized monographs, or remain at a rather superficial level. Some are even misleading, conveying more ideology and unsustained claims than actual scientific contents.

What is missing is an organized set of detailed guidebooks to the relevant literature, that help the students and the newcoming scientist, having some preliminary knowledge of fuzzy sets, get deeper in the field without wasting time, by being guided right away in the heart of the literature relevant for her or his purpose. The ambition of the HANDBOOKS OF FUZZY SETS is to address this need. It will offer, in the compass of several volumes, a full picture of the current state of the art, in terms of the basic concepts, the mathematical developments, and the engineering methodologies that exploit the concept of fuzzy sets.

This collection will propose a series of volumes that aim at becoming a useful source of reference for all those, from graduate students to senior researchers, from pure mathematicians to industrial information engineers as well as life, human and social sciences scholars, interested in or working with fuzzy sets. The original feature of these volumes is that each chapter is written by one or several experts in the concerned topic. It provides introduction to the topic, outlines its development, presents the major results, and supplies an extensive bibliography for further reading.

The core set of volumes are respectively devoted to fundamentals of fuzzy set, mathematics of fuzzy sets, approximate reasoning and information systems, fuzzy models for pattern recognition and image processing, fuzzy sets in decision research and statistics, fuzzy systems modeling and control, and a guide to practical applications of fuzzy technologies.

D. Dubois, H. Prade
Toulouse

Contributing Authors

Jean-Pierre Aubin is Professor at Université de Paris-Dauphine, France. He is interested in complex systems arising in social and life sciences evolving under stochastic and/or contingent uncertainty and subjected to viability constraints. He is the author of several books : *Approximation of Elliptic Boundary-Value Problems,* Wiley-Inerscience *(1972)* , *Applied Abstract Analysis,* Wiley-Interscience *(1977)* , *Applied Functional Analysis,* Wiley-Interscience *(1979), Mathematical Methods of Games and Economic Theory,* North-IIolland *(1979), Explicit Mcthods of Optimization,* Dunod *(1985), Op tima and Equilibria,* Springer-Verlag *(1993), Differential Inclusions (*with A. Cellina *),* Springer-Verlag *(1984), Applied Nonlinear Analysis (* with I. Ekeland *),* Wiley-Interscience *(1984), Exercices d'Analyse Appliquée,* Masson *(1993), Set-Valued Analysis (* with H. Franskowska *),* Birhauser *(1990), Viability Theory,* Birkhauser *(1991), Initiation a l'Analyse Appliquée,* Masson *(1993), Neural Networks and Qualitative Physics : A Viability Approach,* Cambridge University Press *(1996), Dynamic Economic Theory : A Viability Approach* , Springer-Verlag *(1997), Morphological and Mutational Analysis : Tools for Shape Regulation and Optimization,* Birkhauser *(1997).*

Hamid R. Berenji, Ph.D., is a senior research scientist with the Intelligent Inference Systems Corporation at NASA Ames Research Center in Moffett Field, California. He is the principal investigator of the research project on intelligent control, and was a program chairman of the IEEE International Conference on Neural Networks (ICNN'93) in San Francisco. He serves on the editorial board of several technical publications including *IEEE Transactions on Fuzzy Systems* and as an editor for *IEEE Transactions on Neural Networks.* He was a program co-chair for the 1994 IEEE Conference on Fuzzy Systems, Orlando,

Florida. He is a senior member of IEEE and chairs the Neural Networks Council's Technical Committee on Fuzzy Systems.

Phil Diamond is Reader in the Department of Mathematics ,University of Queensland, Autralia. He received his Ph.D.from the University of NSW in 1968. His current interests include dynamical systems, discretization phenomena, H_∞ - control and Fuzzy Set Theory.

Olivier Dordan is Maitre de Conference at the University of Bordeaux II , Laboratory of Applied Mathematics of Bordeaux (ERS 123). His scientific work concerns mainly with the theory of Qualitative Analysis and Population Dynamics. He is the author of the book *Analyse Qualitative* (Masson, 1995).

Dimitar Filev received his Ph.D. in Electrical Engineering from the Czech Technical University in 1979. He is a senior technical specialist with Ford Motor Company specializing in industrial intelligent systems and technologies for control, diagnostics and decision-making. He is conducting research in control theory and applications, modeling of complex systems, fuzzy modeling and control. He was a recipient of the 95' Award for Excellence of MCB University Press and the 96' Henri Ford Technnlogy Award of Ford Motor Company. He is a co-author of three books, including *Essentials of Fuzzy Modeling and Control* (with R. Yager), J.Wiley & Sons (1994).

Laurent Foulloy graduated from the Ecole Normale Superieure de Cachan, France in 1980. He received the Ph.D. and D.Sc. degrees in 1982 and 1990, respectively, both from the University of Paris XI, France. He is currently professor of Electrical Engineering and Head of the Laboratoire d'Automatique et de MicroInformatique Industrielle at the University of Savoie (Ecole Superieure d'Ingenieurs d'Annecy), France. His research interests include intelligent and fuzzy techniques for process control and measurements.

Sylvie Galichet graduated in 1986 and received a Ph.D. degree in 1989, both from the University of Technology of Compiegne, France. She is currently Assistant Professor of Electrical Engineering at the University of Savoie (Ecole Superieure d'Ingenieurs d'Annecy) and works on process control in the Laboratoire d'Automatique et de MicroInformatique Industrielle. Her research interests inlcude fuzzy control and modeling.

Andreas Geyer-Schulz was born on May 19, 1960 in Wiener Neustadt, Austria. He received his Ph.D. from the Vienna University of Economics and Business Administration in 1985. He is currently a Professor in the Department of Applied Computer Science of the above University. He is head of the Department 's research group for Decision Support Systems and of the WU-Linda Group for Distributed Processing. He is the author of the book *Fuzzy Rule-Based Expert Systems and Genetic Machine Learning*, Physica (1995).

Abraham Kandel, a Professor and the Endowed Eminent Scholar in Computer Science and Engineering, is the Chairman of the Computer Science and Engineering at the University of South Florida. He is the Editor of the Fuzzy Track-IEEE MICRO, an Associate Editor of *IEEE Transactions on Systems, Man, and Cybernetics*, and a member of the Editorial Board of the the following Journals : *Fuzzy Sets and Systems, Information Sciences, Expert Systems, Engineering Applications of Artificial Intelligence, The Journal of Grey Systems, Control Engineering Practice, Fuzzy Systems-Reports and Letters, IEEE Transactions on Fuzzy Systems, Book Series on Stidies in Fuzzy Decision and Control, Applied Computing Review Journal, Neural Network World, Fuzzy Mathematics,BUSEFAL.* He is author/co-author/co-editor of *Fuzzy Switching and Automata : Theory and Applications (1979), Fuzzy Techniques in Pattern Recognition (1982), Discrete Mathematics for Computer Scientists (1983), Fuzzy Relational Databases: A key to Expert Systems (1984), Approximate Reasoning in Expert Systems (1985), Fuzzy Mathematical Techniques with Applications (1986), Designing Fuzzy Expert Systems (1986), Digital Logic Design (1988), Engineering Risk and Hazard Assessment (1988), Elements of Computer Organization (1989), Real-Time Expert Systems Computer Architecture (1991), Fuzzy Expert Systems (1992), Hybrid Architectures for Intelligent Systems (1992), Verification and Validation of Rule-Based Expert Systems (1993), Fundamentals of Computer Numerical Analysis (1994), Fuzzy Control Systems (1994), Fuzzy Expert Systems Tools (1996).* Dr Kandel is a Fellow of IEEE, New York Academy of Sciences, AAAS, as well as a member of the ACM, NAFIPS, IFSA, ASEE and Sigma-Xi.

Vladik Kreinovich is Professor of Computer Science. He received his B.S. in Mathematics and Computer Science from Leningrad University, Russia, in 1974, and Ph.D. from the Institute of Mathematics, Soviet Academy of Sciences, Novosibirsk, in 1979. His research interests include representation and processing of uncertainty, especially in Intelligent Control and Interval Computations. He is a member of the Editorial Board of the *International Journal of Reliable Computing*, an co-editor of the monograph *Applications of Interval*

Computations and a co-author of *Applications of Continuous Mathematics to Computer Science.* He was an invited Professor, in Summer 1996, at LAFORIA, University of Paris VI , France.

George Mousouris was born in Cyprus in 1966. He received the B.S. (honors) and M.S. degrees from Brown University, both in 1990, and the Ph.D. degree from the University of Southern California, Los Angeles, in 1996. From 1990 to 1992, he was with the Digital Signal Processing Group of Texas Instruments, Houston, where he dealt with the design and implementation of signal and image processing algorithms on specialized processors. He is currently with the SIPI,USC. His research interests lie in the areas of nonlinear dynamic modeling, fuzzy systems, radial basis functions, neural networks, and financial modeling. Dr. Mousouris is a member of Tau Beta and Sigma Xi.

Hung Nguyen is Professor of Mathematical Sciences at New Mexico State University. His current research interests in Applied Mathematics include Foundations of Random Sets for Data Fusion, Statistics of Stochastic Processes, and Decision Theory and Intelligent Control (Fuzzy Logic). He is an Associate Editor of *IEEE Transactions on Fuzzy Systems, Soft Computing Journal, Approximate Reasoning, Fuzzy Sets and Systems, Journal of Uncertainty, Fuzziness and Knowledge-Based Systems, Fuzzy and Intelligent Systems.* He is a co-author/co-editor of *Uncertainty Models for Knowledge-Based Systems*, North Holland, *(1985), Fuzzy Sets : Theory and Applications-Selected Papers by L.A.Zadeh* , J.Wiley, *(1987), Fundamentals of Mathematical Statistics, volume I: Probability for Statistics*, Springer-Verlag, *(1989), Fundamentals of Mathematical Statistics, volume II: Statistical Inference,* Springer-Verlag, *(1989), Conditioning Logic in Expert Systems*, North Holland, *(1991), Conditioning Inference and Logic for Intelligent Systems : A Theory of Measure-Free Conditioning*, North Holland, *(1991), Fundamentals of Uncertainty Calculi with Applications to Fuzzy Inference*, Kluwer Academic, *(1994), Theoretical Aspects of Fuzzy Control*, J.Wiley, *(1995), Les Incertitudes dans les Systemes Intelligents (Que sais-Je?)*, Presses Universitaires de France, *(1996), A first Course in Fuzzy Logic*, CRC Press, *(1996), A Course in Stochastic Processes : Stochastic Models and Statistical Inference*, Kluwer Academic, *(1996), Mathematics of Data Fusion*, Kluwer Academic, *(1997), Applications of Continuous Mathematics to Computer Science*, Kluwer Academic, *(1997), Random Sets: Theory and Applications*, Springer-Verlag, *(1997).* During 1992-93, he held the LIFE Chair of Fuzzy Theory at Tokyo Institute of Technology, Japan.

Rainer Palm received his degree in Engineering at the Technical University Dresden (1975), Dr.Engineering at Humboldt University,Berlin (1981), Dr.Sc.Tech. at the Academy of Sciences, Berlin (1989). From 1970 to 1982 he was an Electronic Engineer, project leader at the Institute for Automatic Control, Berlin; from 1982-90 : a researcher, project leader and head of research group at the Academy of Sciences, Berlin; in 1991 he was a guest researcher at the Institute for Robotics IPK, Berlin; and from 1991 he is a researcher and project leader at Siemens AG R&D, Munich. He is a co-author of *Model Based Fuzzy Control (1996).* He is an Associate Editor of *IEEE Transactions of Fuzzy Systems.*

Witold Pedrycz is Professor of Computer Engineering in the Department of Electrical and Computer Engineering, University of Manitoba, Winnipeg. He is actively pursuing research in Computational Intelligence, Fuzzy Modeling, Knowledge Discovery and Data Mining, Fuzzy Control including Fuzzy Controllers, Pattern Recognition, Knowledge-Based Neural Networks, and Relational Computation. He is an author of *Fuzzy Control and Fuzzy Systems (1988, 1993), Fuzzy Relation Equations and Their Applications to Knowledge Engineering (1988), Fuzzy Sets Engineering (1995), Computational Intelligence: An Introduction (1997).* He is one of the Editors-in-Chief of *Handbook of Fuzzy Computation.* Dr. Pedrycz is a member of the Edtitorial Board of *IEEE Transactions on Fuzzy Systems, IEEE Transactions on Neural Networks, Fuzzy Sets and Systems, Soft Computing, Intelligent Manufacturing, Pattern Recognition Letters.*

Nadipuram R. Prasad is an Associate Professor of Electrical Engineering at New Mexico State University. He received his Ph.D. in 1989 in Electrical and Computer Engineering. His research interests are in the areas of fuzzy control,neural network control, genetic algorithms, and the integration of soft computing technologies. He was the recipient of the Bromilov Award for teaching excellence at New Mexico State University. He is a member of the Advisory Board for the *Electric Power Systems Research Journal.*

Michio Sugeno received his B.S. degree in physics from the University of Tokyo in 1962. From 1962 to 1965 he worked in nuclear engineering for Mitsubishi Atomic Power Industry. From 1965 to 1976 he was a Research Associate in control engineering at the Tokyo Institute of Technology. He received the Doctor of Engineering degree from TIT in 1974. In 1976 he joined the Graduate School of TIT where he is currently Professor of Systems Science. Since 1989 he has been a Leading Advisor at the Laboratory for International Fuzzy Engineering Research, Yokohama.Dr. Sugeno is member of the Editorial Boards of *Fuzzy Sets and Systems, Intelligent Systems, J. of Uncertainty, Fuzziness and Knowledge-Based Systems, Mathematical Modeling of Systems, Mathware and Soft Computing, IEEE Transactions on Fuzzy Systems.* His research interets include fuzzy computing based on systemic functional grammar and fuzzy logic, fuzzy control of helicopters, and fuzzy measure analysis. Dr. Sugeno received the Honorary Doctor from the Polytechnical University of Madrid in 1997.

Kazuo Tanaka received his Ph.D. degree in Systems Science from Tokyo Institute of Technology, in 1990. He is currently an Associate Professor. He was a Visiting Scientist in Computer Science at the University of North Carolina, Chapel Hill, in 1992 and 1993. He received the Best Young Researchers Award from the Japan Society for Fuzzy Theory and Systems in 1990. He is a member of IEEE Control Systems Society Technical Committee on Intelligent Control. He is the author of two books and co-author of seven books. His research interets include Principle, Analysis and Design of Intelligent Control Systems such as Fuzzy, Neuro, and Evolutionary Control .

Ronald Yager is Director of the Machine Intelligence Institute and Professor of Information Systems at Iona College. He is a fellow of the IEEE and the New York Academy of Sciences. He has served at the National Science Foundation as Program Director in the Information Sciences Program. He was a NASA/Stanford fellow.He is co-president of the International Conference on Information Processing and Management of Uncertainty, Paris. He is Editor and Chief of the International Journal of Intelligent Systems. He also serves on the Editorial Boards of a number of other Journals including *Neural Networks, Data Mining and Knowledge Discovery, Approximate Reasoning, Fuzzy Sets ans Systems, General Systems.* His current research interests include information fusion, multi-criteria decision-making, information retrieval, multi-media information systems, knowledge discovery, uncertainty management, fuzzy set theory, fuzzy-neural modeling and nonmonotonic reasoning.

Yanqing Zhang received the B.S. and M.S. degrees in Computer Science and Engineering from Tianjin University in 1983 and 1986, respectively. He received the Ph.D.degree in the Department of Computer Science and Engineering at the University of South Florida in 1997. His research interests include hybrid intelligent systems, neural networks, fuzzy logic, genetic algorithms, distributed database systems, parallel and distributed processing, nonlinear system modeling and prediction, and fuzzy game theory. He is a student member of IEEE, ACM, and a member of Upsilon Pi Epsilon.

INTRODUCTION: THE REAL CONTRIBUTION OF FUZZY SYSTEMS

Didier Dubois, Hung T. Nguyen, Henri Prade, and Michio Sugeno

Fuzzy Logic and Systems

Fuzzy logic (FL) was invented in the sixties by a leading expert in control engineering who realized that control theory had become beautiful enough to carry on its development on its own, but that there was many real problems it could not solve. Most real complex system control problems involve man. Hence applying control theory to complex control problems may require a formal understanding of how a human operator understands his system, what his goals are, and how he proceeds when controlling it. It requires a dedicated tool for representing human-originated information in a flexible way. And this is where fuzzy logic enters the picture (Zadeh [28]). In a paper that can be considered as the origin of the fuzzy rule-based approach, Zadeh [31] claimed that systems analysis and control requires a trade-off between representations that are very precise and accurate, such as numerical ones stemming from numerical functions, differential equations and the like, and representations that are intelligible, meaningful to humans, hence summarized, possibly in linguistic terms. However these two kinds of representations are sort of antagonistic because the more precise and accurate a representation, the less understandable it is, and conversely. This is Zadeh's principle of incompatibility, and it explains the particular position of fuzzy logic in control and systems engineering, and the misunderstandings and controversies it has created in the control engineering community. The contribution of fuzzy systems may not be at the level where traditional systems engineers expect contributions: where they expect improved performance, they might get improved intelligibility, flexibility and transparency in the representation of dynamic systems and the design of controllers.

The term "fuzzy logic" is itself rather ambiguous because it refers to problems and methods that belong to different fields of investigation. When scanning the literature, it is possible to find three meanings for the expression "fuzzy logic". In its most popular acception, it refers to numerical computations based on fuzzy rules, for the purpose of modeling a numerical function in systems engineering (Mendel [20] for instance). However, in the mathematical literature fuzzy logic means multiple-valued logics (Novak [21]; Hajek [15]), with the purpose of modeling partial truth values and vagueness. Lastly, in Zadeh's papers fuzzy logic is better understood as encompassing fuzzy set-based methods for approximate reasoning at large (Zadeh [32]), and approximate reasoning is a subtopic of Artificial Intelligence.

This state of facts creates miscommunication problems among researchers in the fuzzy set area, and consequently, outside the fuzzy world as well. Indeed the fields concerned by "fuzzy logic", and that use this terminology, include Systems Engineering, Formal Logic and Artificial Intelligence, and some of those fields almost never communicate with one another. Fuzzy logic, as understood by systems engineers, is no logic at all, from the points of view of logicians. Moreover the research programs of Artificial Intelligence and Systems Engineering are quite divergent: the latter is based on numerical methods and is black-box oriented. The former insists on symbolic processing and knowledge representation. Fuzzy logic is devoted to knowledge representation and symbolic/ numeric interface, and its status is rather ambiguous in that respect. Fuzzy set theory has brought together researchers that had little background in common and the temptation exists for each community (Artificial Intelligence, Logic, and Control), to emphasize a narrow view of fuzzy logic that fits its own tradition (and for the most orthodox members of these communities to criticize fuzzy logic on the basis of this narrow view).

As this introduction suggests, while the emphasis was put on knowledge representation issues in the beginning of fuzzy control, the current trend is rather to consider fuzzy rules as one more tool for approximating functions. However fuzzy logic will preserve its specificity only insofar as as a balance between both views is respected and a synergy between intelligibility and performance is exploited.

Early Fuzzy Control: From Expert Knowledge to Control Laws

The classical control engineering paradigm is to build up a numerical model of a the system to be controlled and then to determine a controller from the model, in the form of a numerical function. Traditionally the model of the system is linear, based on systems of linear differential equations, and the control law is linear as well, like in PID control. However not all actual systems are linear. More recently the study of non-linear models of systems has blossomed and significant results on non-linear control do exist. Besides other techniques have come to grip with non-linearities, such as adaptive control where coefficients of linear controllers can be tuned as the system evolves, and robust control where some uncertainty or variability in the coefficients of the linear model are coped with. However it is not so easy to bridge the gap between these new techniques and industrial practice. As often pointed out (Verbruggen and Bruijn [25]) only PID controllers remain widely used in the industry.

Fuzzy control has appeared before the advent of the recent developments of the traditional control methodology (nonlinear control, robust control and the like). A famous paper by Zadeh [31], as well as a former one on fuzzy algorithms [30], triggered some applied control engineering research in Europe, especially at Queen Mary College, London, where Mamdani proposed the first fuzzy controller, in 1974, with his student Assilian (Assilian [2]; Mamdani and Assilian [19]). The basics of this approach are recalled in the chapter by Kreinovich and Nguyen. Its striking feature is that it obviates the need for a mathematical model of the system to be controlled. Indeed instead of modeling the system, the idea is to model an experienced human operator. This is done via a set of control rules that the operator supplies or that are derived from observing his behavior when he controls the system, or even from books containing recommendations about how to control the system. Fuzzy logic then offers a simple way to turn what is nothing more that one or several symbolic decision tables into a numerical continuous control law. Given a numerical input, an inference step is taken that results in a fuzzy output set. The latter is then turned into a precise control value via a selection method dubbed "defuzzification". In practice the set of rules can be "compiled" into a simple look-up table, or even as a mathematical function expressing the control value in terms of the inputs of the controller. Typically these ideas were very early applied to large non-linear systems such as the cement kilns controllers of Holmblad and Afstergaard [16] This empirical methodology was further on developed and exploited widely in Japan.

The fuzzy control approach was at the same time rather heretic from a control-theoretic point of view and appealing for engineers that had to devise the control law of real systems, in an industrial environment. It was heretic because relying on expert knowledge rather than on a mathematical model based on "objective" observations of the system behavior. To the control community, the approach sounded at best heuristic, at worst non-scientific. In particular there were no way of proving the stability of a system controlled by fuzzy logic controllers, since no model was used. On the other hand, the control law described by fuzzy rules is much easier to understand and local improvements of the control policy can be achieved in a modular way by changing rules. Moreover bypassing the need for a system model means that the development of a fuzzy controller is cheaper than running the whole classical approach. Lastly fuzzy controllers encompass non-linear control laws, hence are in theory more general than classical PID controllers. The precise links between PID controllers and fuzzy controllers are made explicit in the chapter by Foulloy and Galichet. Even PID controllers cannot be used as such. They are supplemented by additional features in order to deal with saturations, local exceptions and the like. A fuzzy controller can easily implement these alterations. These remarks highlight the actual complementarity of fuzzy control and classical control. The contribution of fuzzy controllers is to make control laws, even non-linear ones, easier to understand in intuitive terms, easier to specify from human knowledge, easier to implement.

Fuzzy Control Versus Expert Systems and Approximate Reasoning

There is a clear connection between fuzzy controllers and expert systems (Buchanan and Shortliffe [6]). Fuzzy controllers were originally viewed as special cases of expert systems, but the use of overlapping fuzzy categories in the numerical universes of input and control variables produces a smooth interface between numbers and symbol, so that contrary to usual expert systems, fuzzy controllers have an interpolation capability. However the structure of a fuzzy controller is simpler that the one of an expert system since there is generally no chaining between rules: a fuzzy controller is a set of parallel fuzzy rules, in other words, a decision table.

However there is a major difference between a fuzzy controller as Mamdani devised it, and an expert system. In the latter, a rule is sometimes viewed as a kind of logical formula (a material implication, a clause in logic programming approaches, for instance) or as a kind of inference rule (also called a production rule); in any case it is viewed as a constraint on the underlying universes of discourse e. In the logical view, a rule eliminates worlds inconsistent with it and the more there are rules, the less there are worlds remaining, and the more precise the description of the relation between the inputs of the expert system and its output. A fuzzy rule in a Mamdani controller is on the contrary viewed as a fuzzy Cartesian product of its fuzzy inputs and its fuzzy output.

It is a fuzzy point (a fuzzy "granule") in the input output space, and a set of fuzzy rules is understood as a fuzzy graph (Zadeh [33]). Rules are combined *disjunctively* and not conjunctively like in expert systems: here, the more rules the more points in the graph. As a consequence the more fuzzy rules are triggered, the more unspecific the obtained conclusion to be defuzzified. These features of fuzzy controllers make them stand apart from the world of logic, and have created many a misunderstanding between fuzzy logic and logicians proper.

There has been also some confusion between fuzzy rule-based systems and expert systems handling uncertainty weights such as MYCIN [6]. These systems often used minimum and maximum operations to combine the weights and propagate them, and some people might have believed that they were based on fuzzy logic. However these weights in MYCIN-like systems are uncertainty coefficients attached to non-fuzzy rules, or non fuzzy facts, while in a fuzzy controller, the input fact is often precise and certain and the weight carried over to conclusions evaluate the relevance of the rule condition with respect to the input. Fuzzy controllers are devoted to interpolation, not really uncertainty handling.

Moreover, Zadeh's approach to approximate reasoning (see the volume of Approximate Reasoning and Information Systems, in this handbook series) is not in accordance with Mamdani fuzzy controllers: for Zadeh, statements in natural language (such as fuzzy If -Then rules) denote flexible restrictions on the universe of discourse, modeled by possibility distributions, and are thus constraints that are combined in a disjunctive way (using a minimum operation implementing a fuzzy conjunction). On the contrary, rules in a Mamdani controller are viewed as fuzzy data tuples, or kind of prototypical examples of controls to be applied in fuzzily described situations. Nevertheless, it makes sense to take for granted a logical, constraint-oriented point of view of fuzzy rules, and to compare it to the Mamdani approach. This is done by Yager and Filev in their chapter. See also Dubois and Prade [11] for a typology of fuzzy rules.

Why did so many systems engineers follow the road opened by Mamdani, and very few adopted the logical implication-based modeling of fuzzy rules, which some even find as "violating engineering commonsense [20]"? Let us risk some explanations:

A matter of scientific background . While it can be claimed that some fuzzy control papers were written by people not so aware of control theory, it is clear that not so many control engineers are familiar with logic. Control engineers have a data-oriented tradition, and they tend to see a set of fuzzy rules more as the accumulation of fuzzy typical examples. Hence the logical underpinnings of fuzzy rules, the meaning of many-valued implications were not really their concern. Moreover the defuzzification step in fuzzy control inference tends to attenuate the differences between the results obtained using various kinds of rules.

The presence of inconsistency. Using a conjunctive view of fuzzy rule-based systems, and interpreting fuzzy rules as constraints, not all fuzzy rule-based systems will produce an output when triggered by an input. Sometimes the output will be empty. This occurrence denotes a logical inconsistency in the set of rules, a phenomenon which control engineers may see as a drawback when building fuzzy controllers. With Mamdani controllers, any set of fuzzy rules produces an output whatever the input is. It does not mean that this output always makes sense, though, as pointed out by Baldwin and Guild[3] rather early.

Computational simplicity. The inference in Mamdani controllers can always be performed on a rule by rule basis (trigger each enabled rule and compute its modified conclusion; then perform the union of results, what people sometime call FITA approach (First Infer, Then Aggregate results). This method easily carries over to fuzzy inputs. However if rules are viewed as implications expressing constraints, conjunctively combining the results of elementary inferences is generally less accurate that the FATI method (First Aggregate rules, Then Infer). Only when the inputs are precise do the FATI and FITA methods coincide. As a consequence fuzzy implication-based rule systems are computationally less simple than Mamdani controllers.

However, detecting logical inconsistencies in fuzzy rule-based systems may be part of the validation step of a human-originated fuzzy controller. Only an implicative view of fuzzy rules can allow this kind of validation. Moreover the defuzzification step in Mamdani controllers is clearly an additional ad hoc trick to get a precise control value. It is not part of the inference step, but it is necessary, in order to get an interpolative behavior. Some authors have even tried to cope with inconsistency by a suitable tuning of the defuzzification operator. As it turns out, the interpolative behavior can be captured inside the inference machinery using a special kind of rules called "gradual rules" [10]. Moreover Mamdani controllers are not very well suited to rule-chaining. Indeed the output to such a fuzzy rule-based system even under a crisp input is very unspecific; for instance in a typical Mamdani rule-system the modus ponens property (from A and A implies B deduce B) will not hold, generally, unless rule conditions do not overlap [8]. But the overlapping of conditions is the reason for the success of fuzzy logic. Hence chaining Mamdani rules will result in very unspecific results as well, with an increase of imprecision at each inference step, unless defuzzification takes place before each inference step, a very ad hoc remedy. This imprecision contamination effect will not occur for well-conditioned systems of fuzzy implicative rules. These remarks suggest that a study of fuzzy rule-based systems in a more logical tradition may contribute to the future of fuzzy control.

From Fuzzy Control to Fuzzy Modeling

Pushing further the machinery ruling the behavior of Mamdani's fuzzy rule-based systems, Takagi and Sugeno [23] noticed that the use of fuzzy conclusions was not imperatively needed. They proposed fuzzy rules with fuzzy conditions and precise conclusions. The inference step is then reduced to a fuzzy pattern-matching step between precise inputs yielding rule relevance coefficients , that are then exploited as weights in a linear interpolation of outputs. Takagi and Sugeno even went one step further, noticing that precise conclusions of fuzzy rules could depend on the input variables (for instance via linear functions) thus opening the way to a feasible approach to the identification of fuzzy systems, the idea of which has been pioneered by Tong [24]. Sugeno and Tanaka's chapter proposes a refresher on the basics of fuzzy modeling.

While this approach proposes a new family of fuzzy controllers, it is clear indeed that it can be used to interpolate between various mathematical models of the same system each model being valid in a specific region of its state space. Basically each fuzzy rule then means "in such a region the model of the system is given by this equation", where the regions are fuzzily described and may overlap. This proposal to represent a dynamic system via a blending of models, and the identification method, both structural and parametric, that sustained it, had a significant impact on fuzzy systems research, namely:

- It suggested that fuzzy rule-based systems could be used as a tool for modeling non-linear systems (and not only for controlling them)
- It led researchers to consider a fuzzy rule-based system as a universal approximator of functions, hence laying bare a connection between fuzzy systems and neural nets, for instance.
- It also reinstalled fuzzy control inside the control engineering tradition: if fuzzy rule-based models can be identified, fuzzy controllers can obtained from the fuzzy models (the rules of one serving to build the rules of the other), and genuine control engineering issues can be addressed for fuzzy controllers, namely stability, robustness etc... Especially stability of fuzzy systems is a topic that received much attention, as surveyed by Tanaka in this volume.

It also suggested that fuzzy rule-based systems could be derived from objective data, and not only from human expertise. However this evolution has created some epistemologic confusion about the nature of fuzzy rule-based modeling: is a fuzzy model more than just another black-box numerical model? Does a fuzzy model capture articulated knowledge about a system? Can it be viewed as an articulated summary of the system behavior that a layman can understand? These questions lead us into a key-debate in Artificial Intelligence and Information Engineering at large: how to bridge the gap between data and knowledge?

Fuzzy Systems: Modeling Versus Explaining

Formalizing the connection between data and knowledge is indeed one of the most intriguing and challenging issue in the computer era. Especially, what does it mean to produce knowledge from data? There are two (antagonistic) ways of addressing this problem

modelling: build a function that can mimick the data accurately, and gives good results on new data sets;

abstracting: build a system that produces articulated knowledge, possibly in natural language, from the data.

In the first approach, emphasis is put on the ability to *reproduce* what has been observed. Neural nets and similar techniques are well-adapted to this problem. In the second approach, emphasis is put on the ability to ıunderstand and explain the data in a human-friendly way. Clearly this is the realm of Artificial Intelligence. Moreover the term "learning" applied to the optimization of black-box models may seem too strong and even somewhat inappropriate. Learning more naturally reflects the idea of abstracting, and understanding, not just trying to imitate.

On the basis of the above difference between abstracting and modeling, one may consider the epistemological situation of fuzzy rule-based systems. Fuzzy sets aim at representing human-originated knowledge, and more specifically that part of articulated knowledge that refers to numerical universes of discourse; it proposes a natural flexible interface between symbols and numbers. Fuzzy rules are thus capable of expressing both a piece of knowledge and a continuous function. What should make fuzzy rule-based models unique when fitting them to data is their ability to become linguistically interpretable statements that "explain" the data, at least as much as their numerical approximation power, which is shared by many other methods. In that sense fuzzy logic is as strongly related to symbolic AI that it tries to make softer, as to the data-driven tradition of systems analysis.

Interestingly, attempts to build fuzzy models that are based on a faithful encoding of linguistic information can be traced in the early fuzzy literature (the so-called "verbal models" of Wenstop [26, 27]). Perhaps due to the intermediary, bridging position of fuzzy logic between symbolic and numerical processing, this early tradition has been given up and a significant deviation from original motivations and practice of fuzzy logic has been observed in the fuzzy set community since the late eighties. Namely, fuzzy rule-based systems are more and more considered as standard, very powerful universal approximators of functions (Kosko [18]), and less and less as a means of building numerical functions *from* heuristic knowledge, nor as a tool for the linguistic summarization of data. There exist a huge literature on universal approximation properties of fuzzy systems, surveyed by Kreinovitch, Mouzouris and Nguyen in this volume.

This trend raises several questions for fuzzy logic. First, if fuzzy logic is to compete alternative methods in approximation theory, it faces a big challenge because approximation theory is a well-established field in which many results exist. Approximate representation of functions should be general enough to capture a large class of functions, should be simple enough (especially the primitive objects, here the fuzzy rules) to achieve efficient computation and economical storage, and should be capable of extrapolating data. Are fuzzy rules capable of competing on their own with standard approximation methods on such grounds? The answer is far from clear. On the one hand the universal approximation results for fuzzy rule-based systems presuppose a large number of rules. This is good neither for the economy of representation nor for linguistic relevance. On the other hand the identification between fuzzy rule-based systems with neural nets or variants thereof (radial basis functions and the like, see Mendel [20], and the chapter by Kandel, Pedrycz and Zhang) has perhaps created some confusion as to the actual contribution of fuzzy logic. To some extent it is not clear that fuzzy logic-based approximations methods for modeling and control needs fuzzy set theory any longer (Bersini and Bontempi [4]).

Indeed, why should we bother about if-then rules, and about the "readability" of fuzzy rules as knowledge chunks if the aim is to build a numerical function that best fits a data set? Actually, from the point of view of approximation capabilities, the good performance of a fuzzy rule-based system seems to be incompatible with the linguistic relevance of the rules. This incompatibility leads systems engineers into focusing on performance at the expense of explainability, thus cutting off the links between fuzzy logic and knowledge representation, hence with fuzzy set theory itself. This is very surprising a posteriori since the incompatibility between high precision and linguistic meaningfulness in the description of complex systems behavior is exactly what prompted Zadeh [31] into introducing fuzzy sets as a tool for exploiting human knowledge in controlling such systems. It is questionable whether the present trend in fuzzy engineering, that immerses fuzzy logic inside the jungle of function approximation methods will produce path-breaking results that put fuzzy rule-based systems well over already existing tools. It is not clear either that it will accelerate the recognition of fuzzy set theory, since there is a clear trend to keep the name "fuzzy" and forget the contents of the theory. Contrary to this trend, we tend to believe that the specificity of fuzzy logic is its capability of bridging the gap between articulated linguistic descriptions and numerical models of systems, and of generating one from the other.

The reason for the over focus on fuzzy rules as numerical approximation methods is because control engineers are traditionally more concerned with modeling than explaining. Yet, it seems that system engineering practice can benefit from the readability of fuzzy rule-based systems. Fuzzy rules are easier to modify, they can serve as tools for integrating heuristic, symbolic knowledge about systems, and numerical functions issued from mathematical modeling. Fuzzy control should not be regarded as challenging the well-foundedness of classical control theory. The purposes of fuzzy controllers and classical control theory are complementary (Verbruggen and Bruijn, [25]). Whenever mathematical modeling is possible, control theory offers a safer approach, although a lot of work is sometimes necessary to bridge the gap with practical problems. Fuzzy control can be useful to control engineers, since they do employ heuristic knowledge in practice, be it when they specify objectives to attain. Supervision also involves a lot of know-how, despite the existing sophisticated control theory, and some interesting works have also been done in fuzzy rule-based tuning of PID controllers. A Takagi-Sugeno fuzzy controller with linear outputs can be viewed as an adaptive linear controller. Fuzzy logic sounds reasonable when modeling is difficult or costly, but knowledge is available in order to derive fuzzy rules. Fuzzy logic offers a convenient, transparent tool for designing a controller. These design issues are discussed in Palm's chapter.

Curiously, the idea of extracting fuzzy rules from data for the purpose of constructing a summarized explanatory model of system behavior is not so popular. Yet it seems to be one of the future lines of research for fuzzy systems research. Zadeh [34] himself recently advocated the idea of computing with words as being the ultimate purpose of fuzzy logic and he also insists on the role of fuzzy logic for information granulation (Zadeh [35]). In order to achieve this program that repositions fuzzy logic in the perspective of automated explanation tasks, it seems that fuzzy systems research should again serve as a bridge between Systems Engineering and Artificial Intelligence research. Needless to say that in that perspective, control engineers should receive some education in logic, and Artificial Intelligence researchers interested in systems engineering should be aware of control theory. Such a shift in education and concerns would open the road to addressing, in a less ad hoc way, issues in the supervision of complex systems, a problem whose solution requires a blending between knowledge and control engineering, namely computerized tools for automatically explaining the current situation to human operators, and not only tools for approximating real functions, be they non-linear.

Optimizing Fuzzy Systems: The Soft Computing Trend

As stressed above, the design of a fuzzy model or a fuzzy controller may rely on human knowledge or may be derived from data. In general both are needed. Indeed if an expert can sometimes supply a qualitatively correct description of a system behavior or a control policy, the numerical translation offered by fuzzy logic may be quite approximate. It is interesting to have methods that improve the set of fuzzy rules by tuning membership functions for instance. This necessity has led researchers to combine data-driven learning or optimization techniques with fuzzy logic. This combination is often dubbed "soft computing". There are mainly three kinds of techniques to optimize a fuzzy rule-based system: reinforcement learning, connectionist methods or stochastic optimization, especially genetic algorithms, that can be used conjointly. Chapters by Berenji, Kandel, Pedrycz and Zhang, Prasad, and Geyer-Schulz are devoted to these topics that are currently part of soft computing.

The term "soft computing" (Zadeh [36]; Bonissone [5]) is a new keyword in information technologies that refers to a fusion of methodologies mainly bringing together neural networks, fuzzy logic and evolutionary algorithms, among other techniques. For some scientific communities, soft computing is still only a fashionable name with little actual contents. However the current success of the soft computing recipe is apparently due to three facts:

1) the visibility of control-engineering applications of fuzzy logic, that has prompted fuzzy rule-based systems to the forefront of fuzzy set tools, and has occulted other basic fuzzy concepts;

2) the mathematical similarity between the functional form of fuzzy rules and the equation of a formal neuron, that gives a natural method for learning a fuzzy rule-base from data;

3) the fact that genetic algorithms were very soon used to optimize rule-based systems (also called classifier systems see Burke [7]).

Indeed, due to the mathematical similarity between neural nets and fuzzy rule-based systems (RBS) it is natural to exploit complementarity between FL NN, and GA as on the picture below:

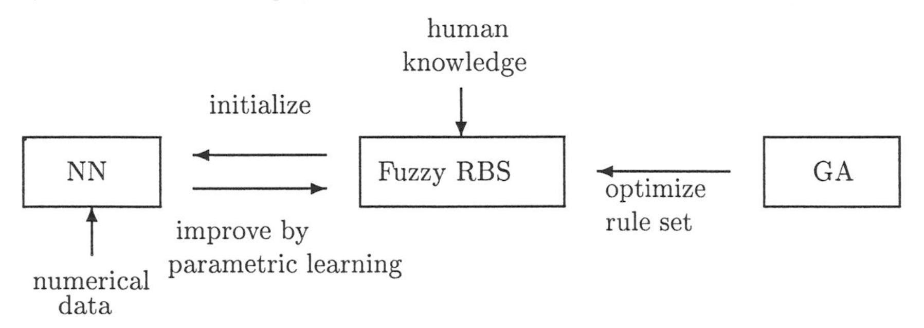

Namely, when fitting a neural network to data in a given problem, the choice of initialization is important for the learning time: a bad initialization may lead to a very long learning stage. The initial architecture of the neural network can be derived from expert knowledge under the form of a fuzzy RBS. Then parametric learning can be carried out by neural net algorithms, and a genetic algorithm can help optimizing the rule set at a more syntactic level, if the human knowledge proved partially erroneous (Geyer-Shulz [13]). The aim of such a scheme is often to build a function that correctly approximates the data set, a purpose that is very specific, and, again, far away from traditional concerns of fuzzy set theory about modeling human imprecise knowledge. In that sense the encounter of fuzzy logic, neural networks and genetic algorithms may seem to be rather contingent and contextual. The complementarity of three techniques in a particular type of application is worth exploiting but does not really creates a new scientific field, nor does it point a would-be intrinsic unity between fuzzy logic, neural nets, and genetic computation. Nevertheless, the merit of soft computing is to point out that modeling human expertise with fuzzy rules is not enough to solve problems and that the possible connection between fuzzy sets and data is important as well (although this has been the concern of fuzzy clustering methods for a long time).

Organization of The Volume

This volume is organized as follows. *Basics of fuzzy rule-based modeling and control* are surveyed in the first four chapters. The fuzzy control methodology in the Mamdani style (Chapter 1 by Kreinovich and Nguyen), that historically came first, and the fuzzy rule based systems with input dependent conclusions (Chapter 2 by Sugeno and Tanaka) are presented. They are the basic building blocks of the major part of the volume. Then Filev and Yager, in Chapter 3, envisage fuzzy systems from the point of view of intelligent systems, casting fuzzy modeling and control in the scope of rule-based approximate reasoning at large, providing variants of fuzzy rules, combination operations and defuzzification operators. Chapter 4, by Kreinovich, Mouzouris and Nguyen, envisages fuzzy rules from the complementary point of view of functional approximation, scanning the huge existing literature.

Advanced fuzzy control is presented in chapters 5 - 7 which are devoted to current techniques in fuzzy control. The links between fuzzy control and traditional PID control methods are laid bare by Foulloy and Galichet in Chapter 5. In the past many authors have shown how to use fuzzy control on very simple linear systems. However such systems are not really the typical ones where fuzzy control is at its best, and the authors of this chapter show that the most usual fuzzy control decision table is nothing but a qualitative encoding of a traditional linear controller. However bridging the gap between fuzzy and linear control naturally brings forth design issues for fuzzy controllers. These are surveyed in Chapter 6 by Palm who combines the scientific and applied engineering points of view. Lastly, in Chapter 7, Tanaka surveys existing results on the theoretical stability of fuzzy controllers applied to fuzzy models. Such results also clearly impinge on design matters.

Optimizations of fuzzy systems is presented in chapters 8 - 11 which are devoted to the huge literature that has appeared in the last ten years and which is devoted to the data-driven improvement or synthesis of fuzzy rule-based systems. Chapter 8 by Berenji can be used as an introductory chapter to these methods. It also emphasizes reinforcement learning schemes. Chapter 9 by Kandel et al. envisages several forms of hybridizations between neural nets and fuzzy rule-based systems. Namely neural nets can be applied to fuzzy systems since the latter can be encoded under the form of a neural architecture and thus naturally amenable to be fitted to a data set. Fuzzy set theoretic connectives can be also be simulated by neural architectures. On the other hand fuzzy logic can be applied to neural networks: the learning effort of a neural net can be decreased by a priori knowledge encoded as fuzzy rules; fuzzy controllers can be used to manage the learning process itself; and fuzzy logic could be envisaged to translate neural networks into articulated knowledge as a set of linguistic rules. The latter kind of application, which is still prospective to-date, highlights the complementarity between black-box methods and fuzzy logic. Chapter 10 by Prasad envisages the use of fuzzy numbers in neural nets, namely the problem of learning from fuzzy data, and the use of fuzzy valued weights. Chapter 11 by Geyer-Schulz gives an authoritative introduction to the use of genetic algorithms for the purpose of optimizing fuzzy rule-based systems.

Advanced topics in fuzzy systems are presented in the last two chapters 12 and 13. These topics are still rather marginal in fuzzy systems engineering, perhaps due to their intrinsic difficulty and mathematical sophistication. There exists a literature which extends set-valued analysis as pioneered by Aumann, to fuzzy set-valued analysis. It is natural to try and envisage a fuzzy system from the point of view of differential equations having fuzzy coefficients and fuzzy solutions. However making any sense of this proposal is not so simple and there exist several ways of posing such a problem. Chapter 12 by Aubin and Dordan suggests one feasible rigorous approach to this problem, via fuzzy differential inclusions. Interestingly they claim that the usual unit interval is not appropriate as a range of membership grades for their purpose and rather use the positive real line in the tradition of convex analysis. Moreover they introduce counterparts to Gaussian probability in fuzzy set theory, as well as transforms such as Cramer and Fenchel transforms that play, in a fuzzy setting the same role as Laplace transform in usual integration theory. Chapter 13 by Diamond offers a particular point of view on an intriguing topic, namely the links between fuzzy set theory and chaos theory. While it is difficult to see any direct connection between the two, Diamond shows that the problem of the possible chaotic behavior of fuzzy systems is worthwhile investigating. However a fuzzy system is here again envisaged from an analytical point of view, namely a fuzzy-valued system. Diamond studies the chaotic behavior of sequences of fuzzy sets.

There are some aspects of fuzzy systems modeling and control research not covered by this volume. First, the pure fuzzy relational approach to fuzzy systems (not based on fuzzy rules) initiated by Zadeh [29], and surveyed by Dubois and Prade [9] (Chap.III.2) is not considered here. Of course, a set of fuzzy rules is an encoding of a fuzzy relation. However some authors have studied fuzzy systems at the mathematical level in terms of abstract fuzzy relations. This theoretical view has produced many mathematically-oriented papers in the seventies, dealing with fuzzy automata, and standards such as controllability, and minimal realization for fuzzy relational systems (Santos [22] ; Arbib and Manes [1]), and this tradition seems to be almost extinct, probably due to a lack of intuitive appeal.

Moreover the earliest applications of fuzzy sets to control were not from the start envisaged as it was to be done by Mamdani and his followers, from 1974 on. The first idea was to introduce fuzzy constraints and fuzzy objectives in the multiple stage control problem, as an alternative to the quadratic criterion of standard optimal control. This trend of research, exemplified by early papers by Gluss [14] and Fung and Fu [12] has ceased to be the forefront of fuzzy control research, despite the continuing interest of some researchers (noticeably Kacprzyk [17]). Due to the extensive use of dynamic programming in this approach, this topic has been dealt with in the volume "Fuzzy Sets in Decision Analysis, Operations Research and Statistics" (R. Slowinski, Ed.) in this series. However we consider that, due to its flexibility, the fuzzy constraint approach in control problems, for expressing practically meaningful control objectives, has great potential.

It is our pleasure to acknowledge those individuals who contributed to the realization of this volume: the contributors, A. Greene at Kluwer Academic, and R. Mines and T. Wang of New Mexico State University (for their technical assistance in designing the book).

References

[1] Arbib, M. A, Manes, E.G.(1975) Fuzzy machines in a category, *J. Austr. Math. Soc.*, **13**, 169-210.

[2] Assilian, S. (1974) *Artificial Intelligence in the Control of Real Dynamic Systems*, PhD Thesis, Queen Mary College, University of London.

[3] Baldwin, J. and Guild, N.C.F. (1980) Modelling controllers using fuzzy relations. *Kybernetes*, **9**, 223-229.

[4] Bersini, H., Bontempi, G. (1997) Now comes the time to defuzzify neuro-fuzzy models, *Fuzzy Sets and Systems*, **90**, 161-169.

[5] Bonissone, P.P. (1997) Soft computing: The convergence of emerging reasoning technologies, *Soft Computing*, **1**(1), 6-18.

[6] Buchanan, B. Shortliffe, E.H. (1984) *Rule-based Expert Systems. The MYCIN Experiments of the Stanford Heuristic Programming Project.* Addison Wesley, Reading, Mass.

[7] Burke, A. W. (1988) The logic of evolution, and the reduction of holistic coherent systems to hierarchical feeback systems. In *Causation in Decision, Belief Change, and Statistics* (W.L. Harper, R. Skyrms, eds.), Kluwer Academic, Dordrecht, pp. 135-199.

[8] Di, Nola A., Pedrycz, W., and Sessa, S. (1989) An aspect of discrepancy in the implementation of modus ponens in the presence of fuzzy quantities, *Int. J. Approximate Reasoning*, **3**, 259-265.

[9] Dubois, D. and Prade, H. (1980) *Fuzzy sets and Systems: Theory and Application* Academic Press.

[10] Dubois, D., Grabisch, M. and Prade, H. (1994) Gradual rules and the approximation of control laws. In *Theoretical Aspects of Fuzzy Control* (H.T. Nguyen, M. Sugeno, R. Tong, R.R. Yager, Eds.), Wiley, New York, pp.147-182.

[11] Dubois, D., Prade, H. (1996) What are fuzzy rules and how to use them. *Fuzzy Sets and Systems*, **84**, 169-185.

[12] Fung, L.W. and Fu, K.S. (1974) The kth optimal policy algorithm for decision-making in a fuzzy environment. In *Identification and System Parameter Estimation* (P. Eykhoff, Ed.), North-Holland, Amsterdam, pp. 1052-1059.

[13] Geyer-Schulz, A. (1996) *Fuzzy Rule-Based Expert Systems and Genetic Machine Learning*, Physica-Verlag, Heidelberg.

[14] Gluss, B. (1973) Fuzzy multistage decision-making, *Int. J. Control*, **17**, 177-192.

[15] Hajek, P. (1995) Fuzzy logic as logic. In *Mathematical Models for Handling Partial Knowledge in Artificial Intelligence* (G. Coletti, D. Dubois, R. Scozzafava, eds.), Plenum Press, New York, pp. 21-30.

[16] Holmblad, L. P. and Afstergaard J.J. (1997) The progression of the first fuzzy logic control application, In *Fuzzy Information Engineering: A Guided Tour of Applications* (D. Dubois, H. Prade, and R.R. Yager eds.), Wiley, New York, pp. 343-356.

[17] Kacprzyk, J. (1983) Multistage Decision-Making under Fuzziness, *ISR Series*, Vol. 79, Verlag Rheinland.

[18] Kosko, B. (1997) *Fuzzy Engineering*. Prentice-Hall, Upper Saddle River, NJ.

[19] Mamdani, E.H. and Assilian, S. (1975) An experiment in linguistic synthesis with a fuzzy logic controller, *Int. J. of Man-Machine Studies*, **7**, 1-13.

[20] Mendel, J. (1995) Fuzzy logic systems for engineering: A tutorial, *Proc. IEEE*, **83**, 345-377.

[21] Novak, V. (1996) Paradigm, formal properties and limits of fuzzy logic, *Int. J. of General Systems*, **24**, 377-406.

[22] Santos E. S. (1972) On reductions of maximin machines, *J. Math. Anal. Appl.*, **40**, 60-78.

[23] Takagi, T. and Sugeno, M. (1985) Fuzzy identification of systems and its application to modelling and control, *IEEE Trans on Systems Man and Cybernetics*, **15**, 116-132.

[24] Tong, R.(1978) Synthesis of fuzzy models for industrial processes: some recent results, *Int. J. General Systems*, **4**, 431-440.

[25] Verbruggen, H.B. and Bruijn, P.M. (1997) Fuzzy control and conventional control: What is (and can be) the real contribution of fuzzy systems? *Fuzzy Sets and Systems*, **90**, 151-160.

[26] Wenstop, F. (1976a) Deductive verbal models of organizations, *Int. J. Man-Machine Studies*, **8**, 293-311.

[27] Wenstop, F.(1976b) Fuzzy set simulation models in a systems dynamic perspective, *Kybernetes*, **6**, 209- 218 (reprinted in Dubois, Prade & Yager's *Readings in Fuzzy sets for Intelligent Systems*, Morgan & Kaufmann, 1993).

[28] Zadeh, L. A. (1965) Fuzzy sets, *Information and Control*, **8**, 338-353.

[29] Zadeh, L. A. (1965) Fuzzy Sets and Systems. In *System Theory* (J. Fox, Ed.), Polytechnic Press, Brooklyn , New York, pp. 29-37 (reprinted in *Int. J. General Systems*, **17**, 129-138, 1990).

[30] Zadeh, L. A. (1968) Fuzzy algorithms, *Information and Control*, **12**, 94-102.

[31] Zadeh, L. A. (1973) Outline of a new approach to the analysis of complex systems and decision processes, *IEEE Trans. Systems, Man and Cybernetics*, **3**, 28-44.

[32] Zadeh, L. A. (1979) A theory of approximate reasoning. In *Machine Intelligence*, Vol. 9 (J.E. Hayes, D. Michie, L.I. Mikulich, eds.), Elsevier, New York, pp. 149-194.

[33] Zadeh, L. A. (1992) The calculus of fuzzy If-Then rules. *AI Expert*, **7**(3), 22-27.

[34] Zadeh, L. A. (1996) Fuzzy logic computing with words, *IEEE Trans. on Fuzzy Systems*, **4**, 103-111.

[35] Zadeh, L. A. (1997a) Toward a theory of fuzzy information granulation and its centrality in human reasoning and fuzzy logic, *Fuzzy Sets and Systems*, **90**, 111-127.

[36] Zadeh, L. A. (1997b) What is soft computing? *Soft Computing*, **1**(1), 1-1.

1 METHODOLOGY OF FUZZY CONTROL: AN INTRODUCTION

Hung T. Nguyen[1] and Vladik Kreinovich[2]

[1]Department of Mathematical Sciences
New Mexico State University
Las Cruces, NM 88003-8001, USA
hunguyen@nmsu.edu

[2]Department of Computer Science
University of Texas at El Paso
El Paso, TX 79968, USA
vladik@cs.utep.edu

1.1 INTRODUCTION: WHY FUZZY CONTROL

Control is necessary. In today's world, many systems can be controlled.

- Some things, such as cars, planes, ships, are actually *designed* to be controlled. In this case, the problem is how to control them so as to achieve certain goals.
- Some systems are not designed by us, but we have learned how to control them: we can, to some extent, control weather, we can control pollution, etc. In these cases, we also need to find out the best ways to control.

Controlling a system means that we *monitor* (i.e., frequently measure) some characteristics $x_1, \ldots, x_n$ of this system, and, depending on the values of these characteristics, we apply different controls $u_1, \ldots, u_m$. An algorithm that transforms the sensor inputs $x_1, \ldots, x_n$ into the corresponding control values $u_i(x_1, \ldots, x_n)$ is called a *control strategy*.

1.1.1 Traditional control methodology and its limitations

Traditional approach to control. The traditional approach to control consists of the following:

- First, we try to describe the behavior of the system in *precise* mathematical terms, i.e., we try to come up with the exact *model* of the system.
  - In some cases, especially when we control a *man-made* system, we may already *know* how exactly this system reacts to different controls. In other words, for each possible initial state $\vec{x} = (x_1, \ldots, x_n)$, and for each possible control values $u_1, \ldots, u_m$, we know how exactly the state will change if we apply this control, i.e., we know how the time derivatives $\dot{x}_i$ depend on x_i and u_j:

    $$\dot{x}_i = f_i(x_1, \ldots, x_n, u_1, \ldots, u_n). \tag{1}$$

    This system of differential equations forms a *model* of the controlled system.
  - In other cases, we *do not know* the differential equation *model*. Therefore, we analyze the system:
    * we apply different controls in different states,
    * determine how the state changes, and thus,
    * determine the desired differential equations model (1).
- Second, we try to describe in precise terms what we want to achieve. We want the control that is the *best*, in the sense of some criterion; for example, we may want to get to the Moon in the shortest possible time, or we may want to design a control strategy for the subway that leads to the most comfortable (smoothest) ride, etc.
  - Some *objectives* (such as "the shortest time to the Moon") are already formulated in well-defined terms.
  - Other *objectives* (such as "the most comfortable ride") need to be re-formulated in these terms.
- Now that the controlled system is described in precise mathematical terms, and the objective function is described in mathematical terms, we can, for each control strategy and for each initial state, determine how exactly the system will change and what the resulting value of the control function will be. Our goal is then to *find the control strategy* for which the resulting value of the objective function is the largest possible. This is a *well-defined mathematical optimization problem*, and traditional control theory has developed many methods for solving these optimization problems and designing the corresponding control strategies.

Traditional control theory is not always applicable: three possible situations. Traditional control theory has many important applications. There are, however, practical cases when this theory is not applicable. Indeed, to apply the traditional control theory, we must:

- know the *model* of the controlled system (described in terms of differential equations);
- know the *objective function* formulated in precise terms;
- *be able to solve* the corresponding mathematical optimization problem.

If one of these conditions is not satisfied, then traditional control methodology is not applicable.

Sometimes, we know the model and the objective function, but we cannot solve the optimization problem. In some cases, we *know the model and the objective function*, but we are *unable to solve* the corresponding optimization problem. There may be three reasons for this inability:

- It could be that the optimization problem is *very complicated* (e.g., highly non-linear), so many attempts to solve it has *not yet* lead to a successful *algorithm*. This lack of success is not necessarily caused by an insufficient effort: It is known that in general, the problem of finding the optimal control is computationally intractable; see, e.g., [21], [41].
- In some complicated cases, methods of traditional control theory do provide an algorithm, but this *algorithm is* so *complicated* that its running time exceeds the time that we have to make a control solution.

  The best example of such a situation is weather prediction. In principle, we know a system of integro-differential equations that describes the weather, and we have pretty accurate measurements of the current state of the atmosphere. However, these computations are very time-consuming:

  * even to predict the weather for *a few hours* ahead, we would need to spend quite some time on the world fastest supercomputer; and
  * if we want to make a *long-term* prediction, e.g., a for a month ahead, we would need much more than one month of computations.

  If computing predictions for next month takes more than a month, then it is no longer a prediction: we will *observe* the next month's weather *before* the *predicting* computations are over.
- The third possibility is that the *problem is new* and unusual for control theory, and so, algorithms for solving it have not yet been developed.

For example, *parking a car* is an example of a problem that traditional control theory has not considered until recently.

Sometimes, we know the model, but we do not know the objective function. In some cases, we *know* the exact *model*, but we *do not know* the exact *objective function*.

For example, if we design a control system for subway, the intended goal is to make the subway ride the most comfortable. Unfortunately, there is no well-accepted formalization of what "comfortable" means.

Sometimes we do not even know the model. Finally, in many control situations, we do not even know the model of the controlled system. There are two reasons for that:

- In many practical applications, especially in applications to manufacturing, we can, *in principle*, measure all the possible variable and determine the model exactly, but this will drastically (and unnecessarily) increase the costs.

  For example, the exact form of the differential equations that describe an airplane or a car depend on its load.

  * A failure of a plane can lead to catastrophic consequences. Therefore, a plane is equipped with lots of sensors. When a plane is designed, a model is formed and tested before the actual plane is allowed to fly.
  * For a car, most possible faults are not that dangerous. Therefore, there are usually much fewer sensors, and models, if any, are only *approximate*. It is, in principle, possible to add sensors to a car and to design its exact model, but this would make the car almost as expensive as a small plane.

- In other practical situations, the main goal of the controlled system is to *explore the unknown*:

  - to control a rover on a new planet, with the terrain of unknown type;
  - to control a catheter during a microsurgery, etc.

  In such situations, the entire objective of the control is to learn as much about the system, and we cannot have a precise model of this system before the control is over.

If we cannot apply traditional control methodology, how can we control? A problem. In the case of exploration of the unknown, there was a *fundamental* reason why we cannot use traditional control methods. In other cases, in principle, traditional control methods may be possible at some later moment of time, but as of now, they do not help us to solve the practical problem of control.

So, what can we do if traditional control methods cannot be used?

1.1.2 What we can use instead of the classical control model

Often, we have an additional expert knowledge. In many practical control situations, we have *expert operators* who successfully control the desired system: expert astronauts know how to dock and land the Space Shuttle; expert operators know how to operate a chemical plant, etc. Therefore, it is desirable to extract the "control rules" from the experts and use this knowledge in an automated controller.

Comment. Control rules do not always come from the top experts, there are two extreme situations when these rules come from common sense:

- One situation is *everyday control*, e.g., control of a car, when almost every person considers himself an expert. In such situations, instead of the "expert rules" coming from top experts, we must have *common sense* rules.
- Another extreme is designing controllers for new systems for which no controllers have been known before, and for which, therefore, there aren't yet any expert controllers. In such situations, we also have to reply on common sense only. As fuzzy control methodology is applied to new areas, such applications become more and more frequent; see, e.g., [16].

Why do we need to extract expert knowledge at all? By definition, an expert is a person who is well knowledgeable about the object. An expert helicopter pilot is a person who can control a helicopter well; an expert doctor is a person who can cure different diseases, etc. If we already have such an expert, why do we need an automated system? There are two reasons for that:

- The first reason is that in every area,
  - there are only a *few* top experts, and
  - there are *many* problems to be solved.

  It is, usually, simply physically impossible to have a top expert for each problem: we cannot have a top surgeon for every surgery, we cannot have a top helicopter pilot for every helicopter in the world. It is therefore desirable to develop a *computer program* that would somehow incorporate the expert knowledge and give advise comparable in quality with the advise of the top experts. In other words, we want to design a computer program that somehow *simulates* the expert.

- Often, it is desirable to avoid using an expert altogether: experts are expensive to use, error-prone, and in some dangerous environments, it is desirable to make completely automatic control systems. E.g., we want to control robots who travel into the volcanos or go to other planets.

In both cases, we need a control program that incorporates expert knowledge. Such programs are called *intelligent control systems.*

Why cannot we simply extract the control strategy $u_i(x_1, \ldots, x_n)$ from an expert? At first glance, the problem seems relatively simple: Since the person is a real expert, we simply ask her multiple questions like "suppose that x_1 is equal to 1.2, x_2 is equal to 1.3, ..., what is u?" After asking all these questions, we will get many patterns, from which we will be able to extrapolate the function $f(x_1, \ldots, x_n)$ using one of the known methods. For example, we may ask the helicopter pilot about all possible combinations of sensor readings and and write down what exactly control the expert recommends. Alas, there are two problems with this idea:

- First, there is a *computational* problem: Since we need to ask a question for each combination of sensor readings, we may end up having to ask *too many* questions. This fact makes our idea *difficult* to implement but *not* necessarily *impossible*: Indeed, we only need to do this lengthy questioning once, so we may afford to spend years, if necessary (actually, the design of the first expert systems did take years).
- However, there is another, more serious problem with this idea that makes it, in most problems, *impossible* to implement. Indeed, this idea may sound reasonable until we try applying it to a skill in which practically all adults consider themselves experts: driving a car. If you ask a driver a question like: "you are driving at 55 miles per hour when the car in 30 feet in front of you slows down to 47 miles per hour, for how many seconds do you hit the brakes?", nobody will give a precise number.

What can we do? We might install measuring devices into a car (or into a driving simulator), and simulate this situation, but the resulting braking times may differ from simulation to simulation.

The problem is not that the expert has some precise number (like 1.453 sec) in mind that he cannot express in words; the problem is that one time it will be 1.39, another time it may be 1.51, etc.

An expert usually expresses his knowledge by using words from natural language. An expert *cannot*, usually, express his knowledge in *precise* numerical terms (such as "hit the brakes for 1.43 sec"), but he *can* formulate his knowledge by using *words* from natural language. For example, an expert can say "hit the brakes for a while".

So, the knowledge that we can extract from an expert consists of statements like "if the velocity is a little bit smaller than maximum, hit the breaks for a while".

There may be also another uncertainty involved, e.g., an expert may say something like "if a patient has a temperature of 100 degrees, and a headache, and ... ⟨several other symptoms⟩, then, most probably, it is a flu". Here, the words "most probably" are the words from natural language that an expert uses to describe his knowledge.

Enters fuzzy control.

- We *know* expert's control rules formulated by words from natural language.
- We *want* to produce the precise control strategy.

The methodology that transforms the informal ("fuzzy") expert control rules into a precise control strategy is called *fuzzy control.* The idea of this methodology was first proposed by L. Zadeh in [10], [11], and the methodology itself was first proposed and applied by E. Mamdani in [3], [4] (see also [6], [2]). In this chapter, we will describe how exactly this transformation is done.

1.2 HOW TO TRANSLATE FUZZY RULES INTO THE ACTUAL CONTROL: GENERAL IDEA

Toy example: a thermostat. We will illustrate the main ideas of fuzzy control methodology on the example of another situation in which everyone feels himself an expert: controlling a *thermostat.*

For simplicity, we will consider a *toy*, simplified thermostat in which a thermometer shows the current temperature and a dial allows us to control the temperature:

- turning the dial to the *left* makes it *cooler*;
- turning it to the *right* makes it *warmer.*

The angle on which we turn the dial will be denoted by u.

We will also assume, for simplicity, that we know the comfort temperature T_0 that we try to achieve. Strictly speaking, the input to our control is the temperature T. However, in reality, our control decisions depend not so much on the *absolute* value of T but rather on the *difference* $T - T_0$ between the actual temperature T and the ideal temperature T_0. Since this difference is important, we will use a special notation for it: $x = T - T_0$. Our goal is to describe the appropriate control u for each possible value x, i.e., to describe a function $u = f(x)$.

For such an easy system, we do not need any expert to formulate reasonable rules; we can immediately describe several reasonable control rules:

- If the temperature T is *close* to T_0, i.e., if the difference $x = T - T_0$ between the actual temperature is *negligible*, then no control is needed, i.e., u should also be negligible.
- If the room is slightly overheated, i.e., if x is positive and small, we must cool it a little bit (i.e., u must be negative and small).
- If the temperature is a little lower than we would like it to be, then we need to heat the room a little bit. In other terms, if x is small negative, then u must be small positive.

We can formulate many similar natural rules. For simplicity, in our toy example, we will restrict ourselves to these three. As a result, we get the following three rules:

- if x is negligible, then u must be negligible;
- if x is small positive, then u must be small negative;
- if x is small negative, then u must be small positive.

Toy example re-formulated. Our goal is to make these and similar rules accessible for the computer. Before describing how to do it, let us first reformulate these rules in a way that will somewhat clarify these rules. Namely, in the formulation of these rules, we want to clearly separate the *properties* (like "x is negligible") and *logical connectives* (like "if ... then"). To achieve this separation, let us introduce a shorthand notation for all the properties, and let us use standard mathematical notations for logical connectives:

- $N(x)$ will indicate that x is negligible;
- $SP(x)$ will indicate that x is small positive;
- $SN(x)$ will indicate that x is small negative;
- $NC(u)$ will indicate that u is negligible;
- $SPC(u)$ will indicate that u is small positive;
- $SNC(u)$ will indicate that u is small negative;
- $A \to B$ is a standard mathematical notation for "if A then B".

In these notations, the above three rules take the form:

$$N(x) \to NC(u); \qquad (2)$$

$$SP(x) \to SNC(u); \qquad (3)$$

$$SN(x) \to SPC(u). \qquad (4)$$

General case. In general, the expert's knowledge about the dependence of u on $x_1, \ldots, x_n$ can be expressed by several rules of the type

$$\text{If } x_1 \text{ is } A_{r1}, \ldots, \text{ and } x_n \text{ is } A_{rn}, \text{ then } u \text{ is } B_r.$$

Here, $r = 1, \ldots, R$ is the rule number, and A_{ri} and B_r are words from natural language that are used in r-th rule, like "small", "medium", "large", "approximately 1", etc.

- In our toy example, we only had *one* input variable: the temperature T (or, to be precise, the difference x between T and the desired temperature T_0).
- In the general case, we have *several* input variables, so, in addition to the logical connective "if ... then", we need another logical connective "and".

In mathematics, "and" is usually denoted by & (or by $\wedge$). If we use the standard mathematical notation for "if ... then" and "and", we can re-formulate the above rules as follows:

$$A_{r1}(x_1) \,\&\, \ldots \,\&\, A_{rn}(x_n) \rightarrow B_r(u), \tag{5}$$

where $r = 1, \ldots, R$. The set of rules is usually called a *rule base*.

Comment. In this text, we will only consider the case when a rule base consists of *straightforward* if then rules. In reality, experts may have some additional information about the object, which may be formulated in a more complicated form. For example, an expert may formulate several different rules bases, and then formulate "meta-rules" that decide which of the rule bases are applicable.

General idea. Our goal is to represent a rule base in a computer. A rule base has a clear structure:

- A rule base consists of *rules.*
- Each rule, in its turn, is obtained from *properties* (expressed by words from natural language) by using *logical connectives.*

In view of this structure, it is reasonable to represent the rule base by first representing the basic elements of the rule base, and then by extending this representation to the rule base as a whole. In other words, it makes sense to follow the following methodology:

- First, we represent the basic properties $A_{ri}(x_i)$ and $B_r(u)$.
- Second, we represent the logical connectives.
- Third, we use the representations of the basic properties and of the logical connectives to get the representations of all the rules.

- Fourth, we combine the representations of different rules into a representation of a rule base.

As a result of these four steps, we get an "advising" (expert) system. For example, if we apply these four steps to the medical knowledge, we ideally, get a system that, given the patient's symptoms, provides the diagnostic and medical advise; e.g., it can say that most probably, the patient has a flu, but it is also possible that he has bronchitis. Such an advice, coming from an expert system, is used by a specialist to make a decision.

In *control*, we want a system to automatically make a decision based on its own conclusions. Therefore, for control situations, we need an *additional*, fifth follow-up step:

- The computer-based system makes a decision.

In the following text, we will describe how these five steps are implemented.

Fuzzy control is indeed very successful. The resulting five-step methodology is indeed very successful in control. The resulting *fuzzy control* is used in various areas ranging from appliances (camcorders, washing machines, etc.) to automatically controlled subway trains in Japan to cement kilns to regulating temperature within the Space Shuttle. The reader who is interested in learning more about fuzzy methods in general and fuzzy control in particular is advised to start, e.g., with one of the following textbooks:

- Klir and Yuan [15] give a more *engineering*-oriented introduction;
- Nguyen and Walker [18] is more oriented towards *mathematical* methods.

1.3 MEMBERSHIP FUNCTIONS AND WHERE THEY COME FROM

Before we represent a property $P(x)$, let us represent the "grade of truth" of this property for each x. To represent a property like "x is positive small" ($SP(x)$), we must be able, for every possible value of the temperature difference x, to represent the expert's opinion of whether this particular value x is indeed small. To represent this opinion, we must solve the following problem:

- All the properties that are traditionally analyzed in mathematics, such as "x is positive", are *crisp*, in the sense that every real number is either positive, or not.
- On the other hand, properties like "x is small" (in which we are interested) are *not crisp*. To be more precise:
  - Some values x are so small that practically everyone would agree that they are small.

- Some values x are so huge that practically everyone will agree that they are not small.
- However, for many intermediate values x, x is small to a degree, i.e., it is neither absolutely small nor absolutely not small, and if human experts have to decide upon the best tag "small" or "not small" for x, they may disagree:
  * For a researcher who performs temperature-sensitive experiments in the lab, the difference of $x = 0.5$ degrees may not be small at all.
  * However, for a living room, even the difference of ± 5 degrees is usually not only small, but even negligible.

Such non-crisp properties are called *fuzzy*. This term was introduced by Lotfi Zadeh in his pioneer paper [9].

How can we represent fuzzy properties?

- If $P(x)$ is a *crisp* property (like "positive"), then for every x, $P(x)$ is either true or false.

  Representing the corresponding truth value in the computer is easy: "true" is usually represented by 1, and "false" by 0.

- If $P(x)$ is a *fuzzy* property (like "small"), then in general, whether a real value x satisfies $P(x)$ is a matter of degree. $P(x)$ may be true only to a limited extent. The compatibility between a value and a fuzzy property is no longer an all or nothing matter. For example, for "small", the larger the value x, the less it is true that x is small. So, for different values x, an expert may have different "grades of truth" ("truth values") that x satisfies this property P:
  - For some values x, the expert considers the statement $P(x)$ to be absolutely true. In this case, the grade of truth corresponds to "true".
  - For some other values x, the expert considers the statement $P(x)$ to be absolutely false. In this case, the grade of truth corresponds to "false".
  - For yet other values x, the expert considers $P(x)$ to be neither absolutely true nor absolutely false. In this case, the expert's grade of truth is *intermediate* between "true" and "false".

  How can we represent these intermediate grades of truth in a computer?

  If 0 corresponds to "false", and 1 to "true", then it is natural to represent grades of truth that are intermediate between "false" and "true" by numbers from the interval $(0, 1)$.

Thus, to describe *arbitrary* degrees of truth, we must describe:

- either absolute truth (represented by 1),
- or absolute falsity (represented by 0),
- or intermediate grades of truth (represented by numbers from the open interval $(0, 1)$).

Therefore, we arrive at the following conclusion:

We must use real numbers from the interval $[0, 1]$.

Historical comments.

- From the *mathematical* viewpoint, what we are doing is a very natural idea: we extend the traditional *2-valued* logic (in which the truth value of each statement is an element of a 2-element set {"true", "false"} $= \{0, 1\}$), to the *interval* $[0, 1]$. Mathematicians have been developing the corresponding mathematical formalisms (called *multiple-valued logics*) since the pioneer works of K. Lukasiewicz in the early 1920s.
- In *computer science*, the idea of using numbers from the interval $[0, 1]$ to describe different grades of truth was also first proposed by Lotfi Zadeh in his pioneer paper [9].

Comments.

- We have used the term *grade of truth*, or *truth value*; other words are also used, such as *degree of belief*, *degree of certainty*, *subjective probability*, etc. From our viewpoint, all of these alternative terms are not completely adequate:
  - On one hand, each of these terms bring with itself some intuitive meaning.
  - On the other hand, the intuitive meaning brought by these words may be sometimes misleading. For example, the terms "belief" and "subjective probability" may lead to a confusion with the "objective probability", i.e., crudely speaking, with a *frequency* of a certain event (like tossing a coin).
  - Our terminology intends to reflect the difference between the two different compatibility degrees:
    * the degree of compatibility between a *precise* number and a *fuzzy* linguistic term, which we call the *degree of truth*, and
    * the degree of compatibility between the *imprecisely located* number (e.g., a random number) and a *crisp* (well-defined) set (e.g.,

an interval); this degree of compatibility is usually called *degree of belief.*

Several different terms can be used for these two notions, but it is important to realize their difference, because these two notions are often mistakenly confused, and such a confusion leads to misunderstandings about fuzzy sets and fuzzy control.

- The interval $[0, 1]$ is probably the most natural to use, but some expert systems use a *different interval* to represent uncertainty. For example, historically the first successful expert system MYCIN [23] developed at Stanford University for diagnosing rare blood diseases, used values from the interval $[-1, 1]$ to represent uncertainty. Another possibility is to use the set $[0, +\infty)$ of all non-negative real numbers; this alternative truth set is used, e.g., in Chapter 12 of this book ("Fuzzy Systems viability theory and toll sets", by Aubin and Dordan).
- The main advantage of using an interval (e.g., $[0, 1]$) is that elements from this interval (i.e., real numbers) can be easily represented and easily processed in a computer. However, sometimes, to get a more adequate representation of the expert's grade of truth, we may need a more sophisticated representation. For example, an expert may not be sure about his truth value. To represent this uncertainty, we may want to describe the expert's truth value not by a *single* number, but by an *interval* of possible numbers (see, e.g., surveys [17], [38]):
  - If an expert has no information about the object, he cannot have any definite truth value. To describe this situation, we will use the entire interval $[0, 1]$ as the description of this expert's grade of truth.
  - If an expert has a certain information, them we may use a narrower interval, or even a number (i.e., a degenerate interval).

  There are several even more sophisticated ways of representing uncertainty. In this chapter, we will only consider the simplest possible way of representing truth values: by using real numbers.

We need to elicit a numerical truth value from the expert. We agreed that for each statement of the type $P(x)$, where:

- P is a fuzzy property (i.e., a property formulated by words from natural language), and
- x is a real number,

a reasonable way to represent the expert's grades of truth is to use numbers from the interval $[0, 1]$. The natural next question is: how can we "extract", elicit the value from an expert?

Direct elicitation is impossible. The ideal situation would be if an expert would directly provide us with this number, but this is, unfortunately, not realistic:

- we want numbers from the interval $[0, 1]$, because they are very *natural* for a *computer*; but
- real numbers from the interval $[0, 1]$ are *not natural* for a *human expert*; it is very difficult for most experts to express their truth value in a given statement by a real number.

Since we cannot elicit these numbers *directly*, we have to use *indirect* elicitation techniques.

There are several dozen different elicitation techniques; in this chapter, we will only describe the most frequently used ones.

First elicitation method: selecting on a scale. If we cannot elicit a *real* number from an expert, maybe we can elicit *some* number from him, and then convert the result into a real number from the interval $[0, 1]$. In many cases, this is indeed possible: many polls ask us to estimate the grade of truth of different statements on a certain integer scale, e.g., on a scale from 0 to 5, or on a scale from 0 to 10. So, we can ask an expert to mark his grade of truth in a given statement $P(x)$ on a given scale 0 to S (0 to 5, 0 to 10, etc). On this scale:

- 0 corresponds to "$P(x)$ is absolutely false";
- S corresponds to "$P(x)$ is absolutely true";
- intermediate marks represent different grades of truth.

As a result of this procedure, we get an integer from 0 to S. The most natural way to convert these integers $0, 1, 2, \ldots, S$ into real numbers from the interval $[0, 1]$ is to *re-scale* the interval $[0, S]$ into the interval $[0, 1]$ by dividing by S. So, if an expert has chosen the mark s on a scale from 0 to S, we will take s/S as the desired truth value.

For example, if an expert selects, 3 on a scale from 0 to 5, then we take $3/5 = 0.6$ as the desired truth value.

The first elicitation method is not always applicable. The above method seems reasonable and is computationally very simple. The problem with this method is that experts often do not feel comfortable expressing their truth values by numbers on an (integer) scale. Usually, these experts only distinguish between three cases:

- a given statement $P(x)$ is true;
- a given statement $P(x)$ is false;

- we do not know whether a given statement $P(x)$ is true or false.

If we have such experts, how can we elicit their grades of truth?

Second elicitation method: polling. We have already mentioned that the first elicitation method is taken from the experience of the polls. In addition to the polls that require us to mark a point on a scale, there are other polls in which we only answer "yes", "no", or "undecided". For example, polls conducted before the elections are usually of this type. As a result of each poll of this type, we get a *percentage* of people who answered "yes" ("60% are planning to vote for candidate X"). This percentage represents exactly what we want: a number from the interval [0, 1] (e.g., 60% means 0.6).

So, we arrive at the following *polling* method of eliciting the truth values from the experts: we ask several experts whether $P(x)$ is true or false. Some of these experts may not answer at all, or give "unknown" as an answer; these experts we need not count. If out of N experts who gave a definite answer, M answered "yes" (i.e., "yes, $P(x)$ is true"), then we take the ratio M/N as the desired truth value.

Problems with the second method. For the polling method to give meaningful results, we need many respondents. From the strict mathematical viewpoint, for a pool to give meaningful results (e.g., to give the percentage with a guaranteed 10% accuracy), we must interview at least 1,000 people.

- For an election poll, interviewing 1,000 people is quite possible: millions participate in the elections, and we want to know the opinion of the representative group of people.
- However, in an expert system, we are trying to formalize the knowledge of the top experts. We may not have that many top experts, and even if we do, top experts are usually very busy people, we may not be able to convince all of them to cooperate with this project. Last but not the least, the time of top experts costs money, and we may not have the money to hire 1,000 of them.

Can we somehow ask a single expert and still get a number?

An alternative elicitation method: using subjective probabilities and bets. Some people feel uncomfortable estimating a probability or marking a value on a scale, but they have a good feel for *bets*. Such people cannot express the probability of their favorite horse winning, but they are absolutely sure that, say, they are willing to bet 4 to 1 but not 5 to 1 that this horse will win.

If we can extract a betting ratio from an expert, then we can easily transform this number into the subjective probability. For example, if an expert is willing to bet 4 to 1 but not any better that $P(x)$ is true (e.g., that most people will agree that $x = 1$ is a small temperature difference), this means that this experts' subjective probability is $4/(4+1) = 0.8$.

The reader should be aware that these "subjective probabilities" are not truth values (membership grades). For example, if we are describing the property "small", then what we *want* to get, for each x, is the degree to which x is small. Instead, we *get* the probability that the value x will be assigned the tag "small" (which is, as a tag, crisp, and not fuzzy). In more mathematical notations, for each value x, we get the conditional probability $P(\text{tag} = \text{"}small'' \mid x)$ that, given the value x, the expert will choose the tag "small". The possibility to extract the actual membership grades for "small" stems from the idea that the shape of the dependence of this subjective probability on x should reflect the shape of the membership function that describes the term "small". This idea leads to the possibility of extracting the membership function from the probability values; for details, see, e.g., [13].

From finitely many truth values to a membership function: extrapolation is needed. We started this section with a problem of describing a fuzzy property $P(x)$ (of the type "x is small"). Namely, for every possible value x, we would like to know the *grade of truth* $t(P(x))$ that this value x satisfies the property P. This grade of truth is usually denoted by $\mu_P(x)$, and the function that transforms a real number x into a value $\mu_P(x) \in [0,1]$ is called a *membership function*, or a *fuzzy set*.

For each value x, we can, using one of the above elicitation techniques, find the value $\mu_P(x)$ for this x. However, this is not sufficient:

- On one hand, we want to know the value $\mu_P(x)$ for all possible real numbers x, i.e., for *infinitely many* different values.
- On the other hand, we can only ask an expert *finitely many* questions and therefore, we can only determine the values of the membership function for *finitely many* different values $x^{(1)}, \ldots, x^{(v)}$.

Thus, after all the elicitation is over, we only know the values $\mu_P(x^{(p)})$ of the desired function $\mu_P(x)$ for v different values $x^{(1)}, \ldots, x^{(v)}$. To reconstruct the desired function, we must use *extrapolation*.

Simple examples of extrapolation. Let us describe a few simple examples of extrapolation. The simplest possible extrapolation is an extrapolation by *piece-wise linear functions*. The simplest case of an extrapolation is when we start with the values x for which the expert is either absolutely sure that x is true, or he is absolutely sure that x is false, i.e., with values for which $\mu_P(x) = 0$ or $\mu_P(x) = 1$. Let us give several examples of such situations.

- Let us first describe the property of the type "x is negligible".

- The only case when we are 100% sure that x is negligible is when $x = 0$. So, we have the value $\mu_P(0) = 1$.
- Usually, we also know the value $\Delta > 0$ after which the difference in temperatures is no longer negligible. For example, for a thermostat that controls the room's temperature, we can take $\Delta = 10$. This means that $\mu_P(x) = 0$ for $x \geq \Delta$ and for $x < -\Delta$.

We know the value of the function $\mu_P(x)$ for $x \leq -\Delta$, for $x = 0$, and for $x \geq \Delta$. We need to use linear extrapolation to find the values of this function for $x \in (-\Delta, 0)$ and for $x \in (0, \Delta)$. In general, the formula for a linear function $f(x)$ that passes through the point $y_1 = f(x_1)$ and $y_2 = f(x_2)$, is

$$f(x) = y_1 + (x - x_1) \cdot \frac{y_1 - y_2}{x_2 - x_1}.$$

For our case, this formula leads to the following membership function:

- To get the values $\mu_P(x)$ for $x \in (-\Delta, x)$, we take $x_1 = -\Delta$, $x_2 = 0$, $y_1 = 0$, and $y_2 = 1$. As a result, we get the expression $\mu_P(x) = (x + \Delta)/\Delta$.
- To get the values $\mu_P(x)$ for $x \in (0, \Delta)$, we take $x_1 = 0$, $x_2 = \Delta$, $y_1 = 1$, and $y_2 = 0$. As a result, we get the expression $\mu_P(x) = 1 - x/\Delta$.

Thus, the function $\mu_P(x)$ takes the following form:

- $\mu_P(x) = 0$ for $x \leq -\Delta$;
- $\mu_P(x) = (x + \Delta)/\Delta$ for $-\Delta \leq x \leq 0$;
- $\mu_P(x) = 1 - x/\Delta$ for $0 \leq x \leq \Delta$; and
- $\mu_P(x) = 0$ for $x > \Delta$.

The graph of this function has the shape of a *triangle* over the x-axis. Therefore, such functions are called *triangular* membership functions.

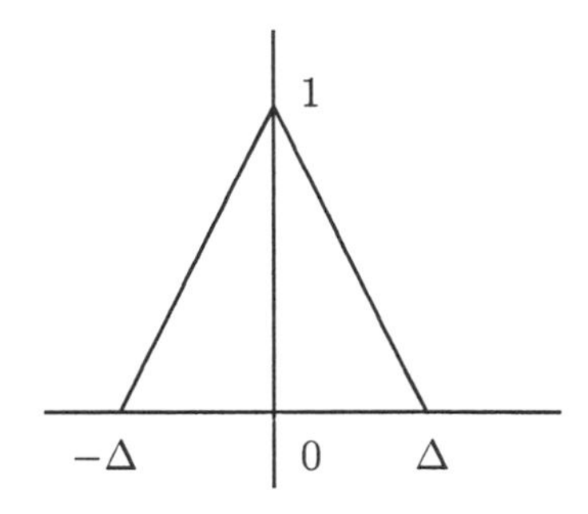

A similar shape describes properties like "close to a", for some fixed a. In this case, the triangular membership function has the form:

- $\mu_P(x) = 0$ for $x \leq a - \Delta$;
- $\mu_P(x) = (x - (a - \Delta))/\Delta$ for $a - \Delta \leq x \leq a$;
- $\mu_P(x) = 1 - (x - a)/\Delta$ for $a \leq x \leq a + \Delta$; and
- $\mu_P(x) = 0$ for $x > a + \Delta$.

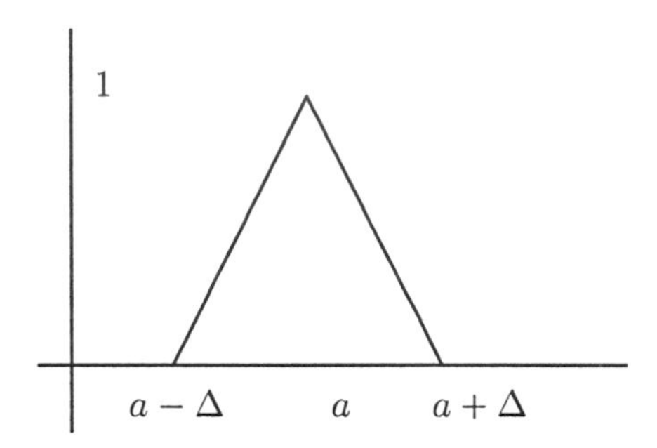

■ For the same property "x is negligible", in some cases, we know the lower bound δ below which the temperature difference x is indeed negligible. In this case, in addition to knowing that $\mu_P(x) = 0$ for $x \leq -\Delta$ and for $x \geq \Delta$, we also know that $\mu_P(x) = 1$ for $-\delta \leq x \leq \delta$. For this function, linear extrapolation to the intervals $(-\Delta, -\delta)$ and (δ, Δ) results in the following membership function:

- $\mu_P(x) = 0$ for $x \leq -\Delta$;
- $\mu_P(x) = (x + \Delta)/(\Delta - \delta)$ for $-\Delta \leq x \leq -\delta$;
- $\mu_P(x) = 1$ for $-\delta \leq x \leq \delta$;
- $\mu_P(x) = 1 - (x - \delta)/(\Delta - \delta)$ for $\delta \leq x \leq \Delta$; and
- $\mu_P(x) = 0$ for $x > \Delta$.

The graph of this function has the shape of a *trapezoid* over the x-axis. Therefore, such functions are called *trapezoidal* membership functions.

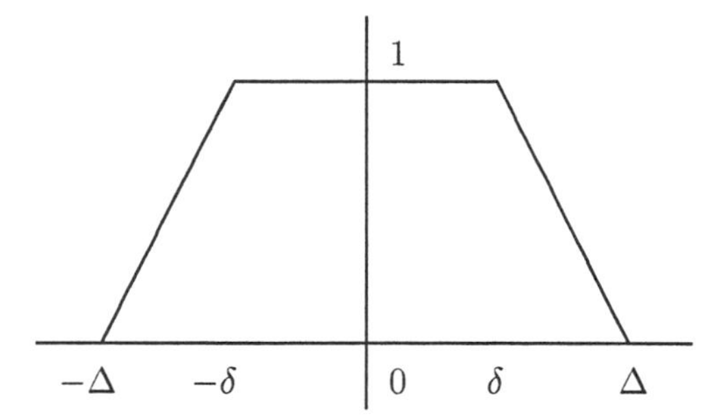

- Another frequent example of piece-wise functions are functions that describe properties like "x is large". For these properties, we usually know that values below a certain δ are definitely not large, and that values above a certain $\Delta \gg \delta$ are definitely large. So, we know that $\mu_P(x) = 0$ for $x \leq \delta$ and $\mu_P(x) = 1$ for $x \geq \Delta$. To get the values of $\mu_P(x)$ for $x \in (\delta, \Delta)$, we use a linear extrapolation. As a result, we get the following function:
  - $\mu_P(x) = 0$ for $x \leq \delta$;
  - $\mu_P(x) = (x - \delta)/(\Delta - \delta)$ for $\delta \leq x \leq \Delta$;
  - $\mu_P(x) = 1$ for $x \geq \Delta$.

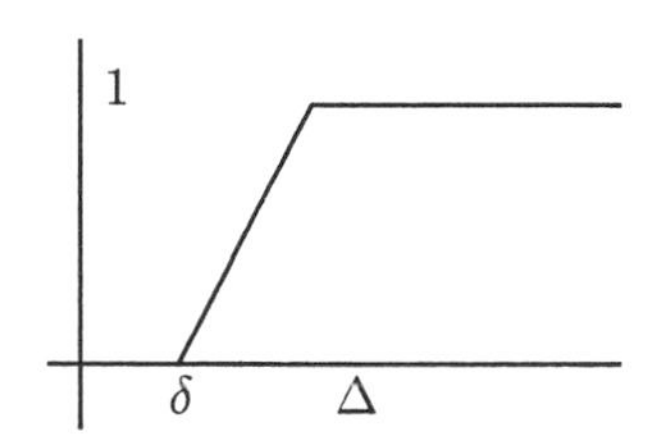

We have described the triangular and trapezoid functions because they are *the simplest.* Interestingly, in fuzzy control, they are also, at present, *the most frequently used membership* functions. This fact has two possible explanations:

- First, fuzzy control is still at its infancy. The first successful application of fuzzy control was produced a little more than 20 years ago (in 1974, by E. Mamdani). Because of that, the potential of simple fuzzy control applications is not yet exhausted. *Maybe, we will need more complicated membership functions* later on, when all simple applications will be used, but so far, we do not seem to need them.

- Second, we are formalizing *approximate* expert knowledge. If all an expert can say about a control is that it should be *small,* or *medium,* or *large,* it may not be reasonable to try to formalize these notions in a too complicated way. Thus, simple models of expert uncertainty, in particular, simple membership functions, may be more reasonable than complicated ones. From this viewpoint, *we may not need more complicated membership functions at all.*

Probably, the truth lies somewhere in between; but anyway, so far, piecewise-linear functions seem to work just fine.

1.4 FUZZY LOGICAL OPERATIONS

How can we represent logical connectives? An ideal solution. Suppose that an expert system contains statements A (e.g., A is "x_1 is small"), and B (e.g., "x_2 is medium"), and we have elicited the truth values $t(A)$ and $t(B)$ from the experts. Suppose now that a user wants to know the truth value of a composite statement $A\&B$.

In principle, knowing only the two numbers $t(A)$ and $t(B)$ is not sufficient to describe the expert's truth value of $A\&B$: e.g., if $t(A) = t(B)$,

- it could be that the experts perceive A and B as *equivalent*, in which case $t(A\&B) = t(A)$, or
- it could also be that the experts perceive A and B as *independent* statements, in which case the possibility of A *and* B being true is smaller than the possibility that one of them is true: $t(A\&B) < t(A)$.

So, the *ideal* situation would be to elicit, from the experts, not only the grades of truth in the *basic* statements from the knowledge base, but also the grades of truth in all possible *logical combinations* of these statements.

This ideal solution is not practically possible. The above described ideal solution is practically impossible: If we have N statements $S_1, ..., S_N$ in the knowledge base, then for each of $2^N - 1$ non-empty subsets $\{S_{i_1}, ..., S_{i_k}\}$ of the knowledge base, we need to elicit the grade of truth in the corresponding AND-statement $S_{i_1}\&...\&S_{i_k}$. For a realistic expert system, N is in hundreds, so asking an expert 2^N questions is impossible.

A practical way to represent logical connectives: logical operations. In view of this practical impossibility, although in *some* cases, we will be able to have the grade of truth $t(A\&B)$ stored in the knowledge base, in *general*, we often have to deal with a following situation:

- we know the grades of truth $t(A)$ and $t(B)$ in statements A and B;
- we know nothing else about A and B; and
- we are interested in the (estimated) truth value of the composite statement $A\&B$.

Since the only information available consists of the values $t(A)$ and $t(B)$, we must compute $t(A\&B)$ based on these values. We must be able to do that for arbitrary values $t(A)$ and $t(B)$. Therefore, we need a *function* that transforms the values $t(A)$ and $t(B)$ into an estimate for $t(A\&B)$. Such a function is called an *AND-operation.* If an AND-operation $f_{\&} : [0,1] \times [0,1] \to [0,1]$ is fixed, then we take $f_{\&}(t(A), t(B))$ as an estimate for $t(A\&B)$.

Similarly:

- to estimate the truth value of $A \vee B$, we need an *OR-operation* $f_{\vee} : [0,1] \times [0,1] \to [0,1]$.
- to estimate the truth value of the negation $\neg A$, we need a *NOT-operation* $f_{\neg} : [0,1] \times [0,1]$.

Terminological comment. AND operations are also called *t-norms*, and OR operations are also called *t-conorms.*

Natural properties of logical operations. The logical operations with fuzzy values must satisfy some *natural conditions.*

- For an expert, $A\&B$ and $B\&A$ mean the same. Therefore, the estimates $f_{\&}(t(A), t(B))$ and $f_{\&}(t(B), t(A))$ for these two statements should coincide. To achieve that, we must require that $f_{\&}(a, b) = f_{\&}(b, a)$ for all a and b; in other words, the operation $f_{\&}$ must be *commutative.*
- Similarly, from the fact that $A\&(B\&C)$ and $(A\&B)\&C$ mean the same, we can deduce the requirement that $f_{\&}$ must be *associative*: $f_{\&}(a, f_{\&}(b, c)) = f_{\&}(f_{\&}(a, b), c)$ for all a, b, and c.
- If A is absolutely false ($t(A) = 0$), then $A\&B$ is also absolutely false, i.e., $f_{\&}(a, 0) = 0$ for all a.
- If A is absolutely true ($t(A) = 1$), then $A\&B$ is true iff B is true, so, the truth value of $A\&B$ must coincide with the truth value of B: $f_{\&}(1, b) = b$ for all b.
- If additional evidence increases our estimates of the truth values of A and B, then our estimated truth value of $A\&B$ must also increase, so $f_{\&}$ must be *monotonic.*
- If the perceived truth value of A changes a little bit, then the truth value of $A\&B$ must also change slightly. In other words, $f_{\&}$ must be *continuous.*

How can we determine AND and OR operations? Since our goal is to describe the expert knowledge, we must elicit these operations from the experts. This can be done in the following manner:

- We form several pairs of statements (A_k, B_k), $k = 1, 2, \ldots$

- For each pair from this set, we elicit, from the experts, the grades of truth $t(A_k)$, $t(B_k)$, and $t(A_k \& B_k)$.
- Then, we use an extrapolation procedure to find a function $f_\&(a,b)$ for which $f_\&(t(A_k), t(B_k)) \approx t(A_k \& B_k)$.

Similar procedures enable us to determine OR and NOT operations.

Historical comment. Empirical evidence shows that in different fields, people use different AND and OR operations. This difference is easy to explain:

- In some applications (e.g., in *medicine*), mistakes can be deadly, so more *cautious* estimates are needed (e.g., $f_\&(a,b) = a \cdot b$).
- In other applications (e.g., in *geology*), we cannot measure as many parameters as in medicine, so, we have to rely more on expertise, and hence, experts must take risks. In these applications, more brave, more optimistic estimates are needed: e.g., a geologist starts to drill in the uncertainty in which a surgeon is not likely to start an incision. Therefore, for such applications, we need more *optimistic* estimates for $t(A\&B)$ (e.g., $f_\&(a,b) = \min(a,b)$).

Simple AND, OR, and NOT operations: an idea. There are many possible AND and OR operations, many of them very complicated. However, in most applications, we do not need this complexity:

- Our goal is to apply these operations to the truth values, that, in their turn, come from the values of membership functions.
- We have already mentioned that in most applications, it is sufficient to use the *simplest* membership functions, that are obtained by applying the *simplest* extrapolation methods to crisp values of the corresponding properties.

Since we start with the simplified expressions for truth values, it makes sense to consider *simple* AND and OR operations for handling these truth values, i.e., to consider AND and OR operations that are obtained by applying some *simple* extrapolation techniques (linear or quadratic) to the *crisp* values of these operations.

Simple AND, OR, and NOT operations: results. To describe negation ("not"), we can have a *linear* operation $f_\neg(a)$. The conditions that $f_\neg(0) = 1$ and $f_\neg(1) = 0$ uniquely determine this operation as $f_\neg(a) = 1 - a$.

The resulting formula $t(\neg A) = f_\neg(d(A)) = 1 - t(A)$ for the *truth value* of $\neg A$ resembles a formula for the *probability* of $\neg A$: If we know the probability $P(A)$ of an event A, then the probability $P(\neg A)$ that this event will not occur is equal to $P(\neg A) = 1 - P(A)$. Thus, the linear NOT operation is consistent with the two probability-like elicitation procedures for truth values: namely, with the procedures that are based on polling and betting.

For AND and OR, linear operations are impossible, but we can have *piecewise linear* or *quadratic* ones. One can show that there are two possible piecewise linear AND operations: $f_\&(a,b) = \min(a,b)$ and $f_\&(a,b) = \max(0, a + b - 1)$. Similarly, we get two possible OR operations: $f_\vee(a,b) = \max(a,b)$ and $f_\vee(a,b) = \min(a+b,1)$. The only *quadratic* AND and OR operations are $f_\&(a,b) = a \cdot b$ and $f_\vee(a,b) = a + b - a \cdot b$ (they are called *algebraic product* and *algebraic sum*).

The operations $\min(a,b)$, $\max(a,b)$, $a \cdot b$, and $a + b - a \cdot b$ were first proposed in the pioneer 1965 paper of L. Zadeh. The operations $\max(0, a + b - 1)$ and $\min(a + b, 1)$ were introduced in [27] under the name of *bold* operations.

Similarly to the linear NOT operation, these six operations are also consistent with the probability-like interpretation: namely, if we know the probabilities $a = P(A)$ and $b = P(B)$ of two events A and B, then:

- The probability $P(A\&B)$ that both events A and B will occur can take any real value from the interval $[\max(a+b-1, 0), \min(a,b)]$. In particular, if we know that A and B are independent events, then $P(A\&B) = a \cdot b$.
- The probability $P(A \vee B)$ that at least one of the events A or B will occur can take any real value from the interval $[\max(a,b), \min(a + b, 1)]$. In particular, if we know that A and B are independent events, then $P(A \vee B) = a + b - a \cdot b$.

More complicated AND and OR operations. The above-described six AND and OR operations are the ones that are most frequently used in expert systems and in fuzzy control. However, these operations are not the only possible ones. Indeed, we can *re-scale* the scale of truth values, i.e., we can represent the truth value $a = t(A)$ of the statement A not by the original value a, but by a new value $a' = t'(A) = r(a) = r(t(A))$, where $r(a)$ is a monotonic function.

How will an AND operation that has the form $c = f_\&(a,b)$ in the old scale look in the new scale $a' = r(a)$? In other words, if we know the truth values $a' = t'(A) = r(t(A))$ and $b' = t'(B) = r(t(B))$ in the new scale, what will be the truth value $c' = t'(A\&B) = r(A\&B))$ in this new scale? To get the expression for c' in terms of a' and b', we must do the following:

- First, we convert the values a' and b' into the old scale by applying the inverse re-scaling r^{-1}: $a = r^{-1}(a')$ and $b = r^{-1}(b')$.
- Then, we apply the AND operation $f_\&(a,b)$ to the values a and b expressed in the old scale. As a result, we get the value $c = f_\&(a,b) = f_\&(r^{-1}(a'), r^{-1}(b'))$.

- Finally, we convert the value c into the new scale by applying the re-scaling $r(a)$: $c' = r(c)$.

As a result of this three-step procedure, we get a new AND operation

$$f'_{\&}(a', b') = r(f_{\&}(r^{-1}(a'), r^{-1}(b'))).$$

Similarly, we get a new OR operation

$$f'_{\vee}(a', b') = r(f_{\vee}(r^{-1}(a'), r^{-1}(b')))$$

and a new NOT operation

$$f'_{\neg}(a') = r(f_{\neg}(r^{-1}(a'))).$$

The new operations are called *isomorphic* to the old ones, because they represent the same operations but on a different scale.

There are special names for operations that are isomorphic to piece-wise linear and quadratic AND and OR operations. (These names may sound somewhat strange for a computer science reader, because they were invented before the computer applications and they describe algebraic properties of the corresponding operations.)

- Operations that are isomorphic to quadratic AND and OR operations, i.e., operations of the type

  $$f'_{\&}(a', b') = r(r^{-1}(a') \cdot r^{-1}(b'))$$

  and

  $$f'_{\vee}(a', b') = r(r^{-1}(a') + r^{-1}(b') - r^{-1}(a') \cdot r^{-1}(b')),$$

  for some strictly increasing continuous function $r(a)$, are called *strictly Archimedean* AND and OR operations.

- Operations that are isomorphic to bold AND and OR, i.e., operations of the type

  $$f'_{\&}(a', b') = r(\max(r^{-1}(a') + r^{-1}(b') - 1, 0))$$

  and

  $$f'_{\vee}(a', b') = r(\min(r^{-1}(a') + r^{-1}(b'), 1)),$$

  for some strictly increasing continuous function $r(a)$, are called *non-strictly Archimedean* AND and OR operations.

- If we use $f_{\&}(a, b) = \min(a, b)$ and $f_{\vee}(a, b) = \max(a, b)$, then, as one can see, for every strictly monotonic function $r(a)$, the isomorphic operations $f'_{\&}(a', b')$ and $f'_{\vee}(a', b')$ have exactly the same form: $f'_{\&}(a', b') = \min(a', b')$ and $f'_{\vee}(a', b') = \max(a', b')$. These two operations are called *idempotent.*

Most of the AND and OR operations that are actually used belong to one of these three types.

In addition of these three classes of operations, we may consider even more complicated AND and OR operations that are, e.g., isomorphic to an idempotent operation on one subinterval of the interval $[0, 1]$ and to a strictly Archimedean one on another subinterval of this interval $[0, 1]$. It turns out that this combination covers *all* possible AND and OR operations: namely, for an arbitrary AND and OR operation that satisfies several reasonable properties (e.g., the ones described above), we can sub-divide the interval $[0, 1]$ into sub-intervals on each of which the operation is isomorphic to an operation from one of the three classes. This general classification result was proven in [32]; see [15], [18] for details.

1.5 MODELING FUZZY RULE BASES

We must formalize implication. We started the description of the expert knowledge by mentioning that the expert knowledge is usually represented by "if-then" rules. Since we want to formalize these rules, it seems reasonable to formalize the statements of the type "if A then B", i.e., using a logical term for such statements, to formalize fuzzy *implication* in the same way as we formalized "and", "or", and "not" operations.

Our goal is to describe the expert knowledge for a computer. So, it is reasonable to generalize the way "crisp" (two-valued) logical operations are implemented in the computer. Most computer languages have built-in logical operations "and", "or", and "not", but usually, not implication. Therefore, even if all the statements are crisp, when we formalize these statements for the computer, we will still need to first *reformulate* implication in terms of other logical operations. Due to this fact, most expert systems and intelligent control systems, traditionally, did not *directly* formalize implication but instead, try first to *reformulate* if-then rules in terms of "and", "or", and "not". In this chapter, we will briefly describe this traditional (and most widely spread) approach.

Comment. In two-valued logic, implication can be proven to be exactly representable in terms of "and", "or", and "not". However, as we go to fuzzy logic, which provides a better description of expert's reasoning, this representation becomes only approximately valid. Thus, it is reasonable to expect that we get a better representation of expert's "if-then" rules if, instead of describing implication in terms of other logical operations, we would formalize implication directly, by using one of the known *fuzzy implication* operations. These expectations were suggested by Zadeh in [11], explicitly formulated in [1], and since then, proven to indeed work well in fuzzy control; see, e.g., [25].

In spite of the naturalness of this idea, most existing fuzzy control systems follow the original Mamdani's idea and use "and" and "or" instead of directly implementing implication. Therefore, in this chapter, we will mainly concentrate on this indirect approach. The logical, implication-based view is described in the chapter by Yager and Filev in this book.

Mamdani's idea: example. Let's first consider our toy thermostat example, with three rules (2)–(4). If we know the difference x between the actual and the desired temperature, what control u should we apply? We have three rules that describe when a control is reasonable. Therefore, u is a reasonable control if one of the the three rules is applicable, i.e., when either:

- the first rule is applicable (i.e., x is negligible) and u is negligible; or
- the second rule is applicable (i.e., x is small positive), and u is small negative; or
- the third rule is applicable (i.e., x is small negative), and u is small positive.

Summarizing, we can say that u is an appropriate choice for a control if and only if either (x is negligible *and* u is negligible), *or* (x is small positive *and* u is small negative), etc.

This approach has the following informal geometric interpretation:

- In the crisp case, when we only consider properties that can be true or false, each property can be described by a crisp set, i.e., normally, by an *interval*. Thus, a general fuzzy property can be described as a fuzzy analogue of an interval, i.e., as a "fuzzy interval".
- In the crisp case, the rule $A \to B$ holds for a pair (x, u) if and only if x belongs to the interval (crisp set) A,a nd u belongs to the interval B. In other words, the set of pairs (x, u) for which the rule holds consists of all possible pairs (x, u) with $x \in A$ and $u \in B$. This set of all possible pairs is called the *Cartesian product* of the sets A and B. Geometrically, the Cartesian product of two intervals is a rectangular-shaped "granule". Thus, we can say that in Mamdani's approach, a rule is viewed as a "fuzzy granule", i.e., a fuzzy Cartesian product of the fuzzy intervals in the input-output space.

Let us describe this approach in more succinct terms. Let us use the following notations:

- $R_k(x, u)$ will indicate that k-th rule is applicable for a given x, and that this rule recommends to use the control value u;
- $C(x, u)$ will indicate that u is a reasonable control for a given input x.

Then, in the above example, we get

$$C(x, u) \equiv R_1(x, u) \lor R_2(x, u) \lor R_2(x, u),$$

where

$$R_1(x,u) \equiv N(x) \,\&\, NC(u); \quad R_2(x,u) \equiv SP(x) \,\&\, SNC(u);$$
$$R_3(x,u) \equiv SN(x) \,\&\, SPC(u).$$

We already know how to formalize the properties (as membership functions, or fuzzy sets), and we know how to formalize the "and" and "or" operations. Thus, for every input x, we can define the truth value $t_r(x,u)$ of r-th rule, i.e., the degree with which this rule will be fired:

$$t_1(x,u) = f_\&(\mu_N(x), \mu_{NC}(u)); \quad t_2(x,u) = f_\&(\mu_{SP}(x), \mu_{SNC}(u));$$
$$t_3(x,u) = f_\&(\mu_{SN}(x), \mu_{SPC}(u)).$$

Reformulating if-then rules in terms of "and", "or", and "not": general case. For a general rule base with R rules of the type (5), we get

$$C(x_1,\ldots,x_n,u) \equiv R_1(x_1,\ldots,x_n,u) \vee \ldots \vee R_R(x_1,\ldots,x_n,u), \tag{6}$$

where

$$R_r(x_1,\ldots,x_n,u) \equiv A_{r1}(x_1) \,\&\, \ldots \,\&\, A_{rn}(x_n) \,\&\, B_r(u). \tag{7}$$

Therefore, for each rule, the "firing degree" is equal to:

$$t_r(x_1,\ldots,x_n,u) = f_\&(\mu_{r1}(x_1),\ldots,\mu_{rn}(x_n),\mu_r(u)), \tag{8}$$

where:

- $\mu_{ri}(x_i)$ is the membership function corresponding to the property A_{ri};
- $\mu_r(u)$ is the membership function corresponding to the property $B_r(u)$;
- $f_\&(a,b,c)$ stands for $f_\&(f_\&(a,b),c)$, and, in general, $f_\&(a_1,\ldots,a_n,a_{n+1}) = f_\&(f_\&(a_1,\ldots,a_n),a_{n+1})$.

Comment. If we use a genuine implication to formalize the if-then rules, then a rule base is modeled as a *disjunction* (and-combination) of rules, not as its conjunction (or-combination). In other words, for a general rule base with R rules of the type (5), we get

$$C(x_1,\ldots,x_n,u) \equiv R_1(x_1,\ldots,x_n,u) \,\&\, \ldots \,\&\, R_R(x_1,\ldots,x_n,u), \tag{6a}$$

where

$$R_r(x_1,\ldots,x_n,u) \equiv A_{r1}(x_1) \,\&\, \ldots \,\&\, A_{rn}(x_n) \rightarrow B_r(u). \tag{7a}$$

Therefore, for each rule, the "firing degree" is equal to:

$$t_r(x_1,\ldots,x_n,u) = f_\rightarrow(f_\&(\mu_{r1}(x_1),\ldots,\mu_{rn}(x_n)),\mu_r(u)). \tag{8a}$$

1.6 INFERENCE FROM SEVERAL FUZZY RULES

General idea. In the previous section, we have shown how to transform the formula (7) into an algorithm that describes the grade of truth (firing degree) $t_r(x_1,\ldots,x_n,u)$ with which each rule is applicable. After we have computed the firing degree of each rule, we can similarly formalize the formula (6) and get a numerical value that describes to what extent each possible control value u is reasonable for a given input $x_1,\ldots,x_n$:

$$\mu_C(x_1,\ldots,x_n,u) = f_\vee(t_1(x_1,\ldots,x_n,u),\ldots,t_R(x_1,\ldots,x_n,u)), \qquad (9)$$

where $f_\vee(a,b,c)$ stands for $f_\vee(f_\vee(a,b),c)$, and, in general, $f_\vee(a_1,\ldots,a_n,a_{n+1}) = f_\vee(f_\vee(a_1,\ldots,a_n),a_{n+1})$.

In particular, in our toy example,

$$\mu_C(x,u) = f_\vee(t_1(x,u),t_2(x,u),t_3(x,u)) =$$

$$f_\vee(f_\&(\mu_N(x),\mu_{NC}(u)),f_\&(\mu_{SP}(x),\mu_{SNC}(u)),f_\&(\mu_{SN}(x),\mu_{SPC}(u))).$$

Comment. If we use a genuine implication to formalize the if-then rules, then we can similarly formalize the formula (6a) and get a numerical value that describes to what extent each possible control value u is reasonable for a given input $x_1,\ldots,x_n$:

$$\mu_C(x_1,\ldots,x_n,u) = f_\&(t_1(x_1,\ldots,x_n,u),\ldots,t_R(x_1,\ldots,x_n,u)). \qquad (9a)$$

In particular, in our toy example,

$$\mu_C(x,u) = f_\&(t_1(x,u),t_2(x,u),t_3(x,u)) =$$

$$f_\&(f_\rightarrow(\mu_N(x),\mu_{NC}(u)),f_\rightarrow(\mu_{SP}(x),\mu_{SNC}(u)),f_\rightarrow(\mu_{SN}(x),\mu_{SPC}(u))).$$

In many cases, this formula leads to a better quality control than Mamdani's approach. For example, if a large number of rules is applicable to the same input, then Mamdani's approach will lead to a fuzzy and imprecise membership function for u, and the more rules are applicable, the more fuzzy and imprecise the output. This features of Mamdani's approach is somewhat counterintuitive, because intuitively, the more rules we know, the more precise should our conclusions be. In the genuine implication approach, we use AND-operation to combine different rules; as a result, in this approach, the more rules we combine, the more precise the conclusion becomes, in good accordance with our intuition.

Numerical example. Let us assume that all six membership functions are piece-wise linear. The functions that describe the temperature difference x are described by the following graph:

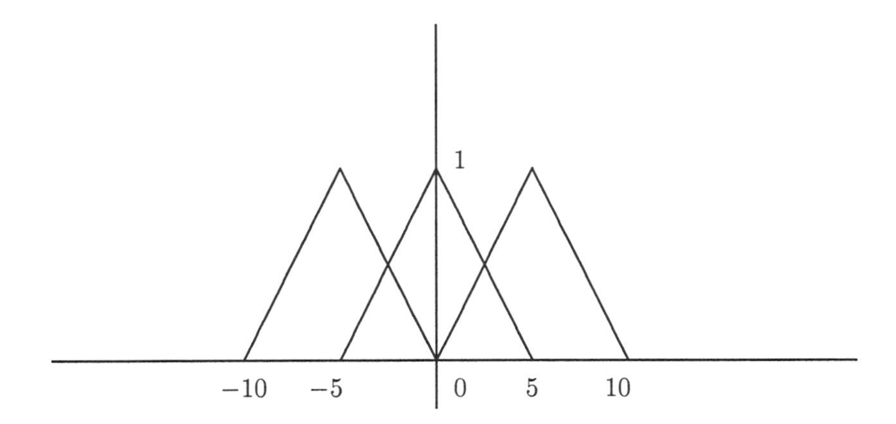

The functions that describe the resulting control u are described by the following graph:

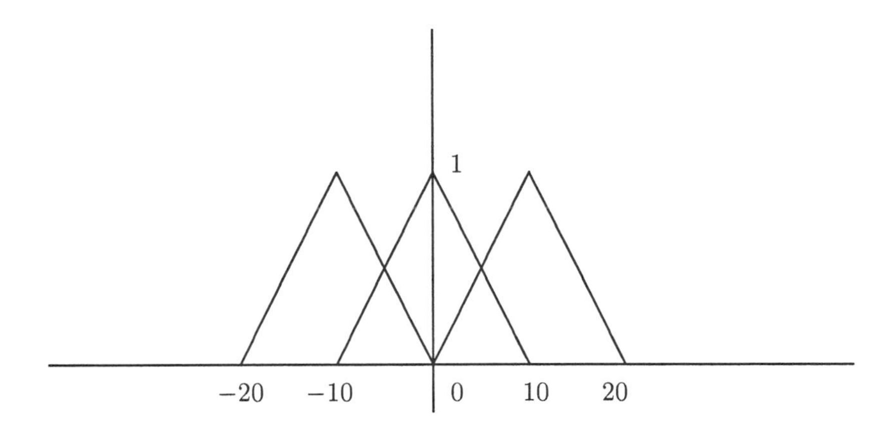

What is the grade of truth $\mu_C(4, -2)$ that for $x = 4°$ the control $u = -4°$ is reasonable? According to our formulas, let us first compute the values of the membership functions. From the above general formula for linear extrapolation, we can find the analytical formulas for these membership functions:

- The term "negligible" is described by the following formulas:

- $\mu_N(x) = 1 + x/5$ for $-5 \le x \le 0$;
- $\mu_N(x) = 1 - x/5$ for $0 \le x \le 5$;
- $\mu_N(x) = 0$ for all other x.
- $\mu_{NC}(u) = 1 + u/10$ for $-10 \le u \le 0$;
- $\mu_{NC}(u) = 1 - u/5$ for $0 \le u \le 10$;
- $\mu_{NC}(u) = 0$ for all other u.

■ The term "small positive" is described by the following formulas:

- $\mu_{SP}(x) = x/5$ for $0 \le x \le 5$;
- $\mu_{SP}(x) = 2 - x/5$ for $5 \le x \le 10$;
- $\mu_{SP}(x) = 0$ for all other x.
- $\mu_{SPC}(u) = u/10$ for $0 \le u \le 10$;
- $\mu_{SPC}(u) = 2 - x/10$ for $10 \le u \le 20$;
- $\mu_{SPC}(u) = 0$ for all other u.

■ The term "small negative" is described by the following formulas:

- $\mu_{SN}(x) = 2 + x/5$ for $-10 \le x \le -5$;
- $\mu_{SN}(x) = -x/5$ for $-5 \le x \le 0$;
- $\mu_{SN}(x) = 0$ for all other x.
- $\mu_{SNC}(u) = 2 + u/10$ for $-20 \le u \le -10$;
- $\mu_{SNC}(u) = -u/10$ for $-10 \le u \le 0$;
- $\mu_{SNC}(u) = 0$ for all other u.

If we use $f_\&(a, b) = \min(a, b)$ and $f_\vee(a, b) = \max(a, b)$, then we get $\mu_C(4, -4) = \max(t_1, t_2, t_3)$, where

$$t_1 = \min(\mu_N(4), \mu_{NC}(-4)); \quad t_2 = \min(\mu_{SP}(4), \mu_{SNC}(-4));$$

$$t_3 = \min(\mu_{SN}(4), \mu_{SPC}(-4)).$$

Here, $\mu_N(4) = 0.2$, $\mu_{NC}(-4) = 0.6$, $\mu_{SP}(4) = 0.8$, $\mu_{SNC}(-4) = 0.4$, and $\mu_{SN}(4) = \mu_{SPC}(-4) = 0$. Hence,

$$t_1 = \min(0.2, 0.6) = 0.2; \quad t_2 = \min(0.8, 0.4) = 0.4; \quad t_3 = \min(0, 0) = 0,$$

and

$$\mu_C(4, -2) = \max(0.2, 0.4, 0) = 0.4.$$

1.7 DEFUZZIFICATION

The problem. As a result of applying the previous steps, we get a *fuzzy set* $\mu(u)$ that describes, for each possible control value u, how reasonable it is to use this particular value. In other words, for every possible control value u, we get a truth value $\mu(u)$ that describes to what extent this value u is reasonable to use. In automatic control applications, we want to transform this *fuzzy* information into a *single* value $\bar{u}$ of the control that will actually be applied.

This transformation from a *fuzzy* set to a (*non-fuzzy*) number is called a *defuzzification*. What defuzzification should we apply?

Main idea. We want to select a value $\bar{u}$ that, on average, would lead to the smallest error. If we choose $\bar{u}$, and the best control is u, then the control error is $\bar{u} - u$. Thus, to determine $\bar{u}$, we can use the least squares method; as weights for each square $(\bar{u} - u)^2$, we can take the grade of truth $\mu(u)$ with which u is a reasonable control value. As a result, we get the following formula for determining $\bar{u}$: $\int \mu(u) \cdot (\bar{u} - u)^2 \, du \to \min$. Differentiating the minimized function with respect to the unknown $\bar{u}$ and equating the derivative to 0, we get the formula

$$\bar{u} = \frac{\int u \cdot \mu(u) \, du}{\int \mu(u) \, du} \tag{10}$$

that is called *centroid defuzzification*. (This and other defuzzification procedures are described in detail in Chapter 2, on Fuzzy Control and Approximate Reasoning, and in [20].)

Warning: caution is required. In most real-life situations, "if-then" rules that form the knowledge base are mutually compatible. In such situations, the above methodology usually leads to a meaningful control. However, there exist real-life situations in which some rules are incompatible.

Let us give a simple example of such a situation. Suppose that we are designing an automatic controller for a car. If the car is traveling on an empty wide road, and there is an obstacle straight ahead (e.g., a box that fell from a truck), then a reasonable idea is to *swerve* to avoid this obstacle. Since the road is empty, there are two possibilities:

- we can swerve to the right; and
- we can swerve to the left.

For swerving, the control variable u is the angle to which we steer the wheel. Based on the distance to the obstacle and on the speed of the car, an experienced driver can describe a reasonable amount of steering u_0.

> In reality, u_0 will probably be a fuzzy value, but for simplicity, we can assume that u_0 is precisely known.

Thus, as a result of formalizing expert knowledge, we conclude that there are two possible control values: u_0 and $-u_0$, both with grade of truth 1. For this $\mu(u)$, formula (10) leads to $\bar{u} = 0$, i.e., the car will run directly into the box.

How can we *detect* such situations? This problem would be detected if genuine implications were used. In Mamdani's method, multimodal outputs indicate logical inconsistency. (see, e.g., [26]).

Once we have detected such a situation, how can we repair it? There are two possible ways of doing it:

- ideally, we should carefully analyze the rules, find out which of them are inconsistent, and modify them so as to avoid this inconsistency;
- if this thorough analysis is not possible, we use heuristic approaches, e.g., we can use special (more complicated) defuzzification methods which enable us to avoid undesirable control decisions; see, e.g., [43], [44], [45].

Simplifications. The formula (10) is sometimes too computationally complicated. To simplify this formula, the membership functions $\mu_r(u)$ (that describe the words $B_r(u)$) can be replaced by their defuzzified values u_r. In this case, instead of the general control rule (5), we get a simplified rule

$$A_{r1}(x_1) \,\&\, \ldots \,\&\, A_{rn}(x_n) \rightarrow u = u_r. \tag{5a}$$

Here, the function $\mu_C(u)$ is different from 0 only for one one of these values u_r, and the integrals in (10) are reduced to easier-to-compute sums:

$$\bar{u} = \frac{t_1 \cdot u_1 + \ldots + t_R \cdot u_R}{t_1 + \ldots + t_R}, \tag{10a}$$

where

$$t_r = f_\&(\mu_{r1}(x_1), \ldots, \mu_{rn}(x_n)). \tag{11}$$

This version of fuzzy control is, computationally, much simpler than the control based on centroid defuzzification, but still leads to a reasonable control.

Takagi-Sugeno version of fuzzy control. In some situations, for each combination of fuzzy inputs, an expert is able not only to describe a single reasonable control value u, but also to describe how exactly, within the corresponding fuzzy area, the control must depend on the inputs $x_1, \ldots, x_n$. In such situations, we get rules of the type

$$A_{r1}(x_1) \,\&\, \ldots \,\&\, A_{rn}(x_n) \rightarrow u = f_r(x_1, \ldots, x_n), \tag{5b}$$

where $f_r(x_1, \ldots, x_n)$ is a function (usually, linear) supplied by the expert. The rule base of the type (5b) was first considered in [42] under the name of a *fuzzy model.* (For a detailed introduction into fuzzy models, see the chapter on fuzzy modeling in this book.) For this fuzzy model, the least squares method similar to the one that we used lead to the following defuzzification formula:

$$\bar{u} = \frac{t_1 \cdot f_1(x_1, \ldots, x_n) + \ldots + t_R \cdot f_R(x_1, \ldots, x_n)}{t_1 + \ldots + t_R}, \qquad (10b)$$

where t_r is determined by the formula (11). This control also has many important applications; see, e.g., [19].

1.8 THE BASIC STEPS OF FUZZY CONTROL: SUMMARY

- We start with the if-then expert rules of the type "If x is small, and ..., then u is small". In general, each of these rules can be represented by a formula (5), where $x_1, \ldots, x_n$ are inputs, and A_{ri} and B_r are words that describe properties of inputs and output.
- For each words w used in these rules, we pick several values $x^{(1)}, \ldots, x^{(k)}$, and use one of the above-described elicitation techniques described to determine the grade of truth $\mu_w(x^{(1)}), \ldots, \mu_w(x^{(n)})$ with which these values satisfy the property w.
- Then, we use some extrapolation technique to determine the membership functions $\mu_w(x)$ (that describe the grade of truth with which different values of x satisfy the property w).
- We choose AND and OR operations $f_{\&}(a, b)$ and $f_{\vee}(a, b)$.
- For each rule r, and for each possible values of input and output, we compute the *firing values* $t_r(x_1, \ldots, x_n, u)$ and then we compute the membership function of the fuzzy control law:
  - In Mamdani's approach, we use the formula (8) to compute the firing values and the formula (9) to combine these values.
  - In the fuzzy implication approach, we use the formula (8a) to compute the firing values and the formula (9a) to combine these values.
- For every input $x_1, \ldots, x_n$, we get a function $\mu_C(u)$ that describes the grade of truth with which this very u is a reasonable control. To get a single recommended control value $\bar{u}$, we use the formula (10) (or one of its modifications).

1.9 TUNING

Why tuning? It is quite possible that the control resulting from the above-described five-step procedure is sometimes inadequate. There are two possible reasons for this:

- First, when an expert formulates the rules, he usually remembers *specific* rules but sometimes forgets to explicitly mention *common sense* rules that are absolutely evident to any human, but that need to be explicitly spelled out for the computer.
- Second, all knowledge elicitation methods are *approximate*, and if we add distortions caused by this approximate character on each step of fuzzy control methodology, we may end up with a rather distorted representation of expert's control.

In view of this possibility, before implementing this control, we must first *test* it. If it turns out that in some situations, the resulting control is inadequate, we have two options:

- If the inadequacy is *huge*, this probably means that we are missing or misinterpreting some of the rules. In this case, we need to confront the experts with these results. Since the rules that the experts have formulated lead to not adequate results, the experts will be able to either *modify* these rules, or *add* new rules that cover these situations.
- If the inadequacy is *small*, then probably, the experts' rules were adequate, and the inadequacy is caused by the approximate character of the expert system methodology. In this case, to make a control better, we can *tune* the parameters of the resulting control. The detailed description of different tuning methods is given in Part C of this book, starting with a general overview in Chapter 8 "Learning and tuning".

Comment. Tuning of fuzzy control was first considered in [7]. Tuning is especially important if we use fuzzy control methodology in novel applications when there aren't yet any expert operators, and the rules that we use are simply coming from common sense (see, e.g., [16]). It turns out [31] that in this case, the optimal control strategy (optimal in some reasonable sense) can be indeed obtained by an appropriate tuning of common sense rules.

Simple iterative tuning. If we *do not know the exact model* of the controlled object, then the only way to predict the quality of different control strategies is to actually test these strategies on the *actual* object. So, if the original fuzzy control does not work well, we try to modify it a little bit and check whether it becomes better or not. This testing takes lots of time.

Tuning is easier in the situation when we *know the exact model* of the controlled object and the exact objective function (and when the only reason for using fuzzy control is that we do not know how to solve the corresponding optimization problem). In this case, we can *simulate* the controlled object, and test different modifications of the original fuzzy control on this computer simulation. This computer testing is fast and easy, and we can, therefore, test many different modifications and find the best one.

Genetic algorithms for tuning. An even better way of tuning fuzzy control is to use randomized optimization techniques instead of the simple search for a maximum. One of these techniques is *genetic algorithms* that simulate the survival of the fittest in nature to generate better and better controls. This technique is described in detail in Chapter 11 "Fuzzy genetic algorithms".

Neural networks for tuning. In addition to *rules*, we can have *records* of the expert's control. The fact that the original fuzzy control is not completely adequate may mean, in particular, that for the recorded inputs $x_1^{(k)}, \ldots, x_n^{(k)}$, the result of applying the fuzzy control methodology may differ from the recorded control $u^{(k)}$ applied by this expert. It is, therefore, desirable to tune the original fuzzy control so that it will fit with the recorded patterns $(x_1^{(k)}, \ldots, x_m^{(k)}, u^{(k)})$.

Neural networks (computer programs that simulate the way our brain operates) are known to be a universal tool for fitting patterns. Therefore, it makes sense to use neural networks for tuning fuzzy control. Details can be found in Chapters 9 and 10 of this book.

Model-based fuzzy control. When we do not have a *crisp* model for the controlled object, i.e., if we do not exactly how different controls will change the state of the system, we may still be able to extract from the experts some information about these changes. Namely, we may be able to extract the information of the type "If x_1 was small, ..., and we apply a small control u, then x_1 will change slightly". This additional information can be formulated in terms of fuzzy logic, and is called a *fuzzy model* of the controlled object (Takagi-Sugeno model mentioned above can be viewed as a particular case of this information).

There are papers in which such models are used to tune the fuzzy control [39], [24], [22].

1.10 METHODOLOGIES OF FUZZY CONTROL: WHICH IS THE BEST?

Which non-linearity should we choose? In the previous section, we have seen that on each stage of fuzzy control methodology, we have several different choices: we can choose different extrapolation techniques, we can choose different AND and OR operations, we can choose different defuzzifications, etc. Different choices often lead to control strategies of different quality; see, e.g., [40]. It is, therefore, important to make choices that lead to the *best* control.

- In some situations, we *know* the exact expression for the *objective function*; in such situation, "the best" simply means the control that leads to the largest possible value of this objective function.
- However, in most applications of fuzzy control, we do not have such a precise expression. In such situations, we must choose the fuzzy control methodology based on some reasonable control criterion.

In this section, we will describe reasonable control characteristics and describe what choices of fuzzy control methodology optimize these characteristics.

Stability. The main objective of the control is that it should control. For example, if we control a car on the road, then, for the largest part of the trip, one of the main objectives of this control is to make sure that it stays in its lane with the desired speed, i.e., that whenever it will accidentally deviate from the straight course, the steering control will return it back on course, and when the speed would deviate, the acceleration or deceleration would bring it back to the optimal cruise speed.

In the general case, we want a control that, after an initial deviation, will bring the controlled system back "on track". This property is called *stability* of the control (and of the controlled system).

Historical comment. For fuzzy control, the problem of stability was first analyzed in [2].

Stability is a generic term. There are many different particular notions of stability, depending on how big initial deviations we allow (usually, only small ones), whether we want the system to be stabilized for a potentially infinite amount of time or only for a given finite interval, etc.

From the practical viewpoint, we can describe the stability of a control strategy by its *relaxation time*: we start with a deviation of a given size $\Delta > 0$, and we measure the time after which the control brings the deviation down to the pre-defined value $\delta \ll \Delta$. This relaxation time depends on the values Δ and δ.

- If the initial deviation Δ is *large*, then we may not be able to return the system to its desired state at all.
- Thus, we will consider only *small* deviations Δ.

In this sense, we can say that a control strategy $u(x_1, \ldots, x_n)$ is *more stable* than the control strategy $u'(x_1, \ldots, x_n)$ if for all sufficiently small Δ and δ, the relaxation time T corresponding to the strategy u is smaller than the relaxation time corresponding to the strategy u'.

It turns out that for for a reasonable class of systems (described by simple smooth differential equations), and for reasonable control rules, the *most stable* control corresponds to $f_{\&}(a, b) = \min(a, b)$ and $f_{\vee}(a, b) = a + b - a \cdot b$ [40], [29].

A detailed analysis of stability of fuzzy control is given in a special chapter by H. Tanaka.

Smoothness. Stability is not all that we expect from a control.

For example, when driving a car, stability means, in particular, that once the car swerved, it should return to the original trajectory. The faster it returns, the more stable is the system. Therefore, from the viewpoint of stability only, the ideal (optimal) control would be the one that brings the car back on track in the shortest possible time (i.e., with the largest possible acceleration). The resulting driving with sudden accelerations may be good on a racetrack or for a car chase, but it is very uncomfortable for passengers. From the passenger viewpoint, we prefer the resulting trajectory to be *smooth.*

Just like stability, smoothness is a generic notion; there are several different understandings (and formalizations) of what "smooth" means. In mathematical terms, *smoothness* usually means that the time derivative $\dot{x}_i(t)$ is small. To compare different control strategies, we must combine the values $\dot{x}_i(t)$ for different moments of time t into a single numerical criterion that characterizes how smooth is the trajectory (i.e., how small are all these derivatives). To form this numerical criterion, it is reasonable to use the idea of the least squares method and to take $I = \sum_i \int (\dot{x}_i(t))^2 \, dt$.

Similarly to stability, it makes sense to compare the values of this smoothness functional for *small* initial deviations Δ. It turns out that from the viewpoint of this comparison, for a reasonable class of systems, the best choice of AND and OR operations is $f_{\&}(a, b) = a \cdot b$ and $f_{\vee}(a, b) = \min(a, b)$ [40].

Comment. At first glance, it may seem that we can require that the control is *both* stable and smooth. However, from the fact that the requirements of smoothness and stability lead to different control strategies, we can conclude that different control requirements are not exactly consistent:

- the most stable control is often not smooth at all;
- the smoothest control is not necessarily very stable.

Therefore, in real life, we must select the control characteristic that is the most adequate for a given control situation, and then choose the fuzzy control methodology that leads to the optimal value of this characteristic.

Computational simplicity. Stability and smoothness are typical examples of the *idealized* goals. When, in mathematical control theory, we look for the optimal control strategy, we look for the optimal mathematical function, without taking into consideration how exactly we are going to implement this function.

In *real life*, however, the computational ability of the processor that actually computes the desired control is limited, so some very good control strategies may be too complicated for this processor. Moreover, in many control situations, we need the control *fast* (e.g., for a car control, if we spend too much time on the computation of the optimal control, the car may, by then, have already wrecked). This necessity is, as we have mentioned, one of the reasons why *fuzzy* control is sometimes used even when the *optimal* control strategy is known.

It can be shown that if we are looking for the control that is the *fastest to compute*, then the best choice is to use $f_\&(a,b) = \min(a,b)$ and $f_\vee(a,b) = \max(a,b)$ [30].

Sensitivity. In the traditional fuzzy control methodology, we assume, among other assumptions, that:

- we know the *exact* values of the inputs x_i; and
- we know the *exact* values of the membership functions.

In reality, both assumptions are idealizations:

- Measurements are never 100% accurate, so, the measurement results $\tilde{x}_i$ of measuring the input variables x_i (that characterize the current state of the controlled system) are, generally speaking, *different* from the *actual* (unknown) values of these variables.
- Similarly, all methods of eliciting the truth values from the experts are only *approximate*, so the (approximate) values $\tilde{\mu}_{ri}(x_i)$ and $\tilde{\mu}(u)$ that are used in the fuzzy control algorithm may be somewhat different from the *ideal* (unknown) values $\mu_{ri}(x_i)$ and $\mu_r(u)$ that characterize the actual expert's grades of truth.

We would like to choose the fuzzy control methodology that would make the resulting control the least sensitive to this uncertainty.

There are two ways to describe the possible errors $\Delta x_i = \tilde{x}_i - x_i$ and $\Delta\mu_{ri}(x_i) = \tilde{\mu}_{ri}(x_i) - \mu_{ri}(x_i)$:

- Usually, we know the *upper bound* Δ on these errors, i.e., we know that the error must belong to the interval $[-\Delta, \Delta]$.
- In some cases, we also know the *probabilities* of different values from this interval. These probabilities are usually described by a normal (Gaussian) distribution with 0 average and a given standard deviation σ.

In the *first* case, the interval uncertainty of the inputs leads to the interval uncertainty in the resulting control. So, the minimal sensitivity corresponds to the smallest width of the resulting interval. Similarly to the cases of smoothness and sensitivity, we will consider these widths for sufficiently small deviations Δ. The following control methodology leads to the smallest sensitivity (i.e., to to the narrowest intervals):

- we must choose piecewise-linear membership functions [17], [35];
- we must choose $f_{\&}(a,b) = \min(a,b)$ and $f_{\vee}(a,b) = \max(a,b)$ [36], [17], [35], [18]; and
- we must choose the centroid defuzzification [29], [28].

In the *second* case, random deviations of the input lead to random deviations of the resulting control. In this case, the minimal sensitivity corresponds to the smallest possible standard deviation of the control value. This smallest value is attained if we use $f_{\&}(a,b) = a \cdot b$ and $f_{\vee}(a,b) = a + b - a \cdot b$ [37], [17], [35], [18].

Comment. These optimization results are in good accordance with the general group-theoretic approach that enables us to classify techniques that are optimal relative to arbitrary reasonable criteria [40], [22].

Conclusion

Contribution of fuzzy control. In many real-life situations, there is a need to automatically control a certain system or object such as a car, a spaceship, or a physical plant. In some situations, we know the exact model of the controlled object, we know exactly how this object will react to different controls, and we can describe precisely the objective of our control – usually, to maximize or minimize a certain characteristic such as the plant's output. In such cases, the search for the optimal control strategy can be re-formulated as a precisely formulated mathematical problem. In many real-life cases, we can explicitly solve this optimization problem and thus, find the desired control.

In many other situations, however, we do not have the exact description of the controlled system (or we may have the exact description, but the corresponding optimization problem is so difficult that we do not know how to solve it). In such situations, we often have an expertise of skilled operators who have the experience of controlling this system: the experience of drivers who control cars, the experience of astronauts who have successfully controlled the spaceships, the experience of chemical engineers who successfully control chemical plants, etc. It is desirable to transform this expert experience into an automatic control strategy.

It is often difficult to come up with such a transformation, because expert operators are often unable to describe their experience in precise terms. Instead, they describe their control by using words of natural language like "small", "a little bit", etc., words that do not have a precise meaning and are, in this sense, *fuzzy.* We therefore need a methodology that would translate such "fuzzy" rules into a precise control strategy. *Fuzzy control* is such a methodology.

The fuzzy control methodology exists for 25 years. During these years, there have been many successful and several spectacular applications of fuzzy control, ranging from the automatic train control in Japan to intelligent appliances (camcorders, washing machines, etc.) to temperature control in the Space Shuttle.

Limitations of fuzzy control. Fuzzy control has been successful in many real-life problems in which traditional control methods failed (or at least were not that successful). Does this mean that traditional control is the thing of the past and only fuzzy methods should be used? Of course not, fuzzy control has its limitations too.

The main limitation of fuzzy control is that it is applicable only in the situations of uncertainty, when we do not have the complete knowledge about the controlled system. Fuzzy control is therefore good but not optimal. Often, as we gain more and more experience of controlling the system, we get a better and better understanding of how the system works. Eventually, this understanding leads to a precise description of the system, which allows us to find the *optimal* control, i.e., a control which is better than any other control and in particular, which is better than the original fuzzy control strategy.

From this viewpoint, for each system, fuzzy control is a temporary phenomenon. Does that mean that as our knowledge grows, fuzzy control will be used less and less? Of course not. As we get more and more knowledge about the systems that we are controlling for a long time, new systems and objects attract our attention, and we need to be able to control them. For example, we may find a precise description of a certain type of camcorder and we may learn how to optimally control camcorders of this type, but then new improved camcorders appear; for these new devices, we do not yet have the exact model, and thus, we have to use fuzzy control (or similar techniques). As the progress intensifies, more and more new objects and systems appear that have to be controlled, and therefore, the relative use of fuzzy control rapidly increases. Fuzzy control methodology has its limitations, yes, but does it have limits? No.

Acknowledgments. This paper was partly supported by the NSF grants EEC-9322370 and DUE-9750858, by NASA grant NCCW-0089, and by the Future Aerospace Science and Technology Program (FAST) Center for Structural Integrity of Aerospace Systems, effort sponsored by the Air Force Office of Scientific Research, Air Force Materiel Command, USAF, under grant number F49620-95-1-0518.

We are thankful to numerous friends and colleagues, especially to Didier Dubois and Henri Prade, for valuable discussions and comments.

References

Historical papers on fuzzy logic and fuzzy control

[1] Baldwin, J. and Guild, N. C. F. (1980). Modelling controllers using fuzzy relations. *Kybernetes*, **9**, 223–229.
[2] Kickert, W. J. M., and Mamdani, E. H. (1978) Analysis of fuzzy logic controller. *Fuzzy Sets and Systems*, **1**, 29–44.
[3] Mamdani, E. H. (1974). Application of fuzzy algorithms for control of simple dynamic plant, *Proceedings of the IEE*, **121** (12), 1585–1588.
[4] Mamdani, E. H. (1977). Application of fuzzy logic to approximate reasoning using linguistic systems, *IEEE Transactions on Computing*, **26**, 1182–1191.
[5] Mamdani, E. H. and S. Assilian, S. (1975). An experiment in linguistic synthesis with a fuzzy logic controller, *Int. J. Man-Mach. Stud.*, **7**, 1–13.
[6] Ostergaard, J. (1977). Fuzzy logic control of a heat exchanger. In: *Fuzzy Automata and decision processes*, Gupta, M. M., Saridis, G. N., and Gaines, B. R. (eds.), North Holland, 285–320.
[7] Procyk, T. J. and Mamdani, E. H.(1979). A linguistic self organizing fuzzy controller, *Automatica*, **15** (1), 15–30.
[8] Tong, R. M. (1977). A control engineering review of fuzzy systems, *Automatica*, **13**, 559–569.
[9] Zadeh, L. A. (1965). Fuzzy sets, *Inform. and Control*, **8**, 338–353.
[10] Zadeh, L. A. (1968). Fuzzy algorithms, *Inform. and Control*, **12** (2), 94–102.
[11] Zadeh, L. A. (1973). Outline of a new approach to the analysis of complex systems and decision processes, *IEEE Trans. on Systems, Man, and Cybernetics*, **1** (1), 28–44.

Textbooks and monographs on fuzzy logic and fuzzy control cited in the chapter

[12] Dubois, D. and Prade, H. (1980). *Fuzzy Sets and Systems: Theory and Applications*, Academic Press, NY.
[13] Dubois, D. and Prade, H. (1988). *Possibility Theory: An Approach to Computerized Processing of Uncertainty*, Plenum Publ., N.Y.
[14] Kandel, A. and Langholtz, G. (eds.) (1994). *Fuzzy Control Systems*, CRC Press, Boca Raton, FL.
[15] Klir, G. and Yuan, B. (1995). *Fuzzy sets and fuzzy logic: theory and applications*, Prentice Hall, Upper Saddle River, NJ.
[16] Kosko, B. (1997). *Fuzzy engineering*, Prentice Hall, Upper Saddle River, NJ.

[17] Nguyen, H. T., Sugeno, M., Tong, R., and Yager, R. (1995) (eds.). *Theoretical aspects of fuzzy control*, J. Wiley, N.Y.

[18] Nguyen, H. T. and Walker, E. A. (1997). *A first course in fuzzy logic*, CRC Press, Boca Raton, FL.

[19] Palm, R., Driankov, D., and Hellendoorn, H. (1997). *Model based fuzzy control*, Springer-Verlag, Berlin, Heidelberg.

[20] Yager, R. R. and Filev, D. (1994) *Essentials of fuzzy control*, Wiley, N.Y.

Supplementary references

Comment. This chapter only deals with the basic issues. For specialized topics, we advise the reader to read other chapters and the books and papers cited in these references. These are the main supplementary references.

The attached list of supplementary references to this chapter is in no way a substitute for the other chapters and for their bibliographies, it simply points to relevant auxiliary issues that were mentioned in this chapter but that are not covered in detail in other chapters.

Since this list quotes quite a few of our own papers, it may give the reader the false impression that we are the only authors active in fuzzy control in the last 5 years. Nothing can be further from the truth. As one can see from the volume as a whole, fuzzy control is an actively growing research field, with dozens of researchers all over the world involved in exciting and fruitful research.

[21] Abello, J. *et al.* (1994). Computing an appropriate control strategy based only on a given plant's rule-based model is NP-hard, In: Hall, L., Ying, H., Langari, R., and Yen, J. (eds.), *NAFIPS/IFIS/NASA'94, Proceedings of the First International Joint Conference of The North American Fuzzy Information Processing Society Biannual Conference, The Industrial Fuzzy Control and Intelligent Systems Conference, and The NASA Joint Technology Workshop on Neural Networks and Fuzzy Logic, San Antonio, December 18–21, 1994*, IEEE, Piscataway, NJ, 331–332.

[22] Bouchon-Meunier, B. *et al.* (1996). On the formulation of optimization under elastic constraints (with control in mind), *Fuzzy Sets and Systems,* **81** (1), 5–29.

[23] Buchanan, B. G. and Shortliffe, E. H. *Rule-based expert systems. The MYCIN experiments of the Stanford Heuristic Programming Project*, Addison-Wesley, Reading, MA, Menlo Park, CA.

[24] Chen, G., Pham, T. T., and Weiss, J. J. (1995). Fuzzy modeling of control systems, *IEEE Transactions on Aerospace and Electronic Systems,* **31** (1), 414–429.

[25] Dubois, D. and H. Prade, H. (1997). What are fuzzy rules and how to use them, *Fuzzy Sets and Systems*, 1997, **84**, 169–185.

[26] Dubois, D., Ughetto, L., and H. Prade, H. (1997). Checking the coherence and redundancy of fuzzy knowledge bases, *IEEE Transactions on Fuzzy Systems*, 1997, **5** (3), 398–417.

[27] Giles, R. (1976). Lukasiewicz logic and fuzzy set theory, *Internat. J. Man-Machine Stud.*, **8**, 313–327.

[28] Kreinovich, V. (1997). Random sets unify, explain, and aid known uncertainty methods in expert systems, in Goutsias, J., Mahler, R. P. S., and Nguyen, H. T. (eds.). *Random Sets: Theory and Applications*, Springer-Verlag, N.Y., 1997, 321–345.

[29] Kreinovich, V., Nguyen, H. T., and Walker, E. A. (1996a). Maximum entropy (MaxEnt) method in expert systems and intelligent control: new possibilities and limitations, In: Hanson, K. and Silver, R. (eds.), *Maximum Entropy and Bayesian Methods, Santa Fe, New Mexico, 1995*, Kluwer Academic Publishers, Dordrecht, Boston, 93–100.

[30] Kreinovich, V. and Tolbert, D. (1994). Minimizing computational complexity as a criterion for choosing fuzzy rules and neural activation functions in intelligent control. In: Jamshidi, M., Nguyen, C., Lumia, R., and Yuh, J. (eds.). *Intelligent Automation and Soft Computing. Trends in Research, Development, and Applications. Proceedings of the First World Automation Congress (WAC'94). August 14–17, 1994, Maui, Hawaii*, TSI Press, Albuquerque, NM, **1**, 545–550.

[31] Lea, R. N. and Kreinovich, V. (1995). Intelligent Control Makes Sense Even Without Expert Knowledge: an Explanation, *Reliable Computing*, 1995, Supplement (Extended Abstracts of APIC'95: International Workshop on Applications of Interval Computations, El Paso, TX, Febr. 23–25, 1995). 140–145.

[32] Ling, C. H. (1965). Representation of associative functions, *Publ. Math. Debrecen*, **12**, 189–212.

[33] Mohler, R. R. (1991). *Nonlinear systems. Vol. 1. Dynamics and control*, Prentice Hall, Englewood Cliff, NJ.

[34] Nguyen, H. T. and Kreinovich, V. (1995). Towards theoretical foundations of soft computing applications, *International Journal on Uncertainty, Fuzziness, and Knowledge-Based Systems*, **3**, 341–373.

[35] Nguyen, H. T. *et al.* (1995a). Interpolation that leads to the narrowest intervals, and its application to expert systems and intelligent control, *Reliable Computing*, 1995, **3** (1), 299–316.

[36] Nguyen, H. T. Kreinovich, V., and Tolbert, D. (1993). On robustness of fuzzy logics. *Proceedings of IEEE-FUZZ International Conference*, San Francisco, CA, March 1993, **1**, 543–547.

[37] Nguyen, H. T. Kreinovich, V., and Tolbert, D. (1994). A measure of average sensitivity for fuzzy logics, *International Journal on Uncertainty, Fuzziness, and Knowledge-Based Systems*, **2** (4), 361–375.

[38] Nguyen, H. T., Kreinovich, V., and Zuo, Q. (1997). Interval-valued degrees of belief: applications of interval computations to expert systems and intelligent control, *International Journal of Uncertainty, Fuzziness, and Knowledge-Based Systems (IJUFKS)*, **5** (3), 313–358.

[39] Pham, T. T., Weiss, J. J., and Chen, G. R. (1992). Optimal fuzzy logic control for docking a boat, *Proc. of the Second International Workshop on Industrial Fuzzy Control and Intelligent Systems, Dec. 2–4, 1992*, College Station, TX, 66–73.

[40] Smith, M. H. and Kreinovich, V. (1995). Optimal strategy of switching reasoning methods in fuzzy control, Chapter 6 in Nguyen, H. T., Sugeno, M., Tong, R., and Yager, R. (eds.). *Theoretical aspects of fuzzy control*, J. Wiley, N.Y., 117–146.

[41] Smith, S. and Kreinovich, V. (1995a). In Case of Interval Uncertainty, Optimal Control is NP-Hard Even for Linear Plants, so Expert Knowledge is Needed, *Reliable Computing*, Supplement (Extended Abstracts of APIC'95: International Workshop on Applications of Interval Computations, El Paso, TX, Febr. 23–25, 1995), 190–193.

[42] Takagi, T. and Sugeno, M. (1985). Fuzzy identification of systems and its applications to modelling and control, *IEEE Transactions on Systems, Man, and Cybernetics*, **15** (1), 116–132.

[43] Yen, J. and Pfluger, N. (1991). Path planning and execution using fuzzy logic, In: *AIAA Guidance, Navigation and Control Conference*, New Orleans, LA, 1991, **3**, 1691–1698.

[44] Yen, J. and Pfluger, N. (1991a). Designing an adaptive path execution system, *IEEE International Conference on Systems, Man and Cybernetics, Charlottesville, VA, 1991.*

[45] Yen, J., Pfluger, N., and Langari, R. (1992). A defuzzification strategy for a fuzzy logic controller employing prohibitive information in command formulation, *Proceedings of IEEE International Conference on Fuzzy Systems, San Diego, CA, March 1992.*

2 INTRODUCTION TO FUZZY MODELING

Kazuo Tanaka[1] and Michio Sugeno[2]

[1]Department of Mechanical and Control Engineering
University of Electo-Communications
1-5-1 Chofugaoka, Chofu, Tokyo 182 Japan
ktanaka@mce.uec.ac.jp

[2]Department of Computational Intelligence
and Systems Science
Tokyo Institute of Technology
4259 Nagatsuta, Midori-ku, Yokohama 227 Japan

2.1 INTRODUCTION

The term "fuzzy modeling" was used in [30]. After that, pioneer works in the field of fuzzy modeling were done in [26, 31]. In 80's, several fuzzy modeling techniques were developed (e.g., [24], [25], and [27]). In particular, Takagi and Sugeno [11] proposed a new type of fuzzy model. The model is called "Takagi-Sugeno fuzzy model (T-S fuzzy model)". Furthermore, they proposed a procedure to identify the T-S fuzzy model from input-output data of systems in [11]. This work has been referred in many papers on fuzzy modeling for a long time. Sugeno and Kang [9, 10] extended the T-S procedure. It consists of parameter identification and structure identification. They stated in [10] that the structure identification is important and the method is better than GMDH[1] in prediction problems for complex systems.

A wide variety of revised procedures for the Sugeno-Kang modeling method (S-K method) have been developed in a nonlinear modeling framework extensively since it was proposed. However, most of the revised procedures have dealt with only small benchmark problems with a few predictor variables. On the other hand, recently, a number of neuro-fuzzy modeling techniques have been proposed. Most of them discusses only parameter adjustment, i.e., parameter identification, of fuzzy models using neural network's learning techniques.

A self-organizing fuzzy identification algorithm (SOFIA) was developed in [15], [16]. It is a revised procedure of the S-K method and is also one of neuro-fuzzy modeling techniques. The SOFIA realizes reduction of computational requirement for identifying a T-S fuzzy model by efficient parameter and structure identification. It was shown in modeling problems [15, 16] of complex systems with a large number of predictor variables that the SOFIA is better than other nonlinear modeling techniques, e.g., the GMDH.

Thus, the S-K method and the SOFIA are practical methods rather than theoretical methods. Because of lack of space, this chapter will present the S-K method and the SOFIA although there are other excellent theoretical and practical works (e.g., [17]-[23]).

Section 2 explains the T-S fuzzy model. Section 3 presents the S-K method and shows the prediction result of water flow rate in the river Dniepr. Section 4 shows the SOFIA. In particular, subsections 4.2 and 4.3 present the prediction results of CO concentration in the air at the busiest traffic intersection in a large city of Japan and O_2 concentration in the flue gas of a municipal refuse incinerator, respectively.

2.2 TAKAGI-SUGENO FUZZY MODEL

A fuzzy model proposed by Takagi and Sugeno [11] is described by fuzzy if-then rules whose consequent parts are represented by linear equations. This fuzzy model is of the following form:

Rule (i): IF x_1 is A_{i1}, ..., x_n is A_{in}, THEN

$$y_i = c_{i0} + c_{i1}x_1 + \cdots + c_{in}x_n, \tag{2.1}$$

where $i = 1, 2, \ldots \ell$, ℓ is the number of if-then rules, c_{ik}'s $(k = 0, 1, \ldots, n)$ are the consequent parameters, y_i is the output from the ith if-then rule, and A_{ik} is a fuzzy set.

Given an input $(x_1, x_2, \ldots, x_n)$, the final output of the fuzzy model is inferred as follows:

$$\begin{aligned} y &= \frac{\sum_{i=1}^{\ell} w_i y_i}{\sum_{i=1}^{\ell} w_i} = \frac{\sum_{i=1}^{\ell} w_i (c_{i0} + c_{i1}x_1 + \cdots + c_{in}x_n)}{\sum_{i=1}^{\ell} w_i} \\ &= \left(\sum_{k=0}^{n} \sum_{i=1}^{\ell} w_i c_{ik} x_k \right) / \left(\sum_{i=1}^{\ell} w_i \right), \end{aligned} \tag{2.2}$$

where $x_0 = 1$, w_i is the weight of the i-th IF-THEN rule for the input and is calculated as

$$w_i = \prod_{k=1}^{n} A_{ik}(x_k), \tag{2.3}$$

where $A_{ik}(x_k)$ is the grade of membership of x_k in A_{ik}.

2.3 SUGENO-KANG METHOD

The S-K method consists of two parts: structure identification and parameters identification. The structure identification consists of premise structure identification and consequent structure identification. The parameters identification also consists of premise parameters identification and consequent parameters identification. The premise parameters are parameters of the fuzzy sets A_{ik} in the premise parts.

Figure 2.1 shows the outline of the S-K method. The structure identification will be mainly discussed in this section. The details of the overall procedure are presented in [10].

2.3.1 Structure identification

The structure identification of fuzzy models consists of two parts:premise structure identification and consequent structure identification. The premise structure identification has two problems. One problem is to select premise variables, i.e., to find which variables are necessary in the premise. The other problem is to find an optimal fuzzy partition of the inputs space, which is essential to fuzzy modeling. The consequent structure identification is reduced to the problem of selecting consequent variables, i.e., a problem to find which variables are necessary in the consequent equations of the if-then rules. To perform the structure identification, a criterion for the verification of an obtained structure should be selected.

A. Criterion for verification of structure.

In the structure identification, there are generally a number of structures which seem to be adequate. Hence a criterion for the verification of a structure is of crucial importance.

The information theoretic criterion AIC proposed by Akaike [29] is well known as a criterion of fitting of a statistical model. It is based on the assumption that residuals obey a normal distribution. There are, however, problems which do not satisfy this assumption in practice.

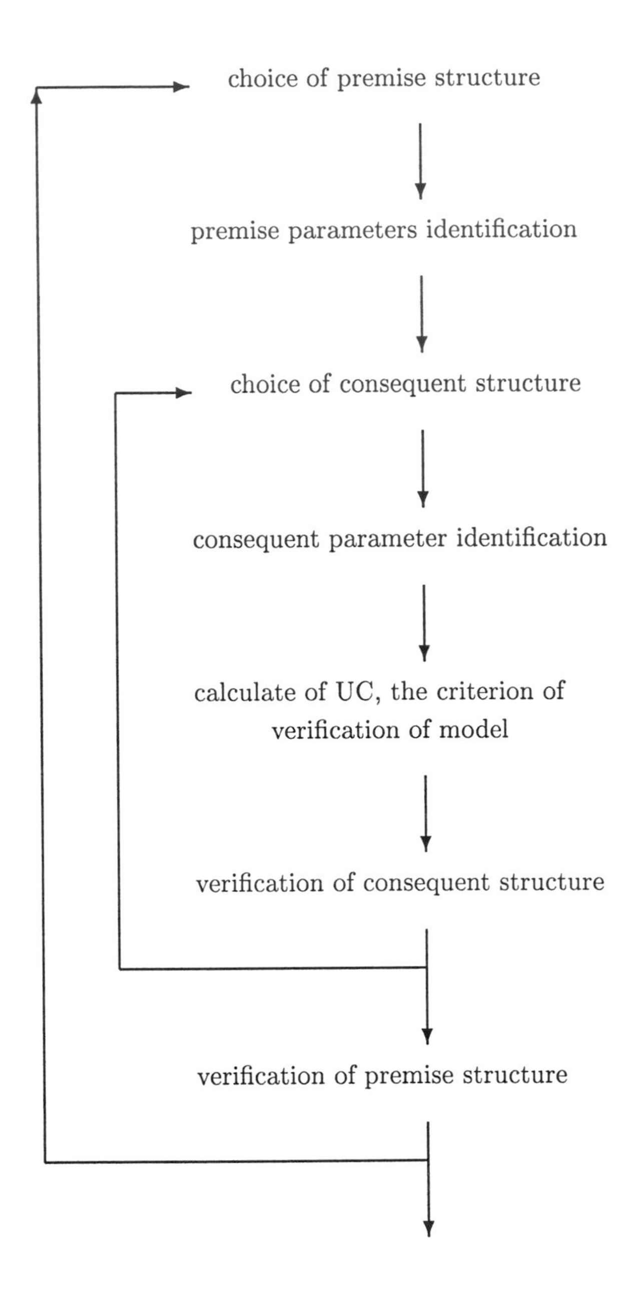

Figure 2.1 Outline of the S-K method.

The S-K method uses the unbiasedness criterion (UC) introduced in GMDH [2]. The basic idea of the UC is as follows: in the presence of moderate noise, the parameters of the model with the true structure are the least sensitive to the changes of the observed data which are used for identifying the parameters. We first divide the observed data into two sets N_A and N_B, and identify two fuzzy models for each data set separately. Then the UC is calculated as

$$\mathrm{UC} = \sqrt{\sum_{i=1}^{n_A} \left(y_i^{AB} - y_i^{AA}\right)^2 + \sum_{i=1}^{n_B} \left(y_i^{BA} - y_i^{BB}\right)^2}, \tag{2.4 }$$

where n_A is the number of the data set N_A, y_i^{AA} the estimated output for the data set N_A from the model identified by using the data set N_A, and y_i^{AB} the estimated output for the data set N_A from the model identified by using the data set N_B. Thus, to calculate the UC, identification of two models is required.

Since a fuzzy model is constructed by partitioning the inputs space, the data for the identification should be distributed uniformly over the space. Therefore, the data of N_A and N_B should be also distributed uniformly over the space. When the number of the observed data is small, we make N_A and N_B have some data in common.

B. Premise structure identification

One of the purposes of the premise structure identification is to partition the inputs space into the fewest fuzzy subspaces. The premise structure identification is based on the following idea. As the number of fuzzy subspaces, i.e. the number of if-then rules, increases, the UC of the fuzzy model decreases. But, if the number of fuzzy subspaces exceeds that of the optimal premise structure, the parameters of the model become sensitive to the changes of the data used for identifying the parameters, and the UC of the model increases.

Here we choose the premise structure which minimizes the UC, and use the following algorithm resembling the forward selection of variables which is a method for finding variables in a linear model. That is, we start the process from the identification of a model with one if-then rule, i.e., a linear model, and increase the number of if-then rules until the UC of the fuzzy model begins to increase. In the process of the premise structure identification, not only the premise parameters but also the consequent parameters of the model are identified as shown in Figure 2.1. Complex method, which is an optimization technique of nonlinear programming problems, is used in the premise parameter identification.

Rule 1:

If w_t is

and y_{t-12} is

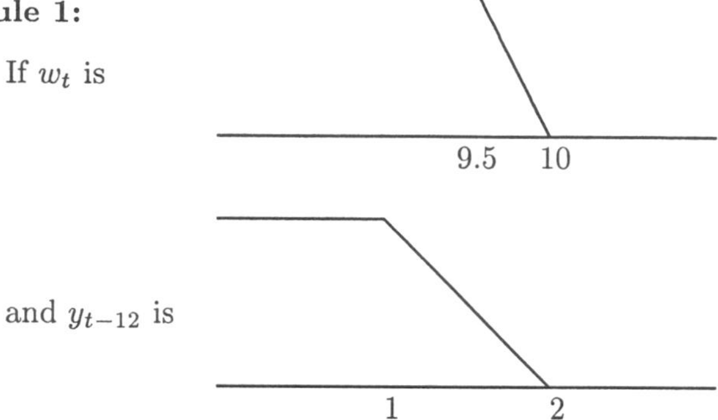

then $\Delta y = -4.9 - 1.9y_{t-1} + 2.3y_{t-12} + 0.48r_{t-1} + 0.34r_{t-2} - 0.3r_{t-8}$
$+0.18r_{t-9} + 0.16w_{t-10}$.

Rule 2:

If w_t is

and y_{t-12} is

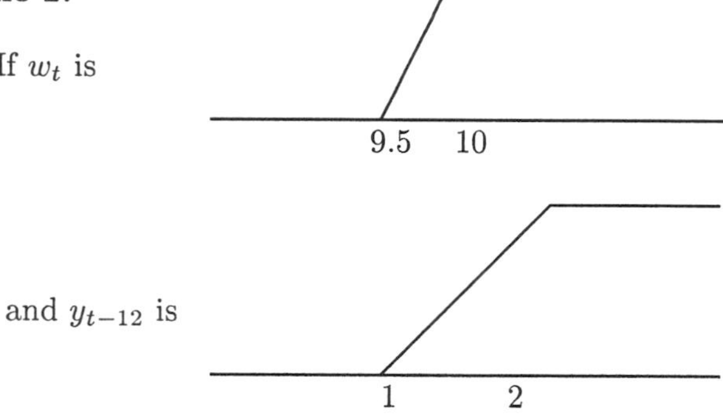

then $\Delta y = 3.5 - 2.4y_{t-1} - 2.4y_{t-7} + 0.13r_{t-1} + 0.34r_{t-2} + 0.3r_{t-8}$
$-0.01r_{t-9} + 0.23w_t - 0.14w_{t-10} - 0.18w_{t-12}$.

Rule 3:

If w_t is

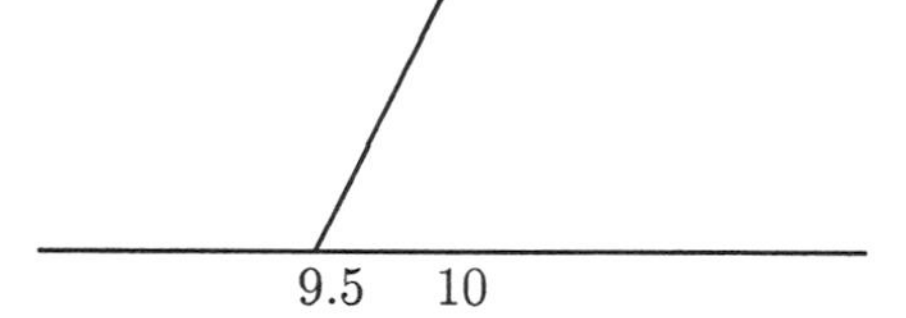

then $\Delta y = -2 + 23y_{t-1} - 7.1y_{t-7} + 1.7y_{t-12} - 4.7r_{t-1} + 0.01r_{t-2}$
$+4.2r_{t-8} + 2.3r_{t-9} - 2.9w_t + 1.4w_{t-10} - 1.2w_{t-12}$.

Figure 2.2 The fuzzy model of remainder.

2.3.2 Prediction of water flow rate in the river Dniepr

The problem is to predict the average annual rate of water flow in the river Dniepr using those values of the river Niemen, a main upper stream of the Dniepr, and Wolf's number concerned with sunspot activity. The prediction duration is 7 years and the data from 1812 to 1971 are given. Ivakhnenko [3] identified a prediction model by using the GMDH method, where the data from 1812 to 1964 are used to identify the model and those from 1965 to 1971 for checking the accuracy of predictions. Ivakhnenko's model consists of two parts, $f_{tr}(t)$, the harmonic trend of time, and $f(y_{t-1}, y_{t-7}, \ldots)$, the prediction of the remainder on the basis of input variables. It is expressed as follows:

$$y_{t+7} = y_{tr}(t) + f\left(y_{t-1}, y_{t-7}, y_{t-12}, t_{t-1}, r_{t-2}, r_{t-8}, r_{t-9}, w_t, w_{t-10}, w_{t-12}\right), \tag{2.5}$$

where y_{t-i} is the water flow rate of Dniepr (100 m^3/sec), r_{t-i} that of Niemen (100 m^3/sec), and w_{t-i} Wolf's number (10 mm^{-2}). First, the harmonic trend of time is constructed. Then, for the remainder, $\Delta y = y_{t+7} - f_{tr}$, a prediction function of input variables is constructed. In Ivakhnenko's model, this prediction function has 76 parameters.

Here we identify a fuzzy prediction model for the remainder. Figure 2.2 shows the identified fuzzy model with 3 if-then rules. To check the prediction accuracy, the mean square error of the checking data is calculated as

$$\Delta = \sum_{t=1965}^{1971} (y_t - y_t^*)^2 \Big/ \sum_{t=1965}^{1971} y_t^2, \tag{2.6}$$

where y_t^* is the predicted value of y_t.

While Δ is 1.3% at the GMDH model, it is 0.93% at the fuzzy model. It should be emphasized that the number of parameters of the fuzzy model is much smaller.

2.4 SOFIA

Subsection 2.3.2 showed effectiveness of the S-K method. However, only disadvantage of the S-K method is a large computational requirement for determining the optimal structure and the optimal parameters of the T-S model in general. The reasons are as follows:

1. In general, a number of patterns of fuzzy partitions must be considered until the optimal fuzzy partition is determined. It should be noted that both the premise and consequent parameter identifications are required whenever a new pattern of fuzzy partition is considered.

2. The complex method is used in the premise parameter identification. The parameter optimization by the complex method is not generally efficient since it does not utilize gradient of a given performance function.

The first problem is serious. Therefore, the best way to reduce computational requirement is to solve the first problem, i.e., to decrease the number of considered fuzzy partition patterns (NFPP) until the optimal fuzzy partition is determined. The main purpose of the SOFIA is to solve the first problem. To solve the first problem, we simplify a procedure for finding the optimal structure of fuzzy partition. In addition, the δ rule is employed in parameter identification. Both the premise and consequent parameters are successively identified by the δ rule. Therefore, the use of complex method can be avoided. We can expect that the second problem is also improved.

To introduce the rule in the parameter identification, the defuzzification process is modified as follows [1, 2] :

$$y = \sum_{i=1}^{\ell} w_i y_i. \tag{2.7}$$

In the SOFIA, the fuzzy sets A_{ik} are assumed to have the following membership functions:

$$A_{ik}(x_k) = \exp\left(-\frac{(x_k - a_{ik})^2}{b_{ik}}\right) \tag{2.8}$$

for $k = 1, 2, \ldots, n$ and $i = 1, 2, \ldots, \ell$.

2.4.1 Algorithm

A. Parameter identification by δ rule.

Successive parameter identification methods of fuzzy models by the δ rule have been reported in [1, 2, 12]. A technique used in [1], [2] is applied to parameter identification in the SOFIA. In the parameter identification, we use normalized data, i.e., the mean and the variance of normalized variable equal 0 and 1, respectively. On the other hand, the raw data (not normalized data) is used in the calculation of the criterion for verification of structure. The detail of the criterion will be given later.

Since each membership function of the fuzzy set A_{ik} is represented by (2.8), we have

$$w_i = \prod_{k=1}^{n} A_{ik}(x_k) = \prod_{k=1}^{n} \exp\left(-\frac{(x_k - a_{ik})^2}{b_{ik}}\right). \tag{2.9}$$

From (2.1) and (2.7),

$$y = \sum_{k=0}^{n} \sum_{i=1}^{\ell} w_i c_{ik} x_k, \tag{2.10 }$$

where $x_0 = 1$. Define a performance function as follows:

$$E = \frac{1}{2}(y^* - y)^2, \tag{2.11 }$$

where y and y^* denote the outputs of a fuzzy model and a real system, respectively. By partially differentiating E with respect to each parameter of the T-S fuzzy model, we obtain

$$\frac{\partial E}{\partial c_{ik}} = \frac{\partial E}{\partial y}\frac{\partial y}{\partial c_{ik}} = -(y^* - y) w_i x_k = -\delta w_i x_k, \tag{2.12 }$$

$$\begin{aligned} \frac{\partial E}{\partial a_{ik}} &= \frac{\partial E}{\partial y}\frac{\partial y}{\partial a_{ik}} = -(y^* - y)\frac{2(x_k - a_{ik})}{b_{ik}} w_i \sum_{k=0}^{n} c_{ik} x_k \\ &= -\delta \frac{2(x_k - a_{ik})}{b_{ik}} w_i \sum_{k=0}^{n} c_{ik} x_k, \end{aligned} \tag{2.13 }$$

$$\begin{aligned} \frac{\partial E}{\partial b_{ik}} &= \frac{\partial E}{\partial y}\frac{\partial y}{\partial b_{ik}} = -(y^* - y)\frac{(x_k - a_{ik})^2}{b_{ik}^2} w_i \sum_{k=0}^{n} c_{ik} x_k \\ &= -\delta \frac{(x_k - a_{ik})^2}{b_{ik}^2} w_i \sum_{k=0}^{n} c_{ik} x_k, \end{aligned} \tag{2.14 }$$

where $\delta = y^* - y$. The final learning law can be defined as

$$c_{ik}^{NEW} = c_{ik}^{OLD} + \varepsilon_1 \delta w_i x_k \tag{2.15 }$$

$$a_{ik}^{NEW} = a_{ik}^{OLD} + \varepsilon_2 \delta \frac{2(x_k - a_{ik}^{OLD})}{b_{ik}^{OLD}} w_i \sum_{k=0}^{n} c_{ik}^{OLD} x_k, \tag{2.16 }$$

$$b_{ik}^{NEW} = b_{ik}^{OLD} + \varepsilon_3 \delta \frac{(x_k - a_{ik}^{OLD})^2}{(b_{ik}^{OLD})^2} w_i \sum_{k=0}^{n} c_{ik}^{OLD} x_k, \tag{2.17 }$$

where ε_1, ε_2, and ε_3 are the learning coefficients and $\varepsilon_1, \varepsilon_2, \varepsilon_3 > 0$. The parameters a_{ik}, b_{ik} and c_{ik} are recursively calculated by (2.15) - (2.17) until the value of the summation of δ for all data points is small enough.

B. A criterion for structure identification.

The S-K method used the unbiasedness criterion (UC) (2.4) introduced in the GMDH as a proper criterion for the verification of structure of a fuzzy model. To calculate the UC, we have to identify two models. This causes a huge computational requirement. To solve the problem, we propose a simplified unbiasedness criterion (SUC). The SUC is calculated as

$$\mathrm{SUC} = \frac{1}{n_B} \sum_{i=1}^{n_B} \frac{|y_i^{BA} - y_i^{B}|}{|y_i^{B}|} \times 100, \tag{2.18}$$

where n_B is the number of data in the data set N_B, y_i^{BA} the estimated output for the data set N_B from the model identified using the data set N_A, y_i^B real output of data set N_B. As mentioned above, the raw data are used in the calculation of the SUC, where it is assumed that $|y_i^B| \neq 0$. We can reduce computational requirement to a half since (2.18) says that it is sufficient to identify a single model. The SUC is used as a criterion for structure identification in the SOFIA.

C. Identification algorithm.

The SOFIA consists of four stages which effectively realize structure identification and parameter identification of a fuzzy model.

- (Stage 1) Identification of a linear system.
- (Stage 2) Determination of the optimal fuzzy partition.
- (Stage 3) Determination of the optimal consequent structure.
- (Stage 4) Parameter identification.

The aim of the first stage is to identify a linear model. The second stage finds the optimal structure of fuzzy partition of the inputs space. The third stage realizes structure identification of consequent parts in the fuzzy model with the optimal structure of fuzzy partition determined in the second stage. The last stage achieves parameter identification of the fuzzy model with the optimal structure.

In subsection D, the procedure of the SOFIA will be concretely demostrated using a simple example which has been used in some modeling exercises. The algorithm will be summarized by using a description like a programming format in Appendix.

(Stage 1) Identification of a linear system.

Forward selection of variables (FSV) based on the value of SUC is employed to find the optimal structure of a linear system. We find a linear model with the least SUC value in this stage. From now on, variables eliminated by the FSV are removed from the consequent parts.

Next,

$$\mathrm{SUC}_1^* \longleftarrow \text{the least SUC value},$$

where the symbol, $\longleftarrow$, means assignment.

$$\mathrm{SUC}^{LOD} \quad \longleftarrow \quad \mathrm{SUC}_1^*$$

and go to (**Stage 2**).

(Stage 2) Determination of the optimal fuzzy partition.

$$p(x_k) \longleftarrow 1, \qquad k = 1 \sim n,$$

$$\mathrm{CHECK}(x_k) \longleftarrow 1, \qquad k = 1 \sim n,$$

where n denotes the number of input variables, i.e., predictor variables. $p(x_k)$ is the number of partition for the k-th premise variable x_k. There is the following relation between the number of rules, , and $p(x_k)$.

$$\ell = \prod_{k=1}^{n} p(x_k).$$

$\mathrm{CHECK}(x_k)$ is used to judge whether $p(x_k)$ should be increased or not. $p(x_k)$ is increased only when CHECK(xk)=1. Next,

$$\begin{aligned} k_0 \quad &\longleftarrow \quad 1, \\ \mathrm{NEPP} \quad &\longleftarrow \quad 0, \end{aligned}$$

where NFPP denotes the number of considered fuzzy partition patterns.

The purpose of this stage is to determine the optimal value of $p(x_k)$ for each premise variable x_k. This stage consists of four parts.

[**Part 1**]

If $\mathrm{CHECK}(x_{k_0}) = 0$, $\mathrm{SUC}^{k_0} \longleftarrow \alpha$ (α is a big value) and go to [**Part 2**]. If $\mathrm{CHECK}(x_{k_0}) = 1$, $p(x_{k_0}) \longleftarrow p(x_{k_0}) + 1$, i.e., increase the number of partition for k_0-th premise variable x_{k_0}. The model with this premise structure is named "model(k_0)". The premise and consequent parameters of the model(k_0) are identified by (2.15) - (2.17). In this parameter identification, variables eliminated in (**Stage 1**) are removed from the consequent parts. After identifying the premise and consequent parameters by (2.15) - (2.17), we calculate SUC value of the model(k_0).

$$\mathrm{SUC}^{k_0} \longleftarrow \text{the SUC value of the model}(k_0).$$

Next,

$$p(x_{k_0}) \longleftarrow p(x_{k_0}) - 1,$$

i.e., decrease the number of partition for the k_0-th premise variable x_{k_0}.

[Part 2]

$k_0 \longleftarrow k_0 + 1$. If $k_0 \leq n$, then go back to **[Part 1]**, else $\mathrm{SUC}^{NEW} \longleftarrow \min_k \mathrm{SUC}^k$ and $k_0 \longleftarrow 1$. If $\mathrm{SUC}^{NEW} \geq \mathrm{SUC}^{OLD}$, finish **(Stage 2)** and select a model with the SUC^{OLD} and go to **(Stage 3)**. Conversely, if $\mathrm{SUC}^{NEW} < \mathrm{SUC}^{OLD}$, go to **[Part 3]**.

[Part 3]

Increase the number of partition, $p(x_k)$, for the k-th premise variable x_k such that $\min_k \mathrm{SUC}^k$. That is, $p(x_k) \longleftarrow p(x_k) + 1$ for the k-th premise variable x_k such that $\min_k \mathrm{SUC}^k$.

[Part 4]

If there exists k such that $\mathrm{SUC}^k \geq \mathrm{SUC}^{OLD}$, $\mathrm{CHECK}(x_k) \longleftarrow 0$. This means that the number of partition for the k-th premise variable x_k never increases from now on.

$\mathrm{SUC}^{OLD} \longleftarrow \mathrm{SUC}^{NEW}$ and NFPP $\longleftarrow$ NFPP+1 and go back to **[Part 1]**.

The number of repetitions from **[Part 1]** to **[Part 4]** corresponds to the NFPP. As mentioned above, the best way to reduce computational requirement is to decrease the NFPP. If the NFPP of the SOFIA is fewer than that of the S-K method, it can be noted from the viewpoint of computational requirement that the SOFIA is better than the S-K method.

(Stage 3) Determination of the optimal consequent structure.

This stage realizes consequent variable selection of a fuzzy model with the optimal fuzzy partition determined in **(Stage 2)**. The FSV is used to select consequent variables in each rules of the fuzzy model. Select a model with the least SUC value and go to **(Stage 4)**.

(Stage 4) Parameter identification.

The parameter identification of the final model selected in **(Stage 3)** is realized using (2.15) - (2.17). The performance indexes for the identification data and the prediction data are calculated by

$$J = \frac{1}{d}\sum_{i=1}^{d}\frac{|y_i^* - y_i|}{|Y_i^*|} \times 100, \tag{2.19 }$$

where d denotes the number of data. y^* and y denote outputs of a real system and an identified fuzzy model, respectively. The values of y and y^* are raw data and are not normalized.

Next, the procedure of the SOFIA will be concretely demonstrated using a simple example which has been used in some modeling exercises. The above algorithm will be summarized by using a description like a programming format in Appendix.

Table 2.1 Identification data and prediction data.

No.	x_1	x_2	x_3	x_4	y	No.	x_1	x_2	x_3	x_4	y
1	1	3	1	1	11.11	21	1	1	5	1	9.545
2	1	5	2	1	6.521	22	1	3	4	1	6.043
3	1	1	3	5	10.19	23	1	5	3	5	5.724
4	1	3	4	5	6.043	24	1	1	2	5	11.25
5	1	5	5	1	5.242	25	1	3	1	1	11.11
6	5	1	4	1	19.02	26	5	5	2	1	14.36
7	5	3	3	5	14.15	27	5	1	3	5	19.61
8	5	5	2	5	14.36	28	5	3	4	5	13.65
9	5	1	1	1	27.42	29	5	5	5	1	12.43
10	5	3	2	1	15.39	30	5	1	4	1	19.02
11	1	5	3	5	5.724	31	1	3	3	5	6.38
12	1	1	4	5	9.766	32	1	5	2	5	6.521
13	1	3	5	1	5.87	33	1	1	1	1	16
14	1	5	4	1	5.406	34	1	3	2	1	7.219
15	1	1	3	5	10.19	35	1	5	3	5	5.724
16	5	3	2	5	15.39	36	5	1	4	5	19.02
17	5	5	1	1	19.68	37	5	3	5	1	13.39
18	5	1	2	1	21.06	38	5	5	4	1	12.68
19	5	3	3	5	14.15	39	5	1	3	5	19.61
20	5	5	4	5	12.68	40	5	3	2	5	15.39

NO.	Combination of input variables x_1	x_2	x_3	x_4	x_0	SUC
1	O	O	O	O	O	14.00
2		O	O	O	O	49.05
3	O		O	O	O	21.33
4	O	O		O	O	26.75
5	O	O	O		O	12.15*
6	O	O	O	O		14.00
7		O	O	-	O	50.73
8	O		O	-	O	20.63
8	O	O		-		17.57
9	O	O	O	-	O	12.17*

Figure 2.3 Forward selection of variables and SU values.

D. Simple Example

This subsection presents a simple example which has been used in many modeling exercises. This example deals with the following nonlinear system:

$$y = (1.0 + x_1^{0.5} + x_2^{-1} + x_3^{-1.5})^2. \qquad (2.20)$$

Table 2.1 shows the input-output data generated from the nonlinear system. x_4 is a dummy variable. No.1 - no.20 are identification data. No.21 - no.40 are prediction data. Of course, the prediction data is never used to identify a fuzzy model. It is used only to check the validity of a fuzzy model identified by using the identification data. The result of the SOFIA will be compared with those of the S-K method [10] and the GMDH [13]. The learning coefficients are set as follows: $\varepsilon_1 = 0.1$, $\varepsilon_2 = 0.01$, $\varepsilon_3 = 0.01$.

(Stage 1)

Figure 2.3 shows the result of variables selections by FSV and the values of SUC. In Figure 2.3, the first line shows that the SUC value for the linear model, $y = f(x_1, x_2, x_3, x_4, x_0)$, was 14.00, where x_0 is the constant term. The second line shows that the SUC value for the linear model, $y = f(x_2, x_3, x_4, x_0)$, was 49.05. It is found from Figure 2.3 that x_4 is eliminated. Therefore, from now on, x_4 is removed from the consequent parts. Since the least SUC is 12.15, $\mathrm{SUC}^{OLD} \longleftarrow 12.15$ and go to **(Stage 2)**.

No.	model (k)	$p(x_1)$	$p(x_2)$	$p(x_3)$	$p(x_4)$	SUC^k	Check (x_k)	NF PP
-	l.m.	1	1	1	1	12.5	-	-
1	(1)	2	1	1	1	17.15	0	1
2	(2)	1	2	1	1	8.33*	1	2
3	(3)	1	1	2	1	9.49	1	3
4	(4)	1	1	1	2	11.00	1	4
5	(1)	-	-	-	-	-	0	-
6	(2)	1	3	1	1	8.55	0	5
7	(3)	1	2	2	1	3.89*	1	6
8	(4)	1	2	1	2	15.72	0	7
9	(1)	-	-	-	-	-	0	-
10	(2)	-	-	-	-	-	0	-
11	(3)	1	2	3	1	4.32*	0	8
12	(4)	-	-	-	-	-	0	-

Figure 2.4 Determination process of fuzzy partition.

(Stage 2)

Figure 2.4 shows the determination process of the optimal fuzzy partition. In this case, NFPP=8. The determined fuzzy partition has $p(x_1) = 1$, $p(x_2) = 2$, $p(x_3) = 2$, and $p(x_4) = 1$. This means that the identified model has 4 rules. The least SUC value is 3.89.

(Stage 3)

This stage performs consequent variable selection of the fuzzy model with the optimal fuzzy partition determined in **(Stage 2)** by the FSV.

(Stage 4)

Figure 2.5 shows the identified fuzzy model. The performance indexes for the identification data and the prediction data are calculated by (2.19). Table 2.2 shows comparison of identification results:linear model, the GMDH, the S-K method and the SOFIA. The prediction data 1 shows performance index for the prediction data (no.21 - no.40 in Table 2.2). The prediction data 2 shows performance index for the 50 input-output data pairs which were randomly generated for x_1, x_2, $x_3 \in [1, 5]$. In the identification data and the prediction

Rule 1 IF x_2 is $\exp\left(-\frac{(x_2+2.165)^2}{5.737}\right)$ and x_3 is $\exp\left(-\frac{(x_3+2.199)^2}{5.742}\right)$
THEN $y = 0.8038x_1 - 1.3777x_3$.

Rule 2 IF x_2 is $\exp\left(-\frac{(x_2+1.999)^2}{5.772}\right)$ and x_3 is $\exp\left(-\frac{(x_3-1.990)^2}{5.778}\right)$
THEN $y = 0.7444x_1 - 0.7609x_3 - 0.3453x_3 + 0.2511$.

Rule 3 IF x_2 is $\exp\left(-\frac{(x_2-1.952)^2}{5.768}\right)$ and x_3 is $\exp\left(-\frac{(x_3+2.113)^2}{5.736}\right)$
THEN $y = 0.7323x_{+}0.7657x_{-}0.7828x_{-}2.0220$.

Rule 4 IF x_2 is $\exp\left(-\frac{(x_2-2.012)^2}{5.769}\right)$ and x_3 is $\exp\left(-\frac{(x_3-2.063)^2}{5.758}\right)$
THEN $y = 0.5030x_1 - 0.0946x_3$.

Figure 2.5 Identified fuzzy model.

data 1, the performance indexes of the SOFIA are much better than those of the linear model and the GMDH. In particular, it should be emphasized that computational requirement of the SOFIA is less than that of the S-K method, i.e., the NFPP of the SOFIA is fewer than that of the S-K method.

Table 2.2 Comparison of identification results.

	Linear model	GMDH	S-K method	SOFIA
Identification data	12.7	4.7	1.5	1.8
Prediction data 1	11.1	5.7	2.1	2.9
Prediction data 2	-	-	4.2	3.4
Number of rules	-	-	3	4
NFPP	-	-	12	8

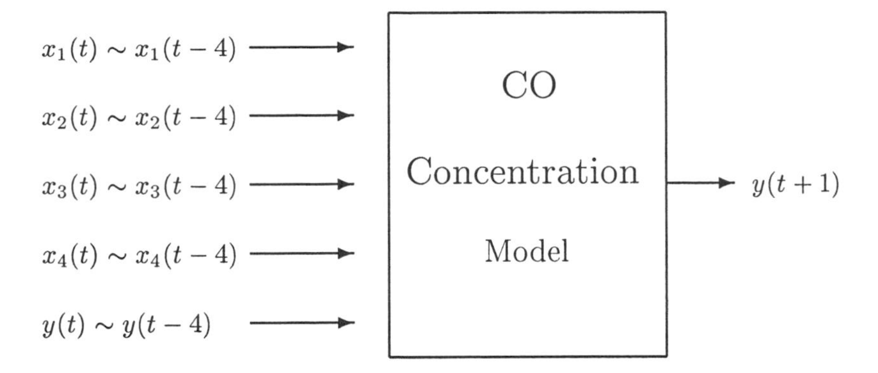

Figure 2.6 Inputs and an output of CO concentration model.

2.4.2 Prediction of CO concentration at a traffic intersection

The effectiveness of the SOFIA was shown in the above example. However, it is a simple example. In this subsection, the SOFIA will be applied to a complex prediction problem. The problem is to predict carbon monoxide concentration in the air at a traffic intersection in a large city of Japan. Carbon monoxide (CO) is one of important factors in air pollution problems.

The data used for identification and prediction are collected at the busiest traffic intersection in a large city of Japan. Input variables and an output variable of CO concentration model are shown in Figure 2.6. x_1 is wind velocity (m/s) , x_2 is volume of traffic, x_3 is temperature (C°), x_4 is amount of sunshine (cal/h), and y is CO concentration (ppm).

The number of data used for identification and prediction are 480 input-output data pairs and 253 input-output data pairs, respectively. The identification data are divided into two data sets N_A (240 input-output data pairs) and N_B (240 input-output data pairs). The sampling interval is 15 minutes. Of course, the prediction data is never used to identify a fuzzy model. It is used only for checking the validity of a fuzzy model identified by using the identification data.

Figures 2.7 and 2.8 show the identified linear model and the identified fuzzy model, respectively. The fuzzy model consists of two rules. Note that x_3 does not appear in the identified models. The reason is that x_3 was eliminated by using the forward selection of variables in (Stage 3).

Table 2.3 shows the values of performance index for the linear model and the fuzzy model. The values are calculated using (2.19). J_1 and J_2 are the

$$\begin{aligned} y(t+1) &= 0.0210x_1(t-2) - 0.0067x_1(t-1) - 0.0176x_1(t) \\ &\quad -0.3325x_2(t-1) + 0.4601x_2(t) - 0.0331x_4(t-4) \\ &\quad +0.0226x_4(t-3) + 0.0062x_4(t-2) + 0.0228x_4(t-1) \\ &\quad -0.0113x_4(t) + 0.0452y(t-4) - 0.0420y(t-3) \\ &\quad -0.0047y(t-2) - 0.3764y(t-1) + 1.2451y(t). \end{aligned}$$

Figure 2.7 Identified linear model.

Rule 1: If $x_2(t-1)$ is $\exp\{-(x_2(t-1)+1.90)^2/5.780\}$ THEN

$$\begin{aligned} y_1(t+1) &= -0.008x_1(t-2) + 0.026x_1(t-1) - 0.032x_1(t) \\ &\quad -0.205x_2(t-1) + 0.249x_2(t) - 0.138x_4(t-4) \\ &\quad -0.181x_4(t-3) + 0.335x_4(t-2) + 0.088x_4(t-1) \\ &\quad -0.112x_4(t) + 0.090y(t-4) - 0.106y(t-3) \\ &\quad +0.011y(t-2) - 0.460y(t-1) + 1.356y(t) - 0.018. \end{aligned}$$

Rule 1: If $x_2(t-1)$ is $\exp\{-(x_2(t-1)-2.06)^2/5.768\}$ THEN

$$\begin{aligned} y_1(t+1) &= -0.013x_1(t-2) + 0.059x_1(t-1) - 0.032x_1(t) \\ &\quad -0.497x_2(t-1) + 0.827x_2(t) + 0.006x_4(t-4) \\ &\quad +0.066x_4(t-3) - 0.130x_4(t-2) + 0.095x_4(t-1) \\ &\quad -0.036x_4(t) + 0.090y(t-4) + 0.007y(t-3) \\ &\quad -0.037y(t-2) - 0.103y(t-1) + 0.748y(t) - 0.002. \end{aligned}$$

Figure 2.8 Identified fuzzy model.

values of performance index for the identification data and the prediction data, respectively. The performance index J_2 of the fuzzy model is superior to that of linear model. Figure 2.9 shows prediction results of the linear model and the fuzzy model for a part of the prediction data. It is, in practice, necessary to obtain good prediction in the high level of CO concentration from the viewpoint of issuing an alarm or warning for dangerous conditions. The prediction values of the linear model do not agree well with the real outputs in the high level of CO concentration. Thus, the linear model is not useful as a prediction model of CO concentration.

Table 1.3 Performances of models.

	Linear model	Fuzzy model
J_1	5.7	4.8
J_2	11.8	5.9

2.4.3 Prediction of O_2 concentration in a municipal refuse incinerator

This subsection deals with fuzzy modeling of a municipal refuse incinerator with 30 predictor variables. It is a prediction problem of O_2 concentration in the flue gas of the municipal refuse incinerator. The prediction problem is difficult because of high nonlinearity and strong interference of many predictor variables. The SOFIA will be compared with linear regression method based on AIC and the S-K method [14].

Inputs and an output of the incinerator are shown in Figure 2.10. The number of data used for identification and prediction are 212 input-output data pairs, respectively. The identification data is divided into two data sets N_A (106 input-output data pairs) and N_B (106 input-output data pairs). The sample interval is one minute. The learning coefficients are set as follows: $\varepsilon_1 = 0.001$, $\varepsilon_2 = 0.0001$, $\varepsilon_3 = 0.0001$. Figure 2.11 Shows the identification result by SOFIA.

Table 2.4 shows comparison of identification results: the linear model (l.m.), the S-K method and the SOFIA. J_1 and J_2 are the values of performance index for the identification data and the prediction data, respectively. The value of J_2 of the linear model is poor because of high nonlinearity and strong interference of many predictor variables. Furthermore, the NFPP of the SOFIA is fewer than that of the S-K method. This means that computational requirement of the SOFIA is less than that of the S-K method.

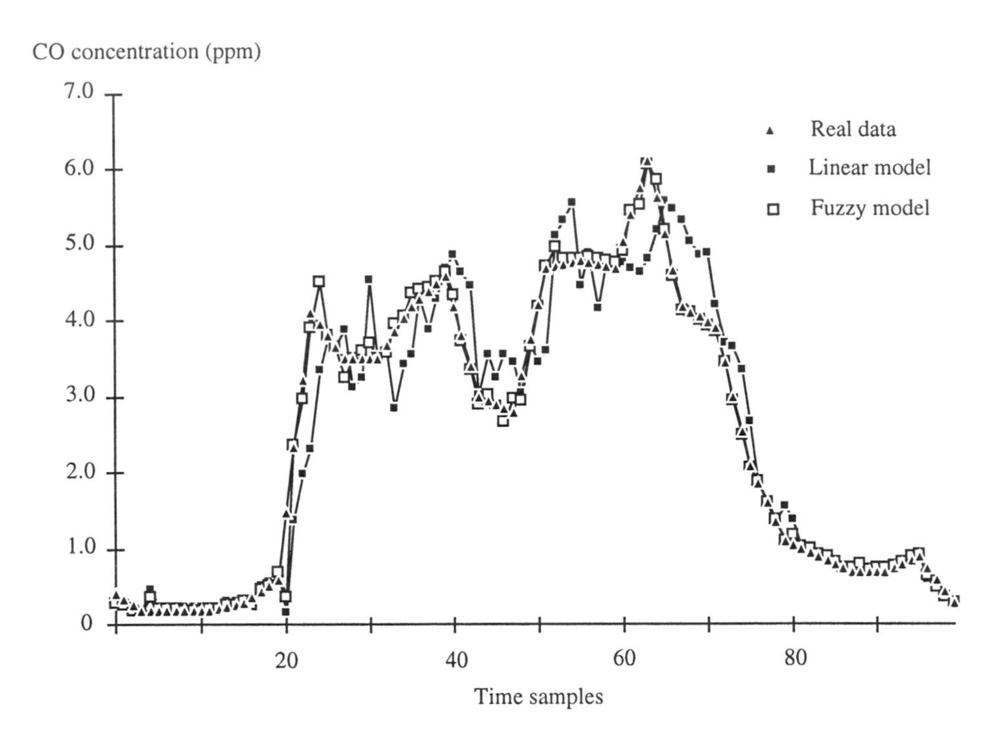

Figure 2.9 Prediction results by linear model and fuzzy model.

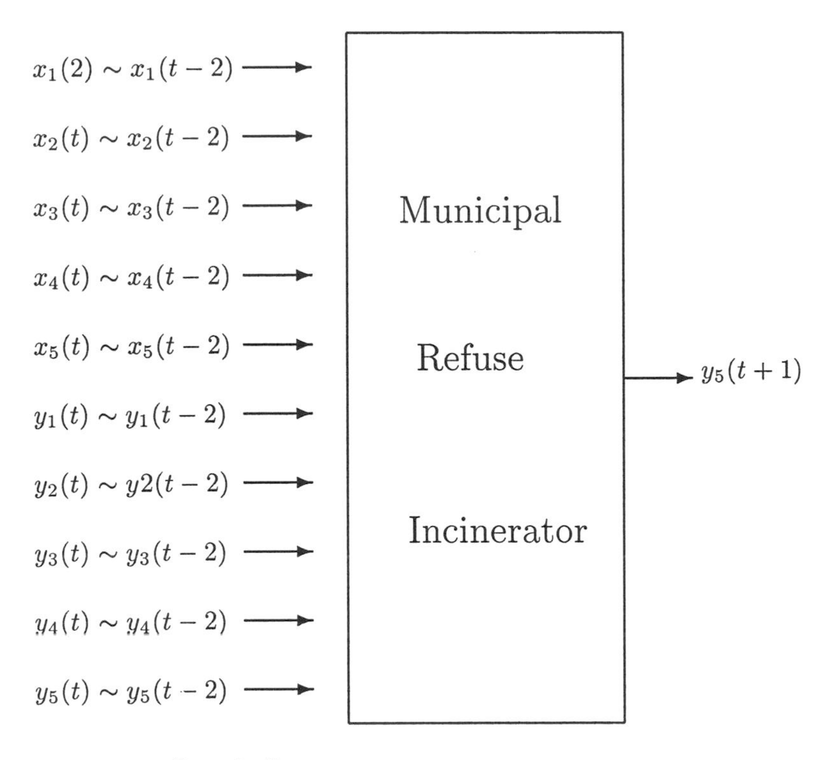

- x_1: cooling air flow rate,
- y_1: temperature of the flue gas at outlet of furnace,
- x_2: overall combustion air flow rate,
- y_2: evaporation rate in the boiler,
- x_3: combustion air flow rate for zone 1,
- y_3: NOx concentration in the flue gas,
- x_4: combustion air flow rate for zone 2,
- y_4: pressure under the stoker,
- x_5: refuse supplying rate,
- y_5: O2 concentration in the flue gas.

Figure 2.10 Inputs and a output of a municipal refuse incinerator.

Rule 1: IF $x_2(t)$ is $\exp\{-(x_2(t)+1.989)^2/5.780\}$, THEN

$$\begin{aligned} y_5(t+1) &= 0.1973x_2(t-2) - 0.1890x_2(t) - 0.0449x_3(t-2) \\ &\quad -0.0372x_3(t-1) - 0.0262x_4(t-2) + 0.0596x_4(t) \\ &\quad -0.0747x_5(t-2) + 0.2264x_5(t-1) - 0.1805x_5(t) \\ &\quad -0.0833y_1(t-1) - 0.0341y_2(t-2) - 0.1261y_2(t) \\ &\quad +0.1117y_3(t-2) - 0.1874y_3(t-1) + 0.8207y_5(t). \end{aligned}$$

Rule 2: IF $x_2(t)$ is $\exp\{-(x_2(t)-2.001)^2/5.768\}$, THEN

$$\begin{aligned} y_5(t+1) &= 0.1868x_2(t-2) - 0.1193x_2(t-1) - 0.1386x_2(t) \\ &\quad +0.0028x_3(t-2) + 0.0957x_3(t-1) - 0.0173x_4(t) \\ &\quad -0.0345x_5(t-2) + 0.0482x_5(t) - 0.4774y_1(t-2) \\ &\quad +1.6530y_1(t-1) - 1.3603y_1(t) + 0.0236y_2(t-2) \\ &\quad -0.1295y_3(t-1) + 0.1903y_3(t) + 0.7277y_5(t). \end{aligned}$$

Figure 2.11 Identified fuzzy model.

Table 2.4 Comparison of identification results

	Linear model	S-K method	SOFIA
J_1	0.610	0.791	0.596
J_2	13.88	1.051	0.921
Number of rules	-	3	2
NFPP	-	64	40

2.5 CONCLUSION

This chapter has presented the procedures and the effectiveness of two fuzzy modeling techniques: the S-K method and the SOFIA. The S-K method has been applied to the prediction problem of water flow rate in the river Dniepr. The SOFIA has been applied to two prediction problems: CO concentration in the air at the busiest traffic intersection at a large city of Japan and O_2 concentration of a municipal refuse incinerator.

Resently, several methods based on systematic identification techniques have been proposed. Wang [17] developed adaptive fuzzy control techniques. Wang and Langari [18, 19] and Yager and Filev [20] presented new algorithms to identify a fuzzy model using input-output data. Furuhashi and Uchikawa [21, 22] proposed useful fuzzy-neuro modeling techniques. Thus a wide variety of fuzzy modeling techniques will be extensively developed in a nonlinear modeling framework in the future.

Appendix

(Stage 1) Identification of a linear system.

- Identification of a linear model. (Select a linear model with the least SUC value.)
- $\mathrm{SUC}_1^* \longleftarrow$ the least SUC value.
- $\mathrm{SUC}^{OLD} \longleftarrow \mathrm{SUC}_1^*$.
- Go to **(Stage 2)**.

(Stage 2) Determination of the optimal fuzzy partition.

- $p(x_k) \longleftarrow 1. \qquad k = 1 \sim n.$
- $\mathrm{CHECK}(x_k) \longleftarrow 1. \qquad k = 1 \sim n.$
- $k_0 \longleftarrow 1.$
- $\mathrm{NFPP} \longleftarrow 0.$

[Part 1]

IF $\mathrm{CHECK}(x_{k_0})=0$

THEN

$\mathrm{SUC}^{k_0} \longleftarrow \alpha$ (α is a big value), and
go to **[Part 2]**,

ELSE

$p(x_{k_0}) \longleftarrow p(x_{k_0}) + 1$,
identification of premise and consequent parameters,
calculation of SUC,
$\mathrm{SUC}^{k_0} \longleftarrow$ the SUC value of the model(k_0),
$p(x_{k_0}) \longleftarrow p(x_{k_0}) - 1$,
go to **[Part 2]**.

[Part 2]

$k_0 \longleftarrow k_0 + 1.$
IF $k_0 \leq n$

THEN
go back to [**Part 1**],
ELSE
$\mathrm{SUC}^{NEW} \longleftarrow \min_k \mathrm{SUC}^k$,
$k_0 \longleftarrow 1$,
If $\mathrm{SUC}^{NEW} \geq \mathrm{SUC}^{OLD}$,
then
select a model with the SUC^{OLD}, and
go to **(Stage 3)**,
else
go to [**Part 3**].

[Part 3]

$p(x_k) \longleftarrow p(x_k) + 1$ for x_k such that $\min_k \mathrm{SUC}^k$.
Go to [**Part 4**].

[Part 4]

If there exists k such that $\mathrm{SUC}^k \geq \mathrm{SUC}^{OLD}$
then CHECK(x_k) $\longleftarrow 0$.
$\mathrm{SUC}^{OLD} \longleftarrow \mathrm{SUC}^{NEW}$.
NFPP $\longleftarrow$ NFPP+1.
Go back to [**Part 1**].

(Stage 3) Determination of the optimal consequent structure.

Consequent variable selection of a fuzzy model determined in **(Stage 2)**.
Select a model with the least SUC value.
go to **(Stage 4)**.

(Stage 4) Parameter identification.

Parameter identification of the final model selected in **(Stage 3)**.
Calculation of performance index.

Notes

1. 1 GMDH [3] is an identification method for complex nonlinear systems.

References

[1] Ichihashi, H. and Watanabe, T. (1990) Learning control by fuzzy models using a simplified fuzzy reasoning, *Journal of Japan Society for Fuzzy Theory and Systems*, **2**(3), 429-437, (in Japanese).

[2] Ichihashi, H. (1991) Iterative fuzzy modeling and a hierarchical network, *Proceedings of IFSA'91*, 49-52.

[3] Ivakhnenko, A. G. at el., (1978) Principle versions of the minimum bias criterion for a model and an investigation of their noise immunity, *Soviet Automat. Control*, **11**, 27-45.

[4] Mamdani, E. H. (1974) Applications of fuzzy algorithms for control of a simple dynamic plant, *Proceedings IEE* **121**(12), 1585-1588.

[5] Sugeno, M. and Tanaka, K. (1991) Successive identification of a fuzzy model and its applications to prediction of a complex system, *Fuzzy Sets and Systems*, **42**(3), 315-334.

[6] Tanaka, K. and Sugeno, M. (1992) Stability analysis and design of fuzzy control systems, *Fuzzy Sets and Systems*, **46**(1), 135-156.

[7] Suzuki, K., Naka, Y., and Bito, K. (1991) Fuzzy Multi-model control of a high-purity distillation system, *Proceedings of the International Fuzzy Engineering Symposium '91*, **2**, 684-693.

[8] Yamaguchi, T. at el. (1991) Fuzzy associative memory system and its application to a helicopter control, *Proceedings of the International Fuzzy Engineering Symposium'91*, **2**, 770-779.

[9] Sugeno, M. and Kang, G. T. (1986) Fuzzy modeling and control of multilayer incinerator, *Fuzzy Sets and Systems*, **18**, 329-346.

[10] Sugeno, M. and Kang, G. T. (1988) Structure identification of fuzzy model, *Fuzzy Sets and Systems*, **28**, 15-33.

[11] Takagi, T. and Sugeno, M. (1985) Fuzzy identification of systems and its applications to modeling and control, *IEEE Trans. SMC*, **15**, 116-132.

[12] Nomura, H. at el. (1991) A self-tuning method of fuzzy control by descent method, *Proceedings of IFSA'91*, 155-158.

[13] Kondo, T. (1986) Revised GMDH algorithm estimating degree of the complete polynomial, *Trans. Soc. Instrument and Control Engrs.*, **22**, 928-934, (in Japanese).

[14] Suzuki, K. and Tanaka, K. (1991) Fuzzy modeling of a municipal refuse incinerator, *7th Fuzzy System Symposium*, 335-338, (in Japanese).

[15] Tanaka, K. at el. (1993) Self-organizing fuzzy identification of a municipal refuse incinerator, *Proceedings of Int. Fuzzy Systems and Intelligent Control Conf.*, 13-22.

[16] Tanaka, K. at el. (1995) Modeling and control of carbon monoxide concentration using a neuro-fuzzy technique, *IEEE Transactions on Fuzzy Systems*, **3**(3), 271-279.

[17] Wang, Li-Xin (1994) *Adaptive Fuzzy Systems and Control: Design and Stability Analysis*, Prentice Hall.

[18] Wang, L. and Langari, R. (1995) Building Sugeno-type models using fuzzy discretization and orthogonal parameter estimation techniques, *IEEE Transactions on Fuzzy Systems*, **3**(4), 454-458.

[19] Wang, L. and Langari, R. (1996) Complex systems modeling via fuzzy logic, *IEEE Transactions on Systems, Man and Cybernetics*, Part B, **20**(1), 100-106.

[20] Yager, R. R. and Filev, D. P. (1993) Unified structure identification and parameter identification of fuzzy models, *IEEE Transactions on Systems, Man and Cybernetics*, **23**, 1198-1205.

[21] Horikawa, S., Furuhashi,T., and Uchikawa, Y. (1992) On fuzzy modeling using fuzzy neural networks with the back-propagation algorithm, *IEEE Trans. on Neural Networks*, **3**(5).

[22] Horikawa, S., Furuhashi, T., and Uchikawa, Y. (1993) On identification of structure in premises of a fuzzy model using a fuzzy neural network, *Proceedings of the 2nd IEEE International Conference on Fuzzy Systems*, 661-666.

[23] Sugeno, M. and Yasukawa, T. (1993) A fuzzy logic based approach to qualitative modeling, *IEEE Transactions on Fuzzy Systems*, **1**, 7-31.

[24] Pedrycz, W. (1984) An identification algorithm in fuzzy relational systems, *Fuzzy Sets and Systems*, **13**, 153-167.

[25] Czogala, E. and Pedrycz, W. (1981) On identification in fuzzy systems and its applications in control problem, *Fuzzy Sets and Systems*, **6**, 73-83.

[26] Tong, R. M. (1978) Synthesis of fuzzy models for industrial processes, *Int. J. General Systems*, **4**, 143-162.

[27] Xu, C. and Lu, Y. (1987) Fuzzy model identification and self learning for dynamic systems, *IEEE Transactions on Systems , Man and Cybernetics*, **17**, 683-689.

[28] Takagi, H. and Hayashi, I. (1991) NN-driven fuzzy reasoning, *International Journal of Approximate Reasoning*, **5**(3), 191-212.

[29] Akaike, H. (1974) A new look at the statistical model identification, *IEEE Trans. Automatic Control*, **19**, 716-723.

[30] Wenstop, F. (1976) Deductive verbal models of organizations, *Int. J. Man-Machine Studies*, **8**, 293-311.

[31] Gaines, B. R. (1979) Sequential fuzzy systems identification, *Fuzzy Sets and Systems*, **2**, 15-24.

[32] Shen, Q. and Leitch, R. (1993) Fuzzy Qualitative Simulation, *IEEE Transactions on Systems, Man and Cybernetics*, **23**, 1038-1061.

3 FUZZY RULE BASED MODELS AND APPROXIMATE REASONING

Ronald R. Yager[1] and Dimitar P. Filev[2]

[1] Machine Intelligence Institute
Iona College
New Rochelle, NY 10801,

[2] Ford Motor Company
24500 Glendale Ave.
Detroit, MI 48239

3.1 INTRODUCTION

Fuzzy systems models form a special class of systems models that use the apparatus of fuzzy logic to represent the essential features of a system. From a formal point of view, fuzzy systems can be regarded as one alternative to the linear, nonlinear and neural modeling paradigms. Fuzzy systems models, however, possess a unique characteristic that is not available in most other types of formal modeling techniques - this is the ability to mimic the mechanism of approximate reasoning performed in the human mind. The most common fuzzy systems models consist of collections of logical IF - THEN rules with vague predicates; these rules along with the reasoning mechanism are the kernel of a fuzzy model.

Depending on the format of the rules, fuzzy systems models fall into two categories which fundamentally differ in their ability to represent different types of information. The first includes *Linguistic Models* (LMs). In these models fuzzy quantities are associated with linguistic labels, and the fuzzy model is essentially a qualitative expression of the system. Models of this type form the basis for qualitative modelling that describes the system behavior by using a natural language [1]. The second category of fuzzy models is based on the Takagi- Sugeno-Kang (TSK) method of reasoning that was proposed by Sugeno and his co-workers [2-5]. These models are formed by logical rules that have a fuzzy antecedent part and functional consequent; essentially they are a combination of fuzzy and nonfuzzy models. Fuzzy models based on the TSK method of reasoning integrate the ability of LMs for qualitative knowledge representation with an effective potential for expressing quantitative information as well. In addition, this type of fuzzy model permits a relatively easy application of powerful learning techniques for their identification from data. We shall refer to models in this category as TSK *Fuzzy Models.*

Central to the fuzzy systems modeling technology is a partitioning of the input/output space into regions in which we understand the performance of the system. Essentially with the aid of a fuzzy model we simplify the representation of complex systems by representing them by a collection simpler models. The fuzziness comes into play in that we allow these simpler models to partially overlap. As shall become apparent in the remainder of this chapter one of the functions of the inference mechanism used in fuzzy systems modeling is to help combine the suggested solutions from these partially overlapping models.

3.2 LINGUISTIC MODELS

One of the main directions in the theory of fuzzy systems is the linguistic approach originally initiated by Zadeh [1]. A linguistic model is a knowledge-based system made up of rules which incorporate fuzzy knowledge about the real-world. As we shall see different interpretations of the knowledge contained in the model rule-base will lead to different reasoning mechanisms and result in different types of linguistic models. We begin with the case of a single input - single output system.

In single-input single-output systems the encoded knowledge can be expressed by a collection of IF - THEN rules of the form

$$\begin{array}{ll} \textbf{IF} \;\; U \;\text{ is }\; B_1 & \textbf{THEN} \;\; V \;\text{ is }\; D_1 \\ & \textbf{ALSO} \\ & \ldots\ldots \\ & \textbf{ALSO} \\ \textbf{IF} \;\; U \;\text{ is }\; B_m & \textbf{THEN} \;\; V \;\text{ is }\; D_m \end{array} \tag{3.1}$$

In the above U is the input variable and V is the output variable of the LM. B_i and D_i are fuzzy subsets of the base sets X and Y of U and V. Usually the fuzzy sets B_i and D_i are associated with linguistic labels (terms). Membership functions of the fuzzy sets B_i, D_i are denoted respectively $B_i(x)$ and $D_i(y)$.

The left-hand side of the rule, is called the antecedent and is related to the input of the system; the right-hand side, called the consequent, is related to the output.

The LM can be considered as an expert system linguistically describing a given complex system. The set of rules can be seen as an analogy to the collection of equations used for presentation of linear and nonlinear systems in classical modeling techniques. The fuzzy sets, the B_i's and D_i's, are the parameters of the LM; the number of the rules determines its structure. The important idea here is that we have partitioned the input space X into fuzzy regions, each of which has associated with it its own particular output. The role of the fuzzy sets is to form granules (bundles) of input-output values. These bundles are associated with the individual rules. In a limiting case when each bundle contains just one reading and the LM coincides with a collection of input-output data. The number of rules of the model rule-base necessary to describe a given system characterizes the ability of the model for generalization.

3.3 INFERENCE WITH FUZZY MODELS

In this section we describe the fundamental mechanism for reasoning with linguistic models by considering the case of a one rule system. The machinery used to develop this reasoning mechanism is based on the theory of approximate reasoning introduced by Zadeh [6]. According to the theory of approximate reasoning a rule

$$\textbf{IF} \quad U \quad \text{is} \quad B \quad \textbf{THEN} \quad V \quad \text{is} \quad D \tag{3.2}$$

can be translated into a proposition of the form (U, V) is R where R is a fuzzy relationship defined on the Cartesian product space $X \times Y$. Given a fuzzy input $\underline{U \text{ is } A}$ one conjuncts this with (3.2) and obtains the relationship $\underline{(U, V) \text{ is } G}$ where

$$G = A \cap R. \tag{3.3}$$

G a fuzzy set also defined on the space $X \times Y$ with membership function

$$G(x, y) = A(x) \wedge R(x, y). \tag{3.4}$$

Applying the projection principle [6] we get as our output $\underline{V \text{ is } F}$ where F is a fuzzy subset of Y,

$$F(y) = \vee_x (G(x, y)) = \vee_x (A(x) \wedge R(x, y)). \tag{3.5}$$

The above operation can be rewritten in the more compact max-min rule of inference form as

$$F = A \circ R. \tag{3.6}$$

In the above the fuzzy intersection of A and R forming G was interpreted by the min($\wedge$) operator. An alternative translation using the multiplicative t-norm operator instead of the min yields the max-product rule of inference:

$$F = A \diamond R. \tag{3.7 }$$

According to this the membership function of the fuzzy set F inferred by the relation R is:

$$F(y) = \vee_x(G(x,y)) = \vee_x(A(x) \cdot R(x,y). \tag{3.8 }$$

In essence, the rule of inference is an operational form of the linguistic model; it is seen as a mapping that defines a transformation of the input fuzzy value to the output. It is a convenient representation of the model, especially in cases where finite universes of discourse are considered. The application of the rule of inference for practical calculations is rather limited, especially this is the case when the input to the system is a crisp value. However, we shall use it to represent in a compact form the transformation defined by the linguistic model.

For finite universes X and Y with the cardinalities p and q we can express the fuzzy sets A and B, D and F by using vector membership functions as row vectors $[A]$ and $[B]$, $[D]$ and $[F]$ of dimensions p and q respectively, and the fuzzy relation R is a $(p \times q)$ matrix $[R]$. Then the max-min and the max-product rules of inference can be rewritten in the matrix forms:

$$[F] = [A] \circ [R], \tag{3.9 }$$
$$[F] = [A] \diamond [R]. \tag{3.10 }$$

A careful inspection of expressions (3.5) and (3.8) shows that the matrix forms of the max-min (3.9) and max-product (3.10) rules of inference are special kinds of inner products of vector $[A]$ and matrix $[R]$ in which the summation is replaced by the max($\vee$) operation and in the case of max-product rule the pairwise multiplication is replaced by a pairwise min($\wedge$) operation.

In the above we indicated that a fuzzy rule is associated with a fuzzy relationship R which was essentially generic. Furthermore no consideration was given to the case where we have multiple rules, the process of combining the outputs of these rules. In the following we shall discuss two specific approaches for the translation of rules into a fuzzy relationships and the associated combination operators. These two approache s result in two alternative reasoning mechanisms that are realized in models. In the next sections we take a careful look at these two methods.

3.4 MAMDANI (CONSTRUCTIVE) AND LOGICAL (DESTRUCTIVE) MODELS

In the **Mamdani type** [7-9] models the relationship associated with a particular rule is obtained via a conjunction of the antecedent and consequent of the rule. Furthermore, in these models the overall systems output from a collection of rulers is constructed by superimposing the outputs of the individual rules. In this approach each rule

$$\textbf{IF} \quad U \quad \text{is} \quad B_i \quad \text{THEN} \quad V \quad \text{is} \quad D_i \tag{3.11}$$

is interpreted as a fuzzy point, 2-tuple, and is expressed as a fuzzy relationship R_i which is obtained as a fuzzy intersection of the fuzzy sets B_i and D_i:

$$R_i = B_i \cap D_i, \tag{3.12}$$

R_i, defined on the Cartesian space $X \times Y$, has membership function

$$R_i(x, y) = B_i(x) \wedge D_i(y). \tag{3.13}$$

From the above expression it is seen that the fuzzy relation R_i forms a rectangular region in the Cartesian product space $X \times Y$ (Figure 3.1) with joint possibility distribution by (3.13).

Under the Mamdani method, where the linguistic connection between the individual rules is interpreted as an **or** connective, the aggregation of the rules is accomplished via a union of the individual fuzzy relationships

$$R = \bigcup_{j=1}^{m} R_i. \tag{3.14}$$

It is this this representation that essentially leads to a constructive nature of this model. The membership function of the overall model output R is given by

$$R(x, y) = \bigvee_{i=1}^{m} R_i(x, y) = \bigvee_{i=1}^{m} (B_i(x) \wedge D_i(y)). \tag{3.15}$$

For a given input $U = A$, the fuzzy output F obtained by this method is defined by the max-min inference rule (3.6):

$$F = A \circ R = A \circ \left(\bigcup_{i=1}^{m} R_i \right) = A \circ \left(\bigcup_{i=1}^{m} (B_i \cap D_i) \right). \tag{3.16}$$

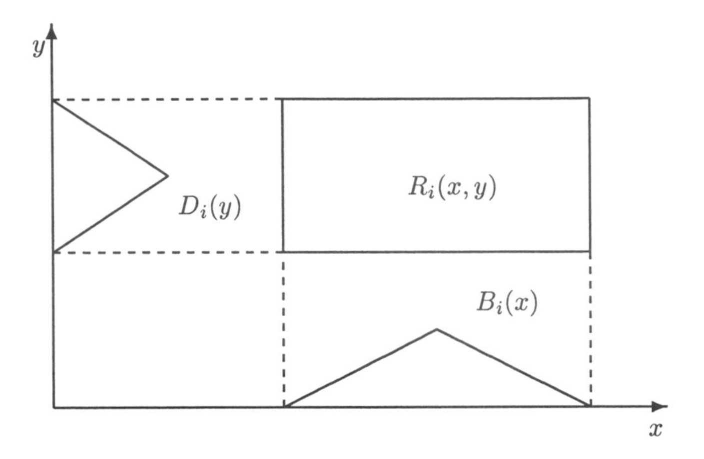

Figure 3.1 Joint possibility distribution $R_1(x,y)$ of the rule IF U is B_i THEN V is D_i

For the membership function of F we have

$$F(y) = \bigvee_x [A(x) \wedge R(x,y)] = \bigvee_x \left[\bigvee_{i=1}^{m} (A(x) \wedge R_i(x,y)) \right],$$

$$F(y) = \bigvee_{i=1}^{m} \left[\bigvee_x (A(x) \wedge B_i(x) \wedge D_i(y)) \right],$$

$$F(y) = \bigvee_{i=1}^{m} [\tau_i \wedge D_i(y)]. \tag{3.17 }$$

where

$$\tau_i = \text{Poss}[B_i|A] = \bigvee_x [B_i(x) \wedge A(x)] \tag{3.18 }$$

denotes the condition possibility B_i of given A. τ_i is the *degree of firing* (DOF) or *firing strength (relevance)* of the i-th rule.

In (3.17) the inherent constructive nature of this approach becomes readily apparent. We see that the output is constructed by a weighted union of the outputs of each of the rules. In particular if none of the rules fire the output is the null set. Thus we see this model is a kind of superposition of the constituents from each of the individual rules.

In special the case, usually occurring in fuzzy logic controllers, where the input is a deterministic value $x^* \in X$, the input fuzzy set A is a fuzzy singleton with membership grade one at x^* and zero elsewhere on the universe X,

$$A(x) = \begin{cases} 0 & \text{if } x \neq x^* \\ 1 & \text{if } x = x^*. \end{cases}$$

In this case the expression (3.18) for the degree of firing (DOF) becomes:

$$\tau_i = \text{Poss}[B_i|A] = B_i(x^*). \quad (3.19)$$

In summary, we obtain the following algorithm for calculation of the output that is inferred by a LM (1) via the Mamdani method assuming given input $U = A$ or $U = x*$:

Algorithm 1.

1. For each rule of the LM (1) - calculate the DOF, the τ_i's, of the rule:

$$\tau_i = \vee_x[B_i(x) \wedge A(x)] \text{ if the input is a fuzzy set } A$$
$$\tau_i = B_i(x*) \text{ if the input is a crisp number } x*.$$

2. Find the fuzzy set F_i inferred by the i-th rule:

$$F_i(y) = \tau_i \wedge D_i(y).$$

3. Aggregate the inferred fuzzy sets F_i by using the max operation:

$$F(y) = \bigvee_{i=1}^{m} F_i(y).$$

Figure 3.2 presents a block-diagram of the internal mechanism of the SI-SO LM which is based on the Mamdani method of reasoning.

Figure 3.3 presents in a graphical form an application of the Mamdani method for calculation of the fuzzy output value F from a system of two rules, assuming that the input is a fuzzy set A.

IF U is B_1 **THEN** V is D_1

ALSO

IF U is B_2 **THEN** V is D_2.

Logical type models are based upon an alternative interpretation for the definition of the relationship R_i obtained from rule and of the aggregation procedure of the individual rules. As we shall see the logical type linguistic models result in a situation in which the overall model output is obtained by

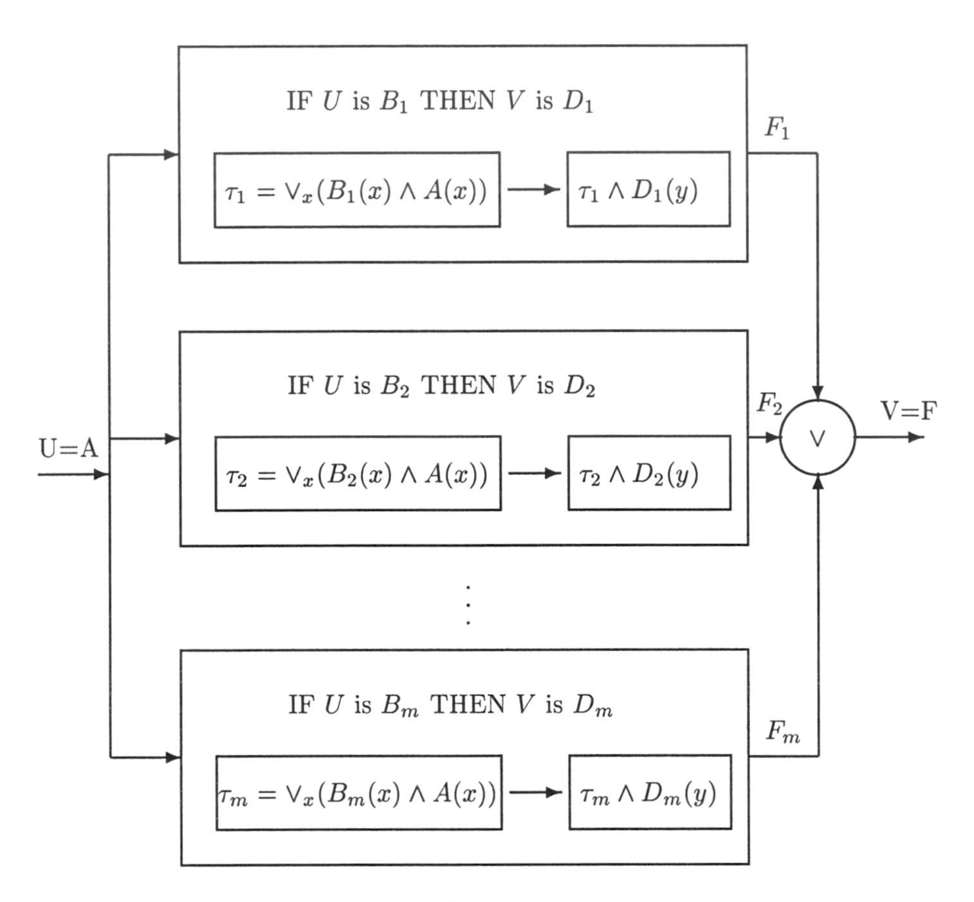

Figure 3.2 Mamdani/Constructive reasoning method.

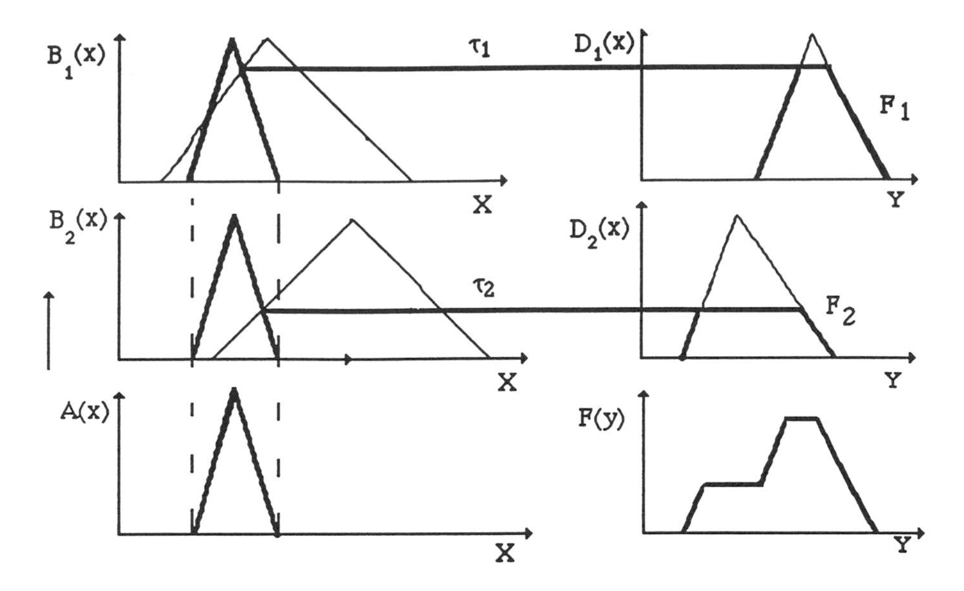

Figure 3.3 Graphical calculation of the output.

removing possibilities not acceptable to the individual rules and as such can be seen as a destructive method. In this approach each individual rule,

$$\text{IF} \quad U \quad \text{is} \quad B_i \quad \text{THEN} \quad V \quad \text{is} \quad D_i.$$

is translated by a logical interpretation of the **IF - THEN** operator and as such results in a fuzzy relationship R_i that is defined by

$$R_i = \bar{B}_i \cup D_i. \tag{3.20}$$

The fuzzy relationship R_i, defined on the Cartesian product space $X \times Y$, has membership function

$$R_i(x, y) = \bar{B}_i(x) \vee D_i(y), \tag{3.21}$$

where $\bar{B}_i(x) = 1 - B_i(x)$.

Under this approach the connection between the individual rules, **ALSO**, is interpreted as an **and** connective resulting in the aggregation of the individual rules being is accomplished by the intersection operator

$$R = \bigcap_{i=1}^{m} R_i. \tag{3.22}$$

Thus in this case the fuzzy relationship R has joint possibility distribution

$$R(x,y) = \bigwedge_{i=1}^{m} R_i(x,y) = \bigwedge_{i=1}^{m} \bar{B}_i(x) \vee D_i(y). \qquad (3.23)$$

For a given input fuzzy set $U = A$, the output inferred using the logical method is obtained by the max-min rule of inference

$$A \circ R = A \circ \left(\bigcap_{i=1}^{m} R_i \right) = A \circ \left[\bigcap_{i=1}^{m} (\bar{B}_i \cup D_i) \right].$$

Let us denote $A \circ R$ as the fuzzy subset G, its membership grade is

$$\begin{aligned} G(y) &= \bigvee_x [A(x) \wedge R(x,y)] = \bigvee_x \left[\bigwedge_{i=1}^{m} (A(x) \wedge R_i(x,y)) \right] \\ &= \bigvee_x \left[\bigwedge_{i=1}^{m} [\bar{B}_i(x) \wedge A(x) \vee D_i(y) \wedge A(x)] \right]. \end{aligned}$$

Consider now the fuzzy subset F where

$$F(y) = \bigwedge_{i=1}^{m} \left[\bigvee_x [\bar{B}_i(x) \wedge A(x) \vee D_i(y) \wedge A(x)] \right].$$

It can be shown that $G(y) \leq F(y)$ for all y, i.e, $G \subseteq F$. Now since the inferred value is V is G and $G \subseteq F$ then by the use of the entailment principle we can infer that V is F. While F is less specific than G, it is a correct inference, and as we shall see, leads a useful formulation for the inference.

Assuming that $A(x)$ is normal, $A(x) = 1$ for some x, we get from the above expression:

$$F(y) = \bigwedge_{i=1}^{m} [\bar{\tau}_i \vee D_i(y)], \qquad (3.24)$$

where

$$\bar{\tau}_i = \text{Poss}[\bar{B}_i | A] = \bigvee_x [\bar{B}_i(x) \wedge A(x)] \qquad (3.25)$$

is the conditional possibility $\bar{B}_i | A$ that represents the DOF of the i-th rule.

Again if we consider the special case of crisp input x^*, the input fuzzy set A is a fuzzy singleton with membership grade

$$A(x) = \begin{cases} 0 & \text{if } x \neq x^* \\ 1 & \text{if } x = x^*. \end{cases}$$

In this case the expression (3.14) for the DOF becomes:

$$\bar{\tau}_i = \text{Poss}[\bar{B}_i|A] = 1 - B_i(x*) = 1 - \tau_i.$$

The above is summarized in the following algorithm for calculation of the fuzzy output inferred by a LM via the logical method:

Algorithm 2.

1. For each rule of the LM (1) calculate

$$\bar{\tau}_i = x[\bar{B}_i(x) \wedge A(x)] \text{ if the input is a fuzzy set } A;$$
$$\bar{\tau}_i = 1 - B_i(u) \text{ if the input is a crisp number } u.$$

2. Find the fuzzy set Ei inferred by the rule

$$F_i(y) = \bar{\tau}_i \wedge D_i(y).$$

3. Aggregate the inferred fuzzy sets E_i by using the *min* operation:

$$F(y) = \bigwedge_{i=1}^{m} F_i(y).$$

From the above algorithm we can see the essential destructive nature of this approach. We first note that if a rule has $\tau_i = 0$, $\bar{\tau}_i = 1$, then $F_i = Y$ and it plays no role in the aggregation process of step three. On the other if $\tau_i = 1$, $\bar{\tau}_i = 0$, then $F_i = D_i$ and in step 3 it essentially eliminates any solutions with low membership grade in D_i.

Figure 2.4 presents a block-diagram of the inference mechanism based on the logical method of reasoning.

The algorithm for calculating the output of the LM (Figure 3.3), via the logical method, is presented in graphical form in Figure 3.5.

Straightforward extensions of the above two approaches to reasoning with linguistic models can be had by realizing that the max and min operators used in (3.17) and (3.24) are only special cases of the t-norm and t-conorm operators. Therefore, we can obtain from expressions (3.17) and (3.24) the more general form of the outputs inferred by the Mamdani type (constructive) and the logical type (destructive) models:

$$F(y) = \bigoplus_{i=1}^{m} [\tau_i \otimes D_i(y)], \tag{3.26}$$

$$E(y) = \bigotimes_{i=1}^{m} [\tau_i \oplus D_i(y)], \tag{3.27}$$

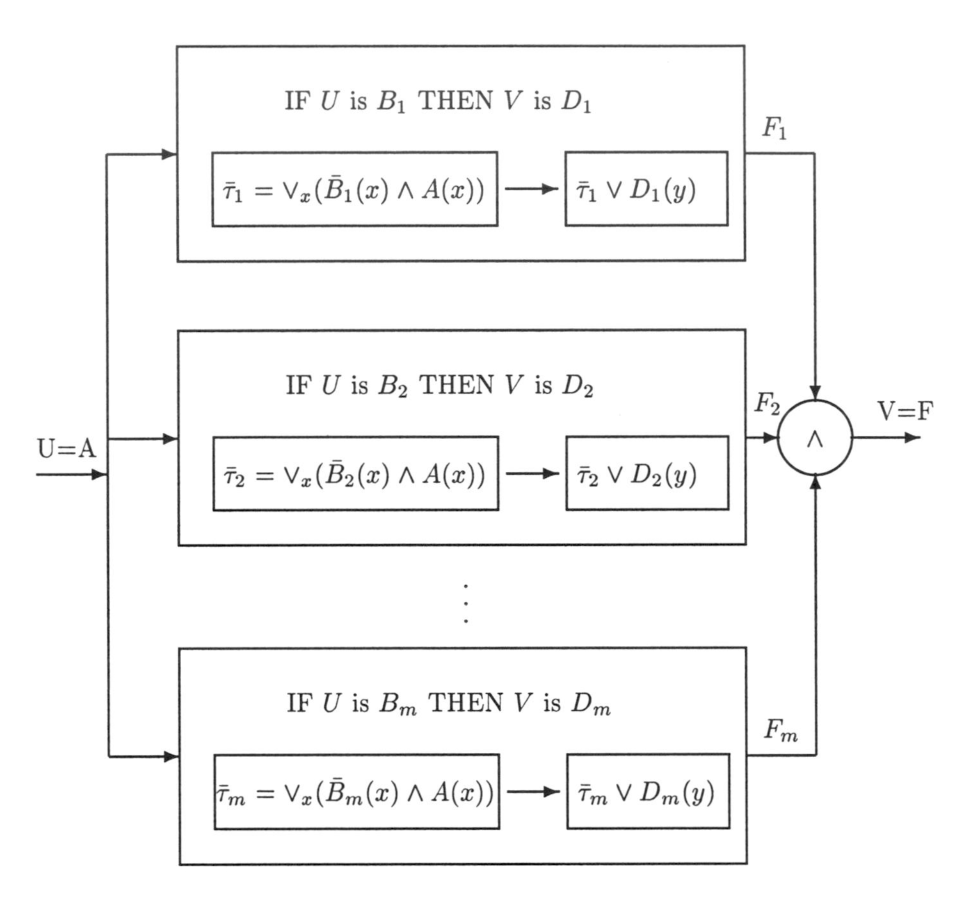

Figure 3.4 Block-diagram of the logical method of reasoning.

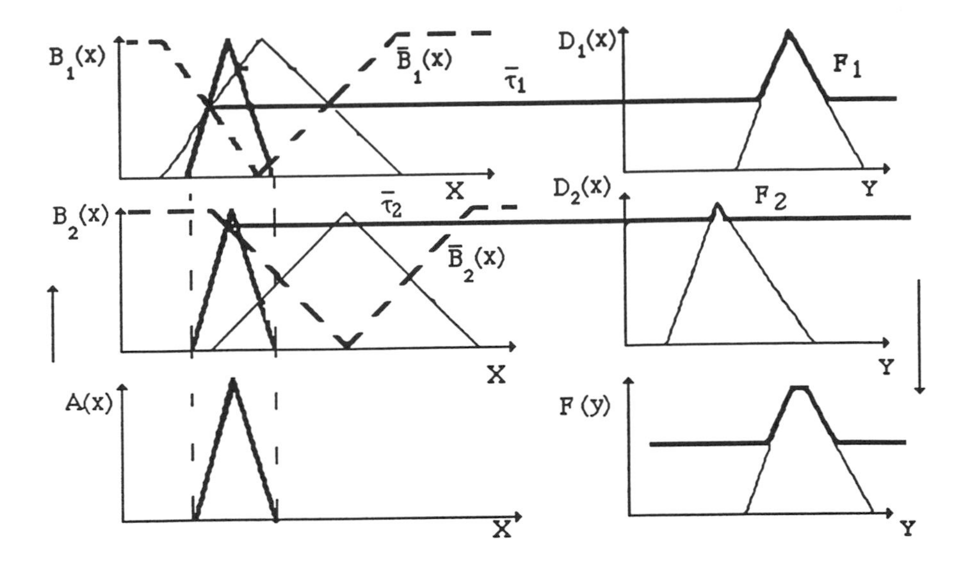

Figure 3.5 Calculation of the output by the logical method (Algorithm 2).

where $\otimes$ and *oplus* are any t-norm and t-conorm operator respectively. For example, if the algebraic product operator is considered in (3.26) as a t-norm operator we obtain Larsen's method of fuzzy reasoning:

$$F(y) = \bigvee_{i=1}^{m} [\tau_i D_i(y)].$$

Geometrically a linguistic model forms a fuzzy region in the space $X \times Y$. Under the Mamdani approach the region is defined by the relationship R formed by taking the union of fuzzy relations R_i, $i = 1, \ldots, m$ which are obtained from individual rules using expression (3.12); fuzzy relations R and R_i are shown in Figure 3.6. Different reasoning methods assign alternative membership functions to the fuzzy relationship R. If the support sets of fuzzy sets B_i and D_i become more narrow, we will get in the limiting case the model of an overall system as a collection of input-output pairs.

3.5 LINGUISTIC MODELS WITH CRISP OUTPUTS

In the preceding we have provided the output of our fuzzy model in terms of a fuzzy subset. In many applications, particularly in fuzzy logic control, we

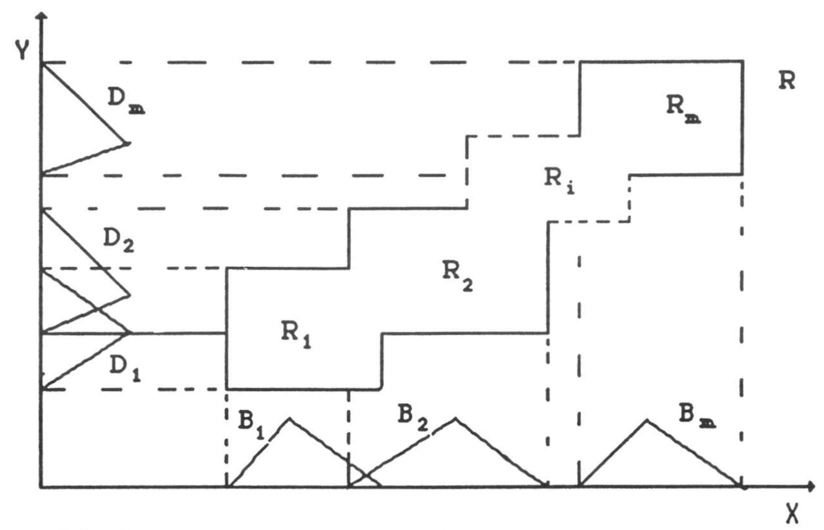

Figure 3.6 Formation of fuzzy relationship R associated with the he Mamdani method.

need our model output in terms of a single crisp value which can be applied to the system being modeled. In order to accomplish this we must defuzzify the fuzzy output value F. In [10] Yager and Filev provide a comprehensive discussion of the defuzzification process. The most commonly used method for defuzzification in fuzzy models is the Center of Area (COA) method. This method defines the defuzzified value as a fuzzy centroid:

$$y^* = \left(\int_Y yF(y)dy\right) / \left(\int_Y F(y)dy\right). \tag{3.28 }$$

For a finite universe Y this become

$$y^* = \left(\sum_{j=1}^{q} y_i F(y_j)\right) / \left(\sum_{j=1}^{q} F(y_j)\right), \tag{3.29 }$$

where q is the cardinality of Y.

Thus in the cases when a crisp expression of the model output V is required, the Algorithms 1 and 2 should be completed by one additional step - the calculation of the crisp value of the output $y*$ based on formulas (3.28) or (3.29).

A careful look at these algorithms shows that they have inherent nonlinearities due to the use of the max and min operators; in addition, the use of these operators provides no analytical expression of the relationship between the input and output variables. In order to reduce the effect of the nonlinearities in the Mamdani type model one can use a multiplicative operator (which is an alternative t-norm) instead of the min operator. As mentioned earlier, this type of fuzzy reasoning is known as Larsen's method. Then the expression for the fuzzy value in this case becomes:

$$F(y) = \bigvee_{i=1}^{m} F_i(y) = \bigvee_{i=1}^{m} \tau_i D_i(y). \quad (3.30)$$

A further simplification, often used in applications, is to replace the max operator used for the aggregation of the F_i's, by a simple summation [11]. In this case we obtain:

$$F(y) = \sum_{i=1}^{m} F_i(y) = \sum_{i=1}^{m} \tau_i D_i(y). \quad (3.31)$$

Obviously the use of the summation can bring the membership grade of $F(y)$ out of the unit interval. However, it doesn't have an effect on the defuzzified value, due to the normalization in (3.28) and (3.29). By substituting for $F(y)$ in (3.29) we get for the COA defuzzified value:

$$y* = \left(\sum_{i=1}^{m} \tau_i y_i^* \right) / \left(\sum_{i=1}^{m} \tau_i \right), \quad (3.32)$$

where the y_i's are the centroids of the individual consequent fuzzy sets, the D_i's. Thus the COA defuzzified value inferred by the model is determined by the weighted average of the centroids of the individual consequent fuzzy sets. The above expression for the defuzzified value is called the Simplified Method of Fuzzy Reasoning. The major advantage of the replacement of the max-aggregation by a summation is that it leads to a simpler form of the reasoning mechanism. This simplified form permits an analytical expression for the relationship between the systems input and output, and it opens the possibility for learning fuzzy models from data. As is seen from expression (3.32), the algorithm for determination of the crisp output value of the Mamdani model consists of two steps: calculation of the DOF, the τ_i's, and substituting into the final formula (3.32). The operations of finding the fuzzy sets inferred by the individual rules, aggregating and defuzzifying are omitted. The simplified method of fuzzy reasoning is presented in the block-diagram in Figure 3.7.

3.6 MULTIPLE VARIABLE LINGUISTIC MODELS

Thus far we have only considered the situation in which we had a single input and output in this section we shall look at the situation in which we allow multiple inputs and outputs. In this section we shall restrict ourselves to just considering the Mamdani type reasoning.

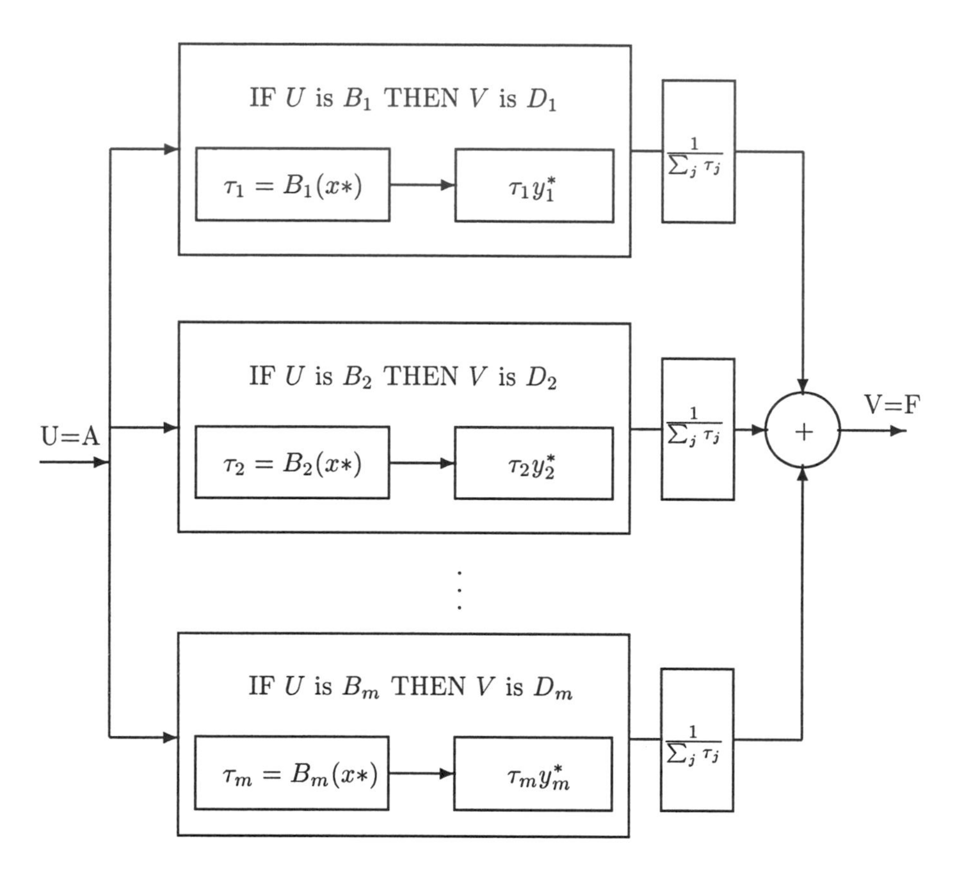

Figure 3.7 Block-diagram of the simplified reasoning mechanism.

We first consider the case of more than one input variable, *Multiple-Input, Single-Output* (MISO) model:

$$\begin{array}{ll} \textbf{IF } U_1 \text{ is } B_{11} \textbf{AND } U_2 \text{ is } B_{12} & \textbf{AND ... AND} U_r \text{ is } B_{1r} \textbf{THEN } V \text{ is } D_1 \\ & \textbf{ALSO} \\ & \dots \\ \textbf{IF } U_m \text{ is } B_{m1} \textbf{AND } U_2 \text{ is } B_{m2} & \textbf{AND ... AND } U_r \text{ is } B_{mr} \textbf{THEN } V \text{ is } D_m. \end{array} \qquad (3.33)$$

In the above $U_1, \ldots, U_r$ are the input variables and V is the output variable. B_{ij} and D_i, $i = 1, \ldots, m$, $j = 1, \ldots, r$ are fuzzy subsets of the universes of discourse $X_1, \ldots, X_r$ and Y of $U_1, \ldots, U_r$ and V. The extension of the results from those obtained for SISO LM in the preceding section is straightforward. Each rule

IF U_m is B_{m1}**AND** U_2 is B_{m2}**AND ... AND** U_r is B_{mr}**THEN** V is D_m.

is associated with a fuzzy relationship R_i,

$$R_i = B_{i1} \cap \cdots \cap B_{ir} \cap D_i. \qquad (3.34)$$

R_i is defined on the Cartesian space $X_1 \times \cdots \times X_r \times Y$ and its membership function is:

$$R_i(x_1, ..., x_r, y) = B_{i1}(x_1) \wedge \cdots \wedge B_{i1}(x_r) \wedge D_i(y). \qquad (3.35)$$

The individual fuzzy relationships are again aggregated using the fuzzy union, resulting in the overall fuzzy relationship R for the MISO LM (3.33):

$$R = \bigcup_i R_i = \bigcup_i (B_{i1} \cap \cdots \cap B_{ir} \cap D_i). \qquad (3.36)$$

The membership function of the fuzzy relationship R is

$$R(x_1, ..., x_r, y) = \bigvee_{i=1}^{m} R_i(x_1, ..., x_r, y) = \bigvee_{i=1}^{m} B_{i1}(x_1) \wedge \cdots \wedge B_{ir}(x_r) \wedge Di(y). \qquad (3.37)$$

In this environment our input is a collection of r fuzzy subsets $U_1 = A_1$, $\ldots, U_r = A_r$, one for each antecedent variable. The output is obtained by extending the max-min rule of inference to the case of r inputs:

$$F = (A_1, ..., A_r) \circ R. \qquad (3.38)$$

The membership function of the inferred fuzzy set F is:

$$F(y) == \bigvee_{i=1}^{m} \tau_i \wedge D_i(y), \qquad (3.39)$$

where τ_i denotes the DOF of the i-th rule and is defined as

$$\tau_i = \tau_{i1} \wedge \tau_{i2} \wedge \cdots \wedge \tau_{ir},$$

where

$$\tau_{ij} = \text{Poss}[B_{ij}|A_j] = Max_{x_i}[B_{ij}(x_i) \wedge A_i(x_i)]). \qquad (3.40)$$

In the special case, where the inputs are crisp values x_1^*, ..., x_r^* the input fuzzy sets, A_1, ..., A_r, become fuzzy singletons, and the DOF of the i-th rule becomes:

$$\tau_i = B_{i1}(x_1^*) \wedge \cdots \wedge B_{ir}(x_r^*). \qquad (3.41)$$

In summary the above yields the following algorithm for calculation of the output inferred by a MISO LM:

Algorithm 3

1. For each rule of the LM (3.33) calculate the DOF, the i's, of the rule:

$$\tau_i = \left(\bigvee_{x_1}[B_{i1}(x_1) \wedge A_1(x_1)]\right) \wedge \cdots \wedge \left(\bigvee_{x_r}[B_{ir}(x_r) \wedge A_r(x_r)]\right)$$

   if the input is a fuzzy set A.

   $\tau_i = B_{i1}(x_1^*) \wedge \cdots \wedge B_{ir}(x_r^*)$ if the inputs are a crisp numbers $x_1^*, ..., x_r^*$.

2. Find the fuzzy set F_i inferred by the i-th rule:

$$F_i(y) = \tau_i \wedge D_i(y).$$

3. Aggregate the inferred fuzzy sets F_i by using fuzzy union:

$$F(y) = \bigvee_{i=1}^{m} F_i(y).$$

In the case, most often used in fuzzy logic control, where in addition to having singleton inputs we desire a crisp output, by applying the simplified method of reasoning (3.32), we formally get the same result as in the case of the SISO LM:

$$y^* = \left(\sum_{i=1}^{m} \tau_i y_i^*\right) / \left(\sum_{i=1}^{m} \tau_i\right), \qquad (3.42)$$

however the τ_i's according to expression (3.41) are

$$\tau_i = B_{i1}(x_1^*) \wedge \cdots \wedge B_{ir}(x_r^*).$$

An alternative expression of the output fuzzy set F can be obtained. It is easy to verify the identity

$$
\begin{aligned}
R &= \bigcup_i R_i = \bigcup_i (B_{i1} \cap D_i \cap B_{i2} \cap D_i \cap \cdots B_{ir} \cap Di) \\
&= \bigcup_i \left(R_i^1 \cap R_i^2 \cap \cdots \cap R_i^r\right), \qquad (3.43)
\end{aligned}
$$

where the R_i^1 R_i^2,..., R_i^r are SISO fuzzy relationships defined by

$$R_i^j = B_{ij} \cap D_i, \qquad j = 1, \ldots, r.$$

Using this formulation the output fuzzy subset can be expressed as

$$
\begin{aligned}
F &= (A_1, ..., A_r) \circ R = \bigcup_{i=1}^{m} (A_1, ..., A_r) \circ R_i \\
&= \bigcup_{i=1}^{m} \left[(A_1 \circ R_i^1) \cap \cdots \cap (A_r \circ R_i^r)\right] \qquad (3.44)
\end{aligned}
$$

This result permits a simplified decomposed expression for the output fuzzy set inferred by the MISO LM:

$$F = (A_1, ..., A_r) \circ R = (A_1, ..., Ar) * \begin{pmatrix} R_i^1 \\ \vdots \\ R_i^r \end{pmatrix}, \qquad (3.45)$$

where the symbol (*) denotes operation $(\circ, \cap)$.

We now turn to the more general case of multiple - input, multiple - output (MIMO) fuzzy if/then rules:

U_1 is B_{11}**AND** U_2 is B_{12} **AND ... AND** U_r is B_{1r}
THEN V_1 is D_{11}; V_2 is D_{12}; ...; V_s is D_{1s}
ALSO
...   (3.46)
ALSO
U_1 is B_{m1}**AND** U_2 is B_{m2} **AND ... AND** U_r is B_{mr}
THEN V_1 is D_{m1}; V_2 is D_{m2}; ...; V_s is D_{ms}.

The input variables $U_1, ..., U_r$ and linguistic labels (terms) B_{ij}, $i = 1, \ldots, m$, $j = 1, \ldots, r$ are defined in the same manner as in the case of MISO LM.

$V_1, ..., V_s$ are output variables and their linguistic labels, the D_{ij}'s, are defined as fuzzy subsets on the universes of discourse $Y_1, ..., Y_s$ of the output variables.

If the outputs $V_1, ..., V_s$ are independent variables, then the MIMO model can be decomposed to a collection of s MISO models

$$U_1 \text{ is } B_{i1} \textbf{AND } U_2 \text{ is } B_{i2} \textbf{AND ... AND } U_r \text{ is } B_{ir} \textbf{THEN} V_1 \text{ is } D_{it},$$

$t = 1, \ldots, s$, each consisting of m rules:

$$\begin{array}{ll} U_1 \text{ is } B_{i1} \textbf{AND } U_2 \text{ is } B_{i2} & \textbf{AND ... AND } U_r \text{ is } B_{ir} \textbf{THEN } V_s \text{ is } D_{it} \\ & \textbf{ALSO} \\ & \ldots \\ & \textbf{ALSO} \\ U_1 \text{ is } B_{i1} \textbf{AND } U_2 \text{ is } B_{i2} & \textbf{AND ... AND } U_r \text{ is } B_{ir} \textbf{THEN } V_s \text{ is } D_{it}, \end{array} \tag{3.47 }$$

$t = 1, \ldots, m$. Using the result of the previous section we obtain for the output fuzzy sets $V_t = F^t, t = 1, \ldots, s$ inferred by the MISO subsystems for the inputs $U_1 = A_1, ..., U_r = A_r$:

$$F^t(y_t) = \bigvee_{i=1}^{m} \tau_i \wedge D_{it}(y_t), \quad t = 1, \ldots, s, \tag{3.48 }$$

where by τ_i, $i = 1, \ldots, m$, is the DOF of the i-th rule of the t-th subsystem:

$$\tau_i = \text{Poss}[B_{i1}|A_1] \wedge \cdots \wedge \text{Poss}[B_{ir}|A_r].$$

To obtain a more compact expression for the inferred output fuzzy sets $F_1, ..., F_s$ we introduce the SISO fuzzy relations defined by the conjunctions of the j-th antecedent and the t-th consequent reference fuzzy sets:

$$R_i^{jt} = B_{ij} \cap D_{it}, \quad i = 1, \ldots, m, \, j = 1, \ldots, r, \, t = 1, \ldots, s.$$

Then by applying the decomposed rule of inference to each of the subsystems we get for the output fuzzy values F^t, $t = 1, \ldots, s$ of the individual outputs $V_1, ..., V_s$ of the MIMO LM:

$$(F^1, F^2, ..., F^s) = \bigcup_{i=1}^{m} (A_1, ..., A_r) * \begin{pmatrix} R_i^{11} & R_i^{12} & \cdots & R_i^{1s} \\ R_i^{21} & R_i^{22} & \cdots & R_i^{2s} \\ \cdot & \cdot & \cdot & cdot \\ R_i^{r1} & R_i^{r2} & \cdots & R_i^{rs} \end{pmatrix}, \tag{3.49 }$$

whereby the symbol $(*)$ denotes operation $(\circ, \cap)$.

An alternative form of the MIMO LM considers the consequents of the fuzzy if/then rules in (3.46) connected by the AND operator. i.e.:

$$
\begin{array}{rl}
U_1 \text{ is } B_{11} \textbf{AND } U_2 \text{ is } B_{12} & \textbf{AND ... AND } U_r \text{ is } B_{1r} \\
\textbf{THEN } V_1 \text{ is } D_{11} \textbf{AND } V_2 \text{ is} & D_{12} \textbf{AND } \cdots \textbf{ AND } V_s \text{ is } D_{1s} \\
 & \textbf{ALSO} \\
 & \ldots \\
 & \textbf{ALSO} \\
U_1 \text{ is } B_{m1} \textbf{AND } U_2 \text{ is } B_{m2} & \textbf{AND ... AND } U_r \text{ is } B_{mr} \\
\textbf{THEN } V_1 \text{ is } D_{m1} \textbf{AND } V_2 \text{ is} & D_{m2} \textbf{AND } \cdots \textbf{ AND } V_s \text{ is } D_{ms}.
\end{array} \tag{3.50}
$$

Extending the concept of fuzzy conjunction to the case of $(r+s)$ Cartesian products space, formed by the universes $X_1, ..., X_r$, $Y_1, ..., Y_s$ we arrive at the multivariable fuzzy relation associated with the collection of rules (3.50):

$$R = \bigcup_{i=1}^{m} R_i = \bigcup_{i=1}^{m} (B_{i1} \cap \cdots \cap B_{ir} \cap D_{i1} \cap \cdots \cap D_{is}) \tag{3.51}$$

with membership function:

$$
\begin{aligned}
R(x_1, ..., x_r, y_1, ..., y_s) &= \bigvee_{i=1}^{m} R_i(x_1, ..., x_r, y_1, ..., y_s) \\
&= \bigvee_{i=1}^{m} B_{i1}(x_1) \wedge \cdots \wedge B_{ir}(x_r) \wedge D_{i1}(y_l) \wedge \cdots \wedge D_{is}(y_s).
\end{aligned} \tag{3.52}
$$

For given input fuzzy sets $U_1 = A_1$, ..., $U_r = A_r$ the LM infers an output fuzzy set F in the output Cartesian product space $Y = Y_1 \times \cdots \times Y_s$ that is defined by the projection of the fuzzy intersection $R \cap A_1 \cap \cdots \cap A_r$ on Y:

$$F(y_1, ..., y_s) = \bigvee_{X_1, ..., x_r} [R(x_1, ...x_r; y_1, ..., y_s) \wedge A_1(x_1) \wedge \cdots \wedge A_r(x_r)]. \tag{3.53}$$

In order to obtain individual output fuzzy sets $F_1, ..., F_s$, fuzzy set F has to be projected on the universes of discourse $Y_1, ..., Y_s$:

$$F^1(y_1) = \bigvee_{y_2, ..., y_s} F(y_1, ..., y_s), \quad ..., \quad F^s(y_s) = \bigvee_{y_1, ..., y_{s-1}} F(y_1, ..., y_s). \tag{3.54}$$

The above result requires considerable computational effort. The problem of its analytical description was studied by Gupta et al.[12, 13] and the matrix-like form (3.49) was found as a reasonable approximation of the fuzzy output F by (3.54).

3.7 TAKAGI-SUGENO-KANG (TSK) MODELS

Sugeno and co-workers [3-5]proposed an alternative type of fuzzy systems modeling, called TSK (Takagi-Sugeno-Kang) type of reasoning. The TSK reasoning method is associated with a rule-base of a special format that is characterized by functional type consequents instead of the fuzzy consequents used in the LM discussed before. In this type of model our rule base is

IF u_1 is B_{11} **AND ... AND** u_r is B_{1r} **THEN** $y_1 = b_{10} + b_{11}u_1 + \cdots + b_{1r}u_r$
ALSO

... (3.55)

ALSO
IF u_1 is B_{m1} **AND ... AND** u_r is B_{mr} **THEN**
$y_m = b_{m0} + b_{m1}u_1 + \cdots + b_{mr}u_r.$

In this model the B_{ij}, $j = 1, \ldots, r$, $i = 1, \ldots, m$ are again linguistic labels defined as reference fuzzy sets over the input spaces $X_1, X_2, ..., X_r$ of a MISO system; $u_1, u_2, ..., u_r$ are the values of input variables. Each of the linear functions in the rule consequents can be regarded as a linear model with crisp inputs $u_1, u_2, ..., u_r$, crisp output $_i$ and parameters b_{ij}, $i = 1, \ldots, m$, $j = 0, \ldots, r$. The crisp output y inferred by the fuzzy model under the TSK method is defined by the weighted average of the crisp outputs y_i of individual linear subsystems

$$y = \sum_{i=1}^{m} \tau_i y_i / \left(\sum_{i=1}^{m} \tau_i \right) = \sum_{i=1}^{m} \tau_i (b_{io} + b_{i1}u_1 + \cdots + b_{ir}u_r) / \left(\sum_{i=1}^{m} \tau_i \right), \quad (3.56)$$

where is the DOF of the i-th rule:

$$\tau_i = B_{i1}(u_1) \wedge \cdots \wedge B_{ir}(u_1). \quad (3.57)$$

Geometrically, the rules of the TSK reasoning model (3.55) correspond to an approximation of the mapping $X_1 \times \cdots \times X_r \to Y$ by a piecewise linear function.

In a more general setting the linear functions in the consequents of the rules can be replaced by nonlinear ones. In this case the TSK models become a collection of rules of the format:

IF u_1 is B_{i1}**AND ... AND** u_r is B_{ir}**THEN** $y_i = f_i(u_1, ..., u_r)$, (3.58)

where the outputs of nonlinear subsystems are combined analogously to the linear case via expression (3.56).

From expression (3.56) is seen that the simplified method of reasoning can be derived as a special case of the TSK where $b_{ij} = 0$ for $i = 1, \ldots, m$, $j = 1, \ldots, r$.

The great advantage of the TSK model lies in its representative power, espe cially for describing complex technological processes. It allows us to decompose a complex system into simpler subsystems (even in some cases linear subsystems). This conception is not new to control engineers; in the 70-s it was developed through some of works of Rajbman and his colleagues [14]. The notion of the approach suggested by Rajbman was to decompose the state space of the system into disjunct regions, and then identify the system dynamics of these regions with different, usually nonlinear models (each with its own structure and parameters). This resulted in the representation of the overall model of a nonlinear system as a collection of subsystems that were combined based on a logical (Boolean) switching function. In realistic situations, however, such disjoint (crisp) decomposition is imPossible due to the inherent lack of natural region boundaries in the system, and also due to the fragmentary nature of available knowledge about the system. The TSK model allows us to replace the crisp decomposition by a fuzzy decomposition, and to replace the Boolean switching function by the interpolative TSK reasoning mechanism. When this is done, the system dynamics for any point in the state space can be dealt with as the (co-temporal) combination of the dynamics of the one or more regions to which this point belongs. Moreover, the TSK model allows us to introduce the expert's knowledge in the partitioning of the input and state space; this can be especially useful in the cases where different regions associated with different operating conditions can be specified by using linguistic labels.

3.8 A GENERAL VIEW OF FUZZY SYSTEMS MODELING

As we have seen fuzzy systems modeling involves the use of a fuzzy rule base to model complex systems by partitioning the input space into fuzzy regions in which the output can be more effectively represented. Typical of these situations are set of n rules of the form

$$\text{if} \quad V \quad \text{is} \quad A_i \quad \text{then} \quad U \quad \text{is} \quad B_i,$$

where A_i and B_i are fuzzy subsets of the input and output spaces X and Y. The problem of finding the value of the output variable U given a value for the input variable V is called the *fuzzy model inference or reasoning process.* This process consists of the following four step algorithm:

1. Determination of the relevance or matching of each rule to the current input value.
2. Determination of the individual output of each rule, denoted U is F_i.

3. Aggregation of the individual rule outputs to obtain the overall fuzzy output, denoted U is F.

4. Selection of some action based upon the output fuzzy set F.

Our purpose here is to describe the class of operators appropriate for the implementation of the third step, the rule output aggregation. Because of the strong interrelationship between all the steps in the process we look at all the steps. A particular close connection exists between the second and third step. A fundamental aspect of this connection is the relationship between the choice of identity used in the rule aggregation process and the type of interpretation one gives to the rules in step two.

3.9 MICA OPERATORS

In this section we introduce a class of aggregation operators called MICA operators. We discuss a few properties of these operators and describe some examples of these operators. As we shall see in the next section these MICA operators provide the appropriate generalization of the aggregation operators necessary for the implementation of the third step in the preceding algorithm. More details about these operators can be found in [15].

Assume I is the unit interval. A bag drawn from I is any collection of elements which is contained in X. A bag is different from a subset in that it allows multiple copies of the same element. A bag is similar to a set in that the ordering of the elements in the bag doesn't matter.

In the following we shall let U^I indicate the set of all bags of the set I. A function $F : U^I \to I$ is called a *bag mapping* from U^I into the unit interval. An important property of bag mappings are that they are commutative in the sense that the ordering of the elements doesn't matter. Thus if $\text{Agg}(x_1, ..., x_n)$ is an aggregation operator representable by a bag mapping then Agg must be a commutative (symmetric) type aggregation.

Definition 3.1 *Assume F is a bag mapping, let A be any bag. An element $u \in I$ is called an* **identity** *for A under F if $F(A) = F(A \oplus < u >)$.*

Definition 3.2 *Assume A and B are two bags of the same cardinality. If the elements in A and B can be indexed in such a way that $a_i \geq b_i$ for all i then we shall denote this as $A \geq B$.*

Definition 3.3 *A bag mapping, $M : U^I \to I$, is called a* **MICA** *(Monotonic Identity Commutative Aggregation) operator if it has the following two properties:*

M1: **Monotonic** - *If $A \geq B$ then $M(A) \geq M(B)$.*

M2: **Identity** – *For every bag A there exists an element, $u \in I$, called the identity of A such that if $D = A \oplus < u >$ then $M(D) = M(A)$.*

Thus the MICA operator is endowed with two properties in addition to the inherent commutativity of the bag operator, monotonicity and identity.

The type of identity element provides a very important means for classifying these types of operators. Two important classes of these MICA operators can be distinguished based upon the type of identity. A MICA operator M is said to have a *fixed identity* if there exists some element $g \in I$ such that $u = g$ for all A. Thus the fixed identity type MICA operator has a unique identity that is the same for all bags. A MICA operator M is said to have a *self identity* if for every A the identity element u is such that $u = M(A)$. For our purposes here we shall consider MICA operators with fixed identity. We shall denote this identity as g. It is shown in [15] that if M is a MICA operator with fixed identity g then $M(\emptyset) = M(< g >)$. We call the operator *normal* if $M(< g >) = g$.

The following theorem characterizes an important property of the identity.

Theorem 3.1 *Assume M is a MICA operator with fixed identity g and let A be any bag and let $k \in I$ it is the case that $M(A) \geq M(A \oplus < k >)$ if $g > k$ and $M(A) \leq M(A \oplus < k >)$ if $g < k$.*

The case when $g = 0$ is particularly notable in that $M(A \oplus < k >) \geq M(A)$ for all k, in [16, 17] Yager called these MOM operators and showed that they constitute *orlike* aggregators. Also when $g = 1$ then $M(A \oplus < k >) \leq M(A)$ for all k, these are called MAM operators [16] and constitute *andlike* aggregators.

We now consider some special cases of these MICA operators. In [15] it is shown that the t-norm aggregation operator is a MICA operator which has fixed *identity* $g = 1$. The t-norm has two additional properties not necessarily required of a MICA operator, it is associative and has natural boundaries, $T(1, x) = x$. The t-conorm is a MICA operator which has fixed *identity* $g = 0$, it is also associative and has natural boundaries, $S(0, x) = x$.

In [17, 18] Yager introduced a class of additive MICA operator operators. Assume f is monotone non-decreasing function from the reals into the unit interval then $M(A) = f\left(\sum_i a_i\right)$ is a MICA operator with $g = 0$. If g is monotone non-increasing function from the reals into the unit interval then $M(A) = g\left(\sum_i (1 - a_i)\right)$ is a MICA operator with $g = 1$.

3.10 AGGREGATION IN FUZZY SYSTEMS MODELING

In this section we investigate the algorithm for fuzzy reasoning given in the first section and show that any MICA operator can be used to implement the aggregation individual rule outputs.

Let us briefly consider the last step in the algorithm. With each y of the output space we can associate some action. For any y, $F(y)$ is a measure of the strength to which the system model, under the current input, suggests y as the solution. The key observation is that the action selection process uses F in such a way that if $F(y_1) > F(y_2)$ we are more inclined to implement the action suggested by y_1 than that of y_2, there is a positive association between the membership grade of an element and the action recommended by that element.

Let us now look at the top of the model. First consider an individual rule,

$$\text{if } V \text{ is } A_i \text{ then } U \text{ is } B_i.$$

The intent of this rule is to say that if the input to the system is consistent with the antecedent then this rule is relevant and it says that the system output is B_i. In indicating that B_i is the fuzzy set output, this is saying that if $B_i(y_1) > B_i(y_2)$ there is a preference for having y_1 as the action to follow rather then y_2.

The first step in the algorithm is the matching step where we determine the degree, τ_i, to which the input to the system and the antecedents of the individual rules are compatible. All that we shall observe about the τ_i is that the larger the τ_i the better the matching, and the more relevant the rule. In particular if $\tau_i = 1$ the matching is perfect, and the rule is completely relevant. If $\tau_i = 0$ the rule has no relevance to the current input.

Having obtained this firing level we then use it to obtain the individual rule output U is F_i. The individual rule output F_i is determined by τ_i, the rule firing level, and the rule consequent B_i. We denote this as $F_i = \tau_i \circ B_i$.

Having obtained the output of each rule we must now combine these individual rule outputs to obtain the overall system output, F. We denote this process $F = F_1 \Delta F_2 \Delta \cdots \Delta F_n$. As we shall subsequently see the process used to obtain the overall system output from the individual F_i and the process used to obtain the individual rule outputs from the rule consequent and firing levels are not independent, that is Δ and $\circ$ are related.

Let us look at the process Δ used to combine the individual rule outputs. A basic assumption we shall make is that the operation is pointwise and likewise. By pointwise we mean that for every $y \in Y$, $F(y)$ just depends upon $F_i(y)$, $i = 1, \ldots, n$. By likewise we shall mean that the process used to combine the F_i is the same for all of the y.

We denote the pointwise process used to combine the individual rule outputs as

$$F(y) = \mathbf{Agg}(F_1(y), F_2(y), \ldots, F_n(y)).$$

Let us look at the minimal requirements associated with the **Agg** operator. We first note that the combination of the individual rule outputs should be independent of the choice of indexing of the rules. This implies that a required property that we must associate with the **Agg** operator is that of *commutativity*. We note that the commutativity property allows us to represent the argument of the **Agg** operator as a bag.

For an individual rule output, F_i, the membership grade $F_i(y)$ indicates the degree to which this rule suggests that y is the appropriate solution. In particular if for a pair of elements y_1 and y_2 it is the case that $F_i(y_1) \geq F_i(y_2)$, then we are saying that rule τ_i is preferring y_1 as the system output over y_2. From this we can conclude that if all rules prefer y_1 over y_2 as output then the overall system output should prefer y_1 over y_2. This observation requires us to impose a *monotonicity* condition on the **Agg** operation. In particular if $F_i(y_1) \geq F_i(y_2)$ for all i, then $F(y_1) \geq F(y_2)$.

One other condition we need to impose upon the Aggregation operator. Assume that there exists some rule whose firing level is zero, it is irrelevant to the current input, and hence should not affect the final F. Without loss of generality we shall assume the nth rule has firing level zero. The first observation we can make is that whatever output this rule provides should not make any distinction between the potential outputs. Thus for zero fired rule $F_n(y_i) = F_n(y_j)$ for all i and j. Let us denote this value as g. In addition to $F_n(y_i)$ being the same for all y_i it appears that a second condition is necessary which affects the Aggregation operator. Consider any two elements y_1 and y_2 in Y. Assume that

$$\mathbf{Agg}(F_1(y_1), F_2(y_1)...F_{n-1}(y_1)) = a_1,$$
$$\mathbf{Agg}(F_1(y_2), F_2(y_2)...F_{n-1}(y_2)) = a_2.$$

Since the rule which has zero firing level should play no rule in determining which is the preferred output, the inclusion of its output F_n should not affect the relationship between the suggested system output value for y_1 and y_2. This observation requires that

$$\mathbf{Agg}(F_1(y_i), F_2(y_i), \ldots, F_{n-1}(y_i)) = \mathbf{Agg}(F_1(y_i), F_2(y_i), \ldots, F_{n-1}(y_i), g).$$

That is, g should be an identity element for the Aggregation process. Thus we see that the Aggregation operator *needs an identity element*. In summary, we see that the Aggregation operator, **Agg**, must satisfy three conditions: commutativity, monotonicity and have a fixed identity. These conditions are based on the three requirements; the indexing of the rules be unimportant, a positive association between individual rule output and total system output, and non-firing rules play no role in the decision process.

As we have indicated earlier an operation that has the above three properties is the MICA operation with fixed identity. Thus the class of MICA operators are the prototypical Aggregation operators for the combination of individual rule output. It should be pointed out that no requirement was made on the choice of the identity g. As we shall subsequently see the only requirement is that it be consistent with the choice of interpretation of the implication.

We now look at the operation used to obtain the individual rule output from the rule firing level and the rule consequent, $F_i = \tau_i \circ B_i$, we shall denote this as h where $F_i(y) = h(\tau_i, B_i(y))$. Let us investigate the minimal requirements we desire the operator h to have. First we require that if the antecedent of the rule is completely satisfied we should get B_i as our output, hence $h(1, b) = b$ for all b. A second requirement on the function h is that if the firing level is zero then the output value should be the identity under the Aggregation operation used in step three hence $h(0, b) = g$ for all b where g is the identity under the Aggregation.

A third condition we require is a monotonicity in the second argument, for any a $h(a, b) \geq h(a, b')$ for $b > b'$. This condition is a reflection of the requirement that for no firing level should we inverse the preference ordering between output values.

A fourth condition we require is a consistency in the antecedent argument. In particular, we desire that for a fixed consequent value the output of h go monotonically from the value for zero firing to the value for complete firing. In particular we desire that if $a > a'$ then

$$\begin{array}{llll} \text{if} & b \geq g & \text{then} & h(a,b) \geq h(a',b), \\ \text{if} & b \leq g & \text{then} & h(a,b) \leq h(a',b). \end{array}$$

We note that in the special case when $g = 1$ it is always the case that $b =\leq g$ and hence we get for $a > a'$, $h(a', b) \leq h(a, b)$. At the other extreme is the case when $g = 0$. In this situation since $b \geq g$ for all b we get $h(a, b) \geq h(a', b)$.

Given the selection of a MICA operator with fixed identity g to implement the individual rule Aggregation we shall say that h is an appropriate implication if it satisfies the four conditions specified with $h(0, b) = g$.

It can be shown that $h(a, b) = ab + g$ satisfies all four conditions for any g and hence is an appropriate implication for every g.

If we restrict ourselves to the values of $g = 0$ or 1 we get some further interesting forms.

Theorem 3.2 *For $g = 0$, $h(a, b) = T(a, b)$, where T is any t-norm, satisfies the four conditions.*

Theorem 3.3 *For $g = 1$, $h(a, b) = S(\bar{a}, b)$, where S is any t-conorm, satisfies the four conditions.*

In the preceding we have shown that the Aggregation of the individual rule outputs can be accomplished by a MICA operator, **Agg**, with fixed identity g,

$$F(y) = \mathbf{Agg}(F_1(y), F_2(y), \ldots, F_n(y))$$

and the transformation or implication is implemented by

$$F_i(y) = h(\tau_i, B_i(y)),$$

where h is an appropriate transformation for **Agg**. In the following we shall look at some forms of fuzzy reasoning that satisfy our requirements. While here we shall concentrate on the cases when $g = 0$ or 1 we significantly note that g can be any value in the unit interval.

We first consider the case when $g = 0$, *orlike* inter-rule Aggregation. As we noted the$t$-conorm are members of this class. The most notable example of this case is the max operator. Since in this case h can be any t-norm, two manifestations of this are $F_i(y) = \min[\tau_i, B_i(y)]$ and $F_i(y) = \tau_i B_i(y)$. If we select the max t-conorm and combine this with the min interpretation of h we get $F(y) = \max_i[\tau_i \wedge B_i(y)]$ which is the formulation used by Mamdani [8] in his original work on fuzzy control. If we combine the max with the product we get $F(y) = \max_i[\tau_i B_i(y)]$.

Earlier we have indicated that for $g = 0$ the addition model $\mathbf{Agg}(a1, a2, \ldots, a_n) = f(\sum_i^m a_i)$, where f is a monotone nondecreasing function from the real line into the unit interval, also satisfies the MICA conditions,. Combining this with the product form of h we get $F(y) = f\left(\sum_{i=1}^m \tau_i B_i(y)\right)$. This can be seen to be similar to the neural type models. In these models f is the activation function, the $B_i(y)$ are the weights and τ_i, the firing levels, are the inputs. (see Figure 3.8.) Thus the neural model can be viewed as an example of the same algorithm used in fuzzy systems modeling.

We now turn to those implementations based on the identity equal to one, *andlike* interrule Aggregations. Since t-norms are MICA operators with $g = 1$ two significant manifestations of this are the Min and product Aggregation operators. As we also noted $S(\bar{a}, b)$ is an appropriate implication for any MICA with identity $g = 1$. Two examples of this are $S_1(a, b) = \max(a, b)$ and $S_2(a, b) = a + b - ab$. If we use these to implement the h function we get $F_i(y) = \bar{\tau}_i \vee B_i(y)$ and $F_i(y) = \bar{\tau}_i + \tau_i B_i(y)$. Combining S_1 with the min we get $F(y) = \min_i(\max(\bar{\tau}_i, B_i(y)))$. This form was suggested by Zadeh in his theory of approximate reasoning [6]. An interesting form arises if we combine S_2 with the product t-norm, $F(y) = \prod_i(1 - t\tau_i \bar{B}_i(y))$. If we combine S_2 with the Min we get $F(y) = \min_i(1 - \tau_i \bar{B}_i(y))$.

We have also showed that for $g = 1$, $M(A) = q\left(\sum(1 - a_i)\right)$ is a MICA operator when q is a non-increasing function. If we combine this with $F_i(y) = \bar{\tau}_i + \tau_i B_i(y)$ we get the inference structure $F(y) = q\left(\sum_{i=1}^n \tau_i \bar{B}_i(y)\right)$ This structure is very amenable to a neural implementation of the type shown in figure 3.1. Here q is the activation function, τ_i the input and $\bar{B}_i(y)$ the neural weights.

It is interesting to see the fundamental difference between the inference mechanism for $g = 0$ and $g = 1$. In the case of $g = 0$ we assume that $F(y)$ has an

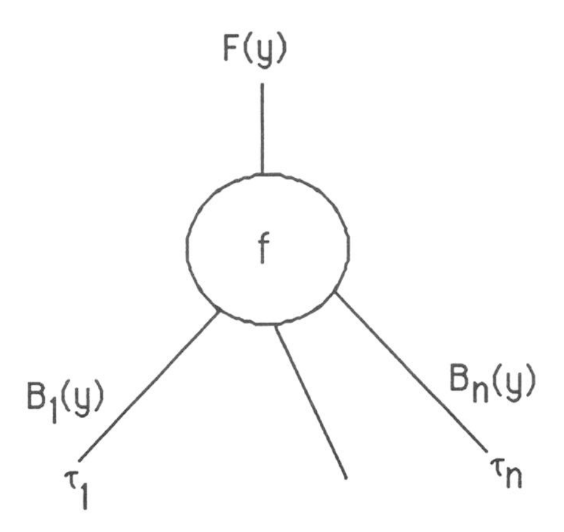

Figure 3.8 Neural Model for the Fuzzy System Inference.

initial value of zero and as rules fire they *add* to $F(y)$ an amount related to the firing level and $B_i(y)$. As noted in [19] this can be viewed as a **constructive** formulation of F. In the case of $g = 1$ we initially assume $F(y)$ is one and as rules fire they reduce $F(y)$ in an amount related to the firing level and $\bar{B}_i(y)$, the difference of Bi(y) from one. As noted in [19] this can be viewed as a **destructive** formulation of F.

3.11 DYNAMIC FUZZY SYSTEMS MODELS

We now turn to the issue of providing a fuzzy description of dynamic systems. Generally a deterministic dynamic system is described by a set of state equations:

$$w(k+1) = f(w(k), u(k)), \tag{3.59 }$$

$$y(k) = g(w(k), u(k)), \tag{3.60 }$$

where $u(k)$ and $y(k)$ are the system's input and output, and $w(k) = [w_1(k), w_2(k), \ldots, w_n(k)]^T$ the vector of state variables at time k. The values of the states $w(k+1)$ and the output $y(k)$ are determined completely by the current values $w(k)$ and $u(k)$, from the mappings f and g. The mappings f and g describe the relationships between state, input and output variables. We shall use the structure of the LM to formulate a linguistic alternative to the equations (3.55) and (3.56) and to define a fuzzy interpretation of the mappings f and g. We shall concentrate our efforts on a general description of the behavior of dynamic systems by an LM from the point of view of input - state and state - output interactions; in effect, therefore, we propose to treat the LM as an alternative nonlinear system model with a specific reasoning mechanism.

We apply the MIMO LM to describe linguistically a SISO dynamic system:

$$
\begin{gathered}
\textbf{IF } U \text{ is } B_1 \textbf{ AND } W_1 \text{ is } A_{11} \textbf{ AND ... AND } W_n \text{ is } A_{1n} \\
\textbf{THEN } \bar{W}_1 \text{ is } \hat{A}_{11}; ...; \bar{W}_n \text{ is } \hat{A}_{1n}; V \text{ is } D_1 \\
\textbf{ALSO} \\
\textbf{IF } U \text{ is } B_2 \textbf{ AND } W_1 \text{ is } A_{21} \textbf{ AND ... AND } W_n \text{ is } A_{2n} \\
\textbf{THEN } \bar{W}_1 \text{ is } \hat{A}_{21}; ...; \bar{W}_n \text{ is } \hat{A}_{2n}; V \text{ is } D_2 \\
\textbf{ALSO} \qquad (3.61) \\
\cdots \\
\textbf{IF } U \text{ is } B_m \textbf{ AND } W_1 \text{ is } A_{m1} \textbf{ AND ... AND } W_n \text{ is } A_{mn} \\
\textbf{THEN } \bar{W}_1 \text{ is } \hat{A}_{m1}; ...; \bar{W}_n \text{ is } \hat{A}_{mn}; V \text{ is } D_m,
\end{gathered}
$$

where U, V are the input and output variables. $W_1, ..., W_n$ are the state variables of the dynamic system and $\bar{W}_1, ..., \bar{W}_n$ denote the updated state variables. B_i, D_i, and A_{ij} are fuzzy subsets associated with the linguistic labels (terms) coding the input, output and state space with respect to the universes X, Y, and $E_1, ..., E_n$. $\hat{A}_{ij}$, $i = 1, \ldots, m$, $j = 1, \ldots, n$, are those linguistic labels coding the state space, the A_{ij}'s, that are assigned to the updated state variables $\bar{W}_1, ..., \bar{W}_n$. The fuzzy coding of the input, output and state variables involves partitioning of the respective spaces X, Y, and $E_1, ..., E_n$. Apparently, the LM generates fuzzy values of the state updates $\bar{W}_1, ..., \bar{W}_n$, and the output V whenever it is supplied with fuzzy or crisp values of U and $W_1, ..., W_n$.

It is assumed that the state variables are independent variables and the output is associated with one of them. Then a single MISO rule updates each of the state variables and the output of the system. The family of MIMO rules forming the LM of a dynamic system (3.61) breaks up into a collection of MISO rules:

$$
\begin{gathered}
\textbf{IF } U \text{ is } B_i \textbf{ AND } W_1 \text{ is } A_{i1} \textbf{ AND ... AND } W_n \text{ is } A_{in} \\
\textbf{THEN } \bar{W}_1 \text{ is } \hat{A}_{i1}, \quad i = 1, \ldots, m \\
\textbf{ALSO} \\
\textbf{IF } U \text{ is } B_i \textbf{ AND } W_1 \text{ is } A_{i1} \textbf{ AND ... AND } W_n \text{ is } A_{in} \\
\textbf{THEN } \bar{W}_2 \text{ is } \hat{A}_{i2}, \quad i = 1, \ldots, m \\
\textbf{ALSO} \\
\cdots \qquad (3.62)
\end{gathered}
$$

IF U is B_i **AND** W_1 is A_{i1} **AND ... AND** W_n is A_{in}
THEN $\bar{W}_n$ is $\hat{A}_{in}$, $i = 1, \ldots, m$
ALSO
IF U is B_i **AND** W_1 is A_{i1} **AND ... AND** W_n is A_{in}
THEN V is D_i, $i = 1, \ldots, m$.

Therefore, by the decomposition of the LM (3.61) we obtained in general $(n+1)$ MISO families of rules (3.62). Each one of these families has as antecedents the input U and state variables $W_1, ..., W_n$; the consequent of each of the first n families is the update $\bar{W}_j$ of j-th state variable W_j. The $(n+1)$-th has as a consequent the output variable V.

For given fuzzy values of the input $U = G$ and states $W_1 = H_1$, $W_2 = H_2, \ldots, W_n = H_n$ the LM infers n fuzzy values $L_1, L_2, ..., L_n$ of the state updates $\bar{W}_1, \bar{W}_2, ..., \bar{W}_n$ and a fuzzy value F of the output V. By using the results for MISO LM (3.39) and (3.40) we get that the inferred fuzzy values $L_1, L_2, ..., L_n$ and F are:

$$L_j(e_j) = \bigvee_{i=1}^{m} \tau_i \wedge \hat{A}_{ij}(e_j), \tag{3.63}$$

$$F(y) = \bigvee_{i=1}^{m} \tau_i \wedge D_i(y), \tag{3.64}$$

where the i-th DOF τ_i is:

$$\begin{aligned}
\tau_i &= \text{Poss}(B_i|G) \wedge \left(\bigwedge_{j=1}^{n} \text{Poss}(A_{ij}|H_j) \right) \qquad (3.65)\\
&= \left[\bigvee_{x} (B_i(x) \wedge G(x)) \right] \wedge \left[\bigvee_{e_1} (A_{i1}(e_1) \wedge H_1(e_1)) \right] \\
&\quad \wedge \cdots \wedge \left[\bigvee_{e_n} (A + in(e_n) \wedge H_1(e_n)) \right]
\end{aligned}$$

For crisp values of the input $U = x*$ and state variables $W_1 = e_1^*, \ldots, W_n = e_n^*$ the above expression for the DOF of the rules, the τ_i's, is simplified:

$$\begin{aligned}
\tau_i &= \text{Poss}(B_i|G) \wedge \left(\bigwedge_{j=1}^{n} \text{Poss}(A_{ij}|H_j) \right) \\
&= B_i(x^*) \wedge A_{i1}(e_1^*) \wedge A_{i2}(e_2^*) \wedge \cdots \wedge (A_{in}(e_n^*). \qquad (3.66)
\end{aligned}$$

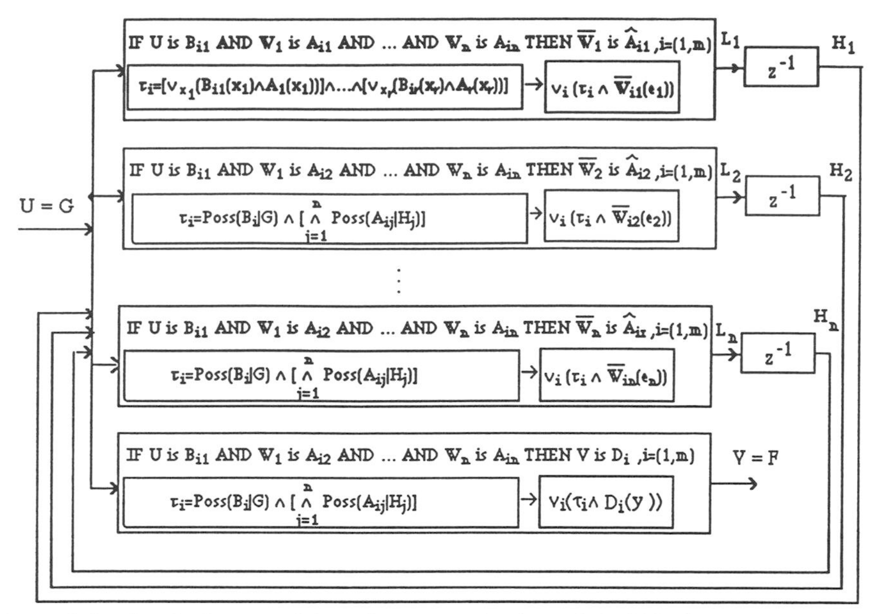

Figure 3.9 Block-diagram of the inference mechanism of the LM of a dynamic system.

The fuzzy values $L_1, L_2, ..., L_n$ of the updated state variables $\bar{W}_1, \bar{W}_2, \ldots, \bar{W}_n$, inferred in the k-th step are fed back in the (k+1) step as fuzzy values of the state variables $W_1, W_2, ..., W_n$; thus in the $(k+1)$ step we have $H_1 = L_1$, $H_2 = L_2, ..., H_n = L_n$.

The internal mechanism of inferring the fuzzy values of the states and the output is illustrated on the block-diagram in Figure 3.9.

Following the mechanism of reasoning in the MISO LM we associate with the first n MISO LM the n MISO fuzzy relations:

$$R_{ij}^x = B_i \cap A_{i1} \cap \cdots \cap A_{in} \cap \hat{A}_{ij}, \quad j = 1, \ldots n$$

defined on the universe $X \times E_1 \times \cdots \times E_n$. The MISO fuzzy relation:

$$R_i^y = B_i \cap A_{i1} \cap \cdots \cap A_{in} \cap D_i$$

defined on the universe $X \times E_1 \times \cdots \times E_n \times Y$ is associated with the last rule family in (3.62). Fuzzy relations R_{ij}^x, R_i^y, are Aggregated by the OR aggregating operator, using fuzzy union:

$$R_j^x = \bigcup_{i=1}^{m} R_{ij}^x = \bigcup_{i=1}^{m} \left(B_i \cap A_{i1} \cap \cdots \cap A_{in} \cap \hat{A}_{ij} \right),$$

$$R^y = \bigcup_{i=1}^{m} R_i^y = \bigcup_{i=1}^{m} (B_i \cap A_{i1} \cap \cdots \cap A_{in} \cap D_i) .$$

For given fuzzy values of the input $U = G$ and states $W_1 = H_1$, $W_2 = H_2$, ..., $W_n = H_n$ the LM infers n fuzzy values $L_1, L_2, ..., L_n$ of the state updates $\bar{W}_1, \bar{W}_2, ..., \bar{W}_n$ and a fuzzy value F of the output V. By application of the rule of inference we get:

$$\begin{aligned} L_j &= [G, H_1, ..., H_n] \circ = [G, H_1, ..., H_n] \circ \bigcup_{i=1}^{m} R_{ij}^x \\ &= \bigcup_{i=1}^{m} [G, H_1, ..., H_n] \circ R_{ij}^x, \quad j = 1, \ldots, n, \end{aligned} \tag{3.67 }$$

$$\begin{aligned} V &= [G, H_1, ..., H_n] \circ R^y = [G, H_1, ..., H_n] \circ \bigcup_{i=1}^{m} R_i^y \\ &= \bigcup_{i=1}^{m} [G, H_1, ..., H_n] \circ R_i^y . \end{aligned} \tag{3.68 }$$

It is rather difficult to directly apply the relational form of LM (3.67) and (3.68) for simulation of LM due to the multidimensionality of the fuzzy relations R_j^x, R^y. Certain simplifications, therefore, are important for the LM implementation; these simplifications are justified by the identity between the MISO and the corresponding SISO relations (see for details expression (3.43)). Introducing the SISO fuzzy relations:

$$R_i^{alj} = A_{il} \cap \hat{A}_{ij}; \quad R_i^{bj} = B_i \cap \hat{A}_{ij}; \quad R^{cl} = A_{il} \cap D_i; \quad R_i^d = B_i \cap D_i, \tag{3.69 }$$

it follows directly from (3.44):

$$L_j = \bigcup_{i=1}^{m} \left[\left(\bigcap_{l=1}^{n} H_l \circ R^{ali} \right) \cap (G \circ R_i^{bj}) \right], \quad j = 1, \ldots, n, \tag{3.70 }$$

$$V = \bigcup_{i=1}^{m} \left[\left(\bigcap_{l=1}^{n} H_l \circ R_i^{cl} \right) \cap (G \circ R_i^d) \right]. \tag{3.71 }$$

If the operator $(\circ, \cap)$ is replaced by the $(*)$ expressions (3.70) and (3.71) yield the matrix-like representation:

$$L = \bigcup_{i=1}^{m} [H * R_i^a \cap G \circ R_i^b], \tag{3.72 }$$

$$V = \bigcup_{i=1}^{m} [H * R_i^c \cap G \circ R_i^d], \tag{3.73 }$$

where $L = (L_1, L_2, ..., L_n)$ and

$$R_i^a = \begin{bmatrix} R_i^{a11} & R_i^{a12} & \dots & R_i^{a1n} \\ R_i^{a21} & R_i^{a22} & \dots & R_i^{a2n} \\ \cdot & \cdot & \cdot & \cdot \\ R_i^{an1} & R_i^{an2} & \dots & R_i^{ann} \end{bmatrix}, \quad R_i^c = \begin{bmatrix} R_i^{c1} \\ R_i^{c2} \\ \cdot \\ R_i^{cn} \end{bmatrix},$$

$$R_i^b = \left[R_i^{b1}, R_i^{b2}, \dots, R_i^{bn}\right], \qquad i = 1, \dots, m.$$

The example below illustrates the inference of fuzzy values of the state and input variables of the LM (3.57) by using the equivalent expressions (3.63) and (3.64) or (3.72) and (3.73). This equivalence is demonstrated in [20].

Geometrically the relations R_j^x, R^y represent the mappings:

$$R_j^x : (X \times E_j) \to E_j\,; \qquad R^y : E \to Y,$$

that are similar to the mappings f and g used in the nonlinear deterministic system covered by the state space model (3.59) and (3.60). The LM models system dynamics by means of a set of rules, each of which describes different strategies following from what might be called "standard situations". Each standard situation (i.e. the premise of a single rule) is in turn characterized by a combination of linguistic variables (labelled fuzzy sets) B_i, $A_{i1}, ..., A_{in}$; geometrically each standard situation is associated with a fuzzy region $B_i \wedge A_{i1} \wedge \cdots \wedge A_{in}$ of the Cartesian product space $X \times E$. As in the case of FLC, the fuzziness in the partitioning of the input X and state space $E = E_1 \times E_2 \times \cdots \times E_n$ results in the fuzziness of the regions $B_i \wedge A_{i1} \wedge \cdots \wedge A_{in}$. The consequents describe the developments of the individual state variables for a particular standard situation. The relevance of an actual combination of fuzzy values G, $H_1, ..., H_n$ of the input and state variables U, $W_1, ..., W_n$ to the different standard situations (fuzzy regions of the $X \times E$ space) is defined by the DOF, the τ_i's, of the rules. Then the fuzzy values $L_1, ..., L_n$ and F of the updated state variables $\bar{W}_1, \bar{W}_2, ..., \bar{W}_n$, and the system output V, depend, roughly speaking, on the (inverse) distance from the Cartesian product of actual fuzzy set $G \wedge H_1 \wedge \cdots \wedge H_n$ to the fuzzy regions $B_i \wedge A_{i1} \wedge \cdots \wedge A_{in}$, $i = 1, \dots, m$.

In the formulation of the LM of dynamic systems we assumed that the individual rules (and the respective fuzzy relations) are not time dependent, i.e. they are constructed by time-invariant reference fuzzy sets. In general, these fuzzy sets can be changed during the time, which gives rise to a nonstationary LM. The concept of a nonstationary LM reflects the reasonable changes in the meaning of the labelled fuzzy sets with time.

The state space LM described by the collection of rules (3.61) is a set of instructions defining how the fuzzy values of the state and output are developed at the moment $k + 1$ on the basis of fuzzy or crisp values of the input and state variables in the moment k. If we equip the LM (3.61) with one more set of instructions:

$$\textbf{IF } W_1 \text{ is } A_{i1} \textbf{AND} \dots \textbf{AND } W_n \text{ is } A_{in} \textbf{THEN } U \text{ is } \hat{B}_i, \quad i = 1, \dots, m, \tag{3.74}$$

where the $\hat{B}_i$'s are linguistic labels from the collection of the linguistic labels B_i, $i = 1, \dots, m$ that partition the input space X , then we obtain an immediate state feedback assessment to the system input. The set of rules (3.74) assigns to each standard situation associated with the fuzzy regions $B_i \wedge A_{i1} \wedge \cdots \wedge A_{in}$ of the space $X \times E$ an action that modifies the system input in a direction determined by the set of rules (3.74). The state feedback is actually a MISO LM and the output U generated by it is defined as follows:

$$G(x) = \bigvee_{i=1}^{m} \tau_i \wedge \hat{B}_i(x), \tag{3.75}$$

where the i-th DOF is:

$$\tau_i = \text{Poss}(B_i|G) \wedge \left(\bigwedge_{j=1}^{n} \text{Poss}(A_{ij}|H_j) \right). \tag{3.76}$$

Thus the LM can be used for representing dynamic knowledge-based systems and control algorithms that are described by using linguistic variables. This strategy is attractive when the entire control problem can be effectively developed as an expert system. By using the rule of inference we can find the model of the closed loop system (such theoretical development can be found, for instance, in [12, 13, 21-25]). It is difficult to imagine the applicability of the concept of fuzzy state feedback in the framework of electrical or mechanical systems. However, we can expect that such types of fuzzy control algorithms can be of significant importance for some "soft" regions, e.g. in economic or social systems. An alternative and more computationally efficient fuzzy model of a dynamic system can be derived by application of the Takagi-Sugeno-Kang Fuzzy Models.

In the above we assumed that we have some knowledge, even incomplete, about the physical nature of the dynamic system represented by a LM; we used the concept of fuzziness to represent this incomplete knowledge in a form that is alternative to a nonlinear state space model of the system. If this is not the case, and we have only quantitative information in the form of input-output data, then the state space LM is no longer an appropriate tool. If we want to find a reasonably qualitative description of the system dynamics, based on an available collection of input-output data $(u(k), y(k))$, $k = 1, \dots, K$, we can use the following simple linguistic model which is derived from the MISO LM (3.33):

$$\begin{aligned}&\textbf{IF } u(k) \text{ is } B_{i0} \textbf{AND } u(k-1) \text{ is } B_{i1} \textbf{AND... AND } u(k-n) \text{ is } B_{in}\\ &\textbf{AND } y(k-1) \text{ is } A_{i1} \textbf{AND } y(k-2) \text{ is } A_{i2} \textbf{AND ... AND } y(k-n)\\ &\text{is } A_{in} \quad \textbf{THEN } y(k) \text{ is } A_{i0}, \quad i=1,\ldots,m, \end{aligned} \tag{3.77}$$

where B_{i0}, $B_{i1}, \ldots, B_{in}$ are the reference fuzzy sets used to partition the input space and $A_{i0}, A_{i1}, ..., A_{in}$ are the reference fuzzy sets that partition the output space. Models of this type are the goal of the qualitative modelling [26]. It is easy to see that the LM (3.77) is the fuzzy counterpart of an n-th order nonlinear model:

$$y(k) = f(u(k), u(k-1), u(k-n), y(k-1), ..., y(k-n)).$$

The crisp output value inferred by the model is obtained by applying the Algorithm 3 and defuzzifying. An alternative form of the internal structure of the input-output dynamic fuzzy model (3.77) follows if the simplified method of reasoning applies. From expression (3.32) we get for the output value $y(k)$:

$$y(k) = \left(\sum_{i=1}^{m} \tau_i a_i^*\right) / \left(\sum_{i=1}^{m} \tau_i\right),$$

where the a_i^*'s are the centroids of the consequent fuzzy sets, the A_{i0}'s, and the τ_i's

$$\tau_i = B_{i0}(u(k)) \wedge B_{i1}(u(k-1)) \wedge B_{in}(u(k-n)) \wedge A_{i1}(y(k-1)) \wedge A_{in}(y(k-n))$$

are the DOF of the rules in the model (3.77). The block-diagram of the internal structure of input-output dynamic fuzzy model by (3.77) is shown in Figure 3.10.

By simple transformation, regarding the past values of the input and the output $u(k-1)$, $u(k-n)$, $y(k-1), ..., y(k-n)$ as state variables, the input-output fuzzy model (3.77) can be rewritten in the state space form (61). An example of an input-output linguistic model of this type the fuzzy transition table, introduced by Tong [27-29]. For instance, the following set of fuzzy rules:

IF $u(k-1)$ is *small* **AND** $y(k-1)$ is *small* **THEN** $y(k)$ is *small*;
IF $u(k-1)$ is *medium* **AND** $y(k-1)$ is *small* **THEN** $y(k)$ is *medium*;
IF $u(k-1)$ is *medium* **AND** $y(k-1)$ is *medium* **THEN** $y(k)$ is *high*;
' **IF** $u(k-1)$ is *gigh* **AND** $y(k-1)$ is *high* **THEN** $y(k)$ is *high*;

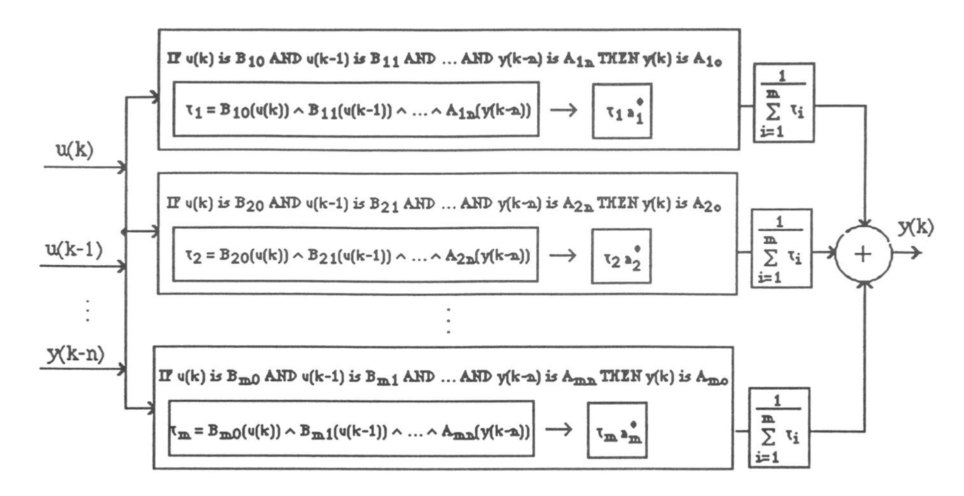

Figure 3.10 Block-diagram of the internal structure of the input-output dynamic model (3.77) via the simplified reasoning method.

describes the dynamics of a first order system. It can be represented by the following state transition table, describing the evolution of the state:

$y(k)$	$u(k-1)\,\&\,y(k-1)$ small	medium	high
small	small		
medium	medium	high	
high			high

The LM (3.77) can be viewed as a fuzzified extension of the AutoRegressive Moving Average (ARMA) models, widely used in digital control [30]. They are characterized with a state vector, formed by past values of the input and output. We have to emphasize here that this new state vector generally doesn't have the physical meaning of the state vector in the case of LM of dynamic systems, but this representation seems to be more convenient for learning the LM on the basis of experimental data.

3.12 TSK MODELS OF DYNAMIC SYSTEMS

A dynamic interpretation of the TSK model is obtained by replacing the input variables in the TSK model with the past input values $y(k-1), y(k-2), ..., y(k-n)$ and also with the current and past input values $u(k), u(k-1), ..., u(k-n)$; the output variable in the TSK modelis replaced by $y(k)$. This substitution in the TSK model (3.55) yields a dynamic form of the TSK model:

$$\begin{aligned} &\textbf{IF } u(k) \text{ is } B_{10} \textbf{ AND } u(k-1) \text{ is } B_{11} \textbf{ AND ... AND } u(k-n) \text{ is } B_{1n} \textbf{ AND} \\ &\qquad y(k-1) \text{ is } A_{i1}\textbf{AND ... AND } y(k-n) \text{ is } A_{in} \qquad (3.78) \\ &\textbf{THEN } y(k) = b_{i0}u(k) + \cdots + b_{in}u(k-n) - a_{i1}y(k-1) - \cdots - a_{in}y(k-n). \end{aligned}$$

Reference fuzzy sets B_{i0}, $B_{i1}, ..., B_{in}$ and $A_{i1}, A_{i2}, ..., A_{in}$ represent linguistic labels that are defined over the input and output spaces. The output y(k) of the overall model is defined from the individual linear subsystems associated with the rules as follows:

$$y(k) = \frac{\sum_{i=1}^{m} \tau_i [b_{i0u}(k) + ... + b_{in}u(k-n) - a_{i1}y(k-1) - ... - a_{in}y(k-n)]}{\sum_{i=1}^{m} \tau_i}, \tag{3.79}$$

where the τ_i's are the DOF of the rules

$$\tau_i = B_{i0}(u(k)) \wedge B_{i1}(u(k-1)) \wedge \cdots \wedge B_{in}(u(k-n)) \wedge A_{i1}(y(k-1)) \wedge \cdots \wedge A_{in}(y(k-n)),$$

where $i = 1, \ldots, m$.

Comparing the dynamic TSK model with the input-output LM (3.77), we see that the only difference between both models is in the consequents. The former has as a consequent the difference equation of the subsystem associated with the i-th region, while the latter has as a consequent the reference fuzzy set A_{i0}, which determines the fuzzy value of the system output in the k-th period. The linear subsystems associated with each rule are only compact expressions of the information contained in the linguistic label that corresponds to the reference fuzzy set A_{i0}. We shall use the relationship between these two alternative models to present a more formalized expression of the FLC.

From expression (3.79) we get the semilinear form of the dynamic TSK model:

$$\begin{aligned} y(k) &= \frac{\sum_{i=1}^{m} \tau_i b_{i0}}{\sum_{i=1}^{m} \tau_i} u(k) + \cdots + \frac{\sum_{i=1}^{m} \tau_i b_{in}}{\sum_{i=1}^{m} \tau_i} u(k-n) \\ &\quad - \frac{\sum_{i=1}^{m} \tau_i a_{i1}}{\sum_{i=1}^{m} \tau_i} y(k-1) - \cdots - \frac{\sum_{i=1}^{m} \tau_i a_{in}}{\sum_{i=1}^{m} \tau_i} y(k-n). \end{aligned}$$

As seen from the above expression all parameters of the dynamic TSK model are dependent on the last n input and $(n-1)$ output readings $u(k), u(k-1), ..., u(k-n)$ and $y(k-1), y(k-2), ...y(k-n)$ through the degrees of firing, the $\tau)_i$'s.

We can express the semilinear form of QLFM (81) in the state space as follows:

$$w(k+1) = Fw(k) + Gu(k), \qquad y(k) = Mw(k) + Nu(k),$$

where matrices F, G, M, N and the state vector w are defined below:

$$F = \begin{bmatrix} 0 & 1 & \cdots & 0 \\ \vdots & \vdots & \vdots & \vdots \\ 0 & 0 & \cdots & 1 \\ -\sum_{i=1}^{m} v_i a_{in} & -\sum_{i=1}^{m} v_i a_{in-1} & \cdots & -\sum_{i=1}^{m} v_i a_{i1} \end{bmatrix}, \quad G = \begin{bmatrix} 0 \\ \vdots \\ 0 \\ 1 \end{bmatrix},$$

$$M = \left[\sum_{i=1}^{m} v_i(b_{in} - a_{in}b_{i0}), \cdots, \sum_{i=1}^{m} v_i(b_{i1} - a_{i1}b_{i0}) \right], \quad N = \sum_{i=1}^{m} v_i b_{i0},$$

$$w(k) = [y(k-n)y(k-n+1)...y(k-1)]^T.$$

In the above we used the variables v_i's to denote the normalized DOF of individual rules:

$$v_i = \tau_i / \left(\sum_{i=1}^{m} \tau_i \right).$$

The structure of the dynamic TSK model can be significantly simplified if only the input is partitioned into fuzzy sets. Then the markers in the antecedents of the rules form a subset of the current and past values of the input $u(k), u(k-1)$, ...; the nonlinear system is represented as a parallel connection of linear subsystems with variable gains, that are dependent on the DOF, the i's [31]. The gains are nonlinear functions of the markers, and therefore of some or all of the current and past input values $u(k), u(k-1)$, ..., through the DOF of the rules i. This model is closely related to the Hammerstain model commonly used to identify nonlinear systems [32].

A straightforward generalization of the concept of a dynamic TSK model can be obtained if nonlinear subsystems associated with the consequents of the rules are considered:

IF $u(k)$ is B_{10} **AND** $u(k-1)$ is B_{11} **AND** ... **AND** $u(k-n)$ is B_{1n}
THEN $w_{(1)}(k+1) = f_1(w_{(1)}(k), k, u(k))$; $y_1 = g_1(w_{(1)}(k), u(k))$
ALSO
...

ALSO

IF $u(k)$ is B_{m0} **AND** $u(k-1)$ is B_{m1} **AND ... AND** $u(k-n)$ is B_{mn}
THEN $w_{(m)}(k+1) = f_m(w_{(m)}(k), k, u(k))$; $y_m = g_m(w_{(m)}(k), u(k))$,

where $w_{(i)}$ denotes the state vector of the i-nonlinear subsystem:

$$w_{(i)}(k+1) = f_1(w_{(i)}(k), k, u),$$

$$y_i(k) = g_1(w_{(i)}(k), u), \qquad i = 1, \ldots, m.$$

The output of the overall system is inferred from the individual outputs yi(k) as follows:

$$y(k) = \left(\sum_{i=1}^{m} \tau_i y_i(k)\right) / \left(\sum_{i=1}^{m} \tau_i\right).$$

3.13 CONCLUSION

Fuzzy systems provide a general methodology for representing complexity of real world problems. The theory of fuzzy system, which is under extensive development, provides a powerful variety of tools for dealing with uncertainty and for representing qualitative and quantitative information.

References

[1] Zadeh, L. A. (1973). Outline of a new approach to the analysis of complex systems and decision processes, *IEEE Trans. Systems, Man, and Cybernetics*, **3**, 28-44.

[2] Sugeno, M. and Takagi, T. (1983). A new approach to design of fuzzy controller, In *Advances in Fuzzy Sets, Possibility Theory and Applications*, Wang, P.P. (Ed.), Plenum Press, New York, pp. 325-334.

[3] Takagi, T. and Sugeno, M. (1983) Derivation of fuzzy control rules from human operators actions, *Proceedings of the IFAC Symposium on Fuzzy Information*, Marseille, pp. 55-60.

[4] Takagi, T. and Sugeno, M. (1985). Fuzzy identification of systems and its application to modeling and control, *IEEE Transactions on Systems, Man and Cybernetics*, **15**, 116-132.

[5] Sugeno, M. and Kang, G. T. (1986). Fuzzy modeling and control of multilayer incinerator, *Fuzzy Sets and Systems*, **18**, 329-346.

[6] Zadeh, L. A. (1979). A theory of approximate reasoning, in *Machine Intelligence*, Vol. 9, Hayes, J., Michie, D., and Mikulich, L.I. (eds.), Halstead Press, New York, pp. 149-194.

[7] Mamdani, E. H. (1974). Application of fuzzy algorithms for control of simple dynamic plant, *Proceedings IEEE 121*, pp. 1585-1588.

[8] Mamdani, E. H. and Assilian, S. (1975). An experiment in linguistic synthesis with a fuzzy logic controller, *International Journal of Man-Machine Studies*, **7**, 1-13.

[9] Mamdani, E. H. and Baaklini, N. (1975). Prescriptive method for deriving control policy in a fuzzy logic control, *Electronic Lett. Ii*, 625-626.

[10] Yager, R. R. and Filev, D. P. (1993). On the issue of defuzzification and selection based on a fuzzy set, *Fuzzy Sets and Systems*, **55**, 255-272.

[11] Mizumoto, M. (1991). Min-max-gravity method versus product-sum-gravity method for fuzzy controls, *Proceedings Fourth IFSA Congress*, Brussels, Engineering Part, pp. 127-130.

[12] Gupta, M. M., Kiszka, J. B. and Trojan, G. J. (1986). Multivariate structure of fuzzy control systems, *IEEE Transactions on Systems, Man and Cybernetics*, **16**, 638-656.

[13] Trojan, G. J., Kiszka, J. B., Gupta, M. M. and Nikiforuk, P. N., (1987). Solution of multivariable fuzzy equations, *Fuzzy Sets and Systems* **22**.

[14] Rajbman, N. S. (1981). Extensions to nonlinear and minimax approaches, in *Trends and Progress in Systems Identification*, Eykhoff, P. (Ed.), Pergamon Press, Oxford, pp. 196-213.

[15] Yager, R. R. (1994). Aggregation operators and fuzzy systems modeling, *Fuzzy Sets and Systems*, **67**, 129-146.

[16] Yager, R. R. (1993). MAM and MOM operators for Aggregation, *Information Sciences*, **69**, 259-273.

[17] Yager, R. R. (1993). Toward a unified approach to Aggregation in fuzzy and neural systems, *Proceedings World Conference on Neural Networks*, Portland, Volume II, pp. 619-622.

[18] Yager, R. R. (1995). A unified approach to Aggregation based upon MOM and MAM operators, *International Journal of Intelligent Systems*, **10**, 809-855.

[19] Yager, R. R. (1990). Using sets as a basis for knowledge representation, *Int. J. of Expert Systems: Research and Applications*, **3**, 147-168.

[20] Yager, R. R. and Filev, D. P. (1994). *Essentials of Fuzzy Modeling and Control*, J. Wiley, New York.

[21] Gupta, M. M., Trojan, G. J. and Kiszka, J. B. (1986). Controllability of fuzzy control systems, *IEEE Transactions on Systems, Man and Cybernetics*, **16**, 576-582.

[22] Pedrycz, W. (1984). Identification in fuzzy systems, *IEEE Trans. on Systems, Man and Cybernetics*, **14**, 361-366.

[23] Pedrycz, W. (1989). *Fuzzy Control and Fuzzy Systems*, J. Wiley, New York.

[24] Czogala, E. and Pedrycz, W. (1982). Control problems in fuzzy systems, *Fuzzy Sets and Systems*, **7**, 257-274.

[25] Tong, R. M. (1980). Some properties of fuzzy feedback systems, *IEEE Transactions on Systems, Man and Cybernetics*, **10**, 327-331.

[26] Sugeno, M. and Yasukawa, T. (1993). A fuzzy-logic based approach to qualitative modeling, *IEEE Transactions on Fuzzy Systems*, **1**, 7-31.

[27] Tong, R. M. (1978). Synthesis of fuzzy models for industrial processes-some recent results, *Int. J. of General Systems*, **4**, 143-163.

[28] Tong, R. M. (1979). The construction and evaluation of fuzzy models, in *Advances in Fuzzy Set Theory and Applications*, Gupta, M. M., Ragade, R. K. and Yager, R. R. (Eds.), North-Holland, Amsterdam, pp. 559-576.

[29] Tong, R. M., Beck, M. B. and Latten, A. (1980). Fuzzy control of the activated sludge waste water treatment process, *Automatica*, **16**, 695-702.

[30] Franklin, G. F., Powell, J. D. and Workman, M. L. (1990). *Digital Control of Dynamic Systems*, Addison-Wesley, Reading, Ma.

[31] Filev, D. P. (1992). System approach to dynamic fuzzy models, *International Journal of General Systems*, **21**, 311-337.

[32] Haber, R. and Unbehauen, H. (1990). Structure identification of nonlinear dynamic systems- a survey on input-output approaches, *Automatica*, **26**, 651-677.

4 FUZZY RULE BASED MODELING AS A UNIVERSAL APPROXIMATION TOOL

Vladik Kreinovich[1], George C. Mouzouris[2] and Hung T. Nguyen[3]

[1]Department of Computer Science
University of Texas at El Paso
El Paso, TX 79968, USA
vladik@cs.utep.edu

[2]Signal and Image Processing Institute
Department of Electrical Engineering — Systems
University of Southern California
Los Angeles, CA 90089-2564, USA
mouzouri@sipi.usc.edu

[3]Department of Mathematical Sciences
New Mexico State University
Las Cruces, NM 88003-8001, USA
hunguyen@nmsu.edu

4.1 INTRODUCTION

4.1.1 Why universal approximation

In some cases, fuzzy rule based model needs tuning. If we have applied some version of fuzzy rule based modeling, and the resulting model is satisfactory, great. But sometimes, the resulting model is not of very high quality:

- we may have misinterpreted some of the expert's rules;

- we may have missed some of the important rules.

So, to improve the quality of the resulting model, we must either:

- re-interpret the existing rules, or
- ask the experts for extra rules.

Before we start tuning, we want to make sure that tuning will help. In order to make sure that this "tuning" will always help, we must guarantee that by appropriate tuning, we will be able to change the initial (not very successful) model into a successful one.

To guarantee that the fuzzy rule based modeling methodology will work in all possible situations, we must make sure that *every possible system can be obtained by applying the fuzzy rule based modeling methodology to appropriate rules.*

Measurements are never 100% accurate, hence, it is sufficient to approximate the desired system. Realistically, measurements are never absolutely precise, hence, we do not need to model the system *exactly*; it is quite sufficient to be able to *approximate* the system, within the measurement accuracy ε, by using appropriate if-then rules.

Mathematical reformulation of the desired property: universal approximation. In more mathematical terms, we want to guarantee that, given an arbitrary function $u(x_1, \ldots, x_n)$, and an arbitrary positive real number $\varepsilon > 0$, we will be able to find a function $\tilde{u}(x_1, \ldots, x_n)$ generated by fuzzy rule based modeling methodology that is ε-close to the original function $u(x_1, \ldots, x_n)$. In mathematics, this ability of approximate an arbitrary function with an arbitrary accuracy is called a *universal approximation* property.

In this chapter, we will briefly review the universal approximation results for fuzzy rule based modeling methodology. This chapter is *not* intended as an exhaustive survey of different universal approximation *results*; our (slightly less ambitious) goal was to overview all possible *approaches*. Because of this, our bibliography is far from being complete.

Also, this chapter is intended to be more an exposition of *ideas*, rather than technical details; the readers who are interested in technical details should refer to the papers that we cite in this survey.

The paper is structured as follows: we start with the main universal approximation results. Then, we describe auxiliary approximation results; in particular, we analyze when it can be guaranteed that the approximating model has the desired properties (such as smoothness). Other auxiliary universal approximation results include:

- systems that use additional expert information (e.g., if-then rules that describe the system and if-then rules that describe the control);
- systems that use unusual logical connectives; and
- approximation results that take the learning process into consideration (i.e., mainly, for fuzzy neural networks).

These universal approximation results are theoretically valid, but these results do not always lead to a practical fuzzy model:

- they do not take into consideration inevitable inaccuracy in the input data, and, besides,
- they may require unrealistically many rules.

So, next, we describe the known method of making the approximation results more realistic.

In this chapter, we describe many different variants of fuzzy rule based modeling methodology. The natural question is: which of these variants should we use? One natural requirements is that the chosen variant must have a universal approximation property. However, since practically all variants have this property, this requirement need to be supplemented by the comparison of the *quality* of the corresponding approximations. We can also go one step back and ask: when should we use fuzzy rule based modeling methodology at all, and when are other intelligent modeling methodologies (such as neural network modeling) better? The answers to these comparison questions is given in the last two sections of this chapter.

4.2 MAIN UNIVERSAL APPROXIMATION RESULTS

4.2.1 *Universal approximation property for the original Mamdani approach*

Before we describe the universal approximation results, let us briefly recall the main formulas of fuzzy rule based modeling methodology (see Chapter 1 for the detailed description).

Rules. Fuzzy rule-based modeling methodology starts with expert "if-then" rules, i.e., with rules of the following type:

$$\text{If } x_1 \text{ is } A_1^j \text{ and } x_2 \text{ is } A_2^j \text{ and } \ldots \text{ and } x_n \text{ is } A_n^j, \text{ then } u \text{ is } B^j,$$

where x_i are parameters that characterize the system's input, u is the output, and A_i^j, B^j are the natural language terms that are used to describe the jth rule (e.g., "small", "medium", etc).

Comment. One of the main applications of fuzzy rule based modeling methodology is *fuzzy control*, when we want to simulate the control applied by expert human controllers. In such control applications, the output u describes applied control.

Mamdani's transformation. The value u is a proper value for the output if and only if one of these rules is applicable.

Thus, the property "u is a proper output" (which we will denote by $C(u)$), can be, therefore, described as follows:

$$\begin{aligned} C(u) \equiv\ & (A_1^1(x_1)\,\&\,A_2^1(x_2)\,\&\,\ldots\,\&\,A_n^1(x_n)\,\&\,B^1(u))\vee \\ & (A_1^2(x_1)\,\&\,A_2^2(x_2)\,\&\,\ldots\,\&\,A_n^2(x_n)\,\&\,B^2(u))\vee \\ & \ldots \\ & (A_1^K(x_1)\,\&\,A_2^K(x_2)\,\&\,\ldots\,\&\,A_n^K(x_n)\,\&\,B^K(u)) \end{aligned}$$

"And" and "or" operations. The natural language terms are described by *membership functions*, i.e., we describe $A_i^j(x)$ as $\mu_i^j(x) \in [0,1]$, and $B^j(u)$ as $\mu_j(u)$.

The logical connectives & and $\vee$ are interpreted, in this context, as operations $f_\&$ and $f_\vee$ on truth values. The most frequent choices of these operations are $\min(a,b)$ and $a \cdot b$ for $f_\&(a,b)$, and $\max(a,b)$ and $a+b-a\cdot b$ for $f_\vee(a,b)$[1].

After these interpretations, we can form the membership function for the output:

$$\mu_C(u) = f_\vee(p_1,\ldots,p_K),$$

where

$$p_j = f_\&(\mu_{j,1}(x_1),\mu_{j,2}(x_2),\ldots,\mu_{j,n}(x_n),\mu_j(u))).$$

Defuzzification. The model must return a single output value u. An operation that transforms a membership function into a single value is called a *defuzzification.* To complete the fuzzy rule based modeling methodology, therefore, we must apply some defuzzification operator F to the membership function $\mu_C(u)$ and thus obtain the crisp output value $\bar{u} = f_C(x)$ that corresponds to $x = (x_1,\ldots,x_n)$. The most widely used defuzzification procedure is *centroid defuzzification*

$$\bar{u} = \frac{\int u\cdot\mu_C(u)\,du}{\int \mu_C(u)\,du}.$$

The above formulas can be further simplified if we only allow fuzziness in the *inputs*, and require the outputs of each rule to be crisp (i.e., to be numbers). In other words, we only allow rules of the type:

If x_1 is A_1^j and x_2 is A_2^j and ... and x_n is A_n^j, then $u = u_j$.

For such rules, the above methodology leads to

$$\bar{u} = \frac{\sum r_j \cdot u_j}{\sum r_j},$$

where

$$r_j = f_{\&}(\mu_{j,1}(x_1), \mu_{j,2}(x_2), \ldots, \mu_{j,n}(x_n)).$$

Who was the first. As successful fuzzy models and fuzzy controls were found for more and more systems, the belief grew that fuzzy rule based modeling methodology is indeed a universal approximation tool.

This belief was explicitly formulated, e.g., by the authors of [18] who showed experimentally that an arbitrary function serving as a control strategy can be well approximated by fuzzy controllers.

Several different universal approximation results expressing this belief were formulated and proved, almost simultaneously, in 1990–92 papers by J. Buckley, Z. Cao, E. Czogala, D. Dubois, M. Grabisch, J. Han, Y. Hayashi, C.-C. Jou, A. Kandel, B. Kosko, J. Mendel, H. Prade, and L.-X. Wang[2]. In the following text, we will describe what each of these authors proved.

Universal approximation result for arbitrary inputs. Bart Kosko, in [20], analyzed the most general case of fuzzy rule based modeling, in which the input x is not necessarily a finite sequence of real numbers (i.e., not necessarily $x \in R^n$), but an element of an arbitrary *compact* set X. He has shown (see also [19], [21], [22], [23], [24], [26], [25]) that for an appropriate choice of an aggregation ($\vee$) operation (namely, for $f_\vee = +$), and for $f_{\&}(a,b) = a \cdot b$, an arbitrary continuous function $f : X \to R$ defined on an arbitrary compact X can be approximated by functions that result from fuzzy if-then rules of the type "if $A(x)$ then $B(u)$", where $x \in X$, $A(x)$ is a fuzzy property (fuzzy subset) of a compact X, and $B(u)$ is a fuzzy property of real numbers.

The choice of $+$ as "or" and $\cdot$ as "and" makes computations simpler: Indeed, since μ_C is the sum and p_j is the product ($p_j = \mu_{j,1}(x) \cdot \mu_j(u)$)), the integrals in Mamdani's approach are simplified into

$$\int u \cdot \mu_C(u)\,du = \int u \cdot \sum(\mu_{j,1}(x) \cdot \mu_j(u))\,du = \sum r_j \cdot U_j,$$

where $r_j = \mu_{j,1}(x)$ and $U_j = \int u \cdot \mu_j(u)\,du$, and similarly,

$$\int \mu_C(u)\,du = \int \sum(\mu_{j,1}(x) \cdot \mu_j(u))\,du = \sum r_j \cdot V_j,$$

where $V_j = \int \mu_j(u)\,du$. Therefore, the Mamdani approach formula is simplified into

$$\bar{u} = \frac{\sum r_j \cdot U_j}{\sum r_j \cdot V_j}.$$

Towards a more realistic universal approximation result. For the practical case of several inputs $x_1, \ldots, x_n$, the compact X is a subset of an n-dimensional space. In this case, Kosko's original theorem requires rules "if A then B" with arbitrary fuzzy properties $A(x_1, \ldots, x_n)$. In the standard fuzzy rule based modeling technique, however, the conditions of the rules are usually of the type $A_1(x)\& \ldots \& A_n(x_n)$. Are rules with such conditions sufficient?

In [31], [29], [32], [33], [30], it was shown that if we use *Gaussian* membership functions, product fuzzy conjunction, and center of average defuzzification, we get a universal approximation[3].

In [27], [28], it was shown that this universal approximation property holds for an *arbitrary* shape of membership functions and for arbitrary "and" and "or" operations, and for an arbitrary defuzzification procedure (see also [86], [213]).

Different universal approximation results were also proven in [2], [9], [14], [15], [16], [17], [3], [4], [8], [10], [5], [7], [6], [34], [35], [36], [13], [37]. (Most of these results will be described in some detail in the following text.)

4.2.2 Generalizations of the standard fuzzy rule based modeling methodology have a universal approximation property

General idea. Since the standard fuzzy rule based modeling methodology has a universal approximation property, any *generalization* of this methodology also has this property.

Several generalizations of the standard methodology have been proposed:

- with more complicated *conclusions* of if-then rules;
- with more complicated *conditions* of if-then rules;
- with more complicated *implication*;
- that take into consideration that experts may assign *different degrees of truth* to different *rules*;
- that take into consideration that different groups of experts may formulate *inconsistent sets of rules.*

Fuzzy rule based modeling with more complicated conclusions of if-then rules: Takagi-Sugeno models are universal approximation tools. In the above approaches, we assumed that for each rule, the output is *fixed* (i.e., does not explicitly depend on the input). In some real-life situations, however, an expert can not only explain that, say, for small x_1 the output should be small, but this expert can also explicitly describe a reasonable dependency of

u on x_1 for the case when x_1 is small. This dependency $u = f_j(x_1, \ldots, x_n)$ is usually described by a linear function, but more complicated dependencies are also possible. In these situations, expert rules are of the type:

If x_1 is A_1^j and x_2 is A_2^j and ... and x_n is A_n^j, then $u = f_j(x_1, \ldots, x_n)$,

and lead to the model of the form

$$\bar{u} = \frac{\sum r_j \cdot f_j(x_1, \ldots, x_n)}{\sum r_j},$$

where

$$r_j = f_\&(\mu_{j,1}(x_1), \mu_{j,2}(x_2), \ldots, \mu_{j,n}(x_n)).$$

This idea is called *Takagi-Sugeno* approach, after its authors [45] (see Chapters 2 and 7 for more detail). For this approach, the universal approximation property was proven in [38], [39], [40], [41], [44].

A natural further generalization of this approach was proposed in [42], [43]. in which in the conclusion of each rule, the desired output u is given not by an explicit formula, but by a (crisp) *dynamical system*, i.e., by a system of differential equations that determine the time derivative of the output variable (i.e., its change in time) as a function of the inputs and of the previous values of the output. This generalization also has a universal approximation property [42], [43].

Fuzzy if-then rules with more complicated conditions: ellipsoid approach. In traditional fuzzy rule based modeling, the condition of each rule is a *conjunction* of several conditions related to different inputs, i.e., each condition is of the type $A_1(x_1)\& \ldots \& A_n(x_n)$. Crudely speaking, we can say that in these conditions, different inputs are *independent* in the sense that the degree with which each input x_i satisfies the condition does not depend on the values of the other inputs. In particular, when each condition $A_i(x_i)$ is described by a Gaussian membership function, and we use product as "and", the resulting membership function for the condition $\mu(x) = \mu_1(x_1) \cdot \ldots \cdot \mu_n(x_n)$ is also Gaussian, and its α-cut (i.e., the set of all values $x = (x_1, \ldots, x_n)$ for which $\mu(x) \geq \alpha$) is an *ellipsoid* with axes coinciding with the coordinate axes.

In some real-life cases, however, there *is* a dependency. As a result, some conditions A_i^j in some expert rules may describe not a single input, but a *combination* of such inputs. The simplest case is when we have a *linear combination.* In this case, each expert rule has the form

If $x_1^{(j)}$ is A_1^j and $x_2^{(j)}$ is A_2^j and ... and $x_n^{(j)}$ is A_n^j, then u is B^j,

where $x_i^{(j)} = w_{i1}^{(j)} \cdot x_1 + \ldots + w_{in}^{(j)} \cdot x_n$ is a linear combination of the (measured) inputs x_i. If we use Gaussian membership functions and $a \cdot b$ as "and", then the resulting membership function is still Gaussian, and its α-cut is an *arbitrarily* oriented ellipsoid. For rules with such ellipsoid-based conditions, a universal approximation property is proven in [46], [47], [50], [51], [52], [48], [49], [53].

Fuzzy if-then rules with a more complicated implication. When an expert pronounces the rule "if x is A, then u is B", e.g., "if an obstacle is close, break", she means not only that if $A(x)$ is true, then $B(u)$ should be true: it also means that if $A(x)$ is *somewhat* true (i.e., if an obstacle is close), then $B(u)$ should also be somewhat true (i.e., we should break some), and the closer the obstacle, the more intensely we need to break. In other words, in addition to the original rule, must we have a "gradual" rule "the more x is A, the more u is B".

Mamdani's approach, in which if-then rules are interpreted by using "and", does not always capture these (implicitly assumed) additional "gradual rule". To capture these meanings, the authors of [54], [55], [56] propose to use a *fuzzy implication* operation $f_{\rightarrow}(a,b)$ instead of "or". In other words, the degree p_j with which j-th rule "if $A^j(x)$ then $B^j(u)$" is true is calculated as $p_j = f_{\rightarrow}(r_j, \mu_j(u))$ (where r_j is the grade to which the condition $A^j(x)$ is true), and the grade to which all K rules from the rule base are true is calculated as $f_{\&}(p_1, \ldots, p_K)$.

In particular, if we use the simplest possible fuzzy implication $f_{\rightarrow}(a,b)$, that is equal to 1 is $a \geq b$ and to 0 else, then the membership function $\mu_C(u)$ describes a *crisp* set of possible output values u for which $\mu_j(u) \geq r_j$ for all j.

It may happen that rules are inconsistent, and there will be no such u; to handle this situation, special interpolation rules are designed. For the resulting methodology, the universal approximation property is proven.

Different grade of truth assigned to different rules. Another generalization of the standard fuzzy rule based methodology is analyzed in [60], [57], [58], [62], [59], [61]: In standard fuzzy rule based modeling methodology, all rules are considered to be equally valid. In reality, however, experts may have *more confidence in some if-then rules* and less confidence in the others. In the spirit of fuzzy approach, the "degree of confidence" (truth value) of each j-th rule can be described by a number w_j from the interval $[0, 1]$. This number w_j determines, crudely speaking, how much "weight" we assign to each rule; therefore, this number is also called the *weight* of j-th rule.

How can we take these weights into consideration when producing a fuzzy rule based model? In *traditional* fuzzy rule based modeling methodology, we are 100% confident in each rule. As a result, we consider a rule "If x is A^j, then u is B^j" to be applicable if both its condition $A^j(x)$ and its conclusion $B^j(u)$ hold, i.e., we take the rule's truth value to be equal to $p_j = f_{\&}(r_j, \mu_j(u))$, where r_j is the grade with which the fuzzy property A^j holds for the given input x.

In the *new* approach, we have some doubts in this rule; as a result, the rule is applicable when not only its conditions and conclusions hold, but also when the rule itself is valid. Thus, the degree of applicability p_j of j-th rule can be obtained by applying the "and" operation to *three* statements instead of two: $p_j = f_{\&}(w_j, r_j, \mu_j(u))$. The resulting formulas for fuzzy rule based modeling becomes correspondingly more complicated.

In particular, when $f_{\&}(a, b) = a \cdot b$, we have $p_j = w_j \cdot f_{\&}(r_j, \mu_j(u))$; in other words, taking the weights into consideration means that we multiply each rule's degree of applicability by this weight. If, in addition, we use + as "or", we get the following formula $\bar{u} = (\sum w_j \cdot r_j \cdot U_j)/(\sum w_j \cdot V_j)$. Even this particular case of "weighted" fuzzy rule based modeling has a universal approximation property.

Several different (inconsistent) sets of rules. Yet another generalization of the standard fuzzy rule based modeling methodology was proposed and analyzed by J. Buckley, B. Kosko, *et al.* Fuzzy rule based modeling methodology starts with a single set of expert rules. In some situations, however, experts cannot agree on the rules. In such situations, we have several different sets of rules; these sets of rules are incompatible and therefore, cannot be easily merged together. Hence, we have to apply fuzzy rule based modeling methodology to each set of rules. As a result, we get several different models $\bar{u}^{(1)}(x_1, \ldots, x_n), \bar{u}^{(2)}(x_1, \ldots, x_n), \ldots$. A (more traditional) approach would be to choose *one* of these sets of rules, based on the quality of the resulting models, and simply disregard all the other sets of rules, i.e., to choose $\bar{u} = \bar{u}^{(k_0)}$, where k_0 is the best model. However, other sets of rules were also proposed by *experts*; this means that they may not be the best *always*, but they definitely describe some expert knowledge that we would want to use in our model. So, it is desirable to take these other models into consideration: e.g., instead of *selecting* the "best" of the models that correspond to different sets of rules (i.e., instead of choosing $\bar{u} = \bar{u}^{(k_0)}$), we can *combine* the values $\bar{u}^{(k)}$ produced by the different models, e.g., into a linear combination $\bar{u} = \sum w^{(k)} \cdot \bar{u}^{(k)}$. In this combination, the weight $w^{(k)}$ of a given model is proportional to the relative quality of corresponding the set of if-then rules: the better the rules, the larger the weight. Such "hierarchical" models are described and analyzed in [65], [68], [69], [63], [64], [66], [70], [67]. Since the resulting model is a generalization of the standard fuzzy rule-based models (namely, the standard rule-based model corresponds to the case when we have only one set of rules), these models also have the universal approximation property.

The resulting combined model is, still, not always of the ideal quality. Ideally, to handle this situation, we should *go back all the way* to the rules, ask the experts to reconcile their rules, and then process the resulting *consistent* combined set of rules. Since this reconciliation is not always possible, the paper [71] proposes the second best thing: *go back one step*, and, instead of combining the *values* $\bar{u}^{(k)}$ that come from different sets of rules, combine the corresponding *membership functions* $\mu_C^{(k)}(u)$, and then apply defuzzification to the combined membership function $\mu_C(u)$. Such a combination often leads (not surprisingly) to a better quality model than the combination of values from different models.

4.2.3 Simplified versions of the standard fuzzy rule based modeling methodology also have a universal approximation property

In the original Mamdani approach, all the stages before defuzzification involve the explicit application of functions, while defuzzification includes *integration* and is, therefore, the most computationally complicated (and thus, time-consuming) stage. For general membership functions, we have to use the general method of computing the integrals: numerical differentiation. For time-sensitive applications such as on-line control, the implementation of this formula can take too much time.

We can, of course, *pre-compute* the values of the model for all possible combinations of inputs. In this case, to find the desired value, we simply need to look up the value of $\bar{u}$ in the corresponding place of the pre-computed *look-up table*. Look-up tables save some time, but require lots of computer memory to store, and, besides, if a table is large, looking up can still take too much time.

For the most widely used membership functions — triangular and trapezoidal — we can find explicit analytical formulas for the integral, and use these formulas instead of the numerical integration [74], [75]. The use of these analytical formulas saves time, but in some practical situations, the resulting formulas may still take too long to compute.

The simplified Mamdani approach is easier to compute, but for some problems, it can still take too much time. For such problems, it is therefore desirable to further simplify this formula. There have been two simplification proposals:

- The simplified Mamdani's defuzzification formula contains several multiplications, additions, and a division. The most time-consuming of them is division. So, to maximally save time, we can *eliminate division*. As a result, we get a formula $\bar{u} = \sum r_j \cdot u_j$ (for Mamdani approach) or $\bar{u} = \sum r_j \cdot f_j(x_1, \ldots, x_n)$ (for Takagi-Sugeno approach). These formulas were proposed in [72], [73], [76], [77]. For Takagi-Sugeno version, with polynomial f_j, the universal approximation property was proven in [72]. In [76], [77], it is shown that we still get the universal approximation property even if we use Mamdani approach (i.e., use constants u_j instead of arbitrary functions $f_j(x_1, \ldots, x_n)$).

- With or without division, the simplified Mamdani approach returns the value $\bar{u}$ that is a linear combination of the values u_j corresponding to different rules. When we use division, we *normalize* this linear combination and thus, make sure that the resulting value $\bar{u}$ is in between the smallest and the largest of the values u_j. Eliminating division makes the result fast to compute, but the downside is that since we eliminated the normalization, the resulting value $\bar{u}$ can be way off, i.e., it can be much larger (or much smaller) than any of the values u_j recommended by the expert rules. So, if we want to value $\bar{u}$ to stay within the reasonable bounds, we must keep a normalization step, and thus, *keep division.*

  In this case, the only way to save computation time is to *save on* two other operations: *multiplication and addition.* The only computer operations with real numbers that are faster than the basic arithmetic operations (addition and multiplication) are minimum and maximum. Therefore, the authors of [78] recommend using max using min instead of multiplication and max instead of addition (from the viewpoint of fuzzy logic, this is a very natural choice, because both max and + are the major examples of "or" operations, and min and product are the main examples of "and" operations). In other words, instead of $\bar{u} = (\sum r_j \cdot u_j)/(\sum r_j)$, the authors of [78] suggest using $\bar{u} = (\max\min(r_j, u_j))/(\max r_j)$; they show that fuzzy rule based modeling with thus simplified defuzzification rule also has a universal approximation property.

- Another possibility to avoid division is to use *Mean of Maximum* defuzzification, i.e., take as the desired value $\bar{u}$ either the value for which $\mu_C(u) \rightarrow \max$, or, if there are several such values, the midpoint of the interval of such values u. The universal approximation property for this defuzzification is proven in [79].

4.2.4 Fuzzy rule-based systems are universal approximation tools for distributed systems

Distributed systems. In the previous section, we explained that fuzzy rule-based systems are universal approximation tools for systems that can be describe by finitely many parameters $x_1, \ldots, x_n$. However, not all real-life systems can be described by finitely many parameters. For example, an appropriate description of a *chemical reaction* requires knowledge of the temperature T, density ρ, and other characteristics at *all* the points $\vec{u}$ inside the reactor. In this case, to describe a state of the plant, we must know the *functions* $T(\vec{u})$, $\rho(\vec{u})$, etc. If we are describing (and/or controlling) a plant which consists of several reactors, we need to know the functions $T(\vec{u}), \rho(\vec{u}), \ldots$, for the first reactor U_1, and also similar functions to describe other reactors $U_2, \ldots$. In general, to describe a state of the system, we need several *functions* $f_1(x), \ldots, f_n(x)$ instead of several *numbers*. Such systems are called *systems with distributed parameters,* or *distributed systems.*

Rule based description of distributed systems. How can we formulate expert rules for such systems? In reality, during a finite period of time, we can only measure the values of a function in *finitely many* points. Therefore, e.g., the actual simulator or controller can only use *finitely many* values $f_i(x_1^j), \ldots, f_i(x_n^j)$ of each function f_i. Thus, reasonable expert rules describe what to do if we know finitely many values, i.e., they are the rules of the type

If $f_1(x_{11}^j)$ is A_{11}^j, $f_1(x_{12}^j)$ is A_{12}^j, ..., $f_1(x_{1m}^j)$ is A_{1m}^j, $f_2(x_{21}^j)$ is A_{21}^j, ..., and $f_n(x_{nm}^j)$ is A_{nm}^j, then u is B^j,

where $f_i(x)$ are functions that characterize the state of the system, x_{kl}^j are the values in which these functions $f_i(x)$ are measured, u is the output, and A_{kl}^j, B^j are the natural language terms that are used to describe the jth rule (e.g., "small", "medium", etc). For these rules, the standard fuzzy rule-based methodology leads to

$$\bar{u} = \frac{\int u \cdot \mu_C(u)\, du}{\int \mu_C(u)\, du},$$

where

$$\mu_C(u) = f_\vee(p_1, \ldots, p_K),$$

and

$$p_j = f_\&(\{\mu_{kl}^j(f_k(x_{kl}^j))\}_{k,l}, \mu_j(u))).$$

A natural question is: do fuzzy rule-based systems of this type universally approximate arbitrary functions $u(f_1, \ldots, f_n)$? The positive answer to this question is given in [81], [82], [80], [83].

Auxiliary result: approximation of universal controllers. The previous results show that the fuzzy rule based methodology is a universal approximation tool. In particular, for control applications, this means that an arbitrary control strategy can be approximated, within an arbitrary accuracy, by an approximate fuzzy controller.

The previous universal approximation results address control of a *single* plant under one *specific* control objective. In real life, however, we may have *different* plants and *different* control objectives. Therefore, the objective of control theory is not only to provide optimal control of *specific* plants under *specific* criteria, but also to develop *general* methods that would help to control an *arbitrary* plant under an *arbitrary* objective. We would, therefore, like fuzzy methodology to provide us with a *universal controller* in the following sense:

we supply it with the description of the system and with an objective (i.e., an optimality criterion), and it will generate the optimal control for this very system under this very criterion.

In [81], [82], [83], it is shown that fuzzy control is a universal tool for such universal controllers as well, i.e., that for every universal controller s of this type, and for every given accuracy $\varepsilon > 0$, we can formulate expert rules that fuzzy control methodology will transform into a universal controller strategy $\tilde{s}$ that is ε-close to s.

4.2.5 Fuzzy rule-based systems are universal approximation tools for discrete systems: Application to expert system design

Fuzzy control methods can be used to control computer processes. In the above example, we considered *continuous* systems, i.e., systems described by one or several *continuous* parameters. It turns out that the same fuzzy rule based modeling and fuzzy control methodology is useful for *discrete* systems, i.e., systems in which state variables only take discrete values.

Computers are a natural example of such discrete systems. So, in this section, we will describe how fuzzy control can be used to control computer processes.

Expert systems. As an example of such a computer process, we can take the process of answering queries in expert systems. An *expert system*, for a certain area of expertize, is a computer system that tries to simulate experts' answers to different questions (*queries*) about this area; e.g., a medical expert system must, given symptoms of a patient, return the possible diagnoses and a reasonable treatment. To be able to do this, an expert system must contain the expert's knowledge; this computer-stored knowledge is called a *knowledge base*.

In addition to the knowledge base, the system must contain a program for answering queries; such a program is called an *inference engine*.

Expert systems are "universal" tools. Designing an inference engine is a very difficult problem: Even when we have *crisp* knowledge, and the knowledge base contains only *propositional* statements F_i — i.e., statements obtained from the elementary statements $S_1, \ldots, S_n$ (like "a patient has a flu") by using "and" (&), "or" ($\vee$), and "not" ($\neg$) — the question of whether a given query follows from the knowledge $F_1, \ldots, F_m$ is, in general, computationally intractable (NP-hard) [84].

NP-hard means that this problem is *universal* in the following sense: any other problem from a very reasonable class (called *NP*) can be reduced to a particular case of this query-answering problem. This universality means that this problem is indeed very difficult to solve: if we could have an algorithm that solves all particular cases of this problem in reasonable time, then we would be thus be able to solve not only this problem, but also *all* reasonable problems. So, this query-answering problem is as computationally difficult as the most complicated of these (realistic) problems.

Heuristic methods are needed. NP-hard means, crudely speaking, that no algorithm can solve *all* particular cases of this problem in reasonable time; thus, *heuristic methods* are needed. In other words, we not only need expert's knowledge about the *domain* to which this expert system is applied, but we also need expert knowledge about the *way* experts answer queries.

It is natural to use fuzzy values (between 0 and 1) to describe heuristic methods. If we ask an expert about a certain query, this expert will often come up with a crisp answer ("yes" or "no"). Producing this answer takes some time. If we ask for an expert's opinion before this time, we will get his *preliminary* opinion; in this preliminary opinion, she is not yet sure whether the answer will be "yes" or "no", but she will probably have some *degree of belief* either in a "yes" answer, or in a "no" answer, or maybe in both. If we ask an expert for the reasons for this degree of belief, she will probably describe some beliefs in the elementary statements $S_1, \ldots, S_n$ and/or their negations.

Therefore, it is natural to simulate this expert reasoning as a step-by-step procedure, in which we start with no beliefs at all (except for the knowledge contained in the knowledge base) and then *modify* our degrees of belief $d(S_i)$ and $d(\neg S_i)$ in the basic statements S_i and their negations $\neg S_i$.

The use of fuzzy rule based modeling methodology. In order to apply fuzzy rule based modeling methodology, we must describe this change in degrees of belief by if-then rules.

A natural way to do this comes from the fact that our knowledge consists of propositional formulas, and it is known that every propositional formula can be reformulated in Conjunctive Normal Form (CNF), i.e., in the form $D_1 \& \ldots \& D_k$, where each of the formulas D_j (called *disjunctions*) is of the type $a \vee b \vee c$, and a, b, and c are *literals* (i.e., elementary statements S_i or their negations $\neg S_i$). Each disjunction D_j, in its turn, can be reformulated as three implications: "if $\neg a$ and $\neg b$, then c"; "if $\neg b$ and $\neg c$, then a"; and "if $\neg a$ and $\neg c$, then b". Thus, the entire knowledge can be represented as a set of such if-then rules.

These rules describing *knowledge* naturally lead to the rules describing *change in degrees of belief*: e.g., the knowledge rule "if $\neg a$ and $\neg b$, then c" leads to the update rule "if $\neg a$ and $\neg b$, then increase the degree of belief in c". Thus, we can apply the standard fuzzy rule based modeling methodology to these rules: at any given moment of time, we know the degree of belief $d(\neg a)$ in $\neg a$ and the degree of belief $d(\neg b)$ in $\neg b$. Therefore, we can compute the degree of belief p_j that this particular rule is applicable as $f_{\&}(d(\neg b), d(\neg c))$, and we can compute the degree of belief $i(c)$ that we should increase $d(c)$ as $f_{\vee}(p_j, \ldots, p_K)$, i.e., as an aggregation for all the rules whose conclusion is this particular increase. Then, for every literal c, we have two conclusions: "update" with degree of belief $i(c)$, and "do not update" with the remaining degree of belief $1 - i(c)$.

If we interpret "update" as adding a constant α to the previous degree of belief, then the standard defuzzification leads to the change from $d(c)$ to the updated value $d(c) + \alpha \cdot i(c)$.

By applying this update procedure again and again, we will get an answer to a query.

This method is successful and related to neural networks. The resulting algorithm coincides with a heuristic method proposed by S. Maslov in [89], [90], [91] (see also [85], [87], [92], [86], [88]). Computer experiments has shown that this is indeed a very successful heuristic for expert systems.

This method was originally proposed based on the idea of simulating biological *neurons*, but it later turned out that exactly these same formulas follow from *fuzzy logic* heuristics, from the ideas of *chemical computing* (i.e., simulating chemical reactions), from heuristics of *numerical optimization* approach, from the ideas of *freedom of choice*, etc. (for a survey and latest results, see [86], [88].

4.3 CAN WE GUARANTEE THAT THE APPROXIMATION FUNCTION HAS THE DESIRED PROPERTIES (SUCH AS SMOOTHNESS, SIMPLICITY, STABILITY OF THE RESULTING CONTROL, ETC.)?

For a model to be good, it must not only approximate the modeled system, but it must also preserve some properties of the modeled system. In the previous sections, we have shown that an arbitrary system can be approximated by a model that originate from the fuzzy if-then rules. To be more precise, an arbitrary function $u(x_1, \ldots, x_n)$ that describes the dependence of the system's output u on the inputs x_i can be approximated by a function $\tilde{u}(x_1, \ldots, x_n)$ that originate from the fuzzy if-then rules. In these sections, to answer the question how good is this approximation, we simply compared the numerical values of the desired and actual output: if these two numerical values are ε-close (where ε is the desired accuracy), then we consider the approximation to be good.

If our only goal is to describe how the system reacts to different inputs, then predicting the system's reaction is all we need. In such applications, all we need is that the simulated output should be ε-close to the actual one. However, in many cases, the model is used to make *decisions*, e.g., to make *control* decisions. From this *control* viewpoint, a better way of comparing the approximate control and the desired control is not only by their values *per se*, but by the resulting behavior of the controlled system. It is therefore desirable to analyze whether we can always approximate a control strategy that has a certain property by a fuzzy control which has the same property. To answer this question, we will first enumerate the basic properties that we can require of the control, and then cite the corresponding universal approximation results.

Most of these results are applicable not only to control, but to the more general case of fuzzy rule based modeling as well.

What properties do we require from control? It is therefore desirable to see if fuzzy control can always approximate the desired control in this (more realistic) sense as well. To answer this question, let us first enumerate the basic properties that we can require of the control.

First property: stability (control must control). The main objective of the control is that it should control. For example, if we control a car on the road, then, for the largest part of the trip, one of the main objectives of this control is to make sure that it stays in its lane with the desired speed, i.e., that whenever it will accidentally deviate from the straight course, the steering control will return it back on course, and when the speed would deviate, the acceleration or deceleration would bring it back to the optimal cruise speed.

In the general case, we want a control that, after an initial deviation, will bring the controlled system back "on track". This property is called *stability* of the control (and of the controlled system).

Stability is a matter of closed loop analysis: one and the same control strategy can be stable for one controlled system (described by a certain set of differential equations), but become unstable for a slightly different system.

Even for a fixed system, stability is a generic term. There are many different particular notions of stability, depending on how big initial deviations we allow (usually, only small ones), whether we want the system to be stabilized for a potentially infinite amount of time or only for a given finite interval, etc.

Second property: smoothness. Stability is not all we expect from a control.

For example, when driving a car, stability means, in particular, that once the car swerved, it should return to the original trajectory. The faster it returns, the more stable is the system. Therefore, from the viewpoint of stability only, the ideal (optimal) control would be the one that brings the car back on track in the shortest possible time (i.e., with the largest possible acceleration). The resulting driving with sudden accelerations may be good on a racetrack or for a car chase, but it is very uncomfortable for passengers. From the passenger viewpoint, we prefer the resulting trajectory to be *smooth*.

The non-smoothness of the optimal control is not a peculiar feature of the car example: in control theory, there are general theorems that show that under certain (reasonably general) conditions, the optimal control is indeed of the above-described "bang-bang" type (see, e.g., [102]; not incidentally, the word "bang-bang" is an "official", well-defined and widely used term in control theory).

Just like stability, smoothness is a generic notion; there are several different understandings (and formalizations) of what "smooth" means. In mathematical terms, *smoothness* is typically formalized as the existence of first, second, or higher order derivatives.

Third property: computational simplicity. Stability and smoothness are typical examples of the *idealized* goals. When, in mathematical control theory, we look for the optimal control strategy, we look for the optimal mathematical function, without taking into consideration how exactly we are going to implement this function.

In *real life*, however, the computational ability of the processor that actually computes the desired control is limited, so some very good control strategies may be too complicated for it. Moreover, in many control situations, we need the control *fast* (e.g., for a car control, if we spend too much time on the computation of the optimal control, the car may, by then, have already wrecked).

In principle, we can always replace the existing processor with a faster one, add extra memory, add an additional processor, etc., but this addition is *not* always *physically possible*:

- For example, in controlling a *space mission*, we are usually very much limited both in terms of the weight and the space required for a processor, and especially, in terms of the power feed; every increase in the computational ability of the controlling processor can only come at the (undesirable) expense of the decrease in the useful payload.

- In commercial applications, e.g., in controlling an *appliance* (which one of the main areas of application for fuzzy control), one of the major considerations is *cost*. Every increase in the computational ability of the processor increases the cost of the simple appliance, and this cost increase is only reasonable if it leads to even better savings in performance[4].

In view of all this, we would like our control to be *computationally simple*[5].

These properties are not exactly consistent: a trade-off is needed. At first glance, it may seem that we should require *all three* properties of our control. However, as we have mentioned, these properties are not exactly consistent:

- the most stable control is often not smooth at all;
- the most stable and the smoothest controls can be computationally complicated.

Therefore, in real life, we must either choose one of these properties that we desire the most, or, even better, look for some trade-off between these three properties.

Can the approximating control preserve the desired property? We know that control strategies that result from applying fuzzy control methodology can be as close to the desired control as possible. For each of the three properties described above, we can ask whether the approximating control can have the desired property:

- If we are interested in the *stable* control, then it is natural to ask whether, for a given controlled system and for a given control u that stabilizes this system, the approximating control $\tilde{u}$ can also be chosen to be stable (it is known in control theory that the very fact that $\tilde{u}$ is close to the stable control u does not necessarily imply that $\tilde{u}$ is stable as well).
- If we are interested in the *smooth* control, then it is natural to ask whether, for an arbitrary smooth function u, the approximating control $\tilde{u}$ can also be chosen to be smooth. For smoothness, it is well known that a function that is close to the smooth one can be not smooth: e.g., the well-known Stone-Weierstrass theorem says that an arbitrary continuous function (not necessarily smooth one) can be, with an arbitrary accuracy, approximated by a polynomial (i.e., by a function that has derivatives of all orders and is thus maximally smooth).
- If we are interested in the *computationally simple* control, then it is natural to ask whether, for a given computationally simple function u, the approximating control $\tilde{u}$ can also be chosen to be computationally simple (this question is not trivial, because it is easy to construct a very complicated function that is close to 0 or to any given simple one).

The answers to these three questions will be given in the following subsections.

4.3.1 Is fuzzy control a universal approximation tool for stable controls?

Stability is one of the most important properties of a control strategy. Therefore, if we want to approximate a *stable* control strategy $u(x_1, \ldots, x_n)$, we would like to be sure that the approximating strategy $\bar{u}(x_1, \ldots, x_n)$ is not only *close* to u, but that this approximating strategy $\bar{u}$ is also *stable.*

The results described above only show that we can always find control rules for which the control strategy $\bar{u}$ resulting from fuzzy control methodology is *close* to u, but this closeness does not automatically guarantee that the system controlled by $\bar{u}$ is also stable.

In [111], [113], [114], [116], it is shown, for Takagi-Sugeno controllers, that, for each controlled system, for each control strategy which is stabilizing for this system, and for each $\varepsilon > 0$, we can find if-then rules for which the resulting fuzzy control strategy is ε-close to the given one and still stabilizing the given system. In these papers, stability is understood in some reasonable *practical* sense. For Mamdani-type control, the possibility of approximation by stable fuzzy controllers is proven in [113], [114], [119] (see also [112], [118]).

In [115], conditions are given under which the approximating control $\bar{u}$ can be chosen as *stable* in a more theoretical sense (i.e, for an unlimited amount of time). In [117], a similar result is proven for *adaptive* fuzzy controllers.

4.3.2 Is fuzzy rule based modeling methodology a universal approximation tool for smooth systems?

Smoothness is often important. In many applications, we want the control to be *smooth.* Smoothness is important, e.g., in the following situations:

- In *aerospace engineering*, if we want to *dock* a spaceship to the space station, then, if we unnecessarily speed up, we'll crash into a space station instead of smoothly approaching it. So here a reasonable criterion is *maximal smoothness.*

- For *public transportation*, we want a smooth control, because abrupt changes make the passengers feel very uncomfortable.

- For *robotic control*: For its movement, the robot is relying on its sensors whose outputs are used to constantly update the robot's image of the environment. For this update to be efficient, the robot must be able to identify obstacles and objects from his previous pictures on the new reading. If a robot moves smoothly, the change is gradual, and it is easier to trace the objects; if a robot make abrupt turns, then the new images are radically different from the old ones, and identification is much more difficult.

Smoothness is often useful not only in control applications, but in general modeling applications as well: if we know that a system is smooth, i.e., that the dependence on its output u on its inputs $x_1, \ldots, x_n$ is differentiable, then we want the model to be smooth too.

Idea: how can we achieve smoothness? According to the fuzzy rule based modeling methodology, the fuzzy model is obtained from the original membership functions (and, for Takagi-Sugeno approach, from the original models) by using:

- first, *&-operations*, to form the degree of firing of a rule;
- then, *aggregation* ($\vee$-)operations, to form the membership function for the output $\mu_C(u)$;
- and, finally, *defuzzification*, to extract a single output value $\bar{u}$ from the membership function $\mu_C(u)$.

In mathematical terms, the resulting fuzzy model is a composition of the corresponding functions and operations.

Therefore, to guarantee the smoothness of the resulting fuzzy model, all these functions and operations must be smooth: we must take smooth membership functions (and, in Takagi-Sugeno approach, smooth models), smooth &- and $\vee$-operations, and a smooth defuzzification procedure.

Splines are optimal. The defuzzification procedure is usually smooth, so we only have to worry about the smoothness of the membership functions and &- and $\vee$-operations. This problem was considered in [100], where it was shown, in particular, that the smoothest control (in some reasonable sense) or, in a more general case, the smoothest fuzzy model, are provided when the membership functions are *splines*, i.e., smooth piece-wise polynomial functions.

How to make a smooth fuzzy model computationally simpler? (Mamdani approach). Smoothness is only one of the desired properties of the fuzzy model (and of fuzzy control), another is computational simplicity. So, it is desirable, within all possible methods that guarantee a certain level of smoothness, to find fuzzy rule based modeling methods that are computationally the simplest possible.

For that purpose, the authors of [109], [108] propose to use the so-called *B-splines* that have an additional advantages of being easy to compute.

If we choose $f_{\&}(a, b) = a \cdot b$ as an &-operation, and, as input membership functions, we choose the splines $\mu_{j,1}(x)$ that form the *partition of unity* (i.e., for which, for every x, $\sum_j \mu_{j,1}(x) = 1$), then the formula

$$\bar{u} = \frac{\sum r_j \cdot u_j}{\sum r_j},$$

where $r_j = \mu_{j,1}(x)$, turns into a *linear* formula $\bar{u}(x) = \sum r_j \cdot u_j$ and thus, becomes *computationally easy*.

Universal approximation results for these easy-to-compute smooth controls (Mamdani approach). For B-splines, the possibility to approximate (and easily approximate) an arbitrary function $f(x_1, \ldots, x_n)$ by linear combinations of basic spline functions is well known; it is used, e.g., in computer graphics and computer-aided design (see, e.g., [95]).

How to make a smooth fuzzy model computationally simpler? (Takagi-Sugeno approach). To make the computations simpler, the authors of [104] use rules in which the right-hand side models are the simplest non-constant splines (i.e., linear functions), and the membership functions are also the simplest splines, i.e., piecewise-linear functions. To be more precise, they use Takagi-Sugeno model, with rules of the type "if x is A^j, then $u = a_j \cdot x + b_j$" and triangular membership functions for fuzzy properties $A^j(x)$.

Universal approximation results for these easy-to-compute smooth fuzzy models (Takagi-Sugeno approach). In [104], it is proven that for Takagi-Sugeno model, with rules of the type "if x is A^j then $u = a_j \cdot x + b_j$", the resulting model $\bar{u}(x) = (\sum(a_j \cdot x + b_j) \cdot \mu_{j,1}(x))/(\sum \mu_{j,1}(x))$ with triangular membership functions $\mu_{j,1}(x)$ can approximate an arbitrary twice differentiable SISO (single input single output) functions $u(x)$ in the sense that not only $\bar{u}(x)$ is close to $u(x)$, but the first and second derivatives $\bar{u}'(x)$ and $\bar{u}''(x)$ are close to the corresponding derivatives $u'(x)$ and $u''(x)$ of the approximated control.

If we only allow *homogeneous* Takagi model, with $b_j = 0$ (i.e., if we allow only models that are typically used in traditional control), then we can only get universal approximation for the *first* derivatives [93].

Smooth easy-to-compute fuzzy models has been successfully applied to real-life problems. The resulting smooth fuzzy models have been successfully applied in control problems, e.g., in robotics [108], [110].

4.3.3 Is fuzzy rule based modeling a universal approximation tool for computationally simple systems?

Computational simplicity is important. In many real-life situations, especially in on-line control, it is important to compute the output as fast as possible.

A fuzzy model is usually computationally simple, but sometimes, an even simpler model is needed. The input-output relation produced by the fuzzy rule based modeling methodology is usually reasonably fast to compute (this is one of the major advantages of fuzzy rule based modeling), but in some situations, the resulting computation time is still too high (see, e.g., [107]). In such situations, it is desirable to approximate the function $\bar{u}(x_1, \ldots, x_n)$ resulting from the fuzzy rule based modeling methodology by a still simpler function $\tilde{u}(x_1, \ldots, x_n)$ that will be even easier to compute.

Idea: second interpolation. To make a fuzzy model computationally simpler, in [98], [94], [103], [101], [99], [96], it is suggested to use *piece-wise linear* or *piece-wise quadratic* functions $\tilde{u}(x_1, \ldots, x_n)$ to approximate the dependence between the inputs $x_1, \ldots, x_n$ and the output u. The author of [96] calls this approximation *second interpolation*.

In this case, the resulting fuzzy model is simply a piece-wise linear (or a piece-wise quadratic) function.

Piece-wise functions are a particular case of splines. Therefore, if piece-wise linear functions do not lead to the desired fuzzy model, we can consider splines (in particular, B-splines) of arbitrary order ([105], [106]).

Universal approximation results for computationally simple fuzzy models. The question of universality of a computationally simple fuzzy rule based modeling can be reformulated as follows: *can we, with a given accuracy, approximate an arbitrary function $u(x_1, \ldots, x_n)$ by a piece-wise linear function?*

The possibility of such an approximation follows from known mathematical results (for details, see [96], [97]; since piece-wise functions are a particular case of splines, this result is a particular case of the general result of approximability by splines.)

4.4 AUXILIARY APPROXIMATION RESULTS

4.4.1 An overview

In the above description of fuzzy rule based modeling methodology, we assumed the following:

- first, that the only expert knowledge that we can use consists of expert rules;
- second, that these rules have the simple "if-then" form ("if ⟨something⟩ and ⟨something⟩, then ⟨something⟩");
- and finally, that we are only interested in the *result* of the tuning, but not in the very *process* of tuning.

In real life, of course, these assumptions may turn out to be false:

- first, in addition to expert rules, we may have additional expert information about the modeled system;

- second, the expert information can be formulated in a more complicated way than simply "if-then" rules;
- finally, we are definitely interested not only in the result of the tuning, but also in the process of tuning.

For these situations, it is also possible to formulate and prove universal approximation results. There results will be described in the following sections.

4.4.2 Using additional expert information (fuzzy models and fuzzy control rules)

In the above text, we showed that the fuzzy rule based modeling methodology has the universal approximation property. The main applications of fuzzy rule based modeling methodology can be divided into two groups:

- designing the *model* of a system, i.e., a function $y(x_1, \ldots, x_n)$ that simulates the way the modeled system functions;
- designing the *control strategy* for a given system, i.e., a function $u(x_1, \ldots, x_n)$ which simulates the way an expert controller controls the system.

In the applications from the first group, the fuzzy rule based modeling methodology transforms the if-then rules that describe how the inputs affect the output of the system, into a system's model. In fuzzy control applications, this methodology transforms the if-then control rules into a control strategy.

In some control situations, however, in addition (or instead of) fuzzy control if-then rules, we have if-then rules that constitute a fuzzy model of the controlled system. For example, an expert in driving, in addition to the rules that say how to best control a car, can also predict how the car will react to different (not necessarily best) controls; this prediction will be formulated in terms of rules like "if you break hard on a slippery road, the car will most probably swerve". How can we use this additional information (i.e., fuzzy model) in designing fuzzy control?

In [123], the universal approximation theorem is used to show that, *in principle*, it is possible to formulate a "universal fuzzy controller", i.e., a device that, given the description of the system and the control objective, describes the optimal control strategy. This proof, however, does not provide us with a feasible algorithm for actually *constructing* such a system. Moreover, as shown in [120], the problem of finding the optimal control strategy is, in general, computationally intractable (NP-hard). Feasible constructions are known for simple fuzzy models ([124], [122]); it is desirable to extend these constructions to more complicated (and thus, more realistic) fuzzy models [121].

4.4.3 Expert rules that use unusual logical connectives

Fuzzy rule based modeling methodology translates the expert's knowledge about the system, that is formulated in "fuzzy" terms of natural language, into a precise model. In many real-life situations, this knowledge is already formulated in terms of if-then fuzzy rules; in fuzzy rule-based modeling methodology, these rules are then, usually, transformed into statements that only use connectives "and", "or", and "not".

However, in many other real-life cases, the fuzzy knowledge about the system can be of much more general type than if-then rules. In these cases, a question appears: can we approximation an arbitrary fuzzy knowledge, that uses *arbitrary* logical *connectives* (including, e.g., different versions of fuzzy implication), by a knowledge described in terms of "and", "or", and "not" fuzzy connectives? In other word, *can we approximate an arbitrary fuzzy logical connective by a combination of these three basic ones?*

It may seem, at first glance, that different unusual connectives are purely mathematical constructions, but, as shown, e.g., in [128], even those connectives that may appear this way actually result from very natural axioms. In view of this result, it is desirable to consider the approximability of *arbitrary* logical connectives.

It turned out ([127], [126]) that in general, such an approximation of an arbitrary connective *is* possible, but only when we, in addition to these three basic connectives, allow *modifiers* such as "very", "slightly", etc. (that without the modifiers, such an approximation is impossible, is also shown in [125]).

4.4.4 Taking the learning process into consideration (fuzzy neural networks)

How can we tune? Traditional universal approximation results only show that *in principle*, we *can* always approximate any given system by a fuzzy model that is obtained by applying fuzzy rule based modeling methodology to appropriate if-then rules. These results does not show *how* to find these rules, i.e., how to tune the original model into the desired one. So, in addition to the universal approximation results, it is desirable to show the universality of some of these *tuning methods.*

Where can these universal tuning methods come from? To answer this question, let us recall that we are talking about tuning methods for *fuzzy rule based modeling.* By definition, fuzzy control is a methodology for *simulating* the expert's knowledge in the computer. Therefore, in order to tune the fuzzy control, it is natural to simulate the way we humans "tune" our knowledge.

Neural networks: simulating human brain. At first glance, it may seem that we can use the same fuzzy rule based modeling methodology to tuning as well: just ask the experts how they tune their rules, and then use this methodology to translate these rules into an actual tuning strategy. The main problem with this idea is that fine-tuning of the knowledge is even harder to explain in words than the knowledge itself. This fine-tuning is done mainly on an *intuitive* level.

Therefore, in order to simulate how this fine-tuning is done, we cannot reply on the experts' statements, we have to actually simulate how this fine-tuning is done in the human brain.

The main "processing units" of human brain are neurons. Therefore, in order to simulate this "brain processing", we must simulate the neurons. Each biological neuron has several inputs (up to 10^4), and a single output. Neural processing starts with sensors sending signals to the neurons. The intensity of the original signal is represented by the frequency of the pulses generated by the sensor. Signals $x_1, \ldots, x_n$ coming from different inputs are "weighted" differently, dependent on the thickness of the connection. As a result, the total number of pulses per time unit that get into the neuron is approximately equal to the weighted sum $x = w_1 \cdot x_1 + \ldots + w_n \cdot x_n$, where $w_1, \ldots, w_n$ are weights of different inputs. The output of a neuron depends on its *threshold* level w_0:

- if the total weighted number of input pulses x is much larger than w_0, the neuron gets maximally excited and emits the largest possible number of pulses;

- if the total weighted number of input pulses x is much smaller than the threshold level w_0, the neuron does not get excited at all, and does not emit any pulses at all;

- if the total weighted number of input pulses x is approximately equal to the threshold value w_0, the neuron gets excited only partly.

Crudely speaking, the output signal y depends on the difference $z = x - w_0$ between the weighted sum x of inputs and the threshold value w_0: $y = s_0(x - w_0)$; if the difference z is $\gg 0$, then $s_0(z) \approx 1$; if $z \ll 0$, then $s_0(z) \approx 0$, and in general, $s_0(z)$ monotonically increases. So, $y = s_0(z) = s_0(x - w_0) = s_0(w_1 \cdot x_1 + \ldots + w_n \cdot x_n - w_0)$.

Experimentally, the input-output relation for biological neurons is best represented by the so-called *logistic* function $s_0(z) = 1/(1+\exp(-z))$. (This choice turns out to be optimal in some reasonable sense, see, e.g., [142], [144]).

Signals coming from the neuron enter other neurons, etc., until we get the final processing results. Artificial (simulated) neurons try to simulate the same structure. In the most widely spread artificial neurons, the output signal y is connected with the input signals $x_1, \ldots, x_n$ by the same formula $y = s_0(w_1 \cdot x_1 + \ldots + w_n \cdot x_n - w_0)$.

It is known that neural networks are *universal approximation* tools (see, e.g., [135], [131], [132], [133], [136], [130], [129], [137], [138], [141], [145], [150], [139], [140], [146], [147], [133], [143], [148], [149], [151], [134]), i.e., that an arbitrary continuous function can be, with an arbitrary accuracy, approximated by a function computed by a neural network[6].

Fuzzy neural networks. Neural networks input well-defined (numerical) inputs and return numerical outputs. They are therefore good in approximating (crisp) functions $f(a_1, \ldots, a_n)$ from real numbers to real numbers. In other words, if we have crisp if-then rules of the type "if $x_1 = a_1^{(1)}, \ldots, x_n = a_n^{(1)}$ then $u = f(a_1^{(1)}, \ldots, a_n^{(1)})$", "if $x_1 = a_1^{(2)}, \ldots, x_n = a_n^{(2)}$ then $u = f(a_1^{(2)}, \ldots, a_n^{(2)})$", ..., then we can use a standard neural network.

Fuzzy rule based modeling is dealing with the situations in which both conditions and conclusions of the if-then rules are not *crisp* numbers, but *fuzzy* sets. Thus, to tune fuzzy rules, we must modify neural networks so that they will be able to process fuzzy sets instead of numbers.

This can be easily done if we allow fuzzy sets in the main formula that defines a neuron, i.e., if we consider a neuron as a device that transforms fuzzy inputs $X_1, \ldots, X_n$ into the fuzzy output

$$Y = s_0(W_1 \cdot X_1 + \ldots + W_n \cdot X_n - W_0),$$

For the resulting *fuzzy neural networks*, we can formulate a similar question: are they universal approximation tools? Can they approximate arbitrary functions from fuzzy sets to fuzzy sets?

This question was first analyzed in [153], [155], with a somewhat unexpected result: fuzzy neural networks are *not* universal approximations tools for functions from fuzzy sets to fuzzy sets. The reason for this negative result is rather intuitively clear: each neuron of the above type (and thus, each neural network formed by such neurons) is *monotonic* in the sense that if we "widen" the inputs, the output fuzzy sets grows "wider", while an arbitrary function from fuzzy sets to fuzzy sets need not necessarily be monotonic. For certain classes of monotonic functions (e.g., for all functions generated by number-to-number ones) fuzzy neural networks are indeed universal approximation tools [155].

To get approximation property in the most general case, it is necessary to replace, in the formulas for a fuzzy neuron, standard fuzzy arithmetic by a more general one; with this replacement, the universal approximation property becomes true [154], [156].

A detailed description of fuzzy neural networks is given in Chapters 8 ("Leaning and tuning"), 9 ("Neuro fuzzy systems"), and 10 ("Neural networks and fuzzy logic").

4.5 HOW TO MAKE THE APPROXIMATION RESULTS MORE REALISTIC

4.5.1 A general overview

The existing universal approximation results are of theoretical nature. The universal approximation results that we have described are of great fundamental importance. These results show that, in principle, for every function $u(x_1, \ldots, x_n)$, and for every accuracy $\varepsilon > 0$, there exists a set of rules for which the application of the fuzzy rule based modeling methodology leads to an output $\bar{u}(x_1, \ldots, x_n)$ that is ε-close to u.

How can we make these results more practical? It is desirable to use the theoretical universal approximation results to actually construct the fuzzy models and use them, e.g., in real-life control. This is, however, not always possible.

Since the universal approximation theorems were mainly motivated by *fundamental, methodological* questions, the authors of these theorems were mainly interested in showing that *some* rules existed, and not so much in looking for *realistic* sets of rules. As a result, the sets of rules presented by these theorems are not always very realistic:

- First, these results often *solve a somewhat oversimplified problem.* To be more precise, the main universal approximation results are based on the assumption that we start with the *exact* values of the input variables $x_1, \ldots, x_n$. In reality, sensors are never 100% accurate. In practical applications, this inaccuracy must be taken into consideration.

- Second, *the solution* provided by the major universal approximation results *is* often *too complicated.* To be more precise, there are often unrealistically many rules in the rule base. This abundance of rules makes the resulting fuzzy model (in particular, fuzzy control) computationally non-realistic.

In the following two sections, we will show how we can solve these two problems and thus, how to make the universal approximation results more realistic (and hence, more practically useful).

4.5.2 Inaccuracies in the input data

Two possible types of inaccuracy: fuzzy and interval. Traditional fuzzy rule based modeling methodology is based on the assumption that we know the *exact* values of the input quantities $x_1, \ldots, x_n$. In real life, these values either come from measurements, or from expert estimates. In both cases, the values are not 100% precise:

- *Measurements* are never 100% accurate, there always is a possibility of a measurement inaccuracy; the best measurements are the ones for which this inaccuracy is the smallest possible. The manufacturer of the measuring instrument usually supplies it with a *guaranteed* accuracy, i.e., with an upper bound Δ on the possible error values. As a result, if, as a result of measuring a certain quantity x, we get the value $\tilde{x}$, the only thing that we can now conclude about the actual value x is that this actual value differs from x by at most Δ, i.e., that x belongs to the interval $[\tilde{x} - \Delta, \tilde{x} + \Delta]$. In other words, instead of the precise value of x_i, we have a (crisp) *set* X_i of possible values of x_i.

- If the values x_i come from *expert estimates*, then these expert estimates also never absolutely accurate. At best, an expert may estimate the actual value of the quantity by saying something like "it is approximately equal to 1". Since we are talking about fuzzy rule based modeling, in fuzzy modeling, such statements are naturally formalized in terms of fuzzy sets. Thus, instead of a precise value x_i, we get a (fuzzy) set X_i of possible values of x_i.

In both cases, we have a set X_i (in general, a fuzzy set) of possible values of x_i. Based on these sets, we must select the output values.

Fuzzy inaccuracy. In case all inputs were supposed to be precisely known, the desired fuzzy model was a function that maps each tuple of real numbers $(x_1, \ldots, x_n)$ (i.e., possible values of the inputs) into a real number (corresponding output).

If we take into consideration that inputs have fuzzy uncertainty, then, in mathematical terms, we need, instead of a mapping from tuples of real numbers into real numbers, we need a mapping that maps each tuple of *fuzzy sets* $(X_1, \ldots, X_n)$ (representing the input) into a real number (the output value). (If we are interested not in the automated control, but in the recommendations for the human controller, then it is sufficient to produce a fuzzy set.)

A methodology that uses fuzzy rules to generate such mappings is proposed and used in [157], [158], [162], [163], [164], [165], [166], [167].

Interval inaccuracy. An important particular case of fuzzy inaccuracy is *interval* inaccuracy. This type of uncertainty occurs if we take measurement inaccuracy into consideration. If we take it into consideration, then the resulting fuzzy model becomes more smooth and the resulting fuzzy control becomes more stable and more smooth (see, e.g., [159], [171], [161], [172], and references therein). Let us illustrate this phenomenon on the example of fuzzy control.

- Traditional fuzzy control techniques, if used appropriately, lead to a control that is stable. However, if we use only a *single* value $\tilde{x}_i$ from the interval $[\tilde{x}_i - \Delta_i, \tilde{x}_i + \Delta_i]$ of possible input values, we will get a control that is *stable for this* particular *value*, and may *not* be *stable* at all for the (unknown) *actual* value. The only way to guarantee that the control is stable for the *actual* (unknown) value is to guarantee that it is stable for *all* values from this interval. This requires, at least, that the algorithm that computes the control values should have this interval at its disposal.

- A measured value $\tilde{x}_i$ is, in general, unpredictably ("randomly") different from the actual value x_i. As a result, the control $\bar{u}$ based on the measured value will "wobble" around the control that correspond to the actual (unknown) x_i. The random wobbling around a smooth process usually makes it less smooth. Thus, the way to avoid this wobbling (and to make control smoother) is to take into consideration that the actual values are within the *intervals*, and then, to choose the *smoothest* possible control within these intervals.

Is the set-valued fuzzy rule based modeling methodology a universal approximation tool? It is interesting to check whether this set-valued fuzzy control methodology is still a universal approximation tool, i.e., whether an arbitrary function from tuples of fuzzy sets into real numbers (or into fuzzy sets) can be approximated by functions stemming from some rule bases.

The universal approximation property was first proven in [158] for the following special type of fuzzy-input situation: When rules are of the type "if x_1 is A_1^j, ..., and x_n is A_n^j, then $B^j(u)$", and the inputs are not known exactly, but represented instead by fuzzy sets $X_1, \ldots, X_n$, then the output $\mu_C(u)$ is defined as a fuzzy set
$(B^{j_1}(u) + \ldots + B^{j_t})/t$, where j_k, $1 \leq k \leq t$, are all the rules j for which $d(X_i, A_i^j)) \geq \alpha$ for some "set distance" d and for some threshold $\alpha > 0$.

A general result, that "non-singleton" (i.e., set-valued) systems can approximate an arbitrary continuous function from compact sets to real numbers with an arbitrary accuracy, was proven in [167].

The universal approximation property for fuzzy inputs is somewhat clarified by results presented in [168], [160], [169], [170]. Namely, in these papers, it is shown that there are functions from fuzzy sets to fuzzy sets (and even from intervals to intervals) that *cannot* be approximated not only by compositions of membership functions and "and" and "or" operations, but also by *any* compositions of functions of one and two variables. However, if we add functions of three or more variables, e.g., the "dot" product ($\sum u \cdot \mu_C(u)$ or $\sum r_j \cdot u_j$) of the type used in centroid defuzzification, then we can approximate arbitrary smooth functions with a reasonable accuracy.

4.5.3 Can we have fewer rules?

Too many rules. We have already mentioned that the fuzzy if-then rule base provided by the major universal approximation results is often too complicated and contains unrealistically many rules. Indeed, to make the model better, we must take into consideration all inputs.

The more inputs we have, the more rules we need to describe: if we have v variables and we only consider 3 possible values of each variable (e.g., negative, negligible, and positive; or small, medium, and large), then we must consider 3^v rules that describe the output for all 3^v combinations. Even for a toy robot, the number of inputs v can be in the dozens. In this case, 3^v becomes unrealistically large.

So, in theory, fuzzy rule based modeling is possible, but in practice, we are often still far from the actual model design or controller design (see, e.g., [180], [181], [182], [185], [179], [184], [197]).

The estimates for an *accuracy* of an approximation with a given number of rules are given also in [194], [195], [196].

In general, we cannot approximate by using fewer rules. One consolation is that this abundance of the rules is not a drawback of fuzzy rule based modeling: it is caused simply by the fact that there are too many functions. Whether we consider arbitrary continuous functions, or smooth (differentiable) functions with a bound on the derivative, or any other reasonable class of functions that we want to approximate, there are simply too many functions in each class.

In precise mathematical terms, if we want to approximate these functions with an accuracy $\varepsilon > 0$, and we are given two functions $f \neq f'$ whose distance is greater than 2ε: $d(f, f') > 2\varepsilon$, then these two functions cannot be approximated by the same functions F: otherwise, we would have $d(f, F) \leq \varepsilon$, $d(f', F) \leq \varepsilon$, and $d(f, f') \leq d(f, F) + d(F, f') \leq 2\varepsilon$. Hence, we need two different functions to approximate f and f' with the desired accuracy.

Similarly, if we have N different functions such that every two of them are more than 2ε apart, we will need N different approximating functions. How many bits do we need to represent this information? If we use k bits, then by using all 2^k combinations of k 0's and 1's, we can represent 2^k different objects. Thus, to represent N different objects, we need at least $\log_2(N)$ bits. For a given class $\mathcal{F}$ of functions, the value of this logarithm for the largest possible N is called an *ε-capacity* of this class ([191], [186]). By explicitly constructing the sets of functions with large N, researchers have shown that the ε-capacity of different classes of functions $f(x_1, \ldots, x_n)$ of n variables grows at least exponentially (i.e., as c^n) with the number of inputs n (see, e.g., Chapter 10 of [191] or [175]).

Thus, in general, there is no way to approximate all the functions from one of these classes without having to spend an unrealistic (exponential) amount of memory just to store the approximating functions.

There is hope. At first glance, the impossibility to approximate an arbitrary function by an appropriate fuzzy rule based model with a realistic number of rules sounds like a negative result. But in many real-life situations, we *do* have successful fuzzy models and successful fuzzy controllers. It is, therefore, desirable to formulate a reasonable class of function for which an approximating fuzzy model with a few rules *is* possible.

An approximation with fewer rules was proposed, for 1-D case, in [189], [190], and for a general case, in [174]: Whereas in the previous approximations, the number of rules increases when we need a better accuracy, in this new approximation scheme, the number of rules remains fixed as the accuracy increases, and in principle, we can get a precise description of the original input-output function, not just an approximation. Moreover, if we allow non-convex membership functions, then we can approximate an arbitrary function by using only two rules ([177], [192], [174], [193], [184]). The downside of these results is that while the standard approximation results use, say, triangular membership functions that are easy to store and to process, this new scheme uses special membership functions that, in essence, code the desired input-output function. So, while we get fewer rules, we do not automatically decrease the total amount of information that needs to be stored.

An idea of actually decreasing this amount of information is described in [180], [181], [182], [185], [176], [184]: it consists of building the rules around the extrema of the input-output function. In these papers, it is shown that if we fix the number of rules, then (under a certain reasonable criterion) locating the membership functions around the extrema indeed leads to the best approximation. Clustering, neural learning, and other learning techniques can then be used to fine-tune the resulting rule bases. An (empirically supported) conjecture is that if an input-output function has few extrema, we will be able to approximate this function with much fewer rules than in the general case.

In [183], another idea is proposed: The expert if-then rules, with which the traditional fuzzy rule based modeling methodology starts, determine the control u based on the *current* values of the inputs $x_1, \ldots, x_n$. In real life, when experts predict the output of a system, they use not only the information about the current state of the modeled system (i.e., about the inputs $x_1, \ldots, x_n$), but also about the *previous* output: e.g., a rule can be "if the inputs did not change much, the output will be almost the same as before". The corresponding expert if-then rules may, therefore, include not only conditions on the current inputs $x_i(t)$, but also conditions on the previous value of the output $u(t-1)$. If we apply the standard fuzzy rule based modeling methodology to these rules, we get an input-output relation of the type $\bar{u}(t) = \bar{u}(x_1(t), \ldots, x_n(t), u(t-1))$. The possibility of such a *feedback* can, often, drastically decrease the number of rules required for an approximation with a given accuracy. The main *drawback* of feedback systems is that the resulting controls are often unstable, so a special care needs to be taken to guarantee the system's stability.

That the hope (of finding reasonable classes of functions for which fewer rules are sufficient) is realistic can be shown on the example of another approximation scheme: neural networks. The negative result formulated above is not only about fuzzy rule based modeling approximations, it is applicable to other means of approximation such as neural networks, wavelet, etc. For neural network, this negative result shows that for some functions, we need exponentially many neurons to approximate. However, in practice, often reasonably few neurons are sufficient for a good approximation. Recently, this empirical fact has been theoretically explained: a reasonable class of functions has been described for which the necessary number of neurons remains quite feasible [178], [173], [187], [188].

4.6 FROM ALL FUZZY RULE BASED MODELING METHODOLOGIES THAT ARE UNIVERSAL APPRIXIMATION TOOLS, WHICH METHODOLOGY SHOULD WE CHOOSE?

At first, there was hope that universal approximation results can help to choose a fuzzy rule based modeling methodology. We have already mentioned that the universal approximation property is an important feature of fuzzy rule based modeling methodology. In view of this importance, when several different variants of this methodology were proposed, variants that use different membership functions, different "and" and "or" operations, and different defuzzification procedures, the hope was that only few of these variants will have the universal approximation property, and that this property will thus help us to select the appropriate methodology.

For example, when Bart Kosko proved his first universal approximation result, that a methodology based on $+$ as aggregation ("or") operation is a universal approximation tool, and proved in a manner that essentially used the properties of $+$, he naturally conjectured that this result may show the advantage of $+$ over other aggregation rules (such as max).

It turned out that the universal approximation property itself is not a good choice criterion; so, how can we choose? As we have seen in the previous sections, most variants of the fuzzy rule based modeling methodology have the universal approximation property. Thus, the very fact that a variant possesses this property does not make it much better than many other variants.

This does not mean, of course, that some variants are not better than the others. Indeed, as we have mentioned, a fuzzy model is not always perfect:

- it can have too many rules, making it computationally non-realistic; or
- the resulting model can be not of very good quality (depending on what we want from the model, it can mean not very smooth, or, in control applications, not very stable, etc.)

So, out of all variants of fuzzy rule based modeling methodology that have the universal approximation properties, it is natural to select the *best* variant, i.e., the variant for which:

- either the *smallest* number of rules is needed, on average, to approximate the given control with a given accuracy; or,
- the resulting model is *the best* according to the chosen criterion (i.e., is the most *smooth*, or, in case of control, the resulting control is the most *stable*, etc.).

In this section, we will briefly describe the situations in which the best choice is known.

4.6.1 Main results. Part I. Best approximation

Choice of membership functions. The authors of [206], [207], compared the quality of the approximation achieved by using different shapes of membership functions. Their numerical experiments have shown that in almost all test situations, the best approximation if we use the "sinc" membership function $\sin(x)/x$.

The paper [199], contains a partial explanation of this result: namely, it is proven that in linear approximation, the function $\sin(x)/x$ is indeed the best (in some reasonable sense). It is desirable to extend this explanation to the general (non-linear) case.

Choice of "and" and "or" operations. In [218], is is shown that the choice of the product $a \cdot b$ as an "and" operation leads to a better approximation than the choice of the minimum $\min(a, b)$.

Choice of defuzzification. In [218], is is shown that the choice of the centroid defuzzification leads to a better approximation than the *Mean of Maximum* defuzzification.

4.6.2 Main results. Part II. Best model

Choice of membership functions. The most *robust* membership functions (i.e., the least sensitive to the inaccuracy of the input data) are piecewise-linear ones [208], [210].

This result explains why the piecewise-linear membership functions are, at present, most frequently used.

Choice of "and" and "or" operations. (These results are (mainly) summarized in [203], [204], [208], [210], [217], [198].)

- If we are looking for the *smoothest* model, then the best choice is to use $f_{\&}(a,b) = a \cdot b$ and $f_{\vee}(a,b) = \max(a,b)$ [203], [204], [217].
- If we are looking for the model that is *most robust* (i.e., least sensitive to the inaccuracy with which we measure the membership functions), then, depending on what exactly we are looking for, we can get two different results:
  - if we are looking for the model that is the most robust *in the the worst case*, then the best choice is to use $f_{\&}(a,b) = \min(a,b)$ and $f_{\vee}(a,b) = \max(a,b)$ [209], [211], [208], [210], [213];
  - if we are looking for the model that is the most robust *in the average*, then the best choice is to use $f_{\&}(a,b) = a \cdot b$ and $f_{\vee}(a,b) = a+b-a \cdot b$ [212], [208], [210], [213];
  - instead of minimizing the *average* error, we can try to minimize the corresponding *entropy* [214], [215], [216], [201], [202], [200]:
    * if we use the *average* entropy (in some reasonable sense), we get the same pair of optimal functions as for average error;
    * for an appropriately defined *worst-case* entropy the optimal operations are $f_{\&}(a,b) = \min(a,b)$ and $f_{\vee}(a,b) = a + b - a \cdot b$.
- If we are looking for the model that is the *fastest to compute*, then the best choice is to use $f_{\&}(a,b) = \min(a,b)$ and $f_{\vee}(a,b) = \max(a,b)$ [205].
- Finally, if, in control applications, we are looking for the *most stable* control for a given system, then the best choice is to use $f_{\&}(a,b) = \min(a,b)$ and $f_{\vee}(a,b) = a + b - a \cdot b$ [203], [204], [217], [202].

Choice of defuzzification. In [203], [204], [202], [200], we show that the optimal defuzzification is given by the centroid formula.

Comment. These optimization results are in good accordance with the general group-theoretic approach that enables us to classify techniques that are optimal relative to arbitrary reasonable criteria [203], [204], [217], [198].

4.7 A NATURAL NEXT QUESTION: WHEN SHOULD WE CHOOSE FUZZY RULE BASED MODELING IN THE FIRST PLACE? AND WHEN IS, SAY, NEURAL MODELING BETTER?

In the previous section, we mentioned that initially, there was hope that only a few variants of the fuzzy rule based modeling methodology would have a universal approximation property and that therefore, this property would be help us select the right variants. Alas, it turns out that most of the variants are universal approximation tools, and thus, we still face the problem of how to choose the best variant. In the previous section, we showed how this problem (of choosing the best fuzzy rule based modeling methodology) can be solved.

Within the scope of this book, it is natural to restrict ourselves to the fuzzy rule based modeling methodology. However, from the practical viewpoint, there is no reason why we should restrict ourselves to this methodology only and not consider other intelligent modeling methodologies that are also known to be universal approximation tools. If we do consider these other methodologies, then we face a similar question: *which of the intelligent modeling methodologies should we choose*? Fuzzy? neural? or some new (yet to be developed) methodology?

There are two ways to compare these two intelligent modeling methodologies:

- First, we can compare the *time* and effort that it takes *to generate* the output of a model.
- Second, we can compare the quality of the resulting model.

For the first way, the answer is easy: neural networks, usually, require lots of time to tune, while fuzzy rule based methodology immediately leads to a reasonable model. So, from the viewpoint of time and effort necessary to generate a model, fuzzy rule based modeling wins hands down.

If, however, we want to compare the *quality* of the resulting models, then the comparison is not so straightforward. The answer depends on which of three quality criteria we use for comparison: smoothness, computational simplicity, or (for control applications) stability. Let us analyze this problem for these three criteria.

4.7.1 Smoothness and stability

Smoothness is the easiest to analyze:

- For each *neuron*, the dependency on the output y on the inputs $x_1, \ldots, x_n$ described by the formula $y = s_0(w_1 \cdot x_1 + \ldots + w_n \cdot x_n - w_0)$ with a smooth function $s_0(z) = 1/(1 + \exp(-z))$. Thus, the dependence between the inputs and the output is smooth. The transformation from the sensor inputs $x_1, \ldots, x_n$ into the output value u that is performed by a neural network is a composition of (smooth) transformations performed by individual neurons, and is, therefore, smooth. Hence, the model generated by a neural network is always smooth.

- The model generated by *fuzzy rule based methodology* is a composition of the membership functions, "and" and "or" operations, and of the defuzzification procedure. The membership functions used in fuzzy rule based modeling are often non-smooth (e.g., triangular); often, non-smooth "and" and "or" operations are also used, such as min for "and" and max for "or". As a result, the model $\tilde{u}(x_1, \ldots, u_n)$ generated by the fuzzy rule based methodology is often non-smooth.

  We have shown (in the above text) how we can gain *some* smoothness by using smooth membership functions and smooth "and" and "or" operations, but typically, the resulting model is *not* infinitely differentiable, and thus, the neural network model is smoother.

Thus, we can conclude that the neural network modeling leads to a smoother model.

This analysis of smoothness can help us to analyze stability of control applications. We have already mentioned that, crudely speaking, stability and smoothness are opposites: the maximally stable control is the least smooth, and vice versa. If we take into consideration that the more stable the control, the less smooth it is, we can then also conclude that the neural network control is the least stable one.

4.7.2 Computational complexity

Often, a model must allow fast output computations. One of the main objectives of modeling is to predict how a modeled system will evolve and how it will react to different changes in the environment. In many real-life applications of modeling (e.g., in many control applications) the system evolves pretty fast, so we need the modeling results fast. In other words, we must be able, given the values $x_1, \ldots, x_n$ of the input variables, to compute the system's simulated output u in the shortest possible time.

The model's output depends on the values of the system's inputs. So, to get a high quality model, we must take into consideration as many parameters that affect the actual system's output as possible. The more inputs we take into consideration, the more numbers we have to process, so, the more computation steps we must perform. So, in many real-life problems, high-quality modeling is a real-time computation problem with a serious time pressure.

Parallel computing is an answer. A natural way to increase the speed of the computations is to perform computations *in parallel* on several processors. To make the computations really fast, we must divide the algorithm into parallelizable steps, each of which requires a small amount of time.

What are these steps?

The fewer variables, the faster. As we have already mentioned, the main reason why modeling algorithms are computationally complicated is that we must process many inputs. For example, modeling or controlling a car is easier than controlling a plane, because the plane (as a 3-D object) has more characteristics to take care of, more characteristics to measure and hence, more characteristics to process. Modeling or controlling a space shuttle, especially during the lift-off and landing, is even a more complicated task, usually performed by several groups of people who control the trajectory, temperature, rotation, etc. In short, the more numbers we need to process, the more complicated the algorithm. Therefore, if we want to decompose our algorithm into fastest possible modules, we must make each module to process as few numbers as possible.

Functions of one variable are not sufficient. Ideally, we should only use the modules that compute functions of one variable. However, if we only have functions of one variables (i.e., procedures with one input and one output), then, no matter how we combine them, we will always end up with functions of one variable. Since our ultimate goal is to compute the modeling function $u = f(x_1, \ldots, x_n)$ that depends on many variables $x_1, \ldots, x_n$, we must therefore enable our processors to compute at least one function of two variables.

What functions of two variables should we choose?

Choosing functions of two variables. Inside the computer, each function is represented as a sequence of hardware implemented operations. The fastest functions are those that are computed by a single hardware operation. The basic hardware supported operations are: arithmetic operations $a + b$, $a - b$, $a \cdot b$, a/b, and $\min(a, b)$ and $\max(a, b)$. The time required for each operation, crudely speaking, corresponds to the number of bits operations that have to be performed:

- Division is done by successive multiplication, comparison and subtraction (basically, in the same way as we do it manually), so, it is a much slower operation than $-$.

- Multiplication is implemented as a sequence of additions (again, basically in the same manner as we do it manually), so it is much slower than $+$.

- $-$ and $+$ are usually implemented in the same way. To add two n-bit binary numbers, we need n bit additions, and also potentially, n bit additions for carries. Totally, we need about $2n$ bit operations.

- Computing the minimum $\min(a, b)$ of two n-bit binary numbers a and b can be done in n binary operations: we compare the bits from the highest to the lowest, and as soon as they differ, the number that has 0 as opposed to 1 is the desired minimum: e.g., the minimum of 0.10101 and 0.10011 is 0.10011, because in the third bit, this number has 0 as opposed to 1.

- Similarly, $\max(a, b)$ is an n-bit operation.

So, the fastest possible functions of two variables are min and max. Similarly fast is computing the minimum and maximum of several (more than two) real numbers. Therefore, we will choose these functions for our modeling-oriented computer.

Summarizing the above-given analysis, we can conclude that our computer will contain modules of two type:

- modules that compute functions of one variable;

- modules that compute min and max of two or several numbers.

How to combine these modules? We want to combine these modules in such a way that the resulting computations are as fast as possible. How can we estimate the computation time? This time is combined from the times of modules on different layers. In principle, different elementary modules take somewhat different time to perform: e.g., minimum is usually computed faster than, say, a square of a number. However, in most computers, all directly hardware supported operations usually take approximately the same time (the largest difference is in the order of 4). Therefore, to get an order-of-magnitude approximation of the computation time, we can assume that all elementary modules require the same computation time.

In this first, crude approximation, the total time required for an algorithm to be performed on a parallel machine is proportional to the number of sequential steps that it takes. We can describe this number of steps in clear geometric terms:

- at the beginning, the input numbers are processed by some processors; these processors form the *first layer* of computations;

- the results of this processing may then go into different processors, that form the *second layer*;

- the results of the second layer of processing go into the *third layer*,
- etc.

In these terms, the fewer layers the computer has, the faster it is:

- functions that can be computed by a single-layer computer require the smallest possible computation time, namely, the time Δt which is equal to the computation time of each processing module;
- functions which can be compare by 2-layer computers (with layers consisting of the above-described standard modules) require the computation time $2 \cdot \Delta t$;
- functions which can be compare by 3-layer computers (with layers consisting of the above-described standard modules) require the computation time $3 \cdot \Delta t$, etc.

So, in the above-described first approximation, looking for the *fastest* model means looking for a combination of processors into the *smallest possible* number of layers.

First result: in the first approximation, fuzzy rule based methodology is the fastest [220]. It turns out that if we restrict ourselves to computers with one or two layers, then there exist functions that cannot be approximated. Actually, these non-approximable functions are not exotic at all: e.g. the function $f(x_1, x_2) = x_1 + x_2$ on the domain $[-1,1]^2$ cannot be approximated by 2-layer computers even with accuracy $\varepsilon = 0.4$.

For *three* layers, we already get a universal approximation tool. The corresponding universal approximation result shows that for every real numbers $T > 0$ and $\varepsilon > 0$, and for every continuous function $f : [-T,T]^n \to R$, there exists a function $\tilde{f}$ that is ε-close to f on $[-T,T]^n$ and that is computable on a 3–layer computer, or, more precisely, a function $\tilde{f}$ of the type $\max(A_1, \ldots, A_k)$, where $A_j = \min(f_{j1}(x_1), \ldots, f_{jn}(x_n))$.

These functions can be explicitly described as the results of applying fuzzy rule based modeling methodology: Indeed, let us define

$$U = \max_{i,j,x_i \in [-T,T]} |f_{ji}(x_i)|,$$

and

$$\mu_{ji}(x_i) = \frac{f_{ji}(x_i) - (-U)}{U - (-U)}.$$

Let us now assume that the experts' rules base consists of exactly two rules:

- "if one of the conditions C_j is true, then $u = U$";
- "else, $u = -U$",

where each condition C_j means that the following n conditions are satisfied:

- x_1 satisfies the property C_{j1} (described by a membership function $\mu_{j1}(x_1)$);
- x_2 satisfies the property C_{j2} (described by a membership function $\mu_{j2}(x_2)$);
- ...
- x_n satisfies the property C_{jn} (described by a membership function $\mu_{jn}(x_n)$).

In logical terms, the condition C for $u = U$ has the form

$$(C_{11} \& \ldots \& C_{1n}) \vee \ldots \vee (C_{k1} \& \ldots \& C_{kn}).$$

If we use min for &, and max for $\vee$ (these are the simplest choices in intelligent modeling methodology), then the degree μ_C with which we believe in a condition $C = C_1 \vee \ldots \vee C_k$ can be expressed as:

$$\mu_C = \max[\min(\mu_{11}(x_1), \ldots, \mu_{1n}), \ldots, \min(\mu_{k1}, \ldots, \mu_{kn})].$$

Correspondingly, the truth value of a condition for $u = -U$ is $1 - \mu_C$. According to fuzzy rule based modeling methodology, we must use a *defuzzification* to determine the actual output, which in this case leads to the choice of

$$u = \frac{U \cdot \mu_C + (-U) \cdot (1 - \mu_C)}{\mu_C + (1 - \mu_C)}.$$

Because of our choice of μ_{ji}, one can easily see that this expression coincides exactly with the function $\max(A_1, \ldots, A_k)$, where $A_j = \min(f_{j1}(x_1), \ldots, f_{jn}(x_n))$. So, we get exactly the expressions that stem from the fuzzy rule based modeling methodology.

Let us summarize our logic:

- we are interested in the modeling methodologies that allow the fastest possible computations of the output u from the given inputs $x_1, \ldots, x_n$;
- the fastest computations are on a parallel computer with simple modules (which are explicitly described in the text);
- in the first approximation, each of the standard modules requires exactly the same computation time;

- in this approximation, the time to perform an algorithm is exactly proportional to the number of layers; so, minimizing computation time means exactly minimizing the total number of layers;

- it turns out that parallel computers with 1 or 2 layers are not enough to get the universal approximation property, and that 3 layers are already enough;

- we also show that functions $u(x_1, \ldots, x_n)$ computed by 3-layer computers are exactly the ones produced by the fuzzy rule based modeling methodology.

So (in the above approximation), *for modeling problems, the fastest possible universal computation scheme corresponds to using fuzzy rule based modeling methodology.*

Second result: neural network modeling methodology is the fastest. We have considered *digital* parallel computers. If we use *analog* processors instead, then min and max stop being the simplest functions. Instead, the sum is the simplest: if we just join the two wires together, then the resulting current is equal to the sum of the two input currents.

In this case, if we use a sum (and more general, linear combination) instead of min and max, 3–layer computers are also universal approximation tools; the corresponding computers correspond to *neural networks* (see, e.g., [219]).

4.7.3 Conclusions

The analysis given above leads us to the following qualitative conclusion:

- If our main objective is to design a model as fast as possible, then we should use fuzzy rule based modeling.

- If our main objective is *smoothness* of a model, then neural network modeling is preferable.

- If our main objective is *computational simplicity*, then:

  - for *software implementations*, fuzzy rule based modeling is better; and

  - for *hardware implementations*, neural network modeling is better.

- In control applications, if our main objective is *stability* of the resulting control, then fuzzy rule based modeling is preferable.

Caution: These conclusions are based on comparing the *potentials* of these two methods, without taking into consideration the imperfection of their implementations. Therefore, these conclusions should be taken as "fuzzy" *guidelines* rather than "crisp" preferences.

For example, the fact that fuzzy rule based modeling has a better potential in achieving stability than neural network modeling does not necessarily mean that the use of the actual fuzzy controller will always lead to more stability than the use of the actual neural network controller.

4.7.4 Collaborate, not compete

We have just seen that each of the two intelligent modeling methodologies (fuzzy rule based modeling and neural network modeling) has its own advantages. It is therefore desirable to *combine* these two methodologies into a single methodology that would use the advantages of both.

Depending on which of the methodologies we start with, we end up with two possible ways of combining these two methodologies:

- we can start with fuzzy rule based modeling methodology, and use a neural network to tune the rules and their parameters; such combined systems, called *neural fuzzy* models, have been described above;

- we can also start with the neural networks; then, the fuzzy inputs and fuzzy tuning rules will enhance the modeling capability of conventional neural networks.

These combined systems are a path to follow.

Acknowledgments. This paper was partly supported by the NSF grant EEC-9322370. We are thankful to numerous researchers, especially to Bernadette Bouchon-Meunier, Piero Bonissone, Jim Buckley, László Kóczy, Bart Kosko, Boris Kovalerchuk, Vera Kůrková, Thomas Runkler, David Sprecher, Li-Xin Wang, Ronald R. Yager, and Lotfi A. Zadeh, for valuable discussions and interesting preprints.

We are especially thankful to Didier Dubois and Henri Prade, the editors of the present Handbook, for their thorough editing and great help.

Notes

1. Sometimes, the term "Mamdani approach" is used only for the case when $f_{\&}(a,b) = \min(a,b)$. The use of $f_{\&}(a,b) = a \cdot b$ is called *Larsen* approach, after the author of the paper [1] that described the first successful applications of this "and" operation.

2. The authors of these papers are listed in *alphabetic*, not in chronological order.

3. The result is actually proven not only for Gaussian functions, but also for functions from any class $\mathcal{F}$ that is closed under shift and multiplication.

4. We are greatly thankful to Piero Bonissone who attracted out attention to this example.

5. Computational simplicity is what sometimes makes engineers use fuzzy control even in the situations when the system is well defined, and the optimal control is known.

6. The paper [152] raises an important concern: Most of the universal approximation results do not take into consideration the precision with which computers represent and process real numbers. The authors of this paper show that in order to get an approximation of arbitrary accuracy, we must have operations with real numbers performed with a much larger precision than in the standard computers. There is, therefore, a need to develop more realistic universal approximation results in which the computer precision is also taken into consideration. Such results are given, e.g., in [149]. A mathematician reader should be warned that the paper [152] is written by non-mathematicians, and is, therefore, somewhat terminologically confusing: e.g., when talking about *polynomials*, the authors of this paper also mean infinite Taylor series, etc.

References

Fuzzy rule-based modeling methodology: an additional reference

[1] Larsen, P. M. (1980). Industrial applications of fuzzy logic control, *Internat. J. Man-Machine Stud.*, **12**, 3–10.

Universal approximation results for the basic fuzzy rule based modeling methodology: basic results

[2] Buckley, J. J. (1992). Universal Fuzzy Controllers, *Automatica*, **28**, 1245–1248.

[3] Buckley, J. J. (1993). Controllable processes and the fuzzy controller, *Fuzzy Sets and Systems*, **53**, 27–31.

[4] Buckley, J. J. (1993a). Sugeno Type Controllers are Universal Controllers, *Fuzzy Sets and Systems,* **53**, 299–303.

[5] Buckley, J. J. (1993d). Approximation paper: Part I, *Proceedings of the Third International Workshop on Neural Networks and Fuzzy Logic, Houston, TX, June 1–3, 1992*, NASA, January 1993, **I** (NASA Conference Publication No. 10111). 170–173.

[6] Buckley, J. J. (1993f). Applicability of the fuzzy controller, In: Wang, P. Z. and Loe, K. F. (eds.), *Advances in Fuzzy Systems: Application and Theory*, World Scientific, Singapore.

[7] Buckley, J. J. and Czogala, E. (1993e). Fuzzy models, fuzzy controllers, and neural nets, *Proc. Polish Academy of Sciences.*

[8] Buckley, J. J. and Hayashi, Y. (1993b). Fuzzy input-output controllers are universal approximators, *Fuzzy Sets and Systems,* **58**, 273–278.

[9] Buckley, J. J., Hayashi, Y., and Czogala, E. (1992a). On the equivalence of neural nets and fuzzy expert systems, *Proc. of Int. Joint Conf. on Neural Networks*, June 7–11, Baltimore, MD, **2**, 691–695.

[10] Buckley, J. J., Hayashi, Y., and Czogala, E. (1993c). On the equivalence of neural nets and fuzzy expert systems, *Fuzzy Sets and Systems,* **53**, 129–134.

[11] Cao, Z. (1990). Mathematical principle of fuzzy reasoning, *Proceedings NAFIPS'90*, Toronto, Canada, June 1990, 362–365.

[12] Cao, Z., Kandel, A., and Han, J. (1990a). Mechanism of fuzzy logic controller, *Proceedings of ISUMA '90*, Univ. of Maryland, December 1990, 603–607.

[13] Castro, J. L. (1995). Fuzzy logic controllers are universal approximators, *IEEE Trans. Syst., Man, Cybern.*, **25** (4), 629–635.

[14] Dubois, D., Grabisch, M., and Prade, H. (1992). Gradual rules and the approximation of functions, *Proceedings of the 2nd International Conference on Fuzzy Logic and Neural Networks*, Iizuka, Japan, July 17–22, 629–632.

[15] Hayashi, Y., Buckley, J. J., and Czogala, E. (1992). Fuzzy expert systems versus neural networks, *Proc. of Int. Joint Conf. on Neural Networks*, June 7–11, Baltimore, MD, **2**, 720–726.

[16] Hayashi, Y., Buckley, J. J., and Czogala, E. (1992a). Approximations between fuzzy expert systems and neural networks, *Proc. of the 2nd Int. Conf. on Fuzzy Logic and Neural Networks*, July 17-22, Iizuka, Japan, 135–139.

[17] Jou, C.-C. (1992). On the mapping capabilities of fuzzy inference systems, *Proceedings of the International Joint Conference on Neural Networks*, Baltimore, Maryland, June 7–11, **2**, 708–713.

[18] Kawamoto, S., Tada, K., Onoe, N., Ishigame, A., and Taniguchi, T. (1992). Construction of exact fuzzy system for nonlinear system and its stability analysis, *8th Fuzzy System Symposium*, Hiroshima, 517–520 (in Japanese).

[19] Kosko, B. (1991). *Neural networks and fuzzy systems: a dynamical systems approach to machine intelligence*, Prentice Hall, 1991.

[20] Kosko, B. (1992). Fuzzy Systems as Universal Approximators, *IEEE Int. Conf. on Fuzzy Systems,* San Diego, CA, March 1992, 1143–1162.

[21] Kosko, B. (1992a). Fuzzy function approximation, *Proceedings of the International Joint Conference on Neural Networks*, Baltimore, Maryland, June 7–11, **1**, 209–213.

[22] Kosko, B. (1994). Fuzzy Systems as Universal Approximators, *IEEE Trans. on Computers,* **43** (11), 1329–1333.

[23] Kosko, B. (1994a). Optimal fuzzy rules cover extrema, *Proceedings of the World Congress on Neural Networks WCNN'94.*

[24] Kosko, B. (1995). Optimal fuzzy rules cover extrema, *International Journal of Intelligent Systems,* **10** (2), 249–255.

[25] Kosko, B. (1996a). Additive fuzzy systems: from function approximation to learning, In: C. H. Chen (ed.), *Fuzzy Logic and Neural Network Handbook*, McGraw-Hill, N.Y., 9-1–9-22.

[26] Kosko, B. and Dickerson, J. A. (1995a). Function approximation with additive fuzzy systems, Chapter 12 in: Nguyen, H. T., Sugeno, M., Tong, R., and Yager, R. (eds.), *Theoretical aspects of fuzzy control*, J. Wiley, N.Y., 313–347.

[27] Nguyen, H. T. and Kreinovich, V. (1992a). *On approximation of controls by fuzzy systems*, Technical Report 92-93/302, LIFE Chair of Fuzzy Theory, Tokyo Institute of Technology.

[28] Nguyen, H. T. and Kreinovich, V. (1993). On Approximation of Controls by Fuzzy Systems, *Proceedings of Fifth IFSA Congress,* Seoul, Korea, **2**, 1414–1417.

[29] Wang, L.-X. (1992). Fuzzy Systems Are Universal Approximators, *Proceedings of Second IEEE International Conference on Fuzzy Systems,* San Diego, CA, March 1992, 1163–1170.

[30] Wang, L.-X. (1994). *Adaptive Fuzzy Systems and Control,* Prentice-Hall, Englewood-Cliffs, NJ.

[31] Wang, L.-X. and Mendel, J. M. (1991). *Generating fuzzy rules from numerical data with applications*, University of Southern California, Signal and Image Processing Institute, Technical Report USC-SIPI # 169.

[32] Wang, L. X. and Mendel, J. M. (1992a). Fuzzy basis functions, universal approximation, and orthogonal least-squares learning, *IEEE Transactions on Neural Networks*, **3**, 807–814.

[33] Wang, L.-X. and Mendel, J. M. (1992b). Generating fuzzy rules by learning from examples, *IEEE Transactions on Systems, Man, and Cybernetics*, **22**, 1414–1417.

[34] Yager, R. R. and Filev, D. P. (1994). *Essentials of fuzzy modeling and control*, J. Wiley & Sons.

[35] Ying, H. (1994). Sufficient conditions on general fuzzy systems as function approximators, *Automatica*, **30** (3), 521–525.

[36] Zeng, X.-J. and Singh, M. G. (1994). Approximation theory of fuzzy systems - SISO case, *IEEE Trans. on Fuzzy Systems,* **2** (2), 162–176.

[37] Zeng, X.-J. and Singh, M. G. (1995). Approximation theory of fuzzy systems - MIMO case, *IEEE Trans. on Fuzzy Systems,* **3**, 219–235.

Takagi-Sugeno methodology, its generalizations, and the corresponding universal approximation results

[38] Buckley, J. J. (1993a). Sugeno Type Controllers are Universal Controllers, *Fuzzy Sets and Systems,* **53**, 299–303.

[39] Buckley, J. J. (1993d). Approximation paper: Part I, *Proceedings of the Third International Workshop on Neural Networks and Fuzzy Logic, Houston, TX, June 1–3, 1992*, NASA, January 1993, **I** (NASA Conference Publication No. 10111). 170–173.

[40] Buckley, J. J. and Hayashi, Y. (1993b). Fuzzy input-output controllers are universal approximators, *Fuzzy Sets and Systems,* **58**, 273–278.

[41] Cao, S. G., Rees, N. W., and Feng, G. (1995). Fuzzy modelling and identification for a class of complex dynamic systems, *Proceedings of the Pacific-Asia Conference on Expert Systems*, Huanshan, China, 212–217.

[42] Cao, S. G., Rees, N. W., and Feng, G. (1996a). Analysis and design of uncertain fuzzy control systems. Part I: Fuzzy modelling and identification, *Proceedings of the Fifth IEEE International Conference on Fuzzy Systems FUZZ-IEEE'96*, New Orleans, September 8–11, **1**, 640–646.

[43] Cao, S. G., Rees, N. W., and Feng, G. (1996b). Analysis and design of uncertain fuzzy control systems. Part II: Fuzzy controller design, *Proceedings of the Fifth IEEE International Conference on Fuzzy Systems FUZZ-IEEE'96*, New Orleans, September 8–11, **1**, 647–653.

[44] Chak, C. K., Feng, G., and Cao, S. G. (1996). Universal fuzzy controllers, *Proceedings of the Fifth IEEE International Conference on Fuzzy Systems FUZZ-IEEE'96*, New Orleans, September 8–11, **3**, 2020–2025.

[45] Takagi, T. and Sugeno, M. (1985). Fuzzy identification of systems and its application to modeling and control, *IEEE Transactions on Systems, Man, and Cybernetics*, **15**, 116–132.

Universal approximation results for the ellipsoid approach

[46] Dickerson, J. A. and Kosko, B. (1993). Fuzzy function learning with covariant ellipsoids, *Proceedings of the IEEE International Conference on Neural Networks*, San Francisco, March 28–April 1, **3**, 1162–1167.

[47] Dickerson, J. A. and Kosko, B. (1993a). Fuzzy function approximation with supervised ellipsoidal learning, *Proceedings of the World Congress on Neural Networks*, Portland, Oregon, July 11–15, **2**, 9–17.

[48] Dickerson, J. A. and Kosko, B. (1996). Fuzzy function approximation with ellipsoidal rules, *IEEE Trans. Syst., Man, Cybern.*, **26** (4), 542–560.

[49] Kim, H. M. and Kosko, B. (1996). Fuzzy prediction and filtering in impulsive noise, *Fuzzy Sets and Systems*, **77**, 15–33.

[50] Kosko, B. (1994). Fuzzy Systems as Universal Approximators, *IEEE Trans. on Computers,* **43** (11), 1329–1333.

[51] Kosko, B. (1994a). Optimal fuzzy rules cover extrema, *Proceedings of the World Congress on Neural Networks WCNN'94*.

[52] Kosko, B. (1995). Optimal fuzzy rules cover extrema, *International Journal of Intelligent Systems*, **10** (2), 249–255.
[53] Kosko, B. (1996a). Additive fuzzy systems: from function approximation to learning, In: Chen, C. H. (ed.), *Fuzzy Logic and Neural Network Handbook*, McGraw-Hill, N.Y., 9-1–9-22.

Universal approximation results for the case when fuzzy if-then rules use a more complicated implication

[54] Dubois, D., Grabisch, M., and Prade, H. (1992). Gradual rules and the approximation of functions, *Proceedings of the 2nd International Conference on Fuzzy Logic and Neural Networks*, Iizuka, Japan, July 17–22, 629–632.
[55] Dubois, D., Grabisch, M., and Prade, H. (1993). Synthesis of real-valued mappings based on gradual rules and interpolative reasoning, *Proc. of the IJCAI'93 Fuzzy Logic in AI Workshop*, Chaméry, France, Aug. 28–Sep. 3, 29–40.
[56] Dubois, D., Prade, H., and Grabisch, M. (1995). Gradual rules and the approximation of control laws, In: Nguyen, H. T., Sugeno, M., Tong, R., and Yager, R. (eds.), *Theoretical aspects of fuzzy control*, J. Wiley, N.Y., 147–181.

Universal approximation results for the case when different grade of truth are assigned to different rules

[57] Dickerson, J. A. and Kosko, B. (1993). Fuzzy function learning with covariant ellipsoids, *Proceedings of the IEEE International Conference on Neural Networks*, San Francisco, March 28–April 1, **3**, 1162–1167.
[58] Dickerson, J. A. and Kosko, B. (1993a). Fuzzy function approximation with supervised ellipsoidal learning, *Proceedings of the World Congress on Neural Networks*, Portland, Oregon, July 11–15, **2**, 9–17.
[59] Dickerson, J. A. and Kosko, B. (1996). Fuzzy function approximation with ellipsoidal rules, *IEEE Trans. Syst., Man, Cybern.*, **26** (4), 542–560.
[60] Kosko, B. (1992). Fuzzy Systems as Universal Approximators, *IEEE Int. Conf. on Fuzzy Systems,* San Diego, CA, March 1992, 1143–1162.
[61] Kosko, B. (1996a). Additive fuzzy systems: from function approximation to learning, In: Chen, C. H. (ed.), *Fuzzy Logic and Neural Network Handbook*, McGraw-Hill, N.Y., 9-1–9-22.
[62] Kosko, B. and Dickerson, J. A. (1995a). Function approximation with additive fuzzy systems, Chapter 12 in: Nguyen, H. T., Sugeno, M., Tong, R., and Yager, R. (eds.), *Theoretical aspects of fuzzy control*, J. Wiley, N.Y., 313–347.

Universal approximation results for hierarchical systems

[63] Buckley, J. J. (1993). Controllable processes and the fuzzy controller, *Fuzzy Sets and Systems*, **53**, 27–31.

[64] Buckley, J. J., and Hayashi, Y. (1993b). Fuzzy input-output controllers are universal approximators, *Fuzzy Sets and Systems,* **58**, 273–278.

[65] Buckley, J. J., Hayashi, Y., and Czogala, E. (1992a). On the equivalence of neural nets and fuzzy expert systems, *Proc. of Int. Joint Conf. on Neural Networks*, June 7–11, Baltimore, MD, **2**, 691–695.

[66] Buckley, J. J., Hayashi, Y., and Czogala, E. (1993c). On the equivalence of neural nets and fuzzy expert systems, *Fuzzy Sets and Systems,* **53**, 129–134.

[67] Hayashi, Y. and Buckley, J. J. (1994). Approximations between fuzzy expert systems and neural networks, *International Journal of Approximate Reasoning*, **10** (1), 63–73.

[68] Hayashi, Y., Buckley, J. J., and Czogala, E. (1992). Fuzzy expert systems versus neural networks, *Proc. of Int. Joint Conf. on Neural Networks*, June 7–11, Baltimore, MD, **2**, 720–726.

[69] Hayashi, Y., Buckley, J. J., and Czogala, E. (1992a). Approximations between fuzzy expert systems and neural networks, *Proc. of the 2nd Int. Conf. on Fuzzy Logic and Neural Networks*, July 17-22, Iizuka, Japan, 135–139.

[70] Jang, J.-S. R. and Sun, C.-T. (1993). Functional equivalence between radial basis function networks and fuzzy inference systems, *IEEE Transactions on Neural Networks*, **4**, 156–159.

[71] Kosko, B. (1995b). Combining fuzzy systems, *Proceedings of the IEEE International Conference on Fuzzy Systems FUZZ-IEEE/IFES'95*, March 1995, **4**, 1855–1863.

Universal approximation results for simplified versions of fuzzy rule based modeling methodology

[72] Buckley, J. J. (1993a). Sugeno Type Controllers are Universal Controllers, *Fuzzy Sets and Systems,* **53**, 299–303.

[73] Buckley, J. J. (1993d). Approximation paper: Part I, *Proceedings of the Third International Workshop on Neural Networks and Fuzzy Logic, Houston, TX, June 1–3, 1992*, NASA, January 1993, **I** (NASA Conference Publication No. 10111). 170–173.

[74] El Hajjaji, A. and Rachid, A. (1994). Explicit formulas for fuzzy controller, *Fuzzy Sets and Systems,* **62**, 135–141.

[75] Kóczy, L. T. and Tikk, D. (1996). Approximation in rule bases, *IPMU'96: Proceedings of the International Conference on Information Processing and*

Management of Uncertainty in Knowledge-Based Systems, Granada, July 1–5, 1996, 489–494.
[76] Su, C.-Y. and Stepanenko, Y. (1994). Adaptive control of a class of nonlinear systems with fuzzy logic, *Proceedings of the Third IEEE International Conference on Fuzzy Systems FUZZ-IEEE'94*, Orlando, FL, June 26–29, **2**, 779–785.
[77] Su, C.-Y. and Stepanenko, Y. (1994a). Adaptive control of a class of nonlinear systems with fuzzy logic, *IEEE Transactions on Fuzzy Systems*, **2** (4), 285–294.
[78] Tan, S., and Vandewalle, J. (1996). Defuzzification, structure transparency, and fuzzy system learning, *Proceedings of the Fifth IEEE International Conference on Fuzzy Systems FUZZ-IEEE'96*, New Orleans, September 8–11, **1**, 470–474.
[79] Zeng, X.-J. and Singh, M. G. (1996). Approximation accuracy analysis of fuzzy systems as function approximators, *IEEE Trans. on Fuzzy Systems,* **4**, 44–63.

Universal approximation results for distributed systems (and for universal controllers)

[80] Cao, S. G., Rees, N. W., and Feng, G. (1996). Fuzzy control of nonlinear discrete-time systems, *Proceedings of the Fifth IEEE International Conference on Fuzzy Systems FUZZ-IEEE'96*, New Orleans, September 8–11, **1**, 265–271.
[81] Kreinovich, V., Nguyen, H. T., and Sirisaengtaksin, O. (1994a). On approximations of controls in distributed systems by fuzzy controllers, *Proceedings of the 5th International Conference on Information Processing and Management of Uncertainty in Knowledge-Based Systems IPMU'94*, Paris, July 4–8, **1**, 79–83.
[82] Kreinovich, V., Nguyen, H. T., and Sirisaengtaksin, O. (1995a). On approximation of controls in distributed systems by fuzzy controllers, In: Bouchon-Meunier, B., Yager, R. R., and Zadeh, L. A. (eds.), *Fuzzy Logic and Soft Computing*, World Scientific, 137–145.
[83] Nguyen, H. T., Kreinovich, V., and Sirisaengtaksin, O. (1996). Fuzzy control as a universal control tool, *Fuzzy Sets and Systems*, **80** (1), 71–86.

Universal approximation results for discrete systems: Application to expert system design

[84] Garey, M. and Johnson, D. (1979). *Computers and intractability: a guide to the theory of NP-completeness*, Freeman, San Francisco.

[85] Kreinovich, V. (1987). Semantics of S. Yu. Maslov's iterative method, *Problems of Cybernetics*, Moscow, 1987, **131**, 30–62 (in Russian); English translation in: Kreinovich, V. and Mints, G. (eds.), *Problems of reducing the exhaustive search*, American Mathematical Society, Providence, RI, 1996.

[86] Kreinovich, V. (1996). S. Maslov's Iterative Method: 15 Years Later (Freedom of Choice, Neural Networks, Numerical Optimization, Uncertainty Reasoning, and Chemical Computing). a chapter in Kreinovich, V. and Mints, G. (eds.), *Problems of reducing the exhaustive search*, American Mathematical Society, Providence, RI.

[87] Kreinovich, V. and Fuentes, L. O. (1991). Simulation of chemical kinetics - a promising approach to inference engines, in: Liebowitz. J. (ed.), *Proceedings of the World Congress on Expert Systems, Orlando, Florida, 1991*, Pergamon Press, N.Y., **3**, 1510–1517.

[88] Kreinovich, V. and Mints, G. (eds.) (1996a). *Problems of reducing the exhaustive search*, American Mathematical Society (AMS Translations — Series 2), **178**, Providence, RI.

[89] Maslov, S. Yu. (1981). Iterative methods in intractable problems as a model of intuitive methods, *Abstracts of the 9th All-Union Symposium on Cybernetics*, 52–56 (in Russian).

[90] Maslov, S. Yu. (1983). Asymmetry of cognitive mechanisms and its implications, *Semiotika i Informatika*, **20**, 3–31 (in Russian).

[91] Maslov, S. Yu. (1987). *Theory of Deductive Systems and its Applications*, MIT Press, Cambridge, MA.

[92] Sirisaengtaksin, O., Fuentes, L. O., and Kreinovich, V. (1995). Non-traditional neural networks that solve one more intractable problem: propositional satisfiability, *Proceedings of the First International Conference on Neural, Parallel, and Scientific Computations*, Atlanta, GA, May 28–31, 1995, **1**, 427–430.

Universal approximation results that describe the possibility of approximating a smooth system by a smooth one, and a simple system by a simple one

[93] Fantuzzi, C. and Rovatti, R. (1996). On the approximation capabilities of the homogeneous Takagi-Sugeno model, *Proceedings of the Fifth IEEE International Conference on Fuzzy Systems FUZZ-IEEE'96*, New Orleans, September 8–11, **2**, 1067–1072.

[94] Galichet, S. and Foulloy, L. (1993). Fuzzy equivalence of classical controllers, *Proceedings of the 1st European Congress on Fuzzy and Intelligent Technologies*, Aachen, Germany, **3**, 1567–1573.

[95] Gordon, W. J. and Riesenfeld, R. F. (1974). B-spline curves and surfaces, In: Barnhill, R. E. and Riesenfeld, R. F. (eds.), *Computer Aided Geometric Design*, Academic Press, N.Y.

[96] Kovalerchuk, B. (1996). Second interpolation in fuzzy control, *Proceedings of the Fifth IEEE International Conference on Fuzzy Systems FUZZ-IEEE'96*, New Orleans, September 8–11, **1**, 150–155.

[97] Kovalerchuk, B. (1996a). *Fuzzy control with the second interpolation*, Working paper, Louisiana State University, Baton Rouge, LA, 1996.

[98] Kovalerchuk, B. and Yusupov, H. (1993). Fuzzy control as interpolation, *Proceedings of the 5th IFSA Congress*, Seoul, Korea, July 4–9, 1151–1154.

[99] Kovalerchuk, B., Yusupov, H., and Kovalerchuk, N. (1994). Comparison of interpolations in fuzzy control, *Proceedings of the 2nd IFAC Workshop on Computer Software Structures Integrating AI/KBS Systems in Process Control*, Lund, Sweden, 76–81.

[100] Kreinovich, V., Quintana, C., Lea, R., Fuentes, O., Lokshin, A., Kumar, S., Boricheva, I., and Reznik, L. (1992). What non-linearity to choose? Mathematical foundations of fuzzy control, *Proceedings of the 1992 International Conference on Fuzzy Systems and Intelligent Control*, Louisville, KY, 349–412.

[101] Meyer-Gramann, K. D. (1993). Easy implementation of fuzzy controller with a smooth control surface, *Proceedings of the 1st European Congress on Fuzzy and Intelligent Technologies*, Aachen, Germany, **1**, 117–123.

[102] Mohler, R. R. (1991). *Nonlinear systems.* **1**. *Dynamics and control*, Prentice Hall, Englewood Cliff, NJ.

[103] Raymond, C., Boverie, S., and Le, J. M. (1993). Practical realization of fuzzy controllers; comparison with conventional methods, *Proceedings of the 1st European Congress on Fuzzy and Intelligent Technologies*, Aachen, Germany, **1**, 149–155.

[104] Rovatti, R. (1996). Takagi-Sugeno models as approximators in Sobolev norms: the SISO case, *Proceedings of the Fifth IEEE International Conference on Fuzzy Systems FUZZ-IEEE'96*, New Orleans, September 8–11, **2**, 1060–1066.

[105] Runkler, T. A. and Glesner, M. (1993). A new approach to fuzzy logic controller realizations using B-splines, *Proceedings of the International Workshop on Current Issues in Fuzzy technology*, Trento, Italy.

[106] Runkler, T. A. and Glesner, M. (1993). Approximative Synthese von Fuzzy–Controllern in Fuzzy Logic, In: Reusch, B. (ed.), *Fuzzy Logic — Theorie und Praxis*, Springer-Verlag, Berlin, 22–31.

[107] Tilli, T. A. (1993). Practical tools for simulation and optimization fuzzy systems with various operators and defuzzification methods, *Proceedings of*

the 1st European Congress on Fuzzy and Intelligent Technologies, Aachen, Germany, **1**, 256–262.
[108] Zhang, J. and Knoll, A. (1996). Constructing fuzzy controllers with B-spline models, *Proceedings of the Fifth IEEE International Conference on Fuzzy Systems FUZZ-IEEE'96*, New Orleans, September 8–11, **1**, 416–421.
[109] Zhang, J., Raczkowsky, J., and Herp, A. (1994). Emulation of spline curves and its application in robot motion control, *Proceedings of the IEEE International Conference on Fuzzy Systems FUZZ-IEEE'94*, Orlando, FL, 831–836.
[110] Zhang, J., Wille, F., and Knoll, A. (1996a). Modular design of fuzzy controller integrating deliberate and reactive strategies, *Proceedings of the IEEE International Conference on Robotics and Automation.*

Universal approximation results and stability issues

[111] Buckley, J. J. (1992). Universal Fuzzy Controllers, *Automatica*, **28**, 1245–1248.
[112] Buckley, J. J. (1993). Controllable processes and the fuzzy controller, *Fuzzy Sets and Systems*, **53**, 27–31.
[113] Buckley, J. J. (1993f). Applicability of the fuzzy controller, In: Wang, P.-Z. and Loe, K. F. (eds.), *Advances in Fuzzy Systems: Application and Theory*, World Scientific, Singapore.
[114] Buckley, J. J. (1995). System stability and the fuzzy controller, In: Nguyen, H. T., Sugeno, M., Tong, R., and Yager, R. (eds.), *Theoretical aspects of fuzzy control*, J. Wiley and Sons, N.Y., 51–63.
[115] Cao, S. G., Rees, N. W., and Feng, G. (1996). Fuzzy control of nonlinear discrete-time systems, *Proceedings of the Fifth IEEE International Conference on Fuzzy Systems FUZZ-IEEE'96*, New Orleans, September 8–11, **1**, 265–271.
[116] Chak, C. K., Feng, G., and Cao, S. G. (1996). Universal fuzzy controllers, *Proceedings of the Fifth IEEE International Conference on Fuzzy Systems FUZZ-IEEE'96*, New Orleans, September 8–11, **3**, 2020–2025.
[117] Jagannathan, S. (1996). Adaptive discrete-time fuzzy logic control of feedback linearizable nonlinear systems, *Proceedings of the Fifth IEEE International Conference on Fuzzy Systems FUZZ-IEEE'96*, New Orleans, September 8–11, **2**, 1273–1278.
[118] Lei, S. and Langari, R. (1996). An approach to synthesis and approximation of stable fuzzy logic controllers, *Proceedings of the Fifth IEEE International Conference on Fuzzy Systems FUZZ-IEEE'96*, New Orleans, September 8–11, **2**, 1446–1452.
[119] Wang, L.-X. (1993). Stable adaptive fuzzy control of nonlinear systems, *IEEE Trans. Fuzzy Syst.*, **1**, 146–155.

Universal approximation results for the case when we have both fuzzy if-then rules describing a model and fuzzy if-then rules describing control

[120] Abello, J., Kreinovich, V., Nguyen, H. T., Sudarsky, S., and Yen, J. (1994). Computing an appropriate control strategy based only on a given plant's rule-based model is NP-hard, In: Hall, L., Ying, H., Langari, R., and Yen, J. (eds.), *NAFIPS/IFIS/NASA'94, Proceedings of the First International Joint Conference of The North American Fuzzy Information Processing Society Biannual Conference, The Industrial Fuzzy Control and Intelligent Systems Conference, and The NASA Joint Technology Workshop on Neural Networks and Fuzzy Logic, San Antonio, December 18–21, 1994*, IEEE, Piscataway, NJ, 331–332.

[121] Bouchon-Meunier, B., Kreinovich, V., Lokshin, A., and Nguyen, H. T. (1996). On the formulation of optimization under elastic constraints (with control in mind). *Fuzzy Sets and Systems,* **81** (1), 5–29.

[122] Chen, G., Pham, T. T., and Weiss, J. J. (1995). Fuzzy modeling of control systems, *IEEE Transactions on Aerospace and Electronic Systems,* **31** (1), 414–429.

[123] Nguyen, H. T., Kreinovich, V., and Sirisaengtaksin, O. (1996). Fuzzy control as a universal control tool, *Fuzzy Sets and Systems*, **80** (1), 71–86.

[124] Pham, T. T., Weiss, J. J., and Chen, G. R. (1992). Optimal fuzzy logic control for docking a boat, *Proc. of the Second International Workshop on Industrial Fuzzy Control and Intelligent Systems, Dec. 2–4, 1992*, College Station, TX, 66–73.

Universal approximation results for the case when expert rules use unusual logical connectives

[125] Kreinovich, V., Nguyen, H. T., and Walker, E. A. (1996b). Maximum entropy (MaxEnt) method in expert systems and intelligent control: new possibilities and limitations, In: Hanson, K. M. and Silver, R. N. (eds.), *Maximum Entropy and Bayesian Methods*, Kluwer Academic Publishers, Dordrecht, 93–100.

[126] Nguyen, H. T., and Kreinovich, V. (1997). Kolmogorov's Theorem and its impact on soft computing, In: Yager, R. R. and Kacprzyk, J. *The Ordered Weighted Averaging Operators: Theory, Methodology, and Applications*, Kluwer, Norwell, MA, 3–17.

[127] Nguyen, H. T., Kreinovich, V., and Sprecher, D. (1996a). Normal forms for fuzzy logic – an application of Kolmogorov's theorem *International Journal on Uncertainty, Fuzziness, and Knowledge-Based Systems,* **4** (4), 331–349.

[128] Türkşen, I. B. and Kreinovich, V. (1996). Fuzzy implication revisited: a new type of fuzzy implication explains Yager's implication operation, In: Dimitrov, V. and Dimitrov, J. (eds.), *Fuzzy Logic and the Management of Complexity (Proceedings of the 1996 International Discourse)*, UTS Publ., Sydney, Australia, **3**, 292–295.

Universal approximation results for neural networks

[129] Blum, E. K. and Li, L. K. (1991). Approximation Theory and feedforward networks, *Neural Networks*, **4**, 511–515.

[130] Cotter, N. E. (1990). The Stone-Weierstrass Theorem and Its Applications to Neural Networks, *IEEE Trans. on Neural Networks,* **1** (4), 290–295.

[131] Cybenko, G. (1989). Approximation by superpositions of a sigmoidal function, *Math. of Control, Signals, and Systems,* **2**, 303–314, 1989.

[132] Funahashi, K. (1989). On the Approximate Realization of Continuous Mappings by Neural Networks, *Neural Networks,* **2**, 183–192.

[133] Funahashi, K. and Nakamura, Y. (1993). Approximation of Dynamical Systems by continuous time recurrent neural networks, *Neural Networks,* **6**, 801–806.

[134] Haykin, S. (1994). *Neural Networks: A Comprehensive Foundation,* Macmillan College Publishing Company, New York, NY.

[135] Hecht-Nielsen, R. (1987). Kolmogorov's Mapping Neural Network Existence Theorem, *Proceedings of First IEEE International Conference on Neural Networks,* 11–14, San Diego, CA.

[136] Hornik, K., Stinchcombe, M., and White, H. (1989). Multilayer Feedforward Neural Networks Are Universal Approximators, *Neural Networks,* **2**, 359–366.

[137] Hornik, K. (1991). Approximation Capabilities of Multilayer Feedforward Neural Networks, *Neural Networks,* **4**, 251–257.

[138] Ito, Y. (1991). Approximation of functions on a compact set by finite sums of a sigmoid function without scaling, *Neural Networks,* **4**, 817–826.

[139] Ito, Y. (1992). Approximation of continuous functions on R^d by linear combinations of shifted rotations of a sigmoid function with and without scaling, *Neural Networks,* **5**, 105–115.

[140] Kosko, B. (1992b). *Neural Networks and Fuzzy Systems*, Prentice Hall, Englewood Cliffs, NJ.

[141] Kreinovich, V. (1991b). Arbitrary Nonlinearity is Sufficient to Represent All Functions by Neural Networks: A Theorem, *Neural Networks,* **4**, 381–383.

[142] Kreinovich, V. and Quintana, C. (1991a). Neural networks: what nonlinearity to choose?, *Proceedings of the 4th University of New Brunswick Artificial Intelligence Workshop*, Fredericton, N.B., Canada, 627–637.

[143] Kreinovich, V. and Sirisaengtaksin, O. (1993). 3-layer neural networks are universal approximators for functionals and for control strategies, *Neural, Parallel, and Scientific Computations*, **1**, 325–346.

[144] Kreinovich, V., Sirisaengtaksin, O., and Nguyen, N. T. (1995). Sigmoid neurons are the safest against additive errors, *Proceedings of the First International Conference on Neural, Parallel, and Scientific Computations*, Atlanta, GA, May 28–31, **1**, 419–423.

[145] Kůrková, V. (1991). Kolmogorov's Theorem is Relevant, *Neural Computation,* **3**, 617–622.

[146] Kůrková, V. (1992). Kolmogorov's Theorem and Multilayer Neural Networks, *Neural Networks,* **5**, 501–506.

[147] Kůrková, V. (1992a). Universal approximation using feedforward neural networks with Gaussian bar units, *Proceedings of ECAI'92, European Conference on Artificial Intelligence*, Wiley, Chichester, 193–197.

[148] Leshno, M., Lin, V. Ya., Pinkus, A., and Schocken, S. (1993). Multilayer Feedforward Networks With a Nonpolynomial Activation Function Can Approximate Any Function, *Neural Networks,* **6**, 861–867.

[149] Nakamura, M., Mines, R., and Kreinovich, V. (1993). Guaranteed intervals for Kolmogorov's theorem (and their possible relation to neural networks). *Interval Computations*, No. 3, 183–199.

[150] Park, J. and Sandberg, J. W. (1991). Universal approximation using radial basis function networks, *Neural Computation,* **3**, 246–257.

[151] Sirisaengtaksin, O. and Kreinovich, V. (1993). Neural networks that are not sensitive to the imprecision of hardware neurons, *Interval Computations*, No. 4, p. 100–113.

[152] Wray, J. and Green, G. G. R. (1995). Neural networks, approximation theory, and finite precision computation, *Neural Networks*, **8** (1), 31–37.

Universal approximation results for fuzzy neural networks

[153] Buckley, J. J. and Hayashi, Y. (1994). Can fuzzy neural networks approximate continuous fuzzy functions?, *Fuzzy Sets and Systems,* **61**, 43–52.

[154] Buckley, J. J. and Hayashi, Y. (1994a). Hybrid fuzzy neural networks are universal approximators, *Proceedings of the Third IEEE International Conference on Fuzzy Systems FUZZ-IEEE'94*, Orlando, FL, June 26–29, **1**, 238–243.

[155] Feuring, T. and Lippe, W.-M. (1995). Fuzzy neural networks are universal approximators, *Proceedings of the IFSA World Congress*, San Paulo, Brazil, 659–662.

[156] Hayashi, Y. and Buckley, J. J. (1994). Approximations between fuzzy expert systems and neural networks, *International Journal of Approximate Reasoning*, **10** (1), 63–73.

Universal approximation results for systems with fuzzy inputs and interval inputs

[157] Buckley, J. J. (1991). Fuzzy I/O controller, *Fuzzy Sets and Systems*, **43**, 127–137.

[158] Buckley, J. J. and Hayashi, Y. (1993b). Fuzzy input-output controllers are universal approximators, *Fuzzy Sets and Systems*, **58**, 273–278.

[159] Kreinovich, V. and Nguyen, H. T. (1994b). Applications of fuzzy intervals: a skeletal outline, In: Hall, L., Ying, H., Langari, R., and Yen, J. (eds.), *NAFIPS/IFIS/NASA'94, Proceedings of the First International Joint Conference of The North American Fuzzy Information Processing Society Biannual Conference, The Industrial Fuzzy Control and Intelligent Systems Conference, and The NASA Joint Technology Workshop on Neural Networks and Fuzzy Logic, San Antonio, December 18–21, 1994*, IEEE, Piscataway, NJ, 461–463.

[160] Kreinovich, V. and Nguyen, H. T. (1995b). On Hilbert's Thirteenth Problem for Soft Computing, *Proceedings of the Joint 4th IEEE Conference on Fuzzy Systems and 2nd IFES*, Yokohama, Japan, March 20–24, 1995, **IV**, 2089–2094.

[161] Lea, R., Kreinovich, V., and Trejo, R. (1996). Optimal interval enclosures for fractionally-linear functions, and their application to intelligent control, *Reliable Computing*, **2** (3), 265–286.

[162] Mouzouris, G. C. and Mendel, J. M. (1994). Non-Singleton Fuzzy Logic Systems, *Proceedings of Third IEEE International Conference on Fuzzy Systems,* June 1994, **1**, 456–461.

[163] Mouzouris, G. C. and Mendel, J. M. (1994a). *Non-Singleton Fuzzy Logic Systems: Theory and Application*, University of Southern California, Signal and Image Processing Institute, Report No. 262.

[164] Mouzouris, G. C. and Mendel, J. M. (1995). Nonlinear Time Series Analysis with Non-Singleton Fuzzy Logic Systems, *Proceedings IEEE/IAFE Conference on Computational Intelligence for Financial Engineering,* New York, NY, April 1995, 47–56.

[165] Mouzouris, G. C. and Mendel, J. M. (1996). Nonlinear predictive modeling using dynamic non-singleton fuzzy logic systems, *Proceedings of the Fifth IEEE International Conference on Fuzzy Systems FUZZ-IEEE'96*, New Orleans, September 8–11, **2**, 1217–1223.

[166] Mouzouris, G. C. and Mendel, J. M. (1996a). Designing fuzzy logic systems for uncertain environments using a singular-value-QR decomposition method, *Proceedings of the Fifth IEEE International Conference on Fuzzy Systems FUZZ-IEEE'96*, New Orleans, September 8–11, **1**, 295–301.

[167] Mouzouris, G. C. and Mendel, J. M. (1997). Non-Singleton Fuzzy Logic Systems: Theory and Application, *IEEE Transactions on Fuzzy Systems*, **5** (1), 56–71.

[168] Nakamura, M. (1994). What fuzzy interval operations should be hardware supported?, In: Hall, L., Ying, H., Langari, R., and Yen, J. (eds.), *NAFIPS/IFIS/NASA'94, Proceedings of the First International Joint Conference of The North American Fuzzy Information Processing Society Biannual Conference, The Industrial Fuzzy Control and Intelligent Systems Conference, and The NASA Joint Technology Workshop on Neural Networks and Fuzzy Logic, San Antonio, December 18–21, 1994*, IEEE, Piscataway, NJ, 393–397.

[169] Nguyen, H. T. and Kreinovich, V. (1995). Towards theoretical foundations of soft computing applications (1995). *International Journal on Uncertainty, Fuzziness, and Knowledge-Based Systems*, **3** (3), 341–373.

[170] Nguyen, H. T., Kreinovich, V., Nesterov, V., and Nakamura, M. (1997a). On hardware support for interval computations and for soft computing: a theorem, *IEEE Transactions on Fuzzy Systems*, **5** (1), 108–127.

[171] Wu, K. C. (1994). A robot must be better than a human driver: an application of fuzzy intervals. In: Hall, L., Ying, H., Langari, R., and Yen, J. (eds.), *NAFIPS/IFIS/NASA'94, Proceedings of the First International Joint Conference of The North American Fuzzy Information Processing Society Biannual Conference, The Industrial Fuzzy Control and Intelligent Systems Conference, and The NASA Joint Technology Workshop on Neural Networks and Fuzzy Logic, San Antonio, December 18–21, 1994*, IEEE, Piscataway, NJ, 171–174.

[172] Wu, K. C. (1996). Fuzzy interval control of mobile robots, *Computers and Electrical Engineering*, **22** (3), 211–229.

Approximation with fewer rules

[173] Barron, A. R. (1993). Universal approximation bounds for superpositions of a sigmoidal function, *IEEE Transactions on Information Theory*, **39**, 930–945.

[174] Bauer, P., Klement, E. P., Moser, B., and Leikermoser, A. (1995). Modeling of fuzzy functions by fuzzy controllers, In: Nguyen, H. T., Sugeno, M., Tong, R., and Yager, R. (eds.), *Theoretical aspects of fuzzy control*, J. Wiley, N.Y.

[175] DeVore, R., Howard, R., and Micchelli, C. A. (1989). Optimal nonlinear approximation, *Manuscripta Mathematica*, **63**, 469–478.

[176] Dickerson, J. A., and Kosko, B. (1996). Fuzzy function approximation with ellipsoidal rules, *IEEE Trans. Syst., Man, Cybern.*, **26** (4), 542–560.

[177] Foo, N. (1994). New perspectives and old problems ..., *Proceedings of the Soft Computing Symposium*, University of New South Wales.

[178] Jones, L. K. (1992). A simple lemma on greedy approximation in Hilbert space and convergence rates for projection pursuit regression and neural network training, *The Annals of Statistics*, **20**, 601–613.

[179] Kóczy, L. T. and Tikk, D. (1996). Approximation in rule bases, *IPMU'96: Proceedings of the International Conference on Information Processing and Management of Uncertainty in Knowledge-Based Systems, Granada, July 1–5, 1996*, 489–494.

[180] Kosko, B. (1994). Fuzzy Systems as Universal Approximators, *IEEE Trans. on Computers,* **43** (11), 1329–1333.

[181] Kosko, B. (1994a). Optimal fuzzy rules cover extrema, *Proceedings of the World Congress on Neural Networks WCNN'94*.

[182] Kosko, B. (1995). Optimal fuzzy rules cover extrema, *International Journal of Intelligent Systems,* **10** (2), 249–255.

[183] Kosko, B. (1996). Asymptotic stability of feedback additive fuzzy systems, *Proceedings of the World Congress on Neural Networks WCNN'96*, September 1996.

[184] Kosko, B. (1996a). Additive fuzzy systems: from function approximation to learning, In: Chen, C. H. (ed.), *Fuzzy Logic and Neural Network Handbook*, McGraw-Hill, N.Y., 9-1–9-22.

[185] Kosko, B. and Dickerson, J. A. (1995a). Function approximation with additive fuzzy systems, Chapter 12 in: Nguyen, H. T., Sugeno, M., Tong, R., and Yager, R. (eds.), *Theoretical aspects of fuzzy control*, J. Wiley, N.Y., 313–347.

[186] Kreinovich, V., Sirisaengtaksin, O., and Cabrera, S. (1994c). Wavelet neural networks are optimal approximators for functions of one variable, *Proceedings of the IEEE International Conference on Neural Networks*, Orlando, FL, July 1994, **1**, 299-303.

[187] Kůrková, V., Kainen, P. C., and Kreinovich, V. (1995). Dimension-independent rates of approximation by neural networks and variation with respect to half-spaces, *Proceedings of World Congress on Neural Networks, WCNN'95, Washington, DC, July 1995*, INNS Press, NJ, **I**, 54–57.

[188] Kůrková, V., Kainen, P. C., and Kreinovich, V. (1997). Estimates of the Number of Hidden Units and Variation with Respect to Half-spaces, *Neural Networks*, **10** (6), 1061–1068.

[189] Lee, J. and Chae, S. (1993). Completeness of fuzzy controller carrying a mapping $f : R \to R$, *Proceedings of the IEEE International Conference on Fuzzy Systems FUZZ-IEEE'93*, **1**, 231–235.
[190] Lee, J. and Chae, S. (1993a). Analysis of function duplicating capabilities of fuzzy controllers, *Fuzzy Sets and Systems*, **56**, 127–143.
[191] Lorentz, G. G. (1966). *Approximation of Functions,* Holt, Reinhart and Winston, New York, 1966.
[192] Watkins, F. A. (1994). *Fuzzy Engineering*, Ph.D. Dissertation, Department of Electrical and Computer Engineering, University of California at Irvine.
[193] Watkins, F. A. (1995). The representation problem for additive fuzzy systems, *Proceedings of the IEEE International Conference on Fuzzy Systems FUZZ-IEEE/IFES'95*, March 1995, **1**, 117–122.
[194] Ying, H. (1994). Sufficient conditions on general fuzzy systems as function approximators, *Automatica*, **30** (3), 521–525.
[195] Zeng, X.-J. and Singh, M. G. (1994). Approximation theory of fuzzy systems - SISO case, *IEEE Trans. on Fuzzy Systems,* **2** (2), 162–176.
[196] Zeng, X. J. and Singh, M. G. (1995a). Approximation accuracy analysis of fuzzy systems with the center–average defuzzifier, *Proceedings Fourth IEEE Int. Conf. on Fuzzy Systems,* 109–116, Yokohama, March 1995.
[197] Zeng, X.-J. and Singh, M. G. (1996). Approximation accuracy analysis of fuzzy systems as function approximators, *IEEE Trans. on Fuzzy Systems,* **4**, 44–63.

How to choose the best variant of fuzzy ruled based modeling methodology

[198] Bouchon-Meunier, B., Kreinovich, V., Lokshin, A., and Nguyen, H. T. (1996). On the formulation of optimization under elastic constraints (with control in mind). *Fuzzy Sets and Systems,* **81** (1), 5–29.
[199] Kosheleva, O. (1996). Why Sinc? Signal Processing Helps Fuzzy Control, *SC-COSMIC, South Central Computational Sciences in Minority Institutions Consortium, Second Student Conference in Computational Sciences,* October 25–27, El Paso, TX, Abstracts, 19–21.
[200] Kreinovich, V. (1997). Random sets unify, explain, and aid known uncertainty methods in expert systems, in Goutsias, J., Mahler, R. P. S., and Nguyen, H. T. (eds.), *Random Sets: Theory and Applications*, Springer-Verlag, N.Y., 321–345.
[201] Kreinovich, V. (1996c). Maximum entropy and interval computations, *Reliable Computing,* **2** (1), 63–79.

[202] Kreinovich, V., Nguyen, H. T., and Walker, E. A. (1996b). Maximum entropy (MaxEnt) method in expert systems and intelligent control: new possibilities and limitations, In: Hanson K. M. and Silver, R. N. (eds.), *Maximum Entropy and Bayesian Methods*, Kluwer Academic Publishers, Dordrecht, 93–100.

[203] Kreinovich, V., Quintana, C., and Lea, R. (1991c). What procedure to choose while designing a fuzzy control? Towards mathematical foundations of fuzzy control, *Working Notes of the 1st International Workshop on Industrial Applications of Fuzzy Control and Intelligent Systems*, College Station, TX, 123–130.

[204] Kreinovich, V., Quintana, C., Lea, R., Fuentes, O., Lokshin, A., Kumar, S., Boricheva, I., and Reznik, L. (1992). What non-linearity to choose? Mathematical foundations of fuzzy control, *Proceedings of the 1992 International Conference on Fuzzy Systems and Intelligent Control*, Louisville, KY, 349–412.

[205] Kreinovich, V. and Tolbert, D. (1994d). Minimizing computational complexity as a criterion for choosing fuzzy rules and neural activation functions in intelligent control. In: Jamshidi, M., Nguyen, C., Lumia, R., and Yuh, J. (eds.), *Intelligent Automation and Soft Computing. Trends in Research, Development, and Applications. Proceedings of the First World Automation Congress (WAC'94). August 14–17, 1994, Maui, Hawaii*, TSI Press, Albuquerque, NM, **1**, 545–550.

[206] Mitaim, S. and Kosko, B. (1996). What is the best shape for a fuzzy set in function approximation?, *Proceedings of the Fifth IEEE International Conference on Fuzzy Systems FUZZ-IEEE'96*, New Orleans, September 8–11, **2**, 1237–1243.

[207] Mitaim, S. and Kosko, B. (1996a). Fuzzy function approximation and the shape of fuzzy sets, *Proceedings of the World Congress on Neural Networks WCNN'96*, September 1996.

[208] Nguyen, H. T. and Kreinovich, V. (1995). Towards theoretical foundations of soft computing applications (1995). *International Journal on Uncertainty, Fuzziness, and Knowledge-Based Systems*, **3** (3), 341–373.

[209] Nguyen, H. T., Kreinovich, V., Lea, B., and Tolbert, D. (1992). How to control if even experts are not sure: robust fuzzy control, *Proceedings of the Second International Workshop on Industrial Applications of Fuzzy Control and Intelligent Systems*, College Station, December 2–4, 153–162.

[210] Nguyen, H. T., Kreinovich, V., Lea, B., and Tolbert, D. (1995a). Interpolation that leads to the narrowest intervals, and its application to expert systems and intelligent control, *Reliable Computing*, 1995, **3** (1), 299–316.

[211] Nguyen, H. T., Kreinovich, V., and Tolbert, D. (1993a). On robustness of fuzzy logics. *Proceedings of IEEE-FUZZ International Conference*, San Francisco, CA, March 1993, **1**, 543–547.

[212] Nguyen, H. T., Kreinovich, V., and Tolbert, D. (1994). A measure of average sensitivity for fuzzy logics, *International Journal on Uncertainty, Fuzziness, and Knowledge-Based Systems*, **2** (4), 361–375.

[213] Nguyen, H. T. and Walker, E. A. (1997b). *A first course in fuzzy logic*, CRC Press, Boca Raton, FL.

[214] Ramer, A. and Kreinovich, V. (1992). Maximum entropy approach to fuzzy control, *Proceedings of the Second International Workshop on Industrial Applications of Fuzzy Control and Intelligent Systems*, College Station, December 2–4, 113–117.

[215] Ramer, A. and Kreinovich, V. (1994). Information complexity and fuzzy control, Chapter 4 in: Kandel, A. and Langholtz, G. (eds.), *Fuzzy Control Systems*, CRC Press, Boca Raton, FL, 75–97.

[216] Ramer, A. and Kreinovich, V. (1994a). Maximum entropy approach to fuzzy control, *Information Sciences*, **81** (3–4), 235–260.

[217] Smith, M. H. and Kreinovich, V. (1995). Optimal strategy of switching reasoning methods in fuzzy control, Chapter 6 in Nguyen, H. T., Sugeno, M., Tong, R., and Yager, R. (eds.), *Theoretical aspects of fuzzy control*, J. Wiley, N.Y., 117–146.

[218] Zeng, X.-J. and Singh, M. G. (1996). Approximation accuracy analysis of fuzzy systems as function approximators, *IEEE Trans. on Fuzzy Systems,* **4**, 44–63.

Comparison between fuzzy rules based modeling methodology and neural network modeling methodology

[219] Kreinovich, V. and Bernat, A. (1994). Parallel algorithms for interval computations: an introduction, *Interval Computations*, No. 3, 6–62.

[220] Lea, R. N. and Kreinovich, V. (1995). Intelligent Control Makes Sense Even Without Expert Knowledge: an Explanation, *Reliable Computing*, 1995, Supplement (Extended Abstracts of APIC'95: International Workshop on Applications of Interval Computations, El Paso, TX, Febr. 23–25, 1995), 140–145.

5 FUZZY AND LINEAR CONTROLLERS

Laurent Foulloy and Sylvie Galichet

Laboratoire d'Automatique et de MicroInformatique Industrielle
LAMII / CESALP, Université de Savoie, BP 806
74016 Annecy Cedex, France

5.1 INTRODUCTION

After the pioneering work of Mamdani [20], [21], [22], the first attempt to provide a design methodology was proposed by MacVicar-Whelan [19] whose antidiagonal table is still used in many applications. The relations between a fuzzy controller using a MacVicar Whelan table and the equivalent linear PI controller have been proposed in [39]. However, ten years after these first results, the study of linear-like or PID-like fuzzy controllers is still active (among many others, see [23], [29], [32], [33] for recent references) and one may wonder whether there is a real interest in implementing linear control with a fuzzy controller. In fact, there are several possible answers corresponding to several points of view.

It is well-known that PID controllers are widely used in industrial plants. However, in many cases, far away from the predefined setpoint, the linear controller no longer works and a human operator substitutes for it. One of the strengths of the fuzzy control approach is to allow to integrate this human expertise into the controller. In such a case, the developer is faced with two problems: the global satisfactory behavior of the linear controller on the one hand and the human expertise in particular situations where the linear controller fails on the other hand. In order to use both types of information, that is the mathematical equation for the linear controller and the linguistic formalism for the human operator, they have to be merged in the same representation. So, the design of a fuzzy controller equivalent to a given linear one provides a rule-based description of the linear control law. The linguistic rules expressed by the human controller can then be easily merged into the rule base. Finally, a fuzzy rule-based representation of the global control strategy is obtained, integrating at the same time the linear behavior and the ability of the human operator. In other words, the synthesis of a fuzzy linear controller may be viewed as an efficient way to convert a mathematical relationship into a knowledge-based model. This approach is well suited for the purpose of information aggregation. A similar methodology has already been proposed for the translation of numerical data ([15], [16], [41]).

It is often argued that a fuzzy controller provides stability of the system under control and ensures robustness with respect to system parameter changes. However, in establishing these properties, especially stability, for industrial applications or for the inverted pendulum system ([3], [14]), available theoretical studies have all been based on the choice of a model of the controlled process and on the assumption of some linear properties of a fuzzy controller. ([2], [4], [7], [18]) show that fuzzy controllers with triangular or more generally unimodal, regularly distributed membership functions with an overlap of 1, have a limit control law which is linear when additive rules are used. By approximating such a fuzzy controller by its limit control law, a stability analysis of the controlled system may be achieved with Lyapunov's functions [3]. [37] shows that the stability of a Sugeno type fuzzy system with polynomial conclusions may be studied by associating a linear subsystem to each rule. Hence, if the fuzzy controllers are said to be qualified to control nonlinear processes, their capacity in terms of stability or robustness is paradoxically analysed with respect to some of their linear properties.

There have been a number of recent papers in the fuzzy literature showing that fuzzy systems are universal approximators. This universal approximation property has been proved in [40] and [15] for two different types of fuzzy systems. In [28], it is shown that many classes of fuzzy systems have the same property. By applying these universality results to real-life control problems, it can be deduced that fuzzy controllers are universal ([5], [6], [17]). In other words, given any controllable process, it can also be controlled by a fuzzy controller. However, although the existence theorem of such a fuzzy controller is an important theoretical result, there is currently no way to analytically determine the existing optimal fuzzy controller. One possible understanding of fuzzy control consists of automatically building a fuzzy controller providing some desired properties. The fuzzy rules are thus expressed by using general parametric functions ([31], [2]) which have to be determined in order for the fuzzy controller to achieve the desired control law or to fit some learning points as best as possible ([30], [35], [36]). Using this type of formulation, assumptions about the fuzzy rules, their number or their type, are avoided.

For all these reasons it seems important to dissociate the linear properties of a fuzzy controller clearly from its nonlinear global behavior. In establishing a methodology for designing fuzzy linear controllers, we will be able to isolate sufficient properties for a fuzzy controller to be linear. An analysis of the nonlinear dynamics of one-dimensional fuzzy controllers is achieved in [38], where the fuzzy control law is decomposed into a linear control law and a superimposed nonlinearity. [45] studies the smallest possible Mamdani type fuzzy controller consisting of two inputs, one output and four fuzzy control rules. The use of a mixed logic (Zadeh AND and Lukasiewicz OR) and a linear defuzzification process is advocated in order for the fuzzy controller to be precisely equivalent to a PI controller. [34] attempts to linearize the same type of fuzzy controller, but in this case described by nine basic rules, expressed by five compacted rules. For the fuzzy controller to achieve exact linearity, different conjunction and disjunction operators have to be used depending on which rule is fired. [42] shows that a fuzzy controller with an arbitrary number of additive rules can be precisely equivalent to a linear controller. A simplified fuzzy reasoning method and mixed fuzzy logics are both considered. [24] determines how to set up the distribution of the membership functions for a nine-rule fuzzy controller in order to fit a PI controller with known gains. By using the *min* conjunction operator, a nonlinearity appears in the control law. This approach of understanding the behavior of a given fuzzy controller can also show that a fuzzy controller based on the product-sum-gravity method is precisely equivalent to a linear controller when a minimal, correctly adjusted rule base is used ([25], [26], [27]).

The aim of this chapter is to show that most of the known results concerning linear-like fuzzy controllers can easily be established using the modal equivalence principle which consists in equating the output of a linear (or a non-linear) controller and a fuzzy one at some particular inputs, called modal values ([11], [12]). As results, the modal equivalence principle provides the structure of the rule base and the tuning of the fuzzy controller.

Section 5.2 introduces the modal equivalence principle for the simplest linear controller, that is the proportional controller, for Mamdani's fuzzy controllers.

Section 5.3 presents its application to PI controllers. It gives the additive structure of the resulting rule base, the tuning and control laws. It is also shown how to choose specific inference operators to obtain exact equivalences with linear controllers.

In section 5.4, the modal equivalence principle is used to synthesize state feedback fuzzy controllers.

Section 5.5 extends the results to Sugeno's fuzzy controllers.

5.2 MODAL EQUIVALENCE PRINCIPLE

The modal equivalence principle consists in computing the output of the fuzzy controller for input modal values, that is values for which membership functions are equal to 1. Then, by equating the obtained result with the output of the classical controller, it is possible to automatically determine the rule base and the tuning of the fuzzy controller.

Let us start with the simplest linear controller, that is the proportional controller, as an introductive example. Let ε be the error between a reference signal and the plant output, and u be the plant input, which is therefore the controller output signal. A proportional controller is represented by

$$u = K\,\varepsilon \tag{5.1}$$

To achieve an equivalent behavior with a fuzzy controller, one may consider using a set of rules whose antecedent part deals with the error and whose conclusion part is concerned with the control leading to

$$\textbf{If}\ \text{error is}\ E_i\ \textbf{then}\ \text{control is}\ U_{f(i)}\,, \tag{5.2}$$

where E_i represents the ith symbol attached to the error. The conclusion is described by the parametric function f such that $f(i)$ represents the index of the conclusion symbol. The function f has therefore to be integer valued.

We assume that the membership functions associated with the input symbols, that is their fuzzy meanings [46], are triangular, regularly distributed with an overlap of 1 (see Figure 5.1). These assumptions induce a strict fuzzy partition of the universes of discourse [1]. Any value such that the membership function of the meaning of a symbol is equal to 1 is called a modal value. Δa represents the distance between two consecutive modal values of the error, so

$$\varepsilon_i = i\,\Delta a + \varepsilon_0\,, \tag{5.3}$$

where ε_i represents the ith modal value of the error .

For the sake of simplicity, it is assumed that the meaning of the symbols E_0 has a membership function centered on the zero of the universe of discourse. In that case, expression (5.3) becomes

$$\varepsilon_i = i\,\Delta a\,. \tag{5.4}$$

It should be noted that the restriction $\varepsilon_0 = 0$ is not necessary. It is only imposed here in order to reduce the volume of further calculations.

We have also to define the membership functions of the meaning of the output symbols U_j. As for the input symbols, unimodal, regularly distributed membership functions are assumed. No particular shape is required, but the membership functions are assumed to be symmetrical around the modal values. By denoting Δb the distance between two consecutive modal values of the control, the jth modal value, u_j, may be expressed as

$$u_j = j\Delta b \,. \tag{5.5}$$

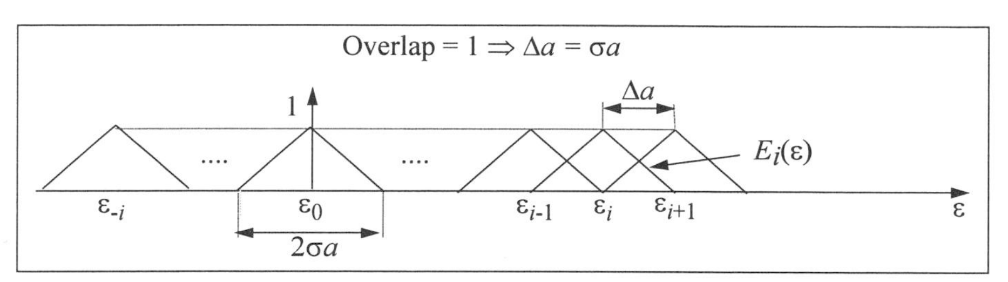

Figure 5.1 Membership functions of the meaning of the input symbols.

For any input value ε, the output of a Mamdani type fuzzy controller ([20], [21], [22], [10]), u_{MAM}, is the defuzzified value of the inferred fuzzy set, F, defined by its membership function

$$F(u) = \max_{i \in \mathbf{Z}} \; \min \left(E_i(\varepsilon), U_{f(i)}(u) \right) . \tag{5.6}$$

It can be noted here that the Mamdani type fuzzy controllers' family is not restricted to the usual "min-max" inference method but includes every inference method based on a conjunctive understanding of the rules, that is every "T -⊥" inference method as proposed in [27], where the t-norm T and the t-conorm ⊥ represent the conjunction and disjunction operators involved in the inference.

According to the properties of t-norms and t-conorms and using the properties of the membership functions for modal values we have

(i) min $(x, 0) = 0$ and min $(x, 1) = x$,
(ii) max $(x, 0) = x$,
(iii) $E_i(\varepsilon_m) = 0$ if $i \neq m$ and $E_i(\varepsilon_m) = 1$ if $i = m$.

Therefore, for the input value $\varepsilon = \varepsilon_m$, equation (5.6) is reduced to

$$F(u) = U_{f(m)}(\delta u) \,. \tag{5.7}$$

Since output membership functions are assumed to be symmetrical , we have the symmetry of F around the modal value $u_{f(m)}$. By using a defuzzification method which preserves this property, the ouput of the Mamdani controller is

$$u_{\mathrm{MAM}} = u_{f(m)} \,. \tag{5.8}$$

According to equation (5.5), the output of the Mamdani controller is

$$u_{\mathrm{MAM}} = f(m)\ \Delta b\ . \tag{5.9}$$

5.2.1 Using the Modal Equivalence Principle

For the same modal input $\varepsilon = \varepsilon_m$, the Proportional controller provides the output

$$u_{\mathrm{P}} = K\,\varepsilon_m = K\,m\,\Delta a\ . \tag{5.10}$$

By equating the values u_{P} and u_{MAM} derived in (5.9) and (5.10), the following relationship is obtained

$$f(m) = m\ K\left(\frac{\Delta a}{\Delta b}\right)\ . \tag{5.11}$$

Since $f(m)$ represents the index of the conclusion symbol of the rule, the parametric function f has to give an integer value. m is the index of a premise symbol and is therefore an integer. Thus, the quantity $K\left(\frac{\Delta a}{\Delta b}\right)$ must also be an integer, that is $K\left(\frac{\Delta a}{\Delta b}\right) = \alpha$, where α is any integer. In order to avoid unused symbols in the conclusion part, it is clear that the best choice is $\alpha = 1$ leading to

$$f(m) = m, \tag{5.12}$$

and the distribution of the membership functions must satisfy the following relationship which provides an equivalent gain

$$K = \frac{\Delta b}{\Delta \mathrm{a}}\ . \tag{5.13}$$

Figure 5.2 shows the output of fuzzy controllers equivalent to a proportional controller whose gain is $K = 1$. The smaller Δb, the closer to the linear controller. We obtain here a classical result stating that the output of a linear fuzzy control becomes a linear function as the number of rules grows [7].

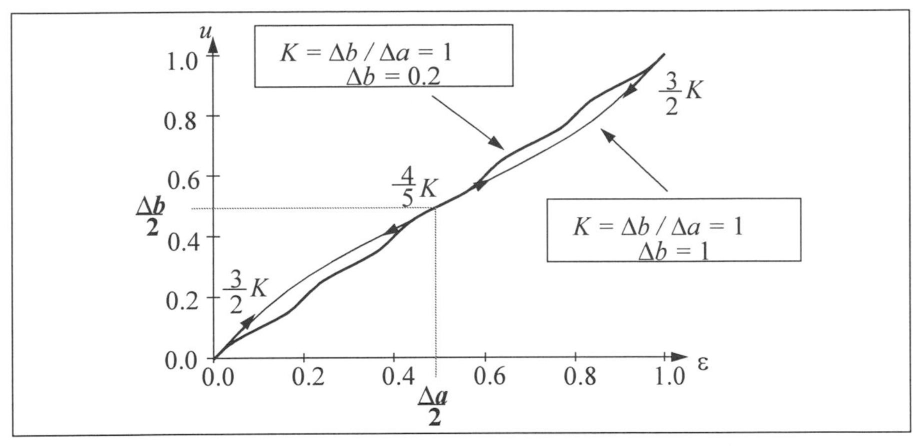

Figure 5.2 Output of the fuzzy proportional controllers.

5.3 APPLICATION TO PI CONTROLLERS

The z-transform of a Proportional-Integral controller is given by

$$C(z) = \frac{u(z)}{\varepsilon(z)} = K_P + K_I\left(\frac{z}{z-1}\right) . \tag{5.14}$$

Let ε be the error between a pre-defined reference signal and the plant output, and $\delta\varepsilon$ (respectively δu) be the change in error (respectively the change in control output signal). In time domain, equation (5.14) can be rewritten as

$$\delta u = K_P \ \delta\varepsilon + K_I \ \varepsilon . \tag{5.15}$$

5.3.1 Rule base

The same inputs and output will be used for the fuzzy controller, that is ε and $\delta\varepsilon$ as inputs and δu as output. In a Mamdani type fuzzy controller, the rules are expressed in a symbolic form. The choice of the input and output variables leads to a definition of the rules in the form

If error is E_i **and** change in error is dE_j **then** change in control is $dU_{f(i,j)}$, (5.16)

where E_i and dE_j represent the ith and j^{th} symbols respectively attached to the error and to the change in error. Each rule is characterized by the pair (i, j) and no assumption is made on the number of symbols. The indices i and j are thus considered as belonging to Z. For a Mamdani type controller, $f(i,j)$ represents the index of the conclusion symbol.

5.3.2 Membership Functions

As for the Proportional controller, we assume that the membership functions of the meaning of the input symbols are triangular, regularly distributed with an overlap of 1 (see Figure 5.1). Δa represents the distance between two consecutive modal values of the error. Similarly, Δb is the distance between two consecutive modal values of the change in error and Δd the distance between two consecutive modal values of the change in control. Once again, it is assumed that the membership functions associated with the symbols E_0, dE_0 and dU_0 are centered on the zero of the universe of discourse leading to

$$\varepsilon_i = i\,\Delta a\,, \tag{5.17}$$

$$\delta\varepsilon_j = j\,\Delta b\,, \tag{5.18}$$

$$\delta u_k = k\,\Delta d\,. \tag{5.19}$$

5.3.3 Outputs for Modal Values

The output of the PI controller, δu_{PI}, for inputs $\varepsilon = \varepsilon_m$ and $\delta\varepsilon = \delta\varepsilon_n$ is obtained by applying equation (5.15) that is

$$\delta u_{\mathrm{PI}} = K_I\,\varepsilon_m + K_p\,\delta\varepsilon_n\,. \tag{5.20}$$

By using the relationships (5.17) and (5.18), δu_{PI} can be reformulated as

$$\delta u_{\mathrm{PI}} = K_I\,m\,\Delta a + K_p\,n\,\Delta b\,. \tag{5.21}$$

For any inputs ε and $\delta\varepsilon$, the output of a Mamdani type fuzzy controller is the defuzzified value of the inferred fuzzy set F given by

$$F(\delta u) = \max_{(i,\,j)\,\in \mathbf{Z}^2}\left(\min\left(E_i(\varepsilon),\, dE_j(\delta\varepsilon),\, dU_{f(i,j)}(\delta u)\right)\right). \tag{5.22}$$

For the same reasons as for the proportional case, for the modal inputs $\varepsilon = \varepsilon_m$ and $\delta\varepsilon = \delta\varepsilon_n$ we have

$$F(\delta u) = dU_{f(m,n)}(\delta u)\,, \tag{5.23}$$

and after defuzzification

$$\delta u_{\mathrm{MAM}} = \delta u_{f(m,n)}\,. \tag{5.24}$$

Finally, from equation (5.19), the output of the Mamdani controller can be reformulated as

$$\delta u_{\mathrm{MAM}} = f(m,\, n)\, \Delta d\,. \tag{5.25}$$

5.3.4 Using the Modal Equivalence Principle

By equating the values δu_{PI} and δu_{MAM} derived in (5.21) and (5.25), the following relationship is obtained

$$K_I\, m\, \Delta a + K_p\, n\, \Delta b = f(m,\, n)\, \Delta d\,, \tag{5.26}$$

$$f(m,\, n) = m\;\; K_I\!\left(\frac{\Delta a}{\Delta d}\right) \;+ n\;\; K_P\!\left(\frac{\Delta b}{\Delta d}\right)\;. \tag{5.27}$$

Since $f(m, n)$ represents the index of the conclusion symbol of the rule, the parametric function f has to give an integer value. Recall that m and n are also integers (indices of the premise symbols). Therefore, the quantities $K_I\left(\frac{\Delta a}{\Delta d}\right)$ and $K_P\left(\frac{\Delta b}{\Delta d}\right)$ must also be integers. In order to use all the possible indices in the numbering of the output symbols, we have to choose $K_I\,\frac{\Delta a}{\Delta d} = \alpha$ and $K_P\,\frac{\Delta b}{\Delta d} = \beta$, where α and β are any integers such that their highest common factor is 1. The expression of the parametric function is finally

$$f(m,\, n) = \alpha\, m + \beta\, n\,, \tag{5.28}$$

and the distribution of the membership functions must satisfy the relationships

$$\frac{\Delta d}{\Delta a} = \frac{K_I}{\alpha} \text{ and } \frac{\Delta d}{\Delta b} = \frac{K_P}{\beta}\,. \tag{5.29}$$

To summarize, for a Mamdani type fuzzy controller to be equivalent to a PI controller at the modal values, the fuzzy rules have to be additive, that is

$$\textbf{If } \text{error is } E_i \textbf{ and } \text{change in error is } dE_j \textbf{ then } \text{change in control is } dU_{\alpha i+\beta j}, \quad (5.30)$$

and the parameters of the fuzzy controller must be tuned in order to verify the relationships (5.29). It can be noted that three parameters (Δa, Δb and Δd) have to be adjusted although only two equations are available. The remaining degree of freedom allows a definition of the number of modal values where the fuzzy control law has to fit the linear law.

5.3.5 *Example with* $\alpha = \beta = 1$

Rule base. The study of this specific case clearly establishes the link between the obtained results and the pre-defined rule base usually assumed for Mamdani's controllers. By regrouping the rules in a table, the additive property of the rules is translated as shown in Figure 5.3.

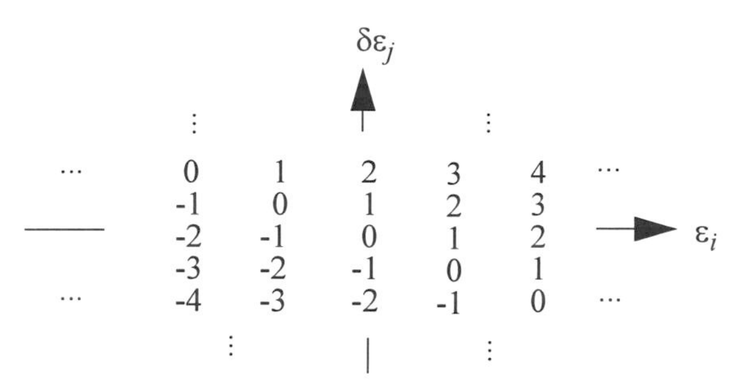

Figure 5.3 Generated rule base for $\alpha = \beta = 1$.

A better understanding is obtained when dealing with linguistic labels instead of the indices. For example, assume that $i \in [-2, 2]$ and $j \in [-2, 2]$. According to (5.28) we have $k = i + j$ and therefore $k \in [-4, 4]$. Let us rename the linguistic terms E_i with more classical labelling, that is

$$E_{-2} = \text{NM},\ E_{-1} = \text{NS},\ E_0 = \text{Z},\ E_1 = \text{PS},\ E_2 = \text{PM}, \quad (5.31)$$

where N, Z, P, S, M stand for Negative, Zero, Positive, Small, Medium, which means that PS is Positive Small.

Let us define the same substitutions for the terms dE_j. New terms have to be introduced for dU_k

$$dU_{-4} = \text{NVB},\ dU_{-3} = \text{NB},\ dU_3 = \text{PB},\ dU_4 = \text{PVB}, \tag{5.32}$$

where V and B stand for Very and Big.
Now, the rule base can be rewritten as shown in Figure 5.4 and one can recognize the classical antidiagonal rule base proposed by MacVicar-Whelan [19]. This assumption has also been used by many authors (see for example [7], [24], [34], [45]). It can also be shown that qualitative control by analysis of the trends in the phase plane leads to the same antidiagonal rule base [8].

		ε				
		NM	NS	Z	PS	PM
δε	PM	Z	PS	PM	PB	PVB
	PS	NS	Z	PS	PM	PB
	Z	NM	NS	Z	PS	PM
	NS	NB	NM	NS	Z	PS
	MM	NVB	NB	NM	NS	Z

Figure 5.4 Generated rule base after re-labelling.

Control laws. Although the use of additive rules is necessary for a Mamdani type fuzzy controller to be equivalent to a PI controller, it is obviously not sufficient to guarantee linear interpolation between modal values. As noted in [34], [45], the usual choice of the inference operators (*min*, *max* t-norm and t-conorm) and of the defuzzification strategy (center of gravity method) leads to nonlinearities in the control law.

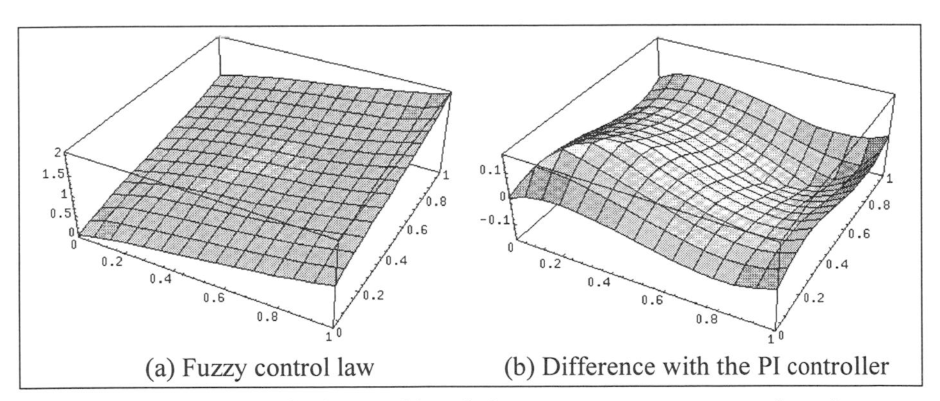

(a) Fuzzy control law (b) Difference with the PI controller

Figure 5.5 Mamdani control law (min-max operators, center of gravity defuzzification).

Figure 5.5a represents the generated fuzzy control law inside the cell defined by the four modal values ε_i, ε_{i+1}, $\delta\varepsilon_j$ and $\delta\varepsilon_{j+1}$ while Figure 5.5b shows the difference between the control law generated by the fuzzy controller and the law generated by the PI controller, which is a plane in this representation. As expected, the difference with the PI controller is zero for the four modal values. It can also be very easily shown that the difference is also zero for input values in the middle of two modal values. For these inputs, the inferred fuzzy subset is a trapezium whose height is 0.5 thus providing, after defuzzification by the center of gravity method, a control action whose value is in the middle of the output modal values. Therefore, inside one cell defined by four modal values, nine points are exactly on the plane defined by the PI controller as shown in Figure 5.6.

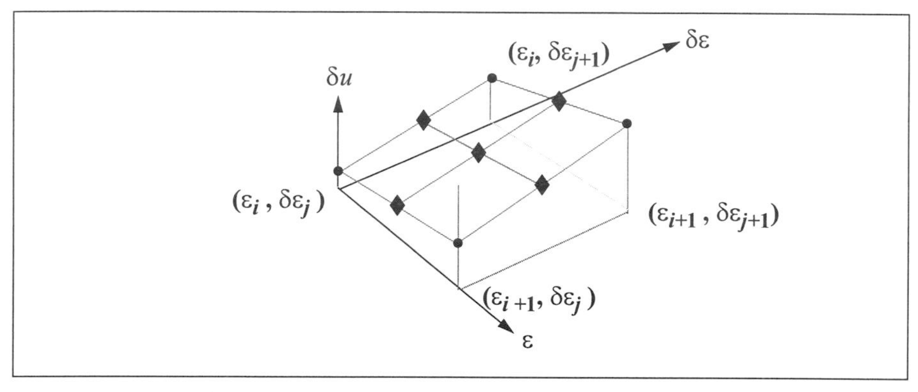

Figure 5.6 Points where the difference with the PI controller is equal to zero.

Tuning. As shown in section 5.3.4, the modal equivalence principle provides the additive structure of the rule base but also the tuning of the fuzzy logic controller with respect to the linear equivalent controller. According to (5.29) we get

$$K_I = \frac{\Delta d}{\Delta a} \text{ and } K_P = \frac{\Delta d}{\Delta b}. \tag{5.33}$$

It means that all the methods developped for tuning PI controllers can be adapted to tune fuzzy PI-like controllers as long as the relationships between Δa, Δb, Δd and the gains K_P and K_I are kept. For given gains K_P and K_I, the smaller Δd the greater the number of points equivalent to the linear controller and therefore the closer to this controller. The relationships (5.29) between the gains of the PI controller and the distances between two consecutive modal values of the error Δa, the change in error Δb, and the change in control Δd, are similar to those expressed in [39] in terms of slope and distance in the phase diagram.

5.3.6 Exact Equivalence

In this section, it is shown how some inference operators and the defuzzification method can be found to provide exact equivalence with PI controllers [13].

The first approach involves selecting some inference operators [43] and seeking a defuzzification strategy providing a linear interpolation. In [44], Yager introduces a parametrized family of defuzzification operators and suggests an algorithm for learning the parameter from a set of desired defuzzified values. In this case, it is possible to use a training set of data (outputs of the PI controller).

The second approach requires choosing a defuzzification strategy and searching for inference operators providing a linear interpolation with the selected defuzzification method. As usual with Mamdani's controllers, the center of gravity defuzzification is assumed.

Theorem [13]: the Mamdani type fuzzy controller defined by

$$F(\delta u) = \perp_{(i,j)\in \mathbf{Z}^2} \left(\mathsf{T} \left(E_i(\varepsilon), dE_j(\delta\varepsilon), dU_{f(i,j)}(\delta u) \right) \right), \tag{5.34}$$

where T is a t-norm and $\perp$ a t-conorm, is precisely equivalent to a PI controller if

(i) input membership functions are triangular, regularly distributed, with an overlap of 1,
(ii) output membership functions are unimodal, symmetrical and regularly distributed,
(iii) fuzzy rules are additive,
(iv) the defuzzification strategy is the center of gravity method,
(v) inference operators are $\mathsf{T}(x, y) = x.y$ and $\perp(x, y) = \min(x+y, 1)$.

This theorem gives sufficient conditions for precise equivalence to be satisfied. In particular, the choice of the inference operators is a main point for the purpose of linear interpolation between modal values. When the partition of the input universes of discourse satisfies the requirement (i), the bounded sum $\perp(x, y) = \min(x+y, 1)$ used in the inference is equivalent to a classical sum and therefore the Mamdani inference is simply reduced to the «Product-Sum» inference proposed by Mizumoto ([25], [27]).

5.4 APPLICATION TO STATE FEEDBACK FUZZY CONTROLLERS

5.4.1 State Feedback Control

The motion of any finite dynamic system can be expressed as a set of first-order ordinary differential equations. This is referred to as the state-variable representation.

When the differential equations are linear, they can be concisely expressed using matrix notation. For a multi-input, multi-output linear system, the state-variable representation is thus given by

$$\begin{aligned} \dot{x} &= Ax + Bu \\ y &= Cx + Du \end{aligned} \quad \text{with} \quad x = \begin{bmatrix} x_1 \\ x_2 \\ \dots \\ x_n \end{bmatrix} \quad u = \begin{bmatrix} u_1 \\ u_2 \\ \dots \\ u_m \end{bmatrix} \quad \text{and} \quad y = \begin{bmatrix} y_1 \\ y_2 \\ \dots \\ y_p \end{bmatrix}, \tag{5.35}$$

where u is the system input, y the system output and x the system state.

The principle of state feedback compensation consists in defining a control law in the form (see Figure 5.7)

$$u = Lr - Kx\,, \tag{5.36}$$

where r is a m-vector corresponding to a new system input, L a square mxm matrix which sets up the static behavior of the compensated system, and K a mxn matrix introduced in the state-vector feedback path that realizes the simple feedback of a linear combination of all the states.

Let us assume that the matrix K has been determined by some classical method. Our purpose is now to design a fuzzy controller equivalent to the classical one. The aim of the next paragraphs is to show how to automatically build a fuzzy controller that approximates the feedback path, that is $v = Kx$.

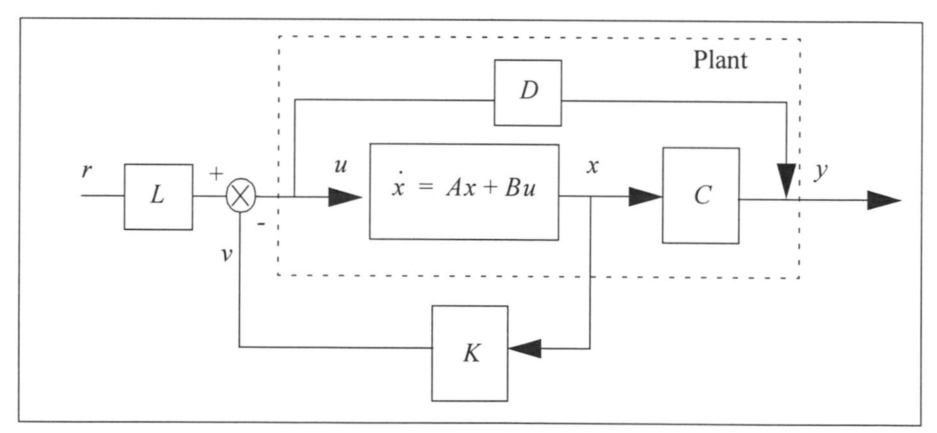

Figure 5.7 State feedback control.

5.4.2 Single-Input, Single-Output Process (m=p=1)

Our purpose is first to generate a fuzzy controller equivalent to the control law $v = Kx$, where v is a real and K a row matrix ($1\text{x}n$). The same inputs and output will be used for the fuzzy controller, that is $x_1, x_2, \ldots., x_n$ as inputs and v as output. In a Mamdani type fuzzy controller, the rules are expressed in a symbolic form. The choice of the input and output variables leads to a definition of the rules in the form

$$\textbf{If } x_1 \text{ is } A_1^{i1} \textbf{ and } x_2 \text{ is } A_2^{i2} \textbf{ and } \cdots \textbf{ and } x_n \text{ is } A_n^{in} \textbf{ then } v \text{ is } B^j . \tag{5.37}$$

The indices correspond to state components and the exponents are used to number the symbols. Thus, A_k^{ik}, $k = 1,2, \ldots, n$, denotes the i_kth symbol defined on the universe of discourse of the k th state variable, that is x_k. The same convention remains valid for the output symbols, where B^j represents the jth symbol attached to the control variable v. A more compact description of the rule base is given by

$$\textbf{If } x \text{ is } (A_1^{i1}, A_2^{i2}, \ldots., A_n^{in}) \textbf{ then } v \text{ is } B^j . \tag{5.38}$$

With regard to linearity, we assume that the membership functions of the meaning of all the symbols are triangular, regularly distributed with an overlap of 1. Figure 5.8 gives an illustration of these assumptions for the symbols A_k^{ik} corresponding to the state variable x_k. We call x_k^{ik} the modal value of x_k for the symbol A_k^{ik}, that is the value of x_k such that

$$A_k^{ik}(x_k) = 1 . \tag{5.39}$$

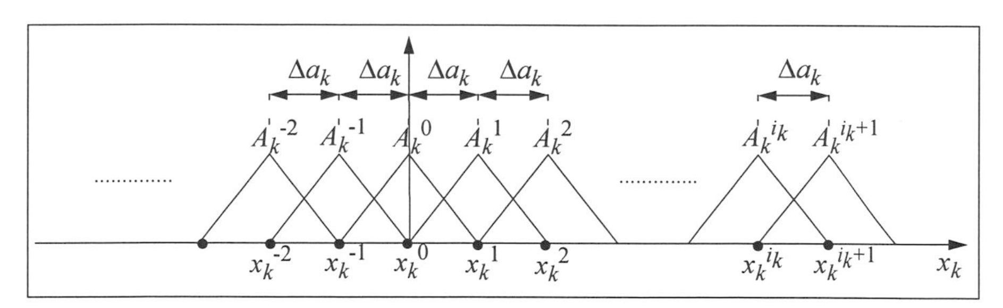

Figure 5.8 Distribution of the membership functions.

We denote Δa_k, $k = 1,2, \ldots, n$, the distance between two consecutive modal values of x_k. In the same way, we denote Δb the distance between two consecutive modal values of v. From these definitions, the following relationships may be deduced

$$x_k^{i_k} = i_k \,.\, \Delta a_k + x_k^0,\; k = 1,2, \ldots, n\,, \tag{5.40}$$

$$v = j \,.\, \Delta b + v^0\,. \tag{5.41}$$

As for the PI case, it is assumed that the meaning of the symbols A_k^0 and B^0 has a membership function centered on the zero of the universes of discourse. In that case, expressions (5.40) and (5.41) become

$$x_k^{i_k} = i_k \,.\, \Delta a_k\,,\; k = 1,2, \ldots, n\,, \tag{5.42}$$

$$v = j \,.\, \Delta b\,. \tag{5.43}$$

We have now to determine the expression of the indices j used in the formulation of the rules conclusion for the fuzzy controller to be equivalent to the classical controller. The correct tuning of the fuzzy controller has also to be determined. That is the values Δa_k and Δb have to be defined with respect to the imposed gains regrouped in the K row matrix. The modal equivalence principle consists in computing the output of the fuzzy controller for input modal values. Then, by equating the obtained result with the output of the classical controller, it will be possible to automatically determine the rule base and the tuning of the fuzzy controller.

The output of the classical controller, $v_{(\mathrm{Cl})}$, for modal inputs $x = (x_1^{i_1}, x_2^{i_2}, \ldots, x_n^{i_n})$ is obtained by applying equation (5.36), that is

$$v_{(\mathrm{Cl})} = K x\,. \tag{5.44}$$

If K_k represents the kth element in the matrix K, equation (5.44) can be reformulated as

$$v_{(\mathrm{Cl})} = \sum_{k=1}^{n} K_k \cdot x_k^{i_k}\,. \tag{5.45}$$

The output of the fuzzy controller, $v_{(\mathrm{Fu})}$, must now be evaluated for the same modal values. For the same reasons as in the PI case, we obtain after defuzzification

$$v_{(\mathrm{Fu})} = v^j\,. \tag{5.46}$$

In fact, the output vector, inferred for an input modal vector, is also modal. By equating the values $v_{(\mathrm{Cl})}$ and $v_{(\mathrm{Fu})}$ derived in (5.45) and (5.46), the following relationship is obtained

$$v^j = \sum_{k=1}^{n} K_k \cdot x_k^{i_k} \; . \tag{5.47}$$

From expressions (5.42), this result may be reformulated as

$$j \cdot \Delta b = \sum_{k=1}^{n} K_k \cdot i_k \cdot \Delta a_k \quad , \tag{5.48}$$

that is

$$j = \sum_{k=1}^{n} \left(\frac{K_k \cdot \Delta a_k}{\Delta b}\right) \cdot i_k \quad . \tag{5.49}$$

Since j represents the index of a conclusion symbol associated with the control variable, it has to be an integer value. Recall that the i_k are also integers (indices of the premise symbols). Therefore, the easiest way to guarantee an integer value for j is to impose that the quantities $\left(\frac{K_k \cdot \Delta a_k}{\Delta b}\right)$ are also integers. In order to use all the possible indices in the numbering of the output symbols, we can simply choose $\frac{K_k \cdot \Delta a_k}{\Delta b} = 1$.

One way to ensure that the Mamdani type fuzzy controller is equivalent to the classical state feedback controller at the modal values is to satisfy the following relationships

$$\begin{aligned} & j = \sum_{k=1}^{n} i_k && \longrightarrow \text{Additive rules} \quad \text{(a)} \\ \forall\; k \in \{1, 2, \ldots, n\} \quad & K_k = \frac{\Delta b}{\Delta a_k} && \longrightarrow \text{Tuning of the fuzzy controller} \quad \text{(b)} \end{aligned} \tag{5.50}$$

5.4.3 Multidimensional Mamdani's controllers

Let us now try to generate a Mamdani type fuzzy controller that implements the feedback control $v = Kx$, where x is a n-vector, v a m-vector and K a mxn matrix.

The first idea consists in choosing the rule base in the form

$$\begin{array}{l} \textbf{If } x_1 \text{ is } A_1^{\,i1} \textbf{ and } x_2 \text{ is } A_2^{\,i2} \textbf{ and } \cdots \textbf{ and } x_n \text{ is } A_n^{\,in} \\ \textbf{then } v_1 \text{ is } B_1^{\,j1} \textbf{ and } v_2 \text{ is } B_2^{\,j2} \textbf{ and } \cdots \textbf{ and } v_m \text{ is } B_m^{\,jm} \end{array} \tag{5.51}$$

By applying the modal equivalence principle, we obtain after computation

$$\begin{array}{lll} \forall\ l \in \{1, 2, \ldots, m\}\ \ j_l = \sum_{k=1}^{n} i_k & \longrightarrow & \text{Additive rules (a)} \\ \left| \begin{array}{l} \forall\ l \in \{1, 2, \ldots, m\} \\ \forall\ k \in \{1, 2, \ldots, n\} \end{array} \right. K_{l,k} = \dfrac{\Delta b_l}{\Delta a_k} & \longrightarrow & \text{Tuning of the fuzzy controller (b)} \end{array} \tag{5.52}$$

where Δb_l represents the distance between two consecutive modal values of the v_l control variable.

The tuning of the fuzzy controller induces the determination of the Δa_k and Δb_l coefficients from the knowledge of the $K_{l,k}$. So, we have to determine m+n parameters satisfying mn equations. Generally, $m+n < mn$ and the system given by (5.52)(b) admits no solution. In that case, there is no way to tune the fuzzy controller. This problem has not appeared in the single-input single-output context, where $m+n > mn$ for $m=1$. In that case, we even have of an extra degree of freedom in the tuning of the parameters. Thus, we can arbitrarily fix Δb and determine the Δa_k, $k = 1, 2, \ldots, n$ from the equations (5.50)(b).

5.4.4 Multi single-output Mamdani's controllers

One way to avoid the problem consists in implementing a mxn control law by using m independent Mamdani's controllers. Thus, each of these controllers deals with n inputs and a single output. The n inputs are the same for all the controllers. Only the output differs from one controller to another. Although the inputs are the same for all the controllers, the partitioning of the corresponding universes of discourse may differ. In this context, the tuning of each controller can be done by using the equations

(5.50) derived in the single output case.

Let us denote $K^{(l)}$, l=1, 2, ..., m, the lth row of the matrix K. The control law $v = Kx$ is thus built as shown in Figure 5.9.

7

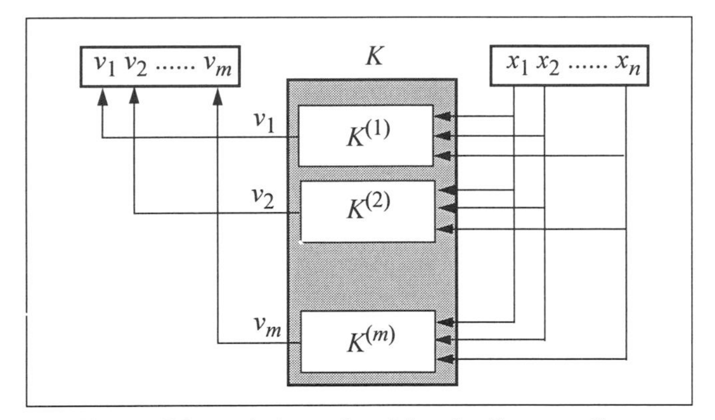

Figure 5.9 *m* independant Mamdani's controllers.

The tuning of the lth controller is realized by using the equations (5.50)(b), that is

$$\left|\begin{array}{l}\forall\ l \in \{1, 2, \ldots, m\} \\ \forall\ k \in \{1, 2, \ldots, n\}\end{array}\right. \qquad K_k^{(l)} - K_{l,k} - \frac{\Delta b^{(l)}}{\Delta a_k^{(l)}} \ . \tag{5.53}$$

The only difference concerning the equations (5.52) obtained in the multidimensional case lies in the fact that the coefficients $\Delta a_k^{(l)}$ now depend on the parameter l.

5.4.5 Application

Let us now synthetize a fuzzy controller that implements the following feedback control law

$$\begin{bmatrix} v_1 \\ v_2 \end{bmatrix} = \begin{bmatrix} 1 & 1/2 \\ 3 & 2 \end{bmatrix} \cdot \begin{bmatrix} x_1 \\ x_2 \end{bmatrix} \ . \tag{5.54}$$

Figure 5.10 represents the corresponding control surfaces for the two control variables v_1 and v_2.

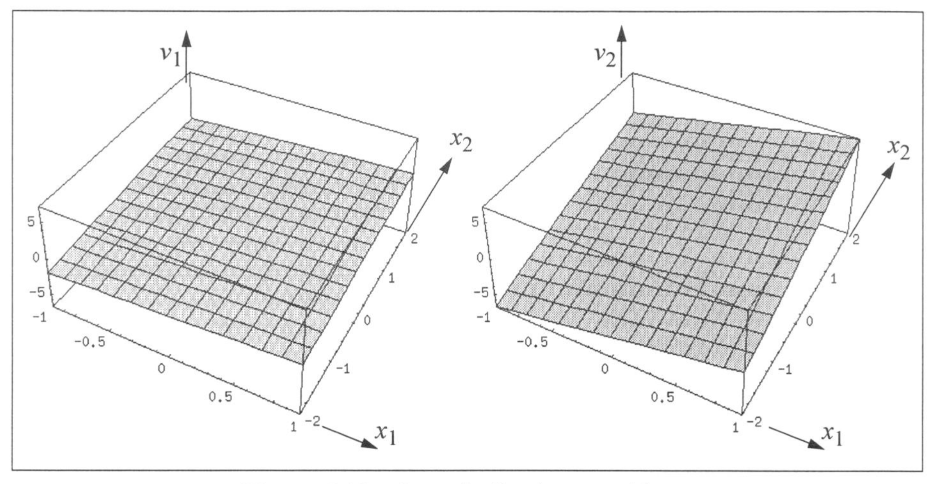

Figure 5.10 State feedback control laws.

The equivalent Mamdani controller is obtained by synthetizing two independent fuzzy controllers as described in section 5.4.4. The rule base of both controllers is the same, that is an additive rule base as advocated in equation (5.50)(a). Both controllers are tuned independently by using equation (5.53), that is

$$
\begin{aligned}
K_{1,1} &= 1 = \frac{\Delta b^{(1)}}{\Delta {a_1}^{(1)}} \qquad & K_{1,2} &= 0.5 = \frac{\Delta b^{(1)}}{\Delta {a_2}^{(1)}} \\
K_{2,1} &= 3 = \frac{\Delta b^{(2)}}{\Delta {a_1}^{(2)}} \qquad & K_{2,2} &= 2 = \frac{\Delta b^{(2)}}{\Delta {a_2}^{(2)}}
\end{aligned}
\tag{5.55}
$$

Figure 5.11 gives an illustration of the difference between the classical control surfaces and the ones obtained with

$$
\begin{aligned}
\Delta {a_1}^{(1)} &= 0.5 \qquad & \Delta {a_2}^{(1)} &= 1 \qquad & \Delta b\ ^{(1)} &= 0.5 \\
\Delta {a_1}^{(2)} &= 0.5 \qquad & \Delta {a_2}^{(2)} &= 0.75 \qquad & \Delta b\ ^{(2)} &= 1.5
\end{aligned}
\tag{5.56}
$$

Nonlinear interpolation is obtained between the modal values. It can be shown that the maximal distance between the two types of controller is proportional to Δb. In our case, we can state a higher error on v_2 than on v_1 stemming from the fact that $\Delta b^{(2)} > \Delta b^{(1)}$.

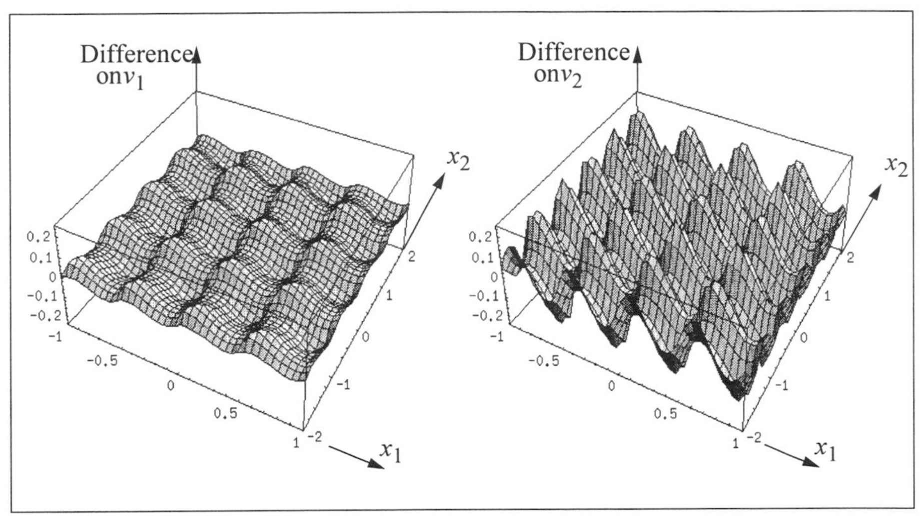

Figure 5.11 Differences between classical control and Mamdani's control.

5.5 EQUIVALENCE FOR SUGENO'S CONTROLLERS

Sugeno's controllers deal with rules whose antecedent parts are symbolic and whose conclusion parts are now functions of the numerical variables associated with the inputs. We will restrict our attention to the case where the functions involved in the conclusion part are reduced to constants. The Sugeno type fuzzy controllers family is thus limited here to the simplified reasoning method proposed in [26]. In this case, the rules for a PI-like controller may be written in the form

$$\textbf{If} \text{ error is } E_i \textbf{ and } \text{change in error is } dE_j \textbf{ then } \delta u = g(i,j)\,, \tag{5.57}$$

where $g(i,j)$ is a constant with respect to the digital inputs ε and $\delta\varepsilon$.

E_i and dE_j represent the ith and j^{th} symbols respectively attached to the error and to the change in error. As for Mamdani's controllers, no assumption is made on the number of symbols. The indices i and j are thus considered as belonging to Z. The function g provides a real constant, that is $g(i,j)$, attributed to the digital output variable δu.

5.5.1 Modal Equivalence Principle

For the inputs ε_m and $\delta\varepsilon_n$, the output of the Sugeno controller, δu_{SUG}, whose rules are defined in (5.57) is described by (see [35], [36])

$$\delta u_{\mathrm{SUG}} = \frac{\sum_{(i,j) \in Z^2} w_{i,j} \cdot g(i,j)}{\sum_{(i,j) \in Z^2} w_{i,j}} \quad \textit{where } w_{i,j} = E_i(\varepsilon) \cdot dE_j(\delta\varepsilon) \quad . \tag{5.58}$$

where $w_{i,j}$ represents the weight associated with the rule whose conclusion is $g(i,j)$.

According to the properties of the membership functions for the modal inputs $\varepsilon = \varepsilon_m$ and $\delta\varepsilon = \delta\varepsilon_n$, we have

(i) $\mu_{E_i}(\varepsilon_m) = 0$ if $i \neq m$ and $\mu_{E_i}(\varepsilon_m) = 1$ if $i = m$,

(ii) $\mu_{dE_j}(\delta\varepsilon_n) = 0$ if $j \neq n$ and $\mu_{dE_j}(\delta\varepsilon_n) = 1$ if $j = n$,

thus, the equation (5.58) is reduced to

$$\delta u_{\mathrm{SUG}} = g(m,n) \ . \tag{5.59}$$

By equating δu_{PI} and δu_{SUG}, we finally obtain

$$g(m,n) = K_I\, m\, \Delta a + K_p\, n\, \Delta b \ . \tag{5.60}$$

To summarize, for a Sugeno type fuzzy controller to be equivalent to a PI controller at the modal values, the parametric function g must verify the relationship (5.60). The fuzzy rules may thus be automatically generated from the following expression

If error is E_i **and** change in error is dE_j **then** $\delta u = K_I\, i\, \Delta a + K_P\, j\, \Delta b$. (5.61)

No particular adjustment of the parameters of the fuzzy controller is required for the modal equivalences to be obtained. The tuning of Δa and Δb allows a simple definition of the number of modal values where the fuzzy controller has to match the PI controller.

It may be noted that the four-rule fuzzy controller studied in ([26], [27]) is a particular case of the fuzzy controller automatically generated here, whose number of rules may be set up to any desired value.

5.5.2 Control law between the modal values

Let us now express the output produced by the fuzzy PI controller for any non-modal inputs, ε and $\delta\varepsilon$. For any error or change in error value, it is always possible to find two consecutive modal values surrounding it. Let us denote ε_m and $\delta\varepsilon_n$ the modal values such that

$$\left|\begin{array}{l}\varepsilon_m < \varepsilon < \varepsilon_{m+1} \text{ and} \\ \delta\varepsilon_n < \delta\varepsilon < \delta\varepsilon_{n+1}.\end{array}\right. \tag{5.62}$$

In that case, we have

$$E_i(\varepsilon) = 0 \text{ if } i \neq m \text{ and } i \neq m+1 , \tag{5.63}$$

$$dE_j(\delta\varepsilon) = 0 \text{ if } j \neq n \text{ and } j \neq n+1 . \tag{5.64}$$

Let us denote I = $\{m, m+1\}$ x $\{n, n+1\}$. Let us also introduce the following notations

$$A = Em(\varepsilon) \quad \text{and} \quad B = dEn(\delta\varepsilon) . \tag{5.65}$$

From the fuzzy partitioning of the universes of discourse we deduce

$$1\text{-}A = Em{+}1(\varepsilon) \text{ and } 1\text{-}B = dEn{+}1(\delta\varepsilon) . \tag{5.66}$$

As the membership functions of the meanings of the symbols are triangular, we have

$$Em(\varepsilon) = -\varepsilon \,/\, \Delta a + m + 1 , \tag{5.67}$$

$$dEn(\delta\varepsilon) = -\delta\varepsilon \,/\, \Delta b + n + 1, \tag{5.68}$$

so

$$\varepsilon = \Delta a \,(-A + m + 1) , \tag{5.69}$$

$$\delta\varepsilon = \Delta b \,(-B + n + 1) . \tag{5.70}$$

From (5.58) and (5.63), the output of the Sugeno controller can be formulated as

$$\delta u_{\text{SUG}} = \frac{\sum_{i,j \in I} w_{i,j} \cdot g(i,j)}{\sum_{i,j \in I} w_{i,j}} \quad \textit{where } w_{i,j} = E_i(\varepsilon) \cdot dE_j(\delta\varepsilon) \quad . \tag{5.71}$$

Let us first evaluate the denominator

$$\sum_{(i,j) \in I} w_{i,j} = w_{m,n} + w_{m+1,n} + w_{m,n+1} + w_{m+1,n+1} \cdot \tag{5.72}$$

By substituting the notations (5.65) and (5.66) in (5.72), we obtain

$$\sum_{(i,j) \in I} w_{i,j} = AB + (1-A)B + (1-B)A + (1-A)(1-B) = 1 \quad . \tag{5.73}$$

So, the denominator of (5.71) vanishes. Let us now examine the numerator.

$$\delta u_{SUG} = ABg(m,n) + (1\text{-}A)Bg(m+1,n) + (1\text{-}B)Ag(m,n+1) + (1\text{-}A)(1\text{-}B)g(m+1,n+1). \tag{5.74}$$

By substituting the expression of the parametric function g derived from the modal equivalence principle (5.60), the output of the Sugeno controller can be rewritten as

$$\begin{aligned} \delta u_{SUG} &= (K_I\, m\, \Delta a + K_p\, n\, \Delta b)\, [\, AB + (1\text{-}A)B + (1\text{-}B)A + (1\text{-}A)(1\text{-}B)\,] \\ &+ K_I\, \Delta a\, [\, (1\text{-}A)B + (1\text{-}A)(1\text{-}B)\,] + K_p\, \Delta b\, [\, (1\text{-}B)A + (1\text{-}A)(1\text{-}B)\,]\, , \end{aligned} \tag{5.75}$$

that is

$$\delta u_{SUG} = K_I\, \Delta a\, (m+1\text{-}A) + K_p\, \Delta b\, (n+1\text{-}B)\, . \tag{5.76}$$

From (5.69) and (5.70), the final output of the Sugeno controller is simply

$$\delta u_{SUG} = K_I\, \varepsilon + K_P\, \delta\varepsilon\, . \tag{5.77}$$

Therefore, the generated Sugeno controller is precisely equivalent to the PI controller. The same result was already obtained in [26] for the particular case of a four-rule controller. Thus, the modal equivalence principle developed in the previous section allows a Sugeno PI controller, which is effectively linear, to be easily synthetized.

5.5.3 Multidimensional Sugeno's controllers

Another way to deal with multi-input multi-output fuzzy controllers consists in choosing the Sugeno formalism. In that case, the rule base is expressed in the form

$$\textbf{If } x_1 \text{ is } A_1^{i_1} \textbf{ and } x_2 \text{ is } A_2^{i_2} \textbf{ and} \cdots \textbf{and } x_n \text{ is } A_n^{i_n}$$
$$\textbf{then } v_1 = f_1(i_1, i_2, .., i_n) \textbf{ and } v_2 = f_2(i_1, i_2, .., i_n) \textbf{ and} \cdots \textbf{and } v_m = f_m(i_1, i_2, .., i_n)\ , \quad (5.78)$$

where $f_l\,(i_1, i_2, .., i_n)$ represents a real value attributed to the lth control variable in the rule labeled $i_1, i_2, .., i_n$.

For input modal values, the outputs of the Sugeno controller are given by

$$\forall \ l \in \{1, 2, \ldots, m\} \ v_{l(\text{Fu})} = f_l\,(i_1, i_2, .., i_n)\ . \quad (5.79)$$

By equating with the outputs of the classical controller, we obtain

$$f_l\ (i_1, i_2, \ldots, i_n) = \sum_{k=1}^{n} K_{l,k} P x_k^{i_k}\ . \quad (5.80)$$

Substituting equation (5.42) in (5.80), the output conclusions are reformulated as

$$v_l = f_l\ (i_1, i_2, \ldots, i_n) = \sum_{k=1}^{n} K_{l,k} \cdot i_k \cdot \Delta a_k\ . \quad (5.81)$$

Finally, the automatic generation of the Sugeno controller is simply based on the application of equation (5.81) for the computation of the real values involved in the conclusion part of the rules.

Let us apply this result to the control law given in Eq. (5.54). We arbitrarily define Δa_1=1 and Δa_2=2. The conclusions of the rules are then computed by using equation (5.81). Some of the obtained rules (the ones necessary for the generation of the control surfaces) are presented in Figure 5.12.

$x_1 \backslash x_2$	A_2^{-1}	A_2^{0}	A_2^{1}
A_1^{-1}	$v_1 = -2$ $v_2 = -7$	$v_1 = -1$ $v_2 = -3$	$v_1 = 0$ $v_2 = 1$
A_1^{0}	$v_1 = -1$ $v_2 = -4$	$v_1 = 0$ $v_2 = 0$	$v_1 = 1$ $v_2 = 4$
A_1^{1}	$v_1 = 0$ $v_2 = -1$	$v_1 = 1$ $v_2 = 3$	$v_1 = 2$ $v_2 = 7$

Figure 5.12 Rule base for the equivalent Sugeno controller.

Figure 5.13 gives an illustration of the difference between the classical control surfaces and the Sugeno ones. A precise equivalence between the two controllers, i.e. linear interpolation between modal values, is obtained.

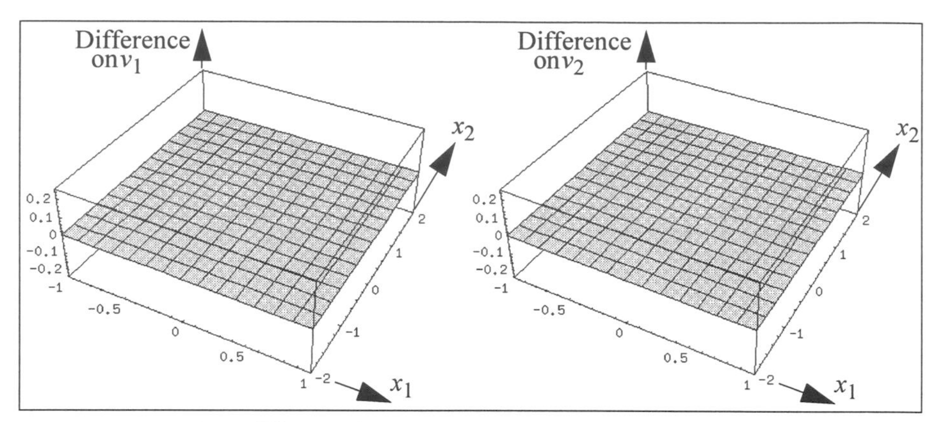

Figure 5.13 Difference between classical control and Sugeno's control.

5.6 CONCLUSION

Any fuzzy approach to process control or modeling requires the following preliminary steps

(i) determination of the rule base,

(ii) determination of the membership functions,

(iii) choice of logical inference operators.

The modal equivalence principle proposed in this chapter provides an efficient way to deal automatically with the first two points when a known linear law is desired. The same method can be easily applied when the desired control law is no longer linear or when only training points are known. In fact, the presented method allows an automatic determination of the rule base and membership functions, which guarantees that some chosen particular points belong to the final control surface. The chosen points, which can be indifferently trained or issued from a mathematical law, are simply taken for being modal values.

For single-output fuzzy controllers, the modal equivalence principle is very efficient. For multi-output fuzzy controllers, the problem is a little bit more complicated. When Mamdani's formalism is used, the application of the modal equivalence principle requires the decomposition of the multi-output problem into a set of single-output laws. However, by using Sugeno's formalism, direct synthesis of multidimensional fuzzy controllers is possible.

References

[1] Bezdek, J.C. (1981) *Pattern Recognition with Fuzzy Objective Function Algorithms*, New-York: Plenum.

[2] Bouslama, F. and Ichikawa, A. (1992) Fuzzy control rules and their natural control laws, *Fuzzy Sets and Systems* **48**, pp. 65-86.

[3] Bouslama, F. and Ichikawa, A. (1992) Application of limit fuzzy controllers to stability analysis, *Fuzzy Sets and Systems* **49**, pp. 103-120.

[4] Buckley, J.J. (1990) Fuzzy controller: Further limit theorems for linear control rules, *Fuzzy Sets and Systems* **36**, pp. 225-233.

[5] Buckley, J.J. (1992) Universal Fuzzy Controllers, *Automatica* **28**, pp. 1245-1248.

[6] Buckley, J.J. (1993) Sugeno type controllers are universal controllers, *Fuzzy Sets and Systems* **53**, pp. 299-303.

[7] Buckley, J.J. and Ying, H. (1989) Fuzzy Controller Theory: Limit Theorems for Linear Fuzzy Control Rules, *Automatica* **25**, pp. 469-472.

[8] Foulloy, L. (1993) Qualitative Control and Fuzzy Control - Towards a Writing Methodology, *AI Communications* **6**, pp. 147-154.

[9] Foulloy, L. and Galichet, S. (1993) Fuzzy Controllers Representation, *Proc. 1st European Congress on Fuzzy and Intelligent Technologies*, Aachen, Germany, pp 142-148.

[10] Foulloy, L. and Galichet, S. (1995) Typology of fuzzy controllers, *Theoretical Aspects of Fuzzy Control*, (Eds H. Nguyen, M. Sugeno, R. Tong and R. Yager), J. Wiley Publishers.

[11] Galichet, S., Dussud, M. and Foulloy, L. (1992) Contrôleurs flous: équivalences et études comparatives, *Actes des Deuxièmes journées nationales sur les applications des ensembles flous*, Nîmes, France, pp. 229-236.

[12] Galichet, S. and Foulloy, L. (1993) Fuzzy Equivalences of Classical Controllers, *Proc. 1st European Congress on Fuzzy and Intelligent Technologies*, Aachen, Germany, pp. 1567-1573.

[13] Galichet, S. and Foulloy, L. (1995) Fuzzy Controllers: Synthesis and Equivalences, *IEEE Transactions on Fuzzy Systems* **3(2)**, pp.140-148.

[14] Hwang, G.C. and Lin, S.C. (1992) A stability approach to fuzzy control design for nonlinear systems, *Fuzzy Sets and Systems* **48**, pp. 279-287.

[15] Kosko, B. (1992) Fuzzy systems as universal approximators, *Proc. 1st IEEE Conference on Fuzzy Systems*, San Diego, CA, pp. 1153-1162.

[16] Kosko, B. (1992), *Neural Networks and Fuzzy Systems: A Dynamical Systems Approach to Machine Intelligence*, Prentice Hall.

[17] Kreinovich, V., Nguyen, H.T. and Sirisaengtaksin, O. (1994) On approximation of controls in distributed systems by fuzzy controllers, *Proc. 5th Int. Conf. on Information Processing and Management of Uncertainty in knowledge-based systems*, Paris, France, pp. 79-83.

[18] Langari, R. (1993) Synthesis of Nonlinear Controllers via Fuzzy Logic, *Proc. 2nd*

IEEE Conference on Fuzzy Systems, San Francisco, CA, pp. 23-28.
[19] MacVicar-Whelan, P.J. (1976) Fuzzy sets for man-machine interaction, *Int. J. Man-Machine Studies* **8**, pp. 687-697.
[20] Mamdani, E.H. (1974) Application of Fuzzy Algorithms for Control of Simple Dynamic Plant, *Control and Science* **121**, pp. 1585-1588.
[21] Mamdani, E.H. and Assilian, S. (1975) An Experiment in Linguistic Synthesis with a Fuzzy Logic Controller, *Int. Journal of Man-Machines Studies* 7, pp.1-13.
[22] Mamdani E.H. (1975) Advances in the Linguistic Synthesis of Fuzzy Controllers, *Int. Journal of Man-Machines Studies* **8**, pp. 669-678.
[23] Mann, G.K.I., Hu, B.G. and Gosine, R.G. (1997) Analysis and Performance Evaluation of Linear-Like Fuzzy PI and PID Controllers, *Proc. 6th IEEE Conference on Fuzzy Systems*, Barcelona, Spain, pp. 383-390.
[24] Matía, F., Jiménez, A., Galán R. and Sanz, R. (1992) Fuzzy controllers: Lifting the linear-nonlinear frontier, *Fuzzy Sets and Systems* **52**, pp. 113-128.
[25] Mizumoto, M; (1990) Fuzzy Controls by Product-sum-gravity Method, *Proc. of Sino-Japan Joint Meeting on Fuzzy Sets and Systems,* Beijing, China, pp. 1-4.
[26] Mizumoto, M. (1992) Realization of PID controls by fuzzy control methods, *Proc. 1st IEEE Conference on Fuzzy Systems*, San Diego, CA, pp. 709-715.
[27] Mizumoto, M. (1993) Fuzzy Controls Under Product-sum-gravity Methods and New Fuzzy Control Methods, *Fuzzy Control Systems,* (Eds A. Kandel and G. Langholz), CRC Press, pp. 276-294.
[28] Nguyen, H.T. and Kreinovich, V. (1993) On approximation of controls by fuzzy systems, *Proc. 5th Int. Fuzzy Systems Association World Congress*, Seoul, Korea, pp. 1414-1417.
[29] Otsubo, A. and Hayashi, K. (1997) Realization of Nonlinear and Linear PID Control Using Simplified Indirect Fuzzy Inference Method, *Proc. 7th International Fuzzy Systems Association World Congress*, Prague, Czeck Republic, pp. 393-397.
[30] Pedrycz, W. (1989) *Fuzzy control and fuzzy systems*, New-York: John Wiley Research Studies Press.
[31] Peng, X.T. (1990) Generating rules for fuzzy logic controllers by functions, *Fuzzy Sets and Systems* **36**, pp. 83-89.
[32] Reznik, L. (1997) Evolution of Fuzzy Controller Design, *Proc. 6th IEEE Conference on Fuzzy Systems*, Barcelona, Spain, pp. 503-508.
[33] Santos, M., Dormido, S. and De La Cruz, J.M. (1996) Fuzzy PID Controllers vs. Fuzzy-PI Controllers, *Proc. 5th IEEE Conference on Fuzzy Systems*, New Orleans, LA, pp. 1598-1604.
[34] Siler, W. and H. Ying, H. (1989) Fuzzy control theory: the linear case, *Fuzzy Sets and Systems* **33**, pp. 275-290.
[35] Sugeno, M. and Kang, G.T. (1988) Structure identification of fuzzy model, *Fuzzy Sets and Systems* **28**, pp. 15-33.
[36] Takagi, T. and Sugeno, M. (1985) Fuzzy identification of systems and its appli-

cations to modeling and control, *IEEE Transactions on Systems, Man and Cybernetics* **15**, pp. 116-132.
[37] Tanaka, K. and Sugeno, M. (1992) Stability analysis and design of fuzzy control systems, *Fuzzy Sets and Systems* **45**, pp.135-156.
[38] Tarn, J.H. and Kuo, L.T. (1993) Analysis of fuzzy nonlinearity - One dimensional case, *Proc. American Control Conference*, San Francisco, CA, pp. 781- 785.
[39] Tang, K.L. and Mulholland, R.J. (1987) Comparing Fuzzy Logic with Classical Controllers Designs, *IEEE Transactions on Systems, Man and Cybernetics* **17**, pp. 1085-1087.
[40] Wang, L.X. (1992) Fuzzy systems are universal approximators, *Proc. 1st IEEE Conference on Fuzzy Systems*, San Diego, CA, pp. 1163-1169.
[41] Wang, L.X. and Mendel, J.M. (1992) Generating Fuzzy Rules by Learning from Examples, *IEEE Transactions on Systems, Man and Cybernetics* **22**, pp. 1414-1427.
[42] Wong, C.C. (1993) Realization of Linear Defuzzified Output via Mixed Fuzzy Logics, *Proc. 2nd IEEE Conference on Fuzzy Systems*, San Francisco, CA, pp. 1167-1172.
[43] Yager, R.R. (1982) Some procedures for selecting fuzzy set-theoretic operators, *Int. J. General Systems* **8**, pp. 115-124.
[44] Yager, R.R. and Filev, D.P. (1993) SLIDE: A Simple Adaptative Defuzzification Method, *IEEE Transactions on Fuzzy Systems* **1**, pp. 69-78.
[45] Ying, H., Siler, W. and Buckley, J.J. (1990) Fuzzy control theory: a nonlinear case, *Automatica* **26**, pp. 513-520.
[46] Zadeh L.A. (1971) Quantitative Fuzzy Semantics, *Information Science* **3**, pp. 159-176.

6 DESIGNS OF FUZZY CONTROLLERS

Rainer Palm

Siemens AG Corporate Research and Development
Dept. ZT IK4
Otto-Hahn-Ring 6
81739 Munich, Germany
Rainer.Palm@mchp.siemens.de

6.1 INTRODUCTION

The fuzzy control paradigm emerged in 1974 when E.H. Mamdani initiated an investigation of fuzzy set theory based algorithms for the control of a simple dynamic plant by using a control law in the form of set of IF-THEN fuzzy rules, i.e., a fuzzy controller [14]. In 1977 Ostergaard [16] reported an application of heat exchanger control by means of a fuzzy controller, and in 1982 Holmblad and Ostergaard presented a cement kiln fuzzy controller [6]. However, only in the late 80's fuzzy control started becoming more and more accepted mainly due to the attention that Japan's industry paid to the new control technology. It is a fact that the commonly used control technique in industrial process control is the PID-controller, though used in a variety of different control schemes, e.g., adaptive , gain scheduling, and supervisory control architectures.

Nowadays, processes and plants under control are so complex that PID controllers are not sufficient even though augmented with additional adaptive, gain scheduling, and supervisory algorithms. Although there is a large number of methods and theories to cope with sufficiently complex control problems in the automation, robotics, consumer and industrial electronics, car, aircraft and ship building industries [36], the restrictions for applying these methods and theories are either too strong, or the methods are too complicated to be dealt with in a practically efficient and inexpensive manner.

Control engineers are in a need of simpler process and plant models and controller design methods being far away from the highly sophisticated mathematical models available and their underlying rigorous assumptions. These simpler design methods should provide good performance characteristics, and they should be robust enough with regard to disturbances, parameter uncertainties, and unmodelled structural properties of the process under control.

Fuzzy control provides a variety of design methods wich can, together with traditional control techniques, cope with modern control problems.

There are mainly three aspects of fuzzy controllers which go beyond the conventional controllers designed via traditional control methods:

1. The use of IF-THEN rules

2. The universal approximation property

3. The property of dealing with vague (fuzzy) values.

The first aspect concerns the human operator's knowledge for controlling a plant. This sort of knowledge reflects heuristic experience and is formulated in terms of IF-THEN fuzzy rules. In the same way, the plant's behavior can also be expressed by a set of IF-THEN fuzzy rules. The major difficulty here is the identification of the fuzzy rules such that they describe the operator's control actions and the systems's response sufficiently well [21, 22, 31]. The identification of this type of fuzzy rules can be done in two ways:

1. knowledge aquisition via the use of interviewing techniques from the area of knowledge based expert systems. This type of identification has been applied successfully to the control of mainly SISO plants and processes, but is difficult to apply and verify for MIMO control problems.

2. black box type of identification via the use of clustering, neural nets, and genetic algorithms based techniques.

In the latter approach one distinguishes between *structure identification* and *parameter identification.*

Structure identification requires *structural a-priori knowledge* about the system to be controlled, e.g. whether the system is assumed to be linear, and what the order of the system might be. The result of identification is a set of fuzzy rules. If one has to identify a plant with only little structural knowledge one has to use algorithms which learn from data. A good identification tool is a *neural net* because of its inherent learning ability. In addition, *genetic algorithms* are used to learn fuzzy rules, as well.

Parameter identification deals with a proper parametrization, scaling, and normalization of physical signals. Parameter identification is a comparetively simple task and can be done by classical methods, e.g., LQ methods and related techniques.

The second aspect from above, the *universal approximation property*, means that a fuzzy system with product-based rule firing, centroid defuzzification , and Gaussian membership functions can approximate any real continuous function on a compact set to arbitrary accuracy [37, 13, 12]. However, in most cases a less accurate approximation of a finite state space, or a nonlinear function with a finite universe of discourse, by a finite number of fuzzy rules is required, while using triangular or trapezoidal membership functions. In this case a certain approximation error must be accepted.

The approximation property is due to the overlap of the membership functions from the IF-parts of the set of fuzzy rules. Because of this overlap, every single rule is influenced by neighboring rules. So, every point in state space is approximated by a subset of fuzzy rules.

The third aspect considers control tasks where the controller inputs are fuzzy values instead of being crisp variables. Fuzzy values are qualitative "numbers" being obtained from different sources. One particular source is a qualitative statement of a human operator while controlling a plant like *Temperature is high*. Another source may originate from a sensor which provides information about the *intensity* of a physical signal with in a certain interval. Here, the *intensity or distribution* of the signal with respect to this interval is expressed by a membership function. In contrast to classical controllers, FCs are able to deal with fuzzy values. Even the mixture of crisp and fuzzy values becomes possible.

Sect. 2 of this chapter deals with fuzzy control techniques including the design goal, the definition of a fuzzy region and the most important FC techniques for systems and controllers.

Sect. 3 deals with the FC as a nonlinear transfer element. In this section the computational structure of a FC, its transfer characteristics and its nonlinearity is discussed.

Sect. 4 presents different heuristic and model based control strategies such as the Mamdani controller, the sliding mode fuzzy controller, the cell mapping control strategy and the Takagi Sugeno control strategy.

Sect. 5 gives a short overview about supervisory control.

Sect. 6 presents the main aspects of adaptive fuzzy control.

6.2 FUZZY CONTROL TECHNIQUES

FC techniques can be devided into experiential (heuristic) and model based techniques. The choice for a special FC technique depends on the way of how the system to be controlled is described. Figure 6.1 shows the most important FC techniques dealing with systems and controllers.

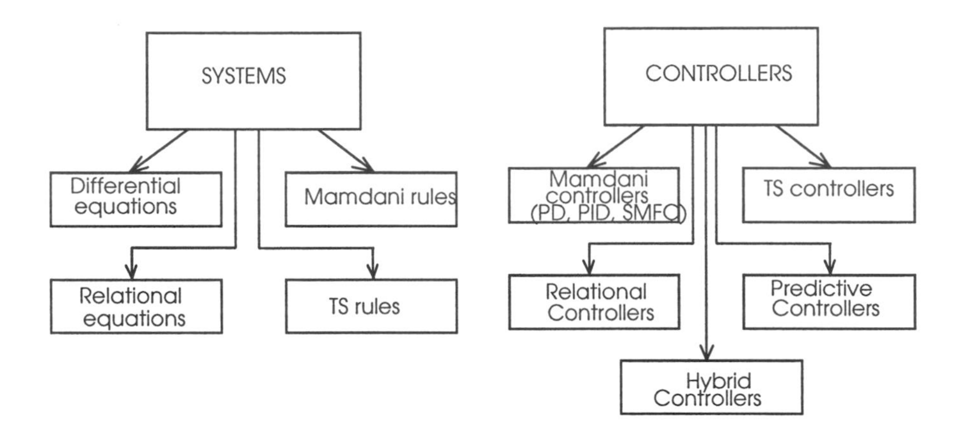

Figure 6.1 FC techniques

In the following subsection we deal with the design goal of fuzzy control. In a subsequent subsection the fuzzy region is defined. Finally, the individual FC techniques for systems and controllers are outlined.

6.2.1 The Design Goal

The objective of the design in fuzzy control can be stated as follows

> Given a model (heuristic or analytical) of the physical system to be controlled and the specifications of its desired behavior. Design a feedback control law in the form of a set of fuzzy rules such that the closed loop system exhibits the desired behavior.

The general control scheme is shown in Fig.(6.2)

Here, we have the following notations

$\mathbf{x}$ is the state vector (also controller input).

$\mathbf{x^d}$ is the desired state vector.

$\mathbf{u}$ is the control input vector (also controller output).

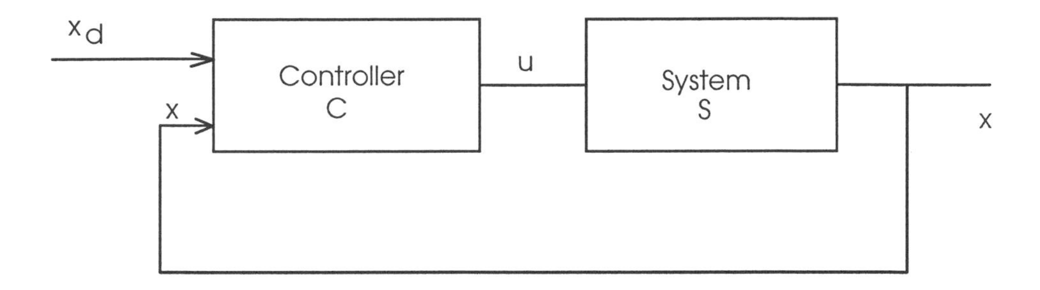

Figure 6.2 General control structure

where the vectors $\mathbf{x}, \mathbf{x^d}, \mathbf{u}$ are continuous functions of time. For simplicity, the output vector $\mathbf{y}$ is set to be equal to the state vector $\mathbf{x}$

$$\mathbf{y} = \mathbf{x}.$$

In the following we define two basic types of nonlinear control problems, namely the *nonlinear regulation (stabilization)* and *nonlinear tracking* [28] (Part II.1). Then, we will briefly discuss the specifications of desired behavior such as *performance, stability, and robustness*, in the context of nonlinear control.

Stabilization And Tracking. In general, the tasks of a control system can be divided in two basic categories:

1. *Stabilization.* In stabilization control problems, a FC, called a stabilizer, or regulator, is to be designed so that the state vector of the closed loop system will be stabilized around a point (operating point, or a set point) of the state space. The *asymptotic stabilization control problem* is to find a control law in terms of a set of fuzzy rules such that, starting anywhere in a region around the setpoint $\mathbf{x^d}$ the state vector $\mathbf{x}$ of the closed loop system goes to the setpoint $\mathbf{x^d}$ as t goes to infinity.

2. *Tracking.* In tracking control problems, a FC is to be designed so that the closed loop system output follows a given time-varying trajectory.The *asymptotic tracking problem* is to find a control law in terms of a set of fuzzy rules such that starting from any initial state $\mathbf{x^0}$ in a region around $\mathbf{x^d}(t)$, the tracking error $\mathbf{x}(t) - \mathbf{x^d}(t)$ tends to $\mathbf{0}$ while the whole state

vector remains bounded. Let us stress here that perfect tracking, i.e., when the initial states imply zero tracking-error, is not possible. Thus the design objective of achieving asymptotic tracking cannnot be achieved. In this case, one should aim at bounded-error tracking, with small tracking-error to be achieved for trajectories of particular interest.

From a theoretical point of view, there is a relationship between the stabilization and the tracking control problems. Stabilization can be regarded as a special case of tracking where the desired trajectory is a constant. On the other hand, if, for example [28] (Part II.1), we are to design a tracker for the open loop system

$$\ddot{y} + f(\dot{y}, y, u) = 0, \tag{6.1}$$

so that $e(t) = y(t) - y_d(t)$ tends to zero, the problem is equivalent to the asymptotic stabilization of the system,

$$\ddot{e} + f(\dot{e}, e, u, y_d, \dot{y}_d, \ddot{y}_d) = 0, \tag{6.2}$$

its state vector components being e and $\dot{e}$. Thus the tracker design problem can be solved if one designs a regulator for the latter nonautonomous open loop system.

Performance. In linear control, the desired behavior of the closed loop system can be systematically specified in exact quantitative terms. For example, the specifications of the desired behavior can be formulated in the time domain in terms of rise time and settling time, overshoot and undershoot, etc. Thus, for this type of control, one first postulates the quantitative specifications of the desired behavior of the closed loop system, and then designs a controller which meets these specifications, for example, by choosing the poles of the closed loop system appropriately.

As observed in [28] (Part II.2), such systematic specifications of the desired behavior of nonlinear closed loop systems, except for those that can be approximated by linear systems, are not obvious at all because the response of a nonlinear system (open or closed loop) to one input vector does not reflect its response to another input vector. Furthermore, a frequency domain description of the behavior of the system is not possible either.

The consequence of the above said is that in specifying the desired behavior of a nonlinear closed loop system one employs some qualitative specifications of performance including stability, accuracy and response speed, and robustness.

Stability. Stability must be guaranteed for the model used for design (the nominal model) either in a local or in a global sense. The regions of stability and convergence are also of interest.

One should however keep in mind that stability does not imply the ability to withstand persistent disturbances of even small magnitude. This is so since the stability of a nonlinear system is defined with respect to initial conditions, and only temporary disturbances may be translated as initial conditions. Thus stability of a nonlinear system is different from stability of a linear system. In the case of a linear system stability always implies the ability to withstand bounded disturbances when of course the system stays in its linear range of operation. The effects of persistent disturbances on the behavior of a nonlinear system are addressed by the notion of robustness.

Accuracy and Response Speed. Accuracy and response speed must be considered for some desired trajectories in the region of operation. For some classes of systems, appropriate design methods can guarantee consistent tracking accuracy independently of the desired trajectory as is the case in sliding mode control and related control methods.

Robustness. Robustness is the sensitivity of the closed loop system to effects which are neglected in the nominal model used for design. These effects can be disturbances, measurement noise, unmodelled dynamics, etc. The closed loop system should be insensitive to these neglected effects in the sense that they should not negatively affect its stability.

We want to stress here that the above specifications of desired behavior are in conflict with each other to some extent, and a good control system can be designed only based on trade-offs in terms of robustness vs. performance, cost vs. performance, etc.

6.2.2 Fuzzy Regions

Before we start with special FC techniques we introduce the definition of a fuzzy region in the state space. In fuzzy control, a crisp state vector $\mathbf{x} = (x_1, \ldots, x_n)^T$ is a state vector the values of which are defined on the closed interval (the domain) X of reals. A crisp control input vector $\mathbf{u} = (u_1, \ldots, u_n)^T$ is a control input vector the values of which are defined on the closed interval (the domain) U of reals. The set of fuzzy values of a component x_i is called the *term set* of x_i denoted as $\mathbf{TX_i} = \{LX_{i1}, \ldots, LX_{im_i}\}$ (e.g. NB, NM, NS, Z, PS, PM, PB with N - negative, P - positive, S - small, M - medium, B - big). LX_{ij} is defined by a membership function $LX_{ij}(x)$. The term set of u_i is likewise denoted as $\mathbf{TU_i} = \{LU_{i1}, \ldots, LU_{ik_i}\}$. LU_{ij} is defined by a membership function $LU_{ij}(u)$.

An arbitrary fuzzy value from $\mathbf{TX_i}$ will be denoted as LX_i where LX_i can be anyone of $LX_{i1}, \ldots, LX_{im_i}$. An arbitrary fuzzy value from $\mathbf{TU_i}$ will be denoted as LU_i where LU_i can be anyone of $LU_{i1}, \ldots, LU_{ik_i}$.
A fuzzy state vector $\mathbf{LX} = (LX_1, \ldots, LX_n)^T$ denotes a vector of fuzzy values. Each component $x_1, \ldots, x_n$ of the state vector $\mathbf{x}$ takes a corresponding fuzzy value $LX_1, \ldots, LX_n$ where $LX_i \in \mathbf{TX_i}$. A fuzzy region $\mathbf{LX^i} = (LX_1^i, \ldots, LX_n^i)^T$ is defined as a fuzzy state vector for which there is a contiguous set of crisp state vectors $\{\mathbf{x}^*\}$, each crisp state vector satisfying the given fuzzy state vector $\mathbf{LX^i}$ to a certain degree different from 0. The fuzzy state space is defined as the set of all fuzzy regions LX^i.
Example
Let $\mathbf{x} = (x_1, x_2)^T$, $\mathbf{TX_1} = \{LX_{11}, LX_{12}, LX_{13}\}$, and $\mathbf{TX_2} = \{LX_{21}, LX_{22}, LX_{23}\}$. Then the total number of different fuzzy state vectors is $\mathbf{M} = 9$ and these state vectors are

1. $\mathbf{LX^1} = (LX_{11}, LX_{21})^T$.
2. $\mathbf{LX^2} = (LX_{11}, LX_{22})^T$.
3. $\mathbf{LX^3} = (LX_{11}, LX_{23})^T$.
4. $\mathbf{LX^4} = (LX_{12}, LX_{21})^T$.
5. $\mathbf{LX^5} = (LX_{12}, LX_{22})^T$.
6. $\mathbf{LX^6} = (LX_{12}, LX_{23})^T$.
7. $\mathbf{LX^7} = (LX_{13}, LX_{21})^T$.
8. $\mathbf{LX^8} = (LX_{13}, LX_{22})^T$.
9. $\mathbf{LX^9} = (LX_{13}, LX_{23})^T$.

6.2.3 FC techniques for systems and controllers

In this subsection we deal with systems and controllers according to the scheme shown in Fig. 6.1.

Heuristic system models. In the case when an analytical model of the plant is not available the control design has to be carried out on the basis of qualitative modeling. This can be done either in terms of a set of Mamdani fuzzy rules or a fuzzy relation [21, 34]. A typical Mamdani rule of a 1st order system is

$$R_{Si}: \quad IF \quad x \quad is \quad PS \quad AND \quad u \quad is \quad NB \quad THEN \quad \dot{x} \quad is \quad NS \tag{6.3}$$

A typical fuzzy relational equation of a 1st order system is

$$\dot{X} = X \circ U \circ \mathbf{S}, \tag{6.4}$$

where X is the fuzzy state, $\dot{X}$ is the fuzzy derivative, U is the fuzzy control variable, and $\mathbf{S}$ is the fuzzy relation. $\circ$ denotes the relational composition (e.g. max-min composition).

Analytical system models. If an analytical model of the plant is available the system's behavior can be described by a set of differential equations or by a set of so-called Takagi Sugeno fuzzy rules (TS rules) [31]. A typical differential equation of an open loop system is

$$\dot{\mathbf{x}} = \mathbf{A} \cdot \mathbf{x} + \mathbf{B} \cdot \mathbf{u} \tag{6.5}$$

On the other hand, a TS fuzzy rule consists of a fuzzy antecedent part and a consequent part consisting of an analytical equation.
A typical TS rule for a 1st order system is

$$R_{Si}: \quad IF \quad x = LX^i \quad THEN \quad \dot{x} = A_i \cdot x + B_i \cdot u \tag{6.6}$$

where LX^i is the ith fuzzy region for x, and A_i and B_i are parameters corresponding to that region.

Fuzzy controllers can be classified as follows.

Mamdani controller. A Mamdani controller works in the following way.

1. A crisp value is scaled into a normalized domain.
2. The normalized value is fuzzified with respect to the input fuzzy sets.
3. By means of a set of fuzzy rules a fuzzy output value is provided.
4. The fuzzy output is defuzzified with the help of an appropriate defuzzification method (center of gravity, height method etc.).
5. The defuzzified value is denormalized into a physical domain.

A typical Mamdani controller of 1st order is

$$R_{Ci}: \quad IF \quad x = LX^i \quad THEN \quad u = LU^i \tag{6.7}$$

where LU^i is the corresponding fuzzy value for the control variable.

Relational controller. According to the description of the system in terms of a relational equation a typical fuzzy relational equation for a 1st order controller is

$$U = X \circ \mathbf{C}, \tag{6.8}$$

where X is the fuzzy state, U is the fuzzy control variable and $\mathbf{C}$ is the fuzzy relation.

Takagi Sugeno controller. A typical TS controller of 1st order is

$$R_{Ci}: \quad IF \quad x = LX^i \quad THEN \quad u = K_i \cdot x \tag{6.9}$$

where LX^i is the ith fuzzy region for x, and K_i is the gain corresponding to that region.

Predictive controller. Predictive fuzzy control was introduced by [40] for automatic train operation. It includes control rules for the time k to predict the behavior of the system for the next time step $k+1$. By means of a performance index $J(k)$, which appears for a specific control action $u(k)$, different features like velocity, riding comfort, energy saving and accuracy of a stop gap are evaluated. By means of going through the whole range of possible control actions $u(k)$ one obtains a range of corresponding performance indices $J(k)$ from which the control action $u(k+1)$ with the highest performance index $J(k)$ is applied to the plant. A typical predictive control rule is

IF the performance index $J(k)$ is LJ^i is obtained
AND a control value $u(k)$ is chosen to be LU^i
THEN the control value to be applied to the plant for the next time step $k+1$
is chosen to be $u(k+1) = LU^i$.

A formal description is

$$IF \quad J(k) = LJ^i \quad AND \quad u(k) = LU^i \quad THEN \quad u(k+1) = LU^i. \tag{6.10}$$

Hybrid controller. A hybrid controller is represented by a mixture of fuzzy controller and conventional controller.
A typical hybrid controller appears if the control law consists of a Mamdani controller C_{fuzz} and an analytical feedforward term C_{comp} which compensates statical or dynamical forces in a mechanical system.

$$u = C_{fuzz}(\mathbf{x}, \mathbf{x^d}) + C_{comp}(\mathbf{x}), \tag{6.11}$$

where $\mathbf{x^d}$ is the desired vector.

6.3 THE FC AS A NONLINEAR TRANSFER ELEMENT

A fuzzy logic controller defines a control law in the form of a static nonlinear *transfer element* (TE) due to the nonlinear nature of the computations performed by a FC. However, the control law of a FC is not represented in an analytic form, but by a set of fuzzy rules. The *antecedent* of a fuzzy rule (IF-part) describes a fuzzy region in the state space. Thus one effectively partitions an otherwise continuous state space by covering it with a finite number of fuzzy regions and consequently, fuzzy rules. The *consequent* of a fuzzy rule (THEN-part) specifies a control law applicable within the fuzzy region from the IF-part of the same fuzzy rule. During control with a FC a point in the state space is affected to a different extent by the control laws associated with all the fuzzy regions to which this particular point in the state space belongs. By using the operations of *aggregation* and *defuzzification*, a specific control law for this particular point is determined. As the point moves in the state space, the control law changes smoothly. This implies that despite of the quantization of the state space in a finite number of fuzzy regions, a FC yields a smooth nonlinear control law.

One goal of this section is to describe computation with a FC and its formal description as a static nonlinear transfer element and thus provide the background knowledge needed for the understanding of control with a FC. Furthermore, we show the relationship between conventional and rule-based transfer elements thus establishing the compatibility between these two conceptually different, in terms of representation, types of transfer elements.

6.3.1 The Computational Structure of a FC

A control law represented in the form of a FC directly depends on the measurements of signals and is thus a static control law. This means that the fuzzy rule-based representation of a FC does not include any dynamics which makes a FC a ***static transfer element***, like a ***state controller***. In addition to that, a FC is, in general, a ***nonlinear static transfer element*** which is due to those computational steps of its computational structure that have nonlinear properties. In what follows we will describe the computational structure of a FC by presenting the computational steps which it involves.

The computational structure of a FC consists of a number of computational steps and is illustrated in Fig. 6.3.

There are five such computational steps constituting the computational structure of a FC. These are the following.

1. *Input scaling (normalization).*

2. *Fuzzification* of controller-input variables.

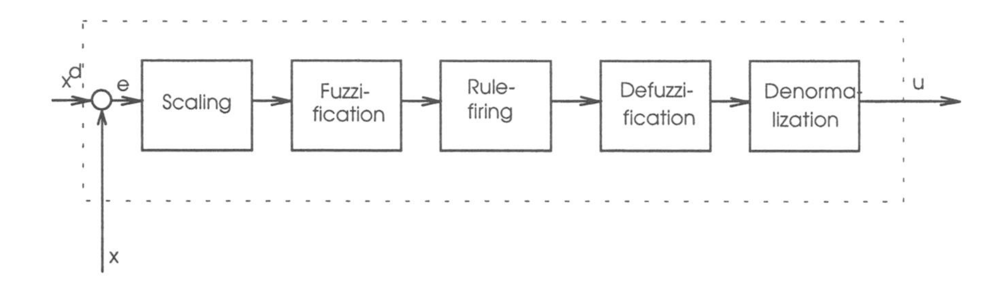

Figure 6.3 The computational structure of a FC.

3. *Inference (rule firing).*

4. *Defuzzification* of controller-output variables.

5. *Output scaling (denormalization).*

We should remark here that the state variables $x_1, x_2, \ldots, x_n$ (or $e, \dot{e}, \ldots, e^{(n)}$) that appear in the IF-part of the fuzzy rules of a FC are also called *controller-inputs.* The controller input variables $u_1, u_2, \ldots, u_m$ that appear in the THEN-part of the fuzzy rules of a FC are also called *controller-outputs.* In the remainder of this chapter we will use controller-inputs and controller-outputs when referring to the state variables $x_1, x_2, \ldots, x_n$ and the input variables $u_1, u_2, \ldots, u_m$ respectively.

We will now consider each of the above computational steps for the case of a *multiple-input/single-output* (MISO) FC. The generalization to the case of multiple-input/multiple-output FC, where there are m controller-outputs u_1, u_2, ..., u_m instead of a single controller-output u, can be easily done.

Input Scaling. There are two principle cases in the context of input scaling.

1. The membership functions defining the fuzzy values of the controller-inputs and controller-outputs are defined off-line on their actual physical domains. In this case the controller-inputs and controller-outputs are

processed only using *fuzzification*, *rule firing* and *defuzzification*. For example, this is the case of Takagi-Sugeno FC-1 and FC-2.

2. The membership functions defining the fuzzy values of controller-inputs and controller-outputs are defined off-line, on a common *normalized domain*. This means that the actual, crisp physical values of the controller-inputs and controller-outputs are mapped onto the same predetermined normalized domain. This mapping, called *normalization*, is done by the so-called *normalization factors*. Input scaling is then the multiplication of a physical, crisp controller-input, with a normalization factor so that it is mapped onto the normalized domain. Output scaling is the multiplication of a normalized controller-output with a *denormalization factor* so that it is mapped back onto the physical domain of the controller-outputs.

The advantage of the second case from the above is that fuzzification, rule firing, and defuzzification can be designed independently of the actual physical domains of the controller-inputs and controller-outputs. For instance, this is the case with SMFC.

To illustrate the notion of input scaling let us consider, for example the state vector $\mathbf{e} = (e_1, e_2, \ldots, e_n)^T = (e, \dot{e}, \ldots, e^{(n)})^T$, where for each i, $e_i = x_i - x_{d_i}$. This vector of physical controller-inputs is normalized with the help of a matrix $\mathbf{N_e}$ containing predetermined normalization factors for each component of $\mathbf{e}$. The normalization is done as

$$\mathbf{e_N} = \mathbf{N_e} \cdot \mathbf{e}, \tag{6.12 }$$

with

$$\mathbf{N_e} = \begin{pmatrix} N_{e_1} & 0 & \ldots & 0 \\ 0 & N_{e_2} & \ldots & 0 \\ \vdots & \vdots & \ddots & \vdots \\ 0 & 0 & \ldots & N_{e_k} \end{pmatrix}, \tag{6.13 }$$

where N_{e_i} are real numbers and the normalized domain for $\mathbf{e}$ is, say $E_N = [-a, +a]$.

Example

Let $\mathbf{e} = (e_1, e_2)^T = (e, \dot{e})^T$ with

$$e = x - x_d \qquad \text{and} \qquad \dot{e} = \dot{x_d} - \dot{x_d}. \tag{6.14 }$$

Then, input-scaling of e into e_N and $\dot{e}$ into $\dot{e}_N$ yields

$$e_N = N_e \cdot e \qquad \text{and} \qquad \dot{e}_N = N_{\dot{e}} \cdot \dot{e}, \tag{6.15 }$$

where N_e and $N_{\dot{e}}$ are the normalization factors for e and $\dot{e}$ respectively.

In the context of a phase plane representation of the dynamic behavior of the controller-inputs, the input scaling affects the angle of the sliding line which divides the phase plane into two semiplanes (see Fig. 6.4).

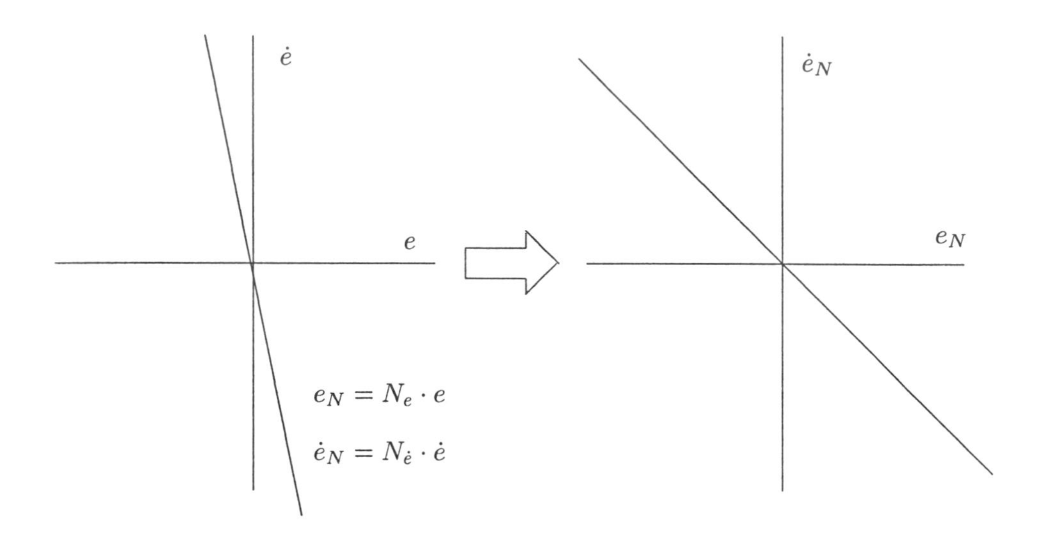

Figure 6.4 Normalization of the phase plane.

In addition to that, we can see how the supports of the membership functions defining the fuzzy values of e and $\dot{e}$ change because of the input scaling of these controller-inputs (see Fig. 6.5).

In the next three subsections on fuzzification, rule firing and defuzzification we consider only the case when the fuzzy values of the controller-inputs and controller-outputs are defined on normalized domains (e.g., E_N and U_N), and in this case we will omit the lower index N from the notation of normalized domains, fuzzy and crisp values. In the subsection on denormalization we will use the lower index N to distinguish between normalized and actual fuzzy and crisp values.

Fuzzification. During fuzzification a crisp controller-input $\mathbf{x}^*$ is assigned a degree of membership to the fuzzy region from the IF-part of a fuzzy rule. Let $LE_1^{\mathbf{i}}, \ldots, LE_n^{\mathbf{i}}$ be some fuzzy values taken by the controller-inputs $e_1, \ldots, e_n$

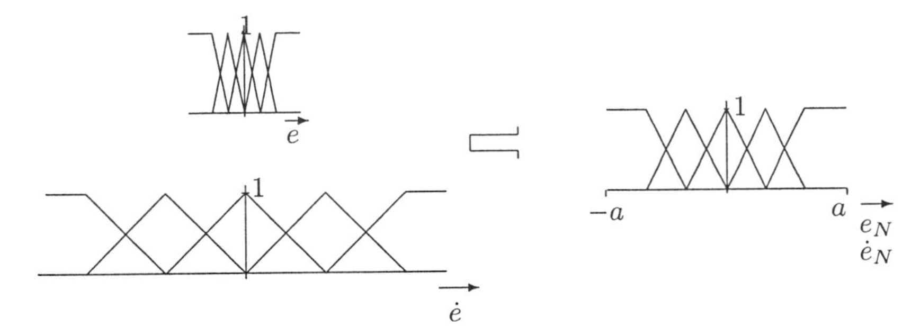

Figure 6.5 Changing of membership functions supports due to input scaling.

in the IF-part of the **i**-th fuzzy rule $R_C^{\mathbf{i}}$ of a FC, i.e., these fuzzy values define the fuzzy region $\mathbf{LE^i} = (LE_1^{\mathbf{i}}, \ldots, LE_n^{\mathbf{i}})^T$.

Each of the above fuzzy values $LE_k^{\mathbf{i}}$ is defined by a membership function on the same (normalized) domain of error E. Thus the fuzzy value $LE_k^{\mathbf{i}}$ is given by the membership function $LE_k^{\mathbf{i}}(e_k)$.

Let us consider now a particular normalized crisp controller-input

$$\mathbf{e}^* = (e_1^*, \ldots, e_n^*)^T, \tag{6.16}$$

from the normalized domain E. Each e_k^* is a normalized crisp value obtained after the input scaling of the current physical controller-input. The *fuzzification* of the crisp normalized controller-input then consists of finding the membership degree of e_k^* in $LE_k^{\mathbf{i}}(e_k)$. This is done for every element of $\mathbf{e}^*$.

Example

Consider the fuzzy rule $R_C^{\mathbf{i}}$ given as

$$R_C^{\mathbf{i}}: \qquad \text{IF} \quad \mathbf{e} = (\text{PS}_e, \text{NM}_{\dot{e}}) \quad \text{THEN} \quad u = \text{PM}_u \tag{6.17}$$

where PS_e is the fuzzy value POSITIVE SMALL of the controller-input e, $\text{NM}_{\dot{e}}$ is the fuzzy value NEGATIVE MEDIUM of the second controller-input $\dot{e}$, and PM_u is the fuzzy value NEGATIVE MEDIUM of the single controller-output u. The membership functions representing these two fuzzy values are given in Fig. 6.6.

In this example we have $\mathbf{e} = (e, \dot{e})^T$ and thus the IF-part of the above rule represents the fuzzy region $\mathbf{LE^i} = (PS_e, NM_{\dot{e}})^T$. Let, furthermore, $e^* = a_1$

and $\dot{e}^* = a_2$ be the current normalized values of the physical controller-inputs e^* and $\dot{e}^*$ respectively as depicted on Fig. 6.6. Then from Fig. 6.6 we obtain the degrees of membership $\text{PS}_e(a_1) = 0.3$ and $\text{NM}_{\dot{e}}(a_2) = 0.65$.

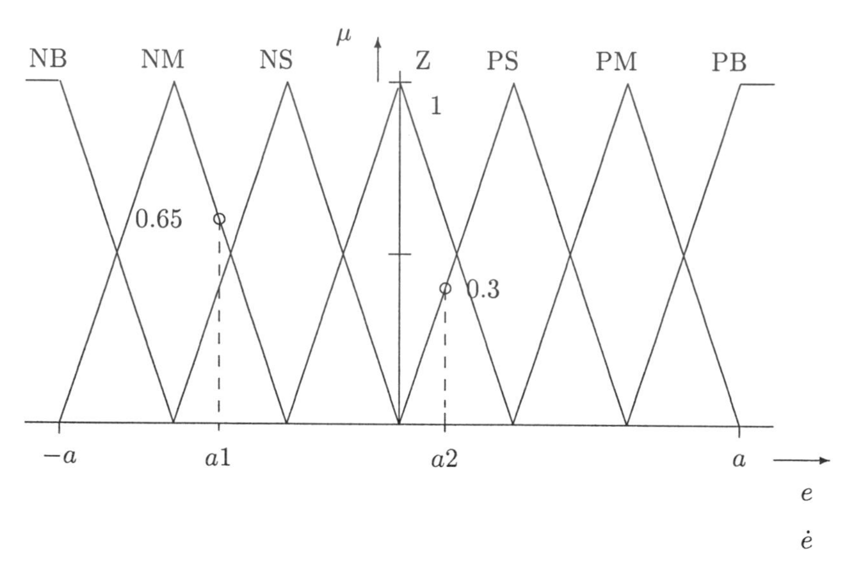

Figure 6.6 Fuzzification of crisp values e^* and $\dot{e}^*$

Rule Firing. For a multi-input/single-output FC the **i**-th fuzzy rule of the set of fuzzy rules has the form

$$R_C^{\mathbf{i}}: \quad \text{IF} \quad \mathbf{e} = \mathbf{LE^i} \quad \text{THEN} \quad u = LU^{\mathbf{i}}, \tag{6.18}$$

where the fuzzy region $\mathbf{LE^i}$ from the IF-part of the above fuzzy rule is given as $\mathbf{LE^i} = (LE_1^{\mathbf{i}}, LE_2^{\mathbf{i}}, \ldots, LE_n^{\mathbf{i}})^T$. Also, $LE_k^{\mathbf{i}}$ denotes the fuzzy value of the k-th normalized controller-input e_k which belongs to the term-set of e_k given as $\mathbf{TE_k} = \{LE_{k1}, LU_{k2}, \ldots, LU_{kn}\}$. Furthermore, $LU^{\mathbf{i}}$ denotes an arbitrary fuzzy value taken by the normalized controller-output u and this fuzzy value belongs to the term-set $\mathbf{TU}$ of u, that is $\mathbf{TU} = \{LU_1, LU_2, \ldots, LU_n\}$.

Let the membership functions defining the fuzzy values from $\mathbf{LE^i}$ and $LU_{\mathbf{i}}$ be denoted by $LE_k^{\mathbf{i}}(e_k)$ $(k = 1, 2, \ldots, n)$ and $LU^{\mathbf{i}}(u)$ respectively. The membership function $LU^{\mathbf{i}}(u)$ is defined on the normalized domain U, and the membership functions $LE_k^{\mathbf{i}}(e_k)$ are defined on the normalized domain E.

Given a controller-input vector $\mathbf{e}^*$ consisting of the normalized crisp values $e_1^*, \ldots, e_n^*$, first the degree of satisfaction $\mathbf{LE^i}(\mathbf{e}^*)$ of the fuzzy region $\mathbf{LE^i}$ is computed as

$$\mathbf{LE^i}(\mathbf{e}^*) = \min\left(LE_1^{\mathbf{i}}(e_1^*), LE_2^{\mathbf{i}}(e_2^*), \ldots, LE_n^{\mathbf{i}}(e_n^*)\right). \qquad (6.19\ )$$

Second, given the degree of satisfaction $\mathbf{LE^i}(\mathbf{e}^*)$ of the fuzzy region $\mathbf{LE^i}$, the normalized controller-output of the **i**-th fuzzy rule is computed as

$$CLU^{\mathbf{i}}(u) = \min\left(\mathbf{LE^i}(\mathbf{e}^*), LU^{\mathbf{i}}(u)\right). \qquad (6.20\ )$$

Thus the controller-output of the **i**-th fuzzy rule is modified by the degree of satisfaction $\mathbf{LE^i}(\mathbf{e}^*)$ of the fuzzy region $\mathbf{LE^i}$ and hence defined as the fuzzy subset $CLU^{\mathbf{i}}(u)$ of $LU^{\mathbf{i}}(u)$. That is

$$\forall u : CLU^{\mathbf{i}}(u) = \begin{cases} LU^{\mathbf{i}}(u) & \text{if } LU^{\mathbf{i}}(u) \leq \mathbf{LE^i}(\mathbf{e}^*), \\ LU^{\mathbf{i}}(u) = \mathbf{LE^i}(\mathbf{e}^*) & \text{otherwise.} \end{cases} \qquad (6.21\ )$$

The fuzzy set $CLU^{\mathbf{i}}(u)$ is called the *clipped controller-output.* It represents the modified version of the controller-output $LU^{\mathbf{i}}(u)$ from the **i**-th fuzzy rule given certain crisp controller-input $e_1^*, \ldots, e_n^*$. The computational step of rule firing is illustrated graphically in Fig. 6.7 for a two-dimensional controller-input vector and a single controller-output.

In the final stage of rule firing, the clipped controller-outputs of all fuzzy rules are combined in a *global controller-output* via *aggregation*

$$\forall u : CU(u) = \max\left(CLU^{\mathbf{1}}, \ldots, CLU^{\mathbf{M}}\right), \qquad (6.22\ )$$

where $CU(u)$ is the fuzzy set defining the fuzzy value of the *global controller-output.*

Defuzzification. The result of rule firing is a fuzzy set CU with a membership function $CU(u)$ as defined in (6.22). The purpose of defuzzification is to obtain a scalar value u from CU. The scalar value u is called a *defuzzified controller-output.* This is done by the *center of gravity method* as follows.

In the continuous case we have

$$u = \frac{\int_U CU(u) \cdot u \cdot du}{\int_U CU(u) \cdot du}, \qquad (6.23\ )$$

and for the discrete case,

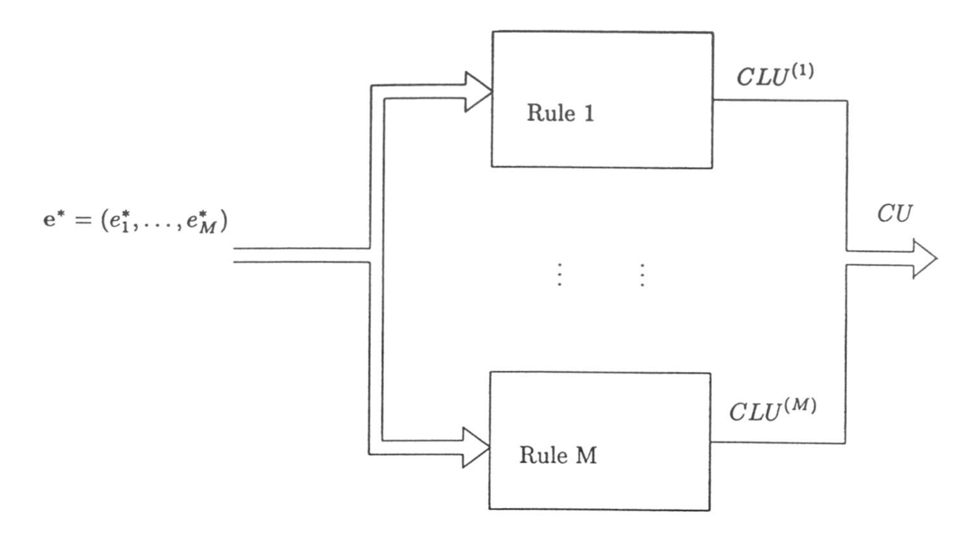

Figure 6.7 The computational step of rule firing.

$$u = \frac{\sum_{U} CU(u) \cdot u}{\sum_{U} CU(u)}. \tag{6.24 }$$

Example
Consider the normalized domain $U = \{1, 2, \ldots, 8\}$ and let the fuzzy set CU be given as

$$CU = \{0.5/3, 0.8/4, 1/5, 0.5/6, 0.2/7\}. \tag{6.25 }$$

Then the defuzzified controller-output u is computed as (see also Fig. 6.8)

$$u = \frac{0.5 \cdot 3 + 0.8 \cdot 4 + 1 \cdot 5 + 0.5 \cdot 6 + 0.2 \cdot 7}{0.5 + 0.8 + 1 + 0.5 + 0.2} = 4.7 \tag{6.26 }$$

Denormalization. In the denormalization procedure the defuzzified normalized controller-output u_N is denormalized with the help of an off-line predetermined scalar denormalization factor N_u^{-1} which is the inverse of the normalization factor N_u .

Let the normalization of the controller-output be performed as:

$$u_N = N_u \cdot u. \tag{6.27 }$$

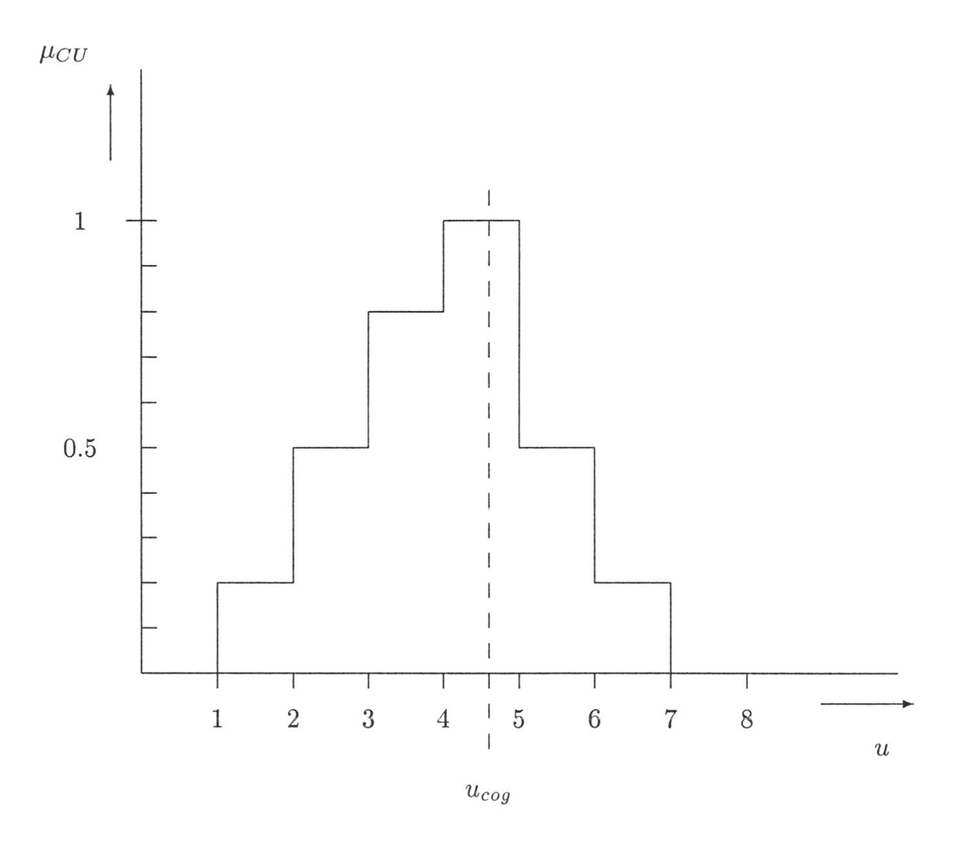

Figure 6.8 Defuzzification of a fuzzy controller-output.

Then, the denormalization of u_N is simply:

$$u = N_u^{-1} \cdot u_N. \tag{6.28 }$$

As we will see in Chap. 3, the choice of N_u essentially determines, together with the rest of the scaling factors, the stability of the system to be controlled. In the case of Takagi-Sugeno FCs the above computational steps are performed on the actual physical domains of the controller-inputs and controller-outputs. Thus the computational steps of normalization and denormalization are not involved in the computational structure of a Takagi-Sugeno FC which in turn eliminates the need for input and output scaling factors.

6.3.2 The Transfer Characteristics

The way to obtain a specific input output transfer characteristics shows the following example (SISO):

1. Suppose there is a set of rules like

   R1: IF x = NB THEN y = PB
   R2: IF x = NS THEN y = PS
   R3: IF x = Z THEN y = Z
   R4: IF x = PS THEN y = NS
   R5: IF x = PB THEN y = NB

   where

   x - input; y - output
   N - negative; P - positive; Z - zero; S - small; B - big.

2. Shape and location of the corresponding membership functions are choosen so that they always overlap at the degree of membership $LX(x) = 0.5$ (see Fig. 6.9).

3. For the specific crisp controller input x_{in} one obtains the degrees of membership $X_{NS}(x_{in}) > 0$ and $X_Z(x_{in}) > 0$ where the remaining degrees of membership $X_{NB}(x_{in})$, $X_{PS}(x_{in})$ and $X_{PB}(x_{in})$ are equal to zero. Hence, only rules R2 and R3 fire. The controller output set is computed by cutting the output set Y_{PS} at the level of $X_{NS}(x_{in})$ and Y_Z at $X_Z(x_{in})$. The resulting output membership function $LY(y)$ takes every rule into account performing the union of the resulting output membership function Y_{Ri} of each rule Ri (i= 1,...,5) which means the maximum operation between them.

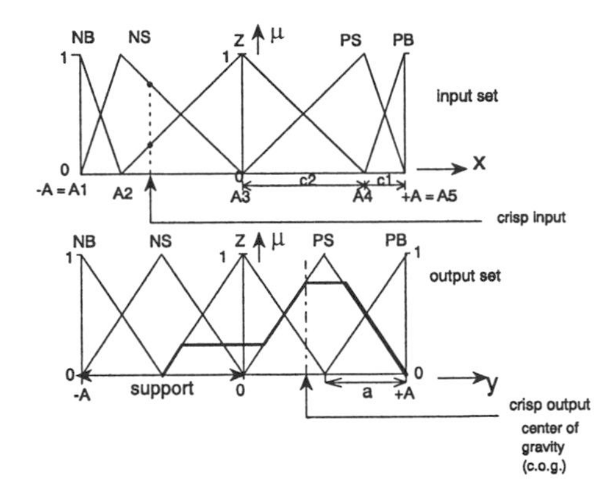

Figure 6.9 Membership functions for input x and output y

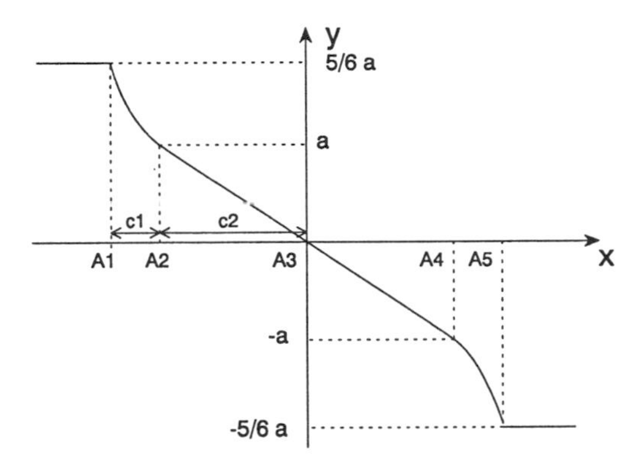

Figure 6.10 Transfer characteristics of a FC

4. The crisp controller output $\bar{y}$ is obtained by calculating the center of gravity of the output set LY:

$$\bar{y} = \frac{\int_{-A}^{+A} Y_{Ri}(y) \cdot y \cdot dy}{\int_{-A}^{+A} Y_{Ri}(y) \cdot dy}. \tag{6.29 }$$

The cut operation (min operation), the max operation over all resulting fuzzy subsets LY_{Ri} and the center of gravity are nonlinear operations

which cause a nonlinear operating line between x and $\bar{y}$. This seems to make a systematic design of a desired transfer function with the help of membership functions difficult.
However, in the x-domain there are operating points A1, A2, A3, A4 and A5 at which only one of the 5 rules fires. At these operating points the center of gravity can be calculated more easily than for the intermediate points. The operating points A1, A2, A3, A4 and A5 form points in the x-y-domain (see Fig. 6.10). The values of the transfer characteristic between the operating points may show a slight nonlinear behavior but from a linear approximation (interpolation) between two operating points one obtains the relation between the supports of the input and output membership functions on the one hand and slopes required of the transfer characteristic on the other hand.

6.3.3 The Nonlinearity of the FC

In this section we will describe the sources of nonlinearity of the transfer characteristic of a FC by relating them to particular computational steps.

System theory distinguishes between two basic types of systems: *linear* and *nonlinear*. A system is linear if and only if it has both the *additivity property* and the *scaling property*, otherwise it is a nonlinear system.

The Additivity Property (Superposition Property):. Let it be the case that

$$y_1 = f(x) \qquad \text{and} \qquad y_2 = f(z). \tag{6.30}$$

Then for the *additivity property* to hold it is required

$$y_1 + y_2 = f(x + z). \tag{6.31}$$

Hence, we obtain

$$f(x) + f(z) = f(x + z). \tag{6.32}$$

Scaling Property (Homogenity Property):. Let it be the case that

$$y = f(x). \tag{6.33}$$

Then for the *scaling property* to hold it is required

$$\alpha \cdot y = f(\alpha \cdot x) \qquad \text{and} \qquad \alpha \cdot f(x) = f(\alpha \cdot x). \tag{6.34}$$

Because of fuzzification and defuzzification, a FC is in fact a crisp TE. This crisp TE has a nonlinear transfer characteristic because of the nonlinear character of fuzzification (when performed on nonlinear membership functions), rule firing, and defuzzification. The argument for this is that if one computational step within the computational structure of the TE is nonlinear then the whole TE is nonlinear as well. Using the additivity and scaling properties of a linear system we will now establish the linearity, or nonlinearity, of each computational step in the computational structure of a FC with respect to these two properties.

In what follows, without any loss of generality, we will use a single MISO fuzzy rule such as

$$R_C: \quad \text{IF} \quad e = LE \quad \text{THEN} \quad u = LU, \tag{6.35}$$

where LE and LU are the fuzzy values taken by the normalized, single controller-input e and the normalized, single controller-output u respectively. These two fuzzy values are determined by the membership functions $LE(e)/e$ and $LU(u)/u$ defined on the normalized domains E and U. Here again we only consider normalized domains, fuzzy and crisp values, and thus the lower index N will be omitted from the notation, unless there is a need to distinguish between normalized and actual crisp and fuzzy values used within the same expression.

Let furthermore e_1^* and e_2^* be two normalized crisp controller-inputs and u_1^* and u_2^* be the defuzzified controller-outputs corresponding to these normalized controller-inputs.

Input Scaling and Output Scaling. Input scaling is linear because it simply multiplies each physical controller-input e_1^* and e_2^* with a predetermined scalar N_e (normalization factor) to obtain their normalized counterparts e_{1N}^* and e_{2N}^*. Thus we have

$$N_e \cdot e_1^* + N_e \cdot e_2^* = N_e \cdot (e_1^* + e_2^*). \tag{6.36}$$

Furthermore, for a given scalar α we have

$$\alpha \cdot N_e \cdot e_1^* = N_e \cdot (\alpha \cdot e_1^*). \tag{6.37}$$

Thus input scaling has the properties of additivity and scaling and is thus a linear computational step. The same is valid for output scaling since it uses N_e^{-1} instead of N_e.

Fuzzification. Let the membership function $LE(e)$ defining the normalized fuzzy value LE be, in general, a nonlinear function, e.g., a triangular membership function. The fuzzification of e_1^* and e_2^* results in finding $LE(e_1^*)$ and $LE(e_2^*)$. Linearity requires

$$LE(e_1^*) + LE(e_2^*) = LE(e_1^* + e_2^*). \tag{6.38}$$

The above equality cannot be fulfilled because the membership function $LE(e)$ is nonlinear. Thus fuzzification in the case of nonlinear membership functions is a nonlinear computational step.

Rule Firing. Let the membership function $LU(u)$ defining the normalized fuzzy value LU be, in general, a nonlinear function. Then the result of rule firing given the normalized crisp controller-input e_1^* will be:

$$\forall u : CLU'(u) = \min\left(LU(e_1^*), LU(u)\right). \tag{6.39}$$

Similarly, for the normalized crisp controller-input e_2^* we obtain

$$\forall u : CLU''(u) = \min\left(LE(e_2^*) LU(u)\right). \tag{6.40}$$

Linearity requires

$$\forall u : CLU'(u) + CLU''(u) = \min\left(LE(e_1^* + e_2^*), LU(u)\right), \tag{6.41}$$

but the above equality does not hold because

- $LU(u)/u$ is a nonlinear membership function,
- $CLU'(u)/u$ and $CLU''(u)/u$ are nonlinear membership functions, (usually defined as piecewise linear functions),
- the function min is a nonlinear function.

Thus rule firing is a nonlinear computational step within the computational structure of a FC.

Defuzzification. Let defuzzification be performed with the center of gravity method. Let furthermore u_1 and u_2 be the normalized defuzzified controller-outputs obtained after defuzzification. That is,

$$u_1 = \frac{\int_U CLU'(u) \cdot u \cdot du}{\int_U CLU'(u) \cdot du}, \tag{6.42}$$

$$u_2 = \frac{\int_U CLU''(u) \cdot u \cdot du}{\int_U CLU''(u) \cdot du}. \tag{6.43}$$

Linearity requires

$$u_1 + u_2 = \frac{\int\limits_U (CLU'(u) + CLU''(u)) \cdot u \cdot du}{\int\limits_U (CLU'(u) + CLU''(u)) \cdot du}. \tag{6.44}$$

However, the above equality cannot be fulfilled since instead of it we have

$$u_1 + u_2 = \frac{\int\limits_U CLU'(u) \cdot u \cdot du}{\int\limits_U CLU'(u) \cdot du} + \frac{\int\limits_U CLU''(u) \cdot u \cdot du}{\int\limits_U CLU''(u) \cdot du}. \tag{6.45}$$

This shows that the nonlinearity of the computational step of defuzzification comes from the normalization of the products $\int_U CLU'(u) \cdot u \cdot du$ and $\int_U CLU''(u) \cdot u \cdot du$.

From all of the above it is readily seen that a FC is a nonlinear TE its sources of nonlinearity being the nonlinearity of membership functions, rule firing, and defuzzification.

However, in the case of a Takagi-Sugeno FC-1 each single fuzzy rule is a linear TE for all controller-inputs (state vectors) that belong to the center of the fuzzy region specified by the IF-part of this rule. At the same time, for controller-inputs outside the center of a fuzzy region this same fuzzy rule is a nonlinear TE. Because of the latter, the set of all fuzzy rules of a Takagi-Sugeno FC-1 defines a nonlinear TE. In the case of a Takagi-Sugeno gain scheduler we have that each fuzzy rule defines a linear TE everywhere in a given fuzzy region.

6.4 HEURISTIC CONTROL AND MODEL BASED CONTROL

Fuzzy control can be classified into the main directions *Heuristic fuzzy control* and *Model based fuzzy control.*

Heuristic control deals with plants which are unsufficiently described from the mathematical point of view while *Model based fuzzy control* deals with plants for which a mathematical model is available. In the following we will shortly describe the following control strategies

Mamdani control (MC)

Sliding mode fuzzy control (SMFC)

Cell mapping control (CM)

Takagi/Sugeno control (TS)

Takagi/Sugeno control (TS) with Lyapunov linearization

6.4.1 The Mamdani Controller

This type of FC obtaines its control strategy from expert knowledge. Since a model of the plant is not available, a simulation of the closed loop cannot be performed. Therefore, the control design bases on trial and error strategies which makes the implementation of the FC critical. The crucial point is that the behavior of the plant to be controlled is only reflected through the operator rules. However, from the control point of view this is not a satisfactory situation. Thus, one seeks methods to build qualitative models in terms of fuzzy rules.

In the context of heuristic control the so-called *Mamdani Control rules* are used where both the antecentent and the consequent include fuzzy values. A typical control rule (operator rule) is

$$R_{Ci}: \qquad IF \quad \mathbf{x} = \mathbf{LX^i} \quad THEN \quad \mathbf{u} = \mathbf{LU^i} \tag{6.46}$$

For a system with two state variables and one control variable we have, for example,

$$R_{Ci}: \qquad IF \quad x = PS \quad AND \quad \dot{x} = NB \quad THEN \quad u = PM \tag{6.47}$$

which can be rewritten into

$$R_{Ci}: \qquad IF \quad (x, \dot{x})^T = (PS, NB)^T \quad THEN \quad u = PM \tag{6.48}$$

with

$$\mathbf{x} = (x, \dot{x})^T.$$

$$\mathbf{LX^i} = (PS, NB)^T.$$

$$\mathbf{u} = u.$$

$$\mathbf{LU^i} = PM.$$

Even if there is only little knowledge about the system to be controlled, one has to have some ideas about the behavior of the system state vector $\mathbf{x}$, its change with time $\dot{\mathbf{x}}$, and the control variable $\mathbf{u}$. This kind of knowledge is structural and can be formulated in terms of fuzzy rules. A typical fuzzy rule for a system is

$$R_{Si}: \qquad IF \quad \mathbf{x} = \mathbf{LX^i} \quad AND \quad \mathbf{u} = \mathbf{LU^i} \quad THEN \quad \dot{\mathbf{x}} = \mathbf{L\dot{X}^i}. \tag{6.49}$$

For the above system with two states and one control variable we have, for example,

$$R_{Si}: \quad IF \quad (x, \dot{x})^T = (PS, NB)^T \quad AND \quad u = PM$$
$$THEN \quad (\dot{x}, \ddot{x})^T = (NM, PM)^T \tag{6.50}$$

with

$$\mathbf{x} = (x, \dot{x})^T.$$

$$\mathbf{LX^i} = (PS, NB)^T.$$

$$\dot{\mathbf{x}} = (\dot{x}, \ddot{x})^T.$$

$$\mathbf{L\dot{X}^i} = (NM, PM)^T.$$

$$\mathbf{u} = u.$$

$$\mathbf{LU^i} = PM.$$

A typical Mamdani controller is the sliding mode fuzzy controller (SMFC) [9, 11, 17].

Once the qualitative system structure is known one has to find the corresponding quantitative knowledge. Quantitative knowledge means the following. In general, both control rules and system rules work with normalized domains. The task is to map inputs and outputs of both the controller and the system to normalized domains. For the system, this task is identical with the identification of the system parameters.

For the controller, this task is identical with the controller design namely to find the proper control gains.

6.4.2 Sliding Mode FC

FCs for a large class of second order nonlinear systems are designed by using the phase plane determined by error e and change of error $\dot{e}$ [23, 24, 33, 37]. The fuzzy rules of these FCs determine a fuzzy value for the input u for each pair of fuzzy values of error and change of error that is, for each fuzzy state vector. The usual heuristic approach to the design of these fuzzy rules is the partitioning of the phase plane into two semi-planes by means of a *sliding (switching) line.* This means that the FC has the so-called *diagonal form.* Another possibility is, instead of using a sliding line, to use a sliding curve like a time optimal trajectory [30].

A typical fuzzy rule for the FC in a diagonal form is

$\dot{e}$ \ e	NB	NM	NS	Z	PS	PM	PB
PB	Z	NS	NS	NM	NM	NB	NB
PM	PS	Z	NS	NS	NM	NM	NB
PS	PS	PS	Z	NS	NS	NM	NM
Z	PM	PS	PS	Z	NS	NS	NM
NS	PM	PM	PS	PS	Z	NS	NS
NM	PB	PM	PM	PS	PS	Z	NS
NB	PB	PB	PM	PM	PS	PS	Z

P - positive
N - negative
Z - zero
S - small
M - medium
B - big

Figure 6.11 A FC in a diagonal form.

$$\text{IF} \quad e = \text{PS} \quad \text{AND} \quad \dot{e} = \text{NB} \quad \text{THEN} \quad u = \text{PS}, \tag{6.51}$$

where PS stands for the fuzzy value of error POSITIVE SMALL, NB stands for the fuzzy value change of error NEGATIVE BIG, and PS stands for the fuzzy value POSITIVE SMALL of the input.

Each semi-plane is used to define only negative or positive fuzzy values of the input u. The magnitude of a specific positive/negative fuzzy value of u is determined on the basis of the distance $|s|$ between its corresponding state vector $\mathbf{e}$ and the sliding line $s = \lambda \cdot e + \dot{e} = 0$. This is normally done in such a way so that the absolute value of the required input u increases/decreases with the increasing/decreasing distance between the state vector $\mathbf{e}$ and the sliding line $s = 0$.

It is easily observed that this design method is very similar to sliding mode control (SMC) with a *boundary layer* (BL) which is a robust control method [35], [28] (Part II, Chap. 7, Sect. 7.2) Sliding mode control is applied especially to control of nonlinear systems in the presence of model uncertainties, parameter fluctuations, and disturbances. The similarity between the diagonal form FC and SMC enables us to redefine a diagonal form FC in terms of a SMC with BL and then to verify its stability, robustness, and performance properties in a manner corresponding to the analysis of a SMC with BL. In the following, the diagonal FC is therefore called SMFC.

However, one is tempted to ask here what does one gain by introducing the SMFC type of controller ? The answer is that SMC with BL is a special case of SMFC. SMC with BL provides a linear transfer characteristic with lower and upper bounds while the transfer characteristic of a SMFC is not nessecarily a straight line between these bounds, but a curve that can be adjusted to reflect given performance requirements. For example, normally a fast rise time and as little overshoot as possible are the required performance characteristics for the closed loop system. These can be achieved by making the controller gains much larger for state space regions far from the sliding line than its gains in state space regions close to the sliding line (see Fig. 6.12).

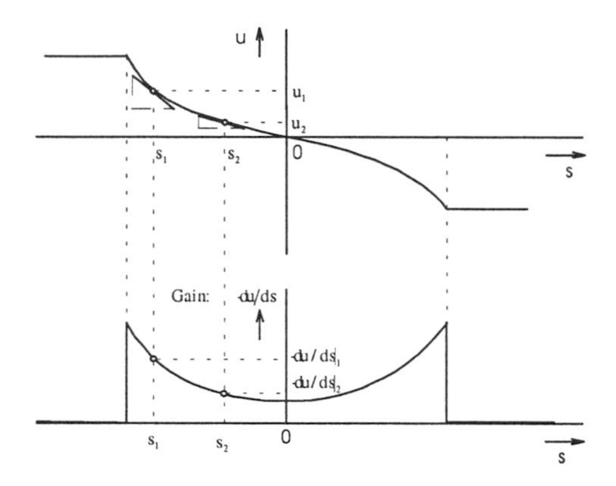

Figure 6.12 The adjustable transfer characteristic of a SMFC.

In this connection it has to be emphasized that a SMFC is a state dependent filter. The slope of its transfer characteristic decides the convergence rate to the sliding line and, at the same time, the bandwidth of the unmodeled disturbances that can be coped with. This means that far from the sliding line higher frequencies are allowed to pass through than in the neighborhood of it. The other function of this state dependent filter is given by the sliding line itself. That is, the velocity with which the origin is approached is determined by the slope λ of the sliding line $s = 0$.

Because of the special form of the rule base of a diagonal form FC (see Fig. 6.11) each fuzzy rule can be redefined in terms of the fuzzy value of the distance $|s|$ between the state vector $\mathbf{e}$ and the sliding line, and the fuzzy value of the input u corresponding to this distance. This helps to reduce the number of fuzzy rules especially in the case of higher order systems. Namely, if the number of state variables is 2 and each state variable has m fuzzy values, the number of fuzzy rules of the diagonal form FC is $M = m^2$. For the same case, the number of fuzzy rules of a SMFC is only m. This is so, because the fuzzy rules of the SMFC only describe the relationship between the distance to the sliding line and the input u corresponding to this distance, rather than the relationship between all possible fuzzy state vectors and the input u corresponding to each fuzzy state vector.

Moreover, the fuzzy rules of a SMFC can be reformulated to include the distance d between the state vector $\mathbf{e}$ and the vector normal to the sliding line and passing through the origin (see Fig. 6.13). This gives an additional opportunity to further affect the rate with which the origin is approached. A fuzzy rule including this distance is of the form

$$\text{IF} \quad s = \text{PS} \quad \text{AND} \quad d = \text{S} \quad \text{THEN} \quad u = \text{NS}. \tag{6.52}$$

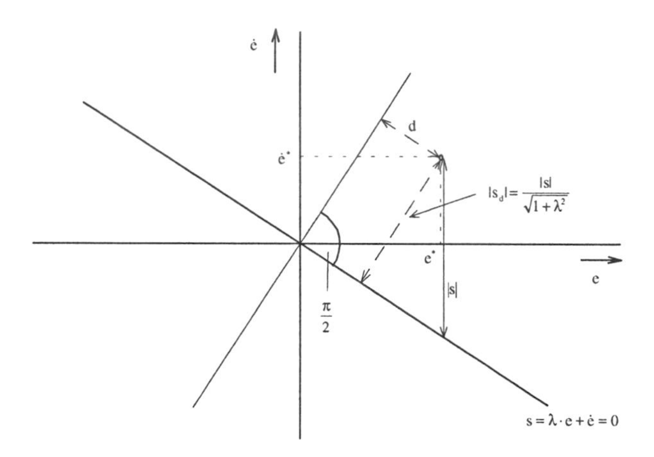

Figure 6.13 The s and d parameters of a SMFC.

Despite of the advantages of a SMFC it poses a number of problems the solutions of which can improve its performance and robustness. One such problem is the addition of an integrator to a SMFC in order to eliminate remaining errors in the presense of disturbances and model uncertainties. There are several ways to accomplish this. One option for example, is to treat the integration term in the same manner as the other parameters of the IF-part of the SMFC's fuzzy rules. This and other available options will be described in Chap. 3 of this report.

Another problem is the so called *scaling* of the SMFC parameters so that the domains on which their fuzzy values are defined are properly determined and optimized with respect to performance. This problem arises in the context of SMFC since the real physical domains of the SMFC parameters are normalized, i.e., their measured values are mapped on their respective normalized domains by the use of *normalization factors.* Thus a normalized input u is the result of the computation with SMFC. This normalized u is then consequently denormalized, i.e., mapped back on its physical domain, by the use of a *denormalization factor*

The determination of the proper scaling factors, via which the normalization and denormalization of the SMFC parameters is performed, is not only part of the design, but also important in the context of adaptation and on-line tuning of the SMFC. The behavior of the closed loop system ultimately depends on the normalized transfer characteristic (control surface) of the SMFC. This control surface is mainly determined by the shape and location of the membership functions defining the fuzzy values of the SMFC's parameters. In this context one need pay attention to the following.

1. The denormalization factor for u influences most stability and oscillations. Because of its impact on stability the determination of this factor has the highest priority in the design.

2. Normalization factors influence most the SMFC sensitivity with respect to the proper choice of the operating regions for s and d. Therefore, normalization factors are second in priority in the design.

3. The proper shape and location of the membership functions and, with this, the transfer characteristic of the SMFC can influence positively the behavior of the closed loop system in different fuzzy regions of the fuzzy state space provided that the operating regions of s and u are properly chosen through a well adjusted normalization factors. Therefore, this aspect comes third in priority.

A third problem is the design of SMFC for MIMO systems. The design for SISO systems can still be utilized though some new aspects and restrictions come into play when this design is extended to the case of MIMO systems. First, we assume that the MIMO system has as many input variables u_i as it has state variables x_i. Second, we assume that the so called *matching condition* holds [28] (Part II, Chap. 7, Sect. 7.2). This condition constrains the so called *parametric uncertainties.* These are, for example, imprecision on the mass or inertia of a mechanical system, inaccuracies on friction functions, etc. *Non-parametric uncertainties* are unmodeled dynamics, neglected time delays, etc.

Let $\dot{\mathbf{x}} = \mathbf{f}(\mathbf{x}) + \mathbf{B} \cdot \mathbf{u}$, be the nonlinear open loop system to be controlled, where $\mathbf{f}$ is a nonlinear vector function of the state vector $\mathbf{x}$, $\mathbf{u}$ is the input vector, and $\mathbf{B}$ is the *input matrix.*

Then, the matching condition requires that the parametric uncertainties have to be within the range of the input matrix $\mathbf{B}$. Since $\mathbf{B}$ is assumed to be a square matrix, this means that $\mathbf{B}$ has to be invertible over the whole state space which is a controllability-like assumption. A second condition required is that the estimate $\hat{\mathbf{B}}$ also is invertible.

6.4.3 Cell Mapping

Cell mapping originates from a computational technique introduced by C.S.Hsu [8] which evaluates the global behavior and the stability of nonlinear systems. It is assumed that the computational (analytical) model of the system is available. Cell mapping has been firstly applied to fuzzy systems by Chen and Tsao [2].

The benefits of using cell mapping for fuzzy controlled systems are

- supporting of self-learning FC strategies
- creating of methodologies for the design of time optimal FC's

The basic idea of Hsu is the following.
Let a nonlinear system be described by the point mapping

$$\mathbf{x}(t_{k+1}) = \mathbf{f}(\mathbf{x}(t_k), \mathbf{u}(t_k)). \tag{6.53 }$$

The t_k represent the discrete time steps over which the point mapping occurs. It has to be emphasized that these time steps need not to be uniform in duration. If one wants to create a map of the state space taking into account all possible states $\mathbf{x}$ and control vectors $\mathbf{u}$, one obtains an infinite number of mappings even for finite domains for $\mathbf{a}$ and $\mathbf{u}$, respectively. Therefore, the (finite) state space is divided into a finite number of cells (see Fig. 6.14). Cells are formed by partitioned the domain of interest of each axis x_i of the state space into intervalls of size s_i which are denotet by an integer valued index z_i. Then, a cell is an n-tuple (a vector) of intervalls $\mathbf{z} = (z_1, z_2, \ldots, z_n)^T$. The remainder of the state space outside the finite state space of interest is lumped together into one so-called *sink cell.* The state of the system (6.53) while in the cell $\mathbf{z}$ is represented by the center point $\mathbf{x}^c$. Now, a cell mapping C is defined by

$$\mathbf{z}(t_{k+1}) = \mathbf{C}(\mathbf{z}(t_k)) \tag{6.54 }$$

which is derived from the point mapping (6.53) by computing the image of a point $\mathbf{x}(t_k)$ and then determining the cell in which the image point is located. It is clear that not all points $\mathbf{x}(t_k)$ in cell $\mathbf{z}(t_k)$ have the same image cell $\mathbf{z}(t_{k+1})$. Therefore, only the image cell of the center $\mathbf{x}^c(t_k)$ is considered. A cell which maps to itself is called an *equilibrium cell.* All cells in the finite state space are called *regular cells.*

The motivation for cell mapping is to obtain an appropriate control action sequence $\mathbf{u}(t_k)$ that drives the system (6.53) to an equilibrium while minimizing a predefined cost function. Therefore, every cell is characterized by

- the Group number $G(\mathbf{z})$ that denotes cells $\mathbf{z}$ belonging to the same periodic domain or domain of attraction,

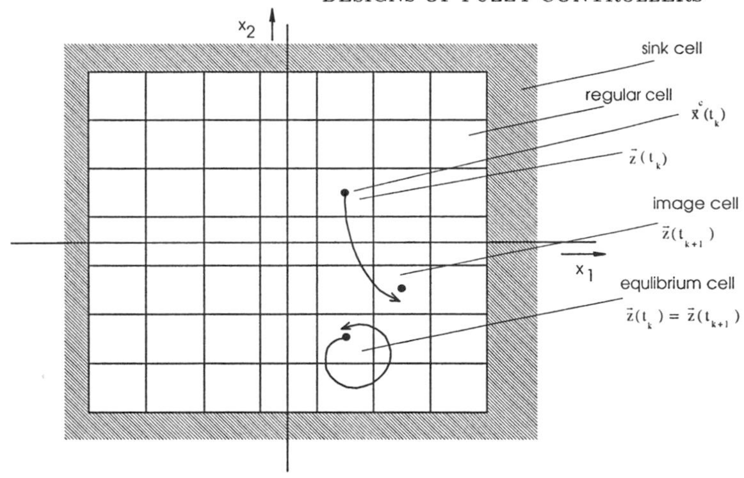

Figure 6.14 Cell mapping principle.

- the step number $S(\mathbf{z})$ that indicates the number of transitions needed to transmit from cell $\mathbf{z}$ to a periodic cell,

- the periodicity number $P(\mathbf{z})$ that indicates the number of cells contributing in the periodic motion.

This characterization is introduced in order to find periodic motions and domain of attractions by a grouping algorithm.

Applied to fuzzy control it is evident that each cell describing the system's behavior belongs a corresponding fuzzy system rule to. Furthermore, each cell describing a particular control action belongs to a corresponding fuzzy control rule.

Smith and Comer developed a fuzzy cell mapping algorithm the aim of which is to calibrate (tune) a FC on the basis of the cell state space concept [29]. Each cell is associated with a control action and a duration which map the cell to a minimum cost trajectory (e.g. minimum time). With a given cost function and a plant simulation model the cell state space algorithm which bases on an optimal control strategy generates a table of desired control actions. The mapping from cell to cell is carried out by a fuzzy controller which smoothes out the control actions while the transitions between the cells. The cell to cell mapping technique has been used to fine tune a Takagi/Sugeno controller (see Fig. 6.15 [20]).

Kang and Vachtsevanos developed a phase portrait assignment algorithm which is related to cell to cell mapping [10]. In this approach states and control variables are partitioned into different cell spaces. The $\mathbf{x}$ -cell space is

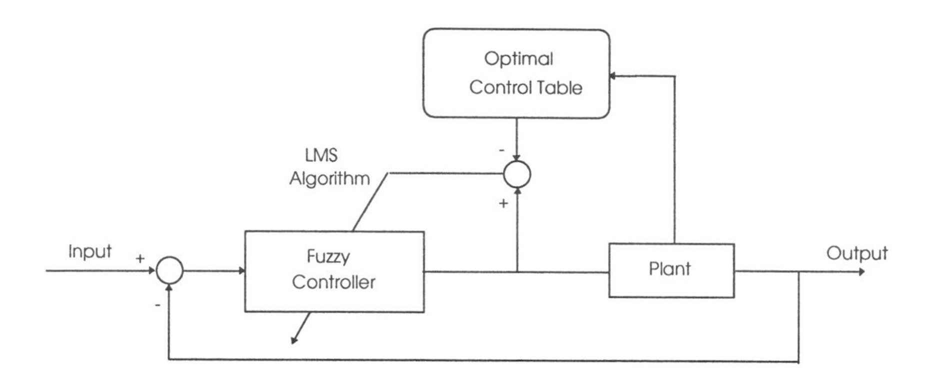

Figure 6.15 Cell mapping by Smith and Comer (Redrawn from Papa et. al.).

recorded by applying a constant input to the system being simulated. Then, by means of a searching algorithm the rule base of the FC is constructed such that asymptotic stability is guaranteed. This is performed by determining the optimal control actions regardless of from which cell the algorithm starts its search (see Fig. 6.16).

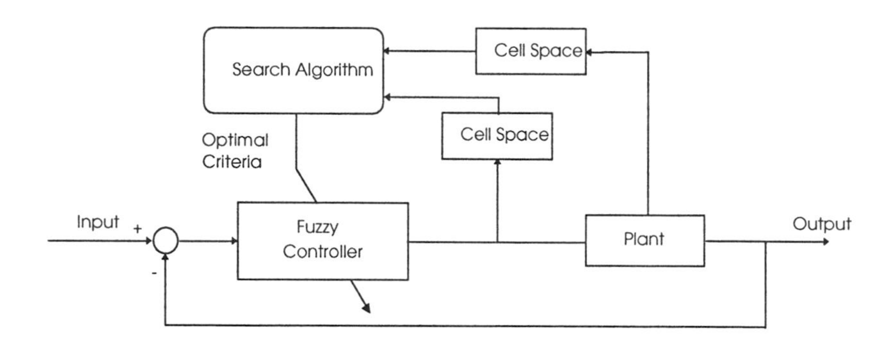

Figure 6.16 Cell mapping by Kang and Vachtsvanos (Redrawn from Papa et. al.)

Hu, Tai, and Shenoi apply genetic algorithms to improve the searching algorithm using cell maps [7]. The aim of this method is to tune a Takagi/Sugeno controller.

6.4.4 TS Model Based Control

Model based FC design starts from the mathematical knowledge of the system to be controlled [31, 32]. In this connection one is tempted to ask why one should use FC in this particular case while conventional control techniques do they work well. The reasons to apply FC in analytical known systems are

1. FC is a user friendly and transparent control method because of its rule based structure,
2. FC provides a nonlinear control strategy which is related to traditional nonlinear control techniques
3. The nonlinear transfer characteristics of a FC can be tuned by changing shape and location of the membership functions so that adaptation procedures can be applied.
4. The approximation property of FC allows the design of a complicated control law with the help of only few rules
5. Gain scheduling techniques can be transfered to FC. In this connection FC is used as an approximator between linear control laws.

The description of the system starts from a fuzzy model of the system which uses both the fuzzy state space and a crisp description of the system.

Let the principle of a Takagi/Sugeno (TS) system be explained by the following example.

Example

Consider a TS system consisting of two rules with x_1 and x_2 as system inputs and y as the system output.

R_1: *if* x_1 is BIG *and* x_2 is MEDIUM *then* $y_1 = x_1 - 3 \cdot x_2$
R_2: *if* x_1 is SMALL *and* x_2 is BIG *then* $y_2 = 4 + 2 \cdot x_1$.

Let the inputs measured be $x_1^* = 4$ and $x_2^* = 60$.

From Fig.6.17 we then obtain:

$$LX_{\mathrm{BIG}}(x_1^*) = 0.3 \qquad LX_{\mathrm{BIG}}(x_2^*) = 0.35$$

and

$$LX_{\mathrm{SMALL}}(x_1^*) = 0.7 \qquad LX_{\mathrm{MED}}(x_2^*) = 0.75.$$

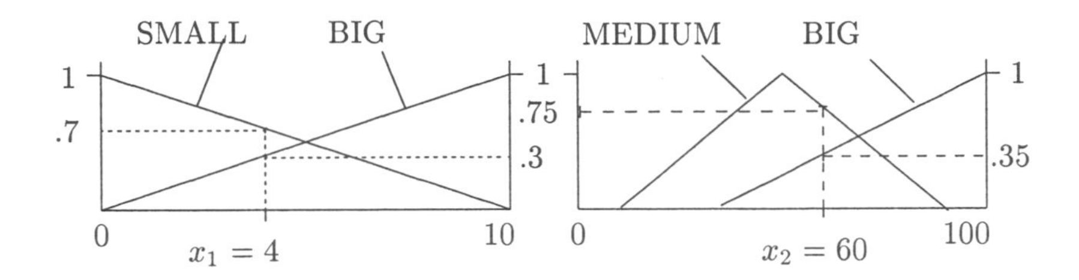

Figure 6.17 Fuzzification procedure for a TS controller

For the degree of satisfaction of R_1 and R_2, respectively, we obtain.

$$\min(0.3, 0.75) = 0.3 \qquad \min(0.7, 0.35) = 0.35.$$

Furthermore, for the consequents of rule R_1 and R_2 we have

$$y_1 = 4 - 3 \cdot 60 = -176 \qquad y_2 = 4 + 2 \cdot 4 = 12.$$

So, the two pairs corresponding to each rule are $(0.3, -176)$ and $(0.35, 12)$ Thus, by taking the weighted normalized sum we get

$$y = \frac{0.3 \cdot (-176) + 0.35 \cdot 12}{0.3 + 0.35} = -74.77.$$

This can be extended to differential equations in the following way.
Let a fuzzy region $\mathbf{LX^i}$ be described by the rule

$$R_{Si}: \qquad IF \quad \mathbf{x} = \mathbf{LX^i} \quad THEN \quad \dot{\mathbf{x}} = \mathbf{A}(\mathbf{x^i}) \cdot \mathbf{x} + \mathbf{B}(\mathbf{x^i}) \cdot \mathbf{u} \qquad (6.55\ )$$

This rule means: IF state vector $\mathbf{x}$ is in fuzzy region $\mathbf{LX^i}$ THEN the system obeys the local differential equation $\dot{\mathbf{x}} = \mathbf{A}(\mathbf{x^i}) \cdot \mathbf{x} + \mathbf{B}(\mathbf{x^i}) \cdot \mathbf{u}$. A summation of all contributing system rules provides the global behavior of the system. In (6.55) $\mathbf{A}(\mathbf{x^i})$ and $\mathbf{B}(\mathbf{x^i})$ are constant system matrices in the center of fuzzy region $\mathbf{LX^i}$ which can be identified by classical identification procedures.

The resulting system equation is

$$\dot{\mathbf{x}} = \sum_{i=1}^{n} w_i(\mathbf{x}) \cdot (\mathbf{A}(\mathbf{x^i}) \cdot \mathbf{x} + \mathbf{B}(\mathbf{x^i}) \cdot \mathbf{u}) \qquad (6.56\ )$$

where $w_i(\mathbf{x}) \in [0,1]$ are the normalized degrees of satisfaction of a fuzzy region $\mathbf{LX^i}$.
The corresponding control rule is

$$R_{Ci}: \qquad IF \quad \mathbf{x} = \mathbf{LX^i} \quad THEN \quad \mathbf{u} = \mathbf{K}(\mathbf{x^i}) \cdot \mathbf{x} \tag{6.57 }$$

and the control law for the whole state space is

$$\mathbf{u} = \sum_{i=1}^{n} w_i(\mathbf{x}) \cdot \mathbf{K}(\mathbf{x^i}) \cdot \mathbf{x}. \tag{6.58 }$$

Together with (6.56) one obtains the closed loop system

$$\dot{\mathbf{x}} = \sum_{i,j=1}^{n} w_i(\mathbf{x}) \cdot w_j(\mathbf{x}) \cdot (\mathbf{A}(\mathbf{x^i}) + \mathbf{B}(\mathbf{x^i}) \cdot \mathbf{K}(\mathbf{x^j})) \cdot \mathbf{x} \tag{6.59 }$$

It has to be emphasized that a system described by a set of rules like (6.55) is nonlinear even in the vicinity of the center of the region. This is due to the fact that $w_i(\mathbf{x})$ depends on the state vector $\mathbf{x}$. Even if $w_i(\mathbf{x})$ is a piecewise linear function of $\mathbf{x}$, the product $w_i(\mathbf{x}) \cdot w_j(\mathbf{x}) \cdot (\mathbf{A}(\mathbf{x^i}) + \mathbf{B}(\mathbf{x^i}) \cdot \mathbf{K}(\mathbf{x^j})) \cdot \mathbf{x}$ in (6.58) will always be a nonlinear function.

6.4.5 Model based Control with Lyapunov linearization

In the following we discuss the case when a mathematical model of the system to be controlled is available and the FC is formulated in terms of fuzzy rules [31, 32, 19, 25]. In this case system and controller are formulated on different semantic levels. Let the system analysis starts from the mathematical model of the system

$$\dot{\mathbf{x}} = \mathbf{f}(\mathbf{x}, \mathbf{u}) \tag{6.60 }$$

and let the FC be formulated in terms of fuzzy rules

$$R_{Ci}: \qquad IF \quad \mathbf{x} = \mathbf{LX^i} \quad THEN \quad \mathbf{u} = \mathbf{LU^i}. \tag{6.61 }$$

In order to study stability, robustnes, and performance of the closed loop system one has to bring system and controller onto the same semantic level. Thus, formally we translate the set of control rules into an analytical structure

$$\mathbf{u} = \mathbf{g}(\mathbf{x}) \tag{6.62 }$$

where, in general the function $\mathbf{g}(\mathbf{x})$ is a nonlinear control surface being a static mapping of the state vector $\mathbf{x}$ to the control vector $\mathbf{u}$ (see Fig. 6.18)

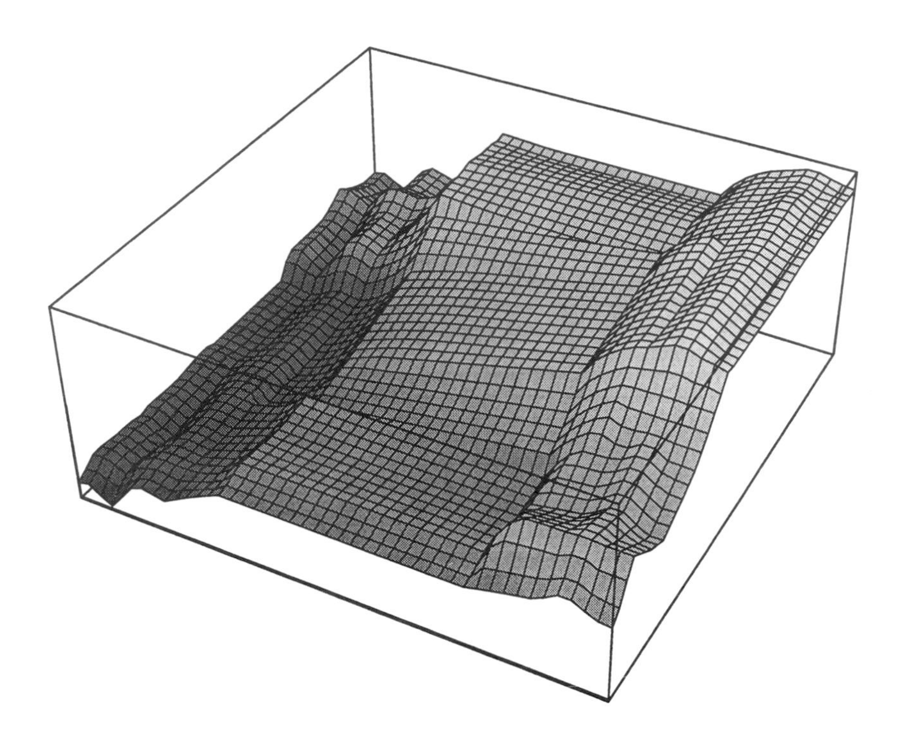

Figure 6.18 Nonlinear control surface $\mathbf{u} = \mathbf{g}(\mathbf{x})$.

The control surface provides information about local and global properties of the controller. For example, the local gain for a specific state vector can be obtained by means of the tangential plane being attached to the corresponding point in state space. From this information one can conclude whether the controlled system is locally stable or not. Furthermore, one obtains a geometrical insight into the way of how the control gain changes as the state trajectory moves in the state space.
Another aspect is the following. In order to study the local behavior of the system around specific points in state space we linearize the system around them and study the closed loop behavior in the linearized region. Let, for example, system (6.60) be linearized around a desired state $\mathbf{x}^{\mathbf{d}}$ and a corresponding state vector $\mathbf{u}^{\mathbf{d}}$

$$\dot{\mathbf{x}} = \mathbf{f}(\mathbf{x}^{\mathbf{d}}, \mathbf{u}^{\mathbf{d}}) + \mathbf{A}(\mathbf{x}^{\mathbf{d}}, \mathbf{u}^{\mathbf{d}}) \cdot (\mathbf{x} - \mathbf{x}^{\mathbf{d}}) + \mathbf{B}(\mathbf{x}^{\mathbf{d}}, \mathbf{u}^{\mathbf{d}}) \cdot (\mathbf{u} - \mathbf{u}^{\mathbf{d}}) \tag{6.63 }$$

where $\mathbf{A}(\mathbf{x^d}, \mathbf{u^d}) = \left.\frac{\partial \mathbf{f}(\mathbf{x},\mathbf{u})}{\partial \mathbf{x}}\right|_{\mathbf{x^d},\mathbf{u^d}}$ and $\mathbf{B}(\mathbf{x^d}, \mathbf{u^d}) = \left.\frac{\partial \mathbf{f}(\mathbf{x},\mathbf{u})}{\partial \mathbf{u}}\right|_{\mathbf{x^d},\mathbf{u^d}}$ are Jacobi matrices.

An appropriate control law is

$$\mathbf{u} = \mathbf{u^d} + \mathbf{K}(\mathbf{x^d}) \cdot (\mathbf{x} - \mathbf{x^d}), \tag{6.64 }$$

where $\mathbf{K}(\mathbf{x^d})$ is the gain matrix. Since the system (6.60) changes its behavior with the setpoint $\mathbf{x^d}$, the control law (6.64) changes with the setpoint $\mathbf{x^d}$ as well. In order to design the controller for the closed loop system at any arbitrary point $\mathbf{x^d}$ in advance we approximate (6.63) by a set of TS fuzzy rules

$$\begin{aligned} R_{Si}: \quad & IF \quad \mathbf{x^d} = \mathbf{LX^i} \\ & THEN \quad \dot{\mathbf{x}} = \mathbf{f}(\mathbf{x^d}, \mathbf{u^d}) + \mathbf{A}(\mathbf{x^i}, \mathbf{u^i}) \cdot (\mathbf{x} - \mathbf{x^d}) + \mathbf{B}(\mathbf{x^i}, \mathbf{u^i}) \cdot (\mathbf{u} - \mathbf{u^d}). \end{aligned} \tag{6.65 }$$

The resulting system equation is

$$\dot{\mathbf{x}} = \mathbf{f}(\mathbf{x^d}, \mathbf{u^d}) + \sum_{i=1}^{n} w_i(\mathbf{x^d}) \cdot (\mathbf{A}(\mathbf{x^i}, \mathbf{u^i}) \cdot (\mathbf{x} - \mathbf{x^d}) + \mathbf{B}(\mathbf{x^i}, \mathbf{u^i}) \cdot (\mathbf{u} - \mathbf{u^d})). \tag{6.66 }$$

This is a linear differential equation because the weights w_i depend on the desired state vector $\mathbf{x^d}$ instead on $\mathbf{x}$.
The corresponding set of control rules is

$$R_{Ci}: \quad IF \quad \mathbf{x^d} = \mathbf{LX^i} \quad THEN \quad \mathbf{u} = \mathbf{u^d} + \mathbf{K}(\mathbf{x^i}) \cdot (\mathbf{x} - \mathbf{x^d}) \tag{6.67 }$$

with the resulting control law

$$\mathbf{u} = \sum_{i=1}^{n} w_i(\mathbf{x^d}) \cdot (\mathbf{u^d} + \mathbf{K}(\mathbf{x^i}) \cdot (\mathbf{x} - \mathbf{x^d}). \tag{6.68 }$$

Substituting (6.68) into (6.66) we obtain the equation for the closed loop system

$$\dot{\mathbf{x}} = \mathbf{f}(\mathbf{x^d}, \mathbf{u^d}) + \sum_{i,j=1}^{n} w_i(\mathbf{x^d}) \cdot w_j(\mathbf{x^d}) \cdot (\mathbf{A}(\mathbf{x^i}, \mathbf{u^i}) + \mathbf{B}(\mathbf{x^i}, \mathbf{u^i}) \cdot \mathbf{K}(\mathbf{x^j})) \cdot (\mathbf{x} - \mathbf{x^d}). \tag{6.69 }$$

Denoting $\mathbf{A}(\mathbf{x^i}, \mathbf{u^i}) + \mathbf{B}(\mathbf{x^i}, \mathbf{u^i}) \cdot \mathbf{K}(\mathbf{x^j})$ by $\mathbf{A_{ij}}$, asymptotic stability of $\mathbf{x} - \mathbf{x^d}$ is guaranteed if there exists a common positive definite matrix $\mathbf{P}$ such that the Lyapunov inequalities

$$\mathbf{A_{ij}^T P} + \mathbf{P A_{ij}} < \mathbf{0} \tag{6.70 }$$

hold where $\mathbf{A_{ij}}$ are Hurwitz matrices [32]. With this result one is able to study stability, robustnes, and performance of the closed loop system around an arbitrary setpoint $\mathbf{x^d}$ just by considering the system at predefined operating points $\mathbf{x^i}$.

6.5 SUPERVISORY CONTROL

A commonly used control technique is supervisory control which is a method to connect conventional control methods and so-called intelligent control methods. This control technique works in such a way that one or more controllers are supervised by a control law on a higher level. Applications to supervisory control for a milling machine and a steam turbine are reported in [5] and [1]. Normally, the low level controllers perform a specific task under certain conditions. These conditions can be

- keeping a predefined error between desired state and current state
- performing a specific control task (e.g. approaching a solid surface by a robot arm)
- being at a specific location of the state space.

Usually, supervisors intervene only if some of the predefined conditions fail. If so, the supervisor both changes the set of control parameters or switches from one control strategy to another.
Very often, supervisory algorithms are formulated in terms of IF-THEN rules. Fuzzy IF-THEN rules avoid hard switching between set of parameters or between control structures. It is therefore useful to build fuzzy supervisors in the cases when "soft supervision" is required.

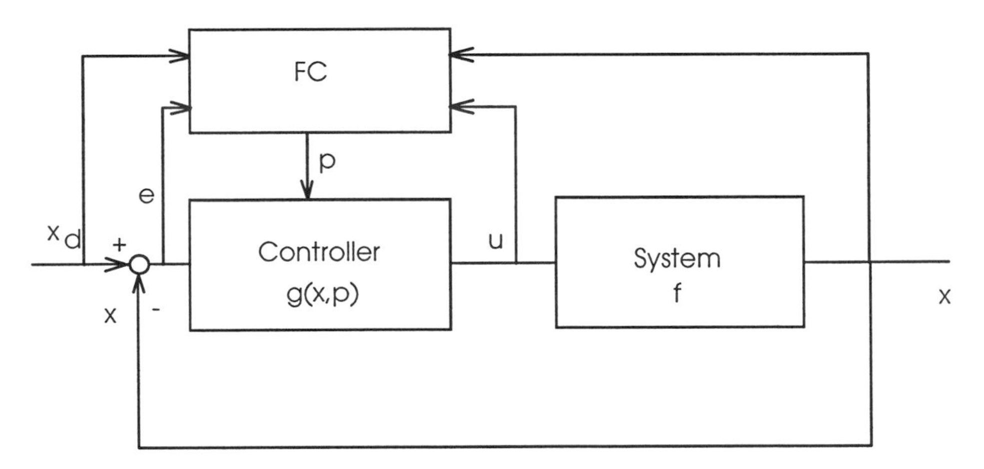

Figure 6.19 Supervisory Control

A formal approach may be the following. Let

$$\dot{\mathbf{x}} = \mathbf{f}(\mathbf{x}, \mathbf{u}) \tag{6.71}$$

be the model of the system and

$$\mathbf{u} = \mathbf{g_p}(\mathbf{x}, \mathbf{p}) \tag{6.72 }$$

be the control law where $\mathbf{p}$ is a parameter vector which has to be determined by the supervisor. The subindex '$\mathbf{p}$' means that with the change of $\mathbf{p}$ the structure of the control law may change. Then, the supervisory law can be written as

$$\mathbf{p} = \mathbf{h}(\mathbf{x}, \mathbf{c}) \tag{6.73 }$$

where $\mathbf{c}$ is the vector of conditions. For example,

$$\mathbf{c} = (|\mathbf{x} - \mathbf{x^d}| > K1; |\dot{\mathbf{x}}^\mathbf{d}| < K2)^T$$

where $K1$ and $K2$ are constant bounds. The corresponding supervisor fuzzy rule is

$$IF \quad \mathbf{x} = \mathbf{LX^i} \quad AND \quad \mathbf{c} = \mathbf{LC^i} \quad THEN \quad \mathbf{p} = \mathbf{p^i}$$

with $\mathbf{LC^i} = (|\mathbf{x} - \mathbf{x^d}| > K1^i; |\dot{\mathbf{x}}^\mathbf{d}| < K2^i)^T$.

Supervision is related to gain scheduling. The distinction between the two is that gain scheduling changes the controller gains with respect to a slowly time varying scheduling variable while the control structure is preserved [26, 15, 27]. On the other hand, supervision can both change the control gains and the control structure, and can deal with fast changing system parameters as well [39].

6.6 ADAPTIVE CONTROL

Many dynamic systems have a known structure, but uncertain or slowly varying parameters. Adaptive control is an approach to the control of such systems. Adaptive controllers, whether designed for linear or nonlinear systems, are inherently nonlinear. We distinguish between *direct* and *indirect* adaptive control methods. Direct adaptive methods start with sufficient knowledge about the system structure and its parameters. Direct change of controller parameters optimize the system's behavior with respect to a given criterion.

In contrast, the basic idea of indirect adaptive control methods is to estimate the uncertain parameters of the system under control (or, equivalently, the controller parameters) on-line, and use the estimated parameters in the computation of the control law. Thus an indirect adaptive controller can be regarded as a controller with on-line parameter estimation. There do exist systematic methods for the design of adaptive controllers for the control of linear systems. There also exist adaptive control methods that can be applied to the control of nonlinear systems. However, the latter methods require measurable states and a linear parametrization of the dynamics of the system under control, i.e., that parametric uncertainty be expressed linearly in terms of a number of adjustable parameters. This is required in order to guarantee stability and tracking convergence. However, when adaptive control of nonlinear systems is concerned, most of the adaptive control methods can only be applied to SISO nonlinear systems.

Since robust control methods are also used to deal with parameter uncertainty, adaptive control methods can be considered as an alternative and complimentary to robust control methods. In principle, adaptive control is superior to robust control in dealing with uncertainties in uncertain or slowly varying parameters.

The reason for this is the learning behavior of the adaptive controller: such a controller improves its performance in the process of adaptation. On the other hand, a robust controller simply attempts to keep a consistent performance. Furthermore, an indirect adaptive controller requires little a priori information about the unknown parameters. A robust controller usually requires reasonable a priori estimates of the parameter bounds.

Conversely, a robust controller has features which an adaptive controller does not possess, such as the ability to deal with disturbances, quickly varying parameters, and unmodelled dynamics.

In control with a FC there exist a number of direct adaptive control methods aimed at improving the FC's performance on-line. The FC's parameters that can be altered on-line are: the scaling factors for the input and output signals, the input and output membership functions, and the fuzzy IF-THEN rules. An adaptive FC, its adjustable parameters being the fuzzy values and their membership functions, is called a *self-tuning* FC. An adaptive FC which can modify its fuzzy IF-THEN rules is called a *self-organizing* FC. Detailed description of the design methods for these two types of direct adaptive FCs can be found in [3] (Chap. 5). Descriptions of indirect adaptive FCs can be found in [38].

The methods for the design of a self-tuning FC can be applied independently of whether its fuzzy IF-THEN rules are derived using model based fuzzy control, or a heuristic design approach and are thus applicable to the different types of FCs.

Since tuning and optimization of controllers is related to adaptive control we use Fig. 6.20 to illustrate this relationship.

In this scheme an adaptation block is arranged above the controller to force the closed loop system behaving according to a parallel installed reference model. The task is to change the parameters of the controller by means of the adaptation block. Tuning or optimization is performed with the following steps:

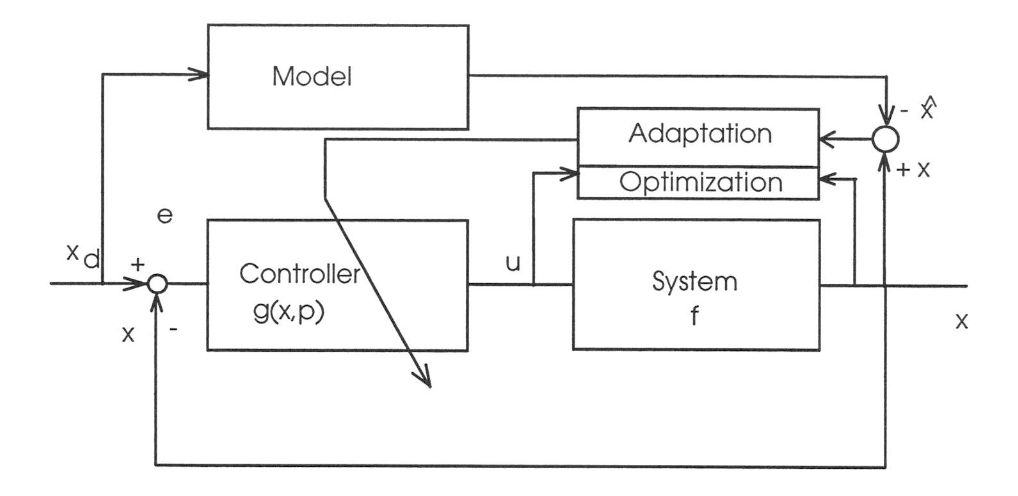

Figure 6.20 Adaptation Control

1. First of all one needs optimization criteria being sufficient criteria for a relevant improvement of the behavior of the system under control. One criterion mostly used is the integral criterion

$$J = \int_0^T (\mathbf{e}^{\mathrm{T}}\mathbf{Q}\mathbf{e} + \mathbf{u}^{\mathrm{T}}\mathbf{R}\mathbf{u})dt \tag{6.74 }$$

   where $\mathbf{e} = \mathbf{x} - \mathbf{x}^{\mathrm{d}}$ is the error, and $\mathbf{Q}, \mathbf{R}$ are weighting matrices. Another performance criterion can be formulated by fuzzy rules, for example,

   IF rise time = SMALL AND settling time = MEDIUM THEN
   performance = HIGH

2. The next point is to choose an appropriate optimization technique, e.g. gradient decent with constant searching step width, *Rosenbrock* method with variable searching step widths.

3. A crucial point is to choose a tuning hierarchy [19] which considers the different impacts of the control parameters on stability, performance, and robustnes of the closed loop system:

   - Tune the output scaling factors
   - Tune the input scaling factors
   - Tune the membership functions

References

[1] Badami, V. V., Chiang, K. H., Houpt, P. K., and Bonisonne, P. P. (1994). Fuzzy Logic Supervisory control for steam turbine prewarming automation. *3rd IEEE Intern. Conference on Fuzzy Systems*, Orlando, pp. 1045-1050.

[2] Chen, Y. Y. and Tsao, T. C. (1989). A Description of the Dynamical Behavior of Fuzzy Systems. *IEEE Trans. on Syst. Man, and Cyb.*, **19**(4), 745-755.

[3] Driankov,D. Hellendoorn, H., and Reinfrank, M. (1993). *An Introduction to Fuzzy Control*, Springer Verlag, Berlin.

[4] Foellinger, O. (1990). *Regelungstechnik*, Huethig Buch Verlag GMBH, Heildelberg.

[5] Haber, R. H., Peres, C. R., Alique, J. R., Ros, S. T., and Alique, A. (1995). Two Approaches for a fuzzy supervisory control system of a vertical milling machine. *VI IFSA Congress*, Sao Paulo, Brazil, pp. 397-400.

[6] Holmblad, L. and Ostergaard, J. J. (1982). Control of a cement kiln by fuzzy logic, In *Fuzzy Information and Decision Processes*. Gupta, M.M. and Sanchez, E.E.(eds.), North-Holland, Amsterdam, pp. 389-399.

[7] Hu, H-T., Tai, H-M., and Shenoi, S. (1994). Incorporating cell map information in fuzzy controller design. *3rd IEEE Intern. Conference on Fuzzy Systems*, Orlando. pp. 394-399.

[8] Hsu, C. S. (1980). A theory of cell-to-cell dynamical systems. *Journal of Applied Mechanics*, **47**, 940-948.

[9] Hwang, G. C., and Li, S. C. (1992). A stability approach to fuzzy control design for nonlinear systems *Fuzzy Sets and Systems*, **48**, 279-287, North-Holland.

[10] Kang, H. and Vachtsevanos, G. (1984). Nonlinear fuzzy control based on the vector field of the phase portrait assignment algorithm. *Proceedings of the American Control Conference*, pp. 1479-1484.

[11] Kawaji, S. and Matsunaga, N. (1991). Fuzzy control of VSS type and its robustness. *IFSA '91 Brussels*, preprints vol. "engineering", 81-88.

[12] Koczy, L. and Kovacs, K. (1994). Linearity and the cnf property in linear fuzzy rule interpolation. *IEEE International Conference on Fuzzy Systems, Fuzz-IEEE'94 - Proceedings*, Orlando, pp.870-875.

[13] Kosko, B. (1992) Fuzzy systems as universal approximators, *IEEE International Conference on Fuzzy Systems, Fuzz-IEEE'92 - Proceedings*, San Diego, pp.1153-1162

[14] Mamdani, E. M. and Assilian, S. (1975). An Experiment in linguistic synthesis with a fuzzy logic controller, *Int. J. Man-Machine Studies*, **7**(1), 1-13.

[15] Nichols, R. A., Reichert, R. T., and Rugh, W. J. (1993). Gain scheduling for H-infinity controllers: a flight control example. *IEEE Trans. on Control Systems Technology* **1**(2), 69-79.

[16] Ostergaard, J. J. (1977). Fuzzy logic control of a heat exchanger process, in *Fuzzy Automata and Decision Processes*, Gupta, M.M.(ed.) North-Holland, New York, pp. 285-320.

[17] Palm, P. (1992). Sliding mode fuzzy control, *IEEE International Conference on Fuzzy Systems, Fuzz-IEEE'92 - Proceedings*, San Diego, pp.519-526.

[18] Palm, R., Driankov, D., and Rehfuess, U., (1995). Lyapunov linearization based design of Takagi-Sugeno controllers, *Proceedings FUZZ-IEEE'95*, Sao Paolo, Brasil, pp. 513–516.

[19] Palm, R., (1993). Tuning of scaling factors in fuzzy controllers using correlation functions, *Proceedings FUZZ-IEEE'93*, san Francisko, California, pp. 691–696.

[20] Papa, M., Tai, H-M., and Shenoi, S. (1995). Design and evaluation of fuzzy control systems using cell mapping, *VI IFSA World Congress*, Sao Paulo, Brazil, pp. 361-364.

[21] W. Pedrycz. (1992). *Fuzzy Control and Fuzzy Systems*, 2nd revised edition, Research Studies Publ.

[22] Pedrycz, W. (1995). Fuzzy control engineering, reality and challenges. *IEEE International Conference on Fuzzy Systems, Fuzz-IEEE/IFES'95 - Proceedings*, Yokohama, pp.437-446

[23] Ray, K. S., and Majumder, D. D., (1984). Application of circle criteria for stability analysis of linear SISO and MIMO systems associated with fuzzy logic controller, *IEEE Transactions on Systems, Man, Cybernetics*, **14**(2), 345–349.

[24] Ray, K. S., Ananda, S., and Majumder, D. D. (1992). L-stability and the related design concept for SISO linear systems associated with fuzzy logic controller, *IEEE Transactions on Systems, Man, Cybernetics*, **14**, 932–939.

[25] Rehfuess, U., and Palm, R. (1995). Design of Takagi-Sugeno controllers based on linear quadratic Control, *Proceedings First International Symposium on Fuzzy Logic*, Zurich, Switzerland, pp. C10–C15.

[26] Rugh, W. J. (1991). Analytical framework for gain scheduling, *IEEE Control Systems Magazine*, **11**(1), 79–84.

[27] Shamma J. S. (1988). *Analysis and Design of Gain Scheduled Control Systems* PhD Thesis No. LIDS-TH-1770 Lab. for Information and Decision Sciences MIT Cambridge MA 02139.

[28] Slotine, J-J. E. and Li, W. (1991). *Applied Nonlinear Control*, Prentice Hall, New Jersey.

[29] Smith, S. M. and Comer, D. J. (1992). An algorithm for automated fuzzy logic controller tuning. *Procedings of the IEEE Intern. Conference on Fuzzy Systems*, 615-622.

[30] Smith, S. M. (1994). A variable structure fuzzy logic controller with runtime adaptation, *Proceedings FUZZ-IEEE'94*, Orlando, Florida, 983–988.

[31] Takagi, T. and Sugeno, M. (1985). Fuzzy identification of systems and its applications to modelling and control. *IEEE Trans. on Syst., Man, and Cyb.*, **SMC-15**(1), 116-132.

[32] Tanaka, T. and Sugeno, M. (1992). Stability analysis and design of fuzzy control cystems. *Fuzzy Sets and Systems*, **45**, 135-156.

[33] Tang, K. L. and Mulholland, R. J. (1987). Comparing fuzzy logic with classical control designs, *IEEE transactions on Systems, Man, Cybernetics*, **SMC-17**(6), 1085–1087.

[34] Tong, R. M. (1980). Some properties of fuzzy feedback systems.*IEEE Trans. on Syst. Man, and Cyb.* **SMC-10**(6), 327-330.

[35] Utkin, V. J. (1977). Variable structure systems, a survey, *IEEE Transactions Automatic Control*, **22**, 212–222.

[36] Vidyasagar, M. (1993). *Nonlinear Systems Analysis*, Prentice Hall, Englewood Cliffs, New Jersey.

[37] Wakileh, B. A. M. and Gill, K. F. (1988). Use of fuzzy logic in robotics, *Computers in Industry*, **10**, 35-46.

[37] Wang, L. (1992). Fuzzy systems are universal approximators, *IEEE International Conference on Fuzzy Systems, Fuzz-IEEE'92 Proceedings*, San Diego, 1163-1169.

[39] Wang, L-X. (1994). Supervisory controller for fuzzy control systems that guarantees stability. *3rd IEEE Intern. Conference on Fuzzy Systems*, Orlando, 1035-1039.

[40] Yasunobu, S. and Miyamoto, S. (1985). Automatic train operation system by predictive fuzzy control. *Industrial Applications of Fuzzy Control*, M. Sugeno (ed.), Elsevier Science Publishers, pp. 1-18.

7 STABILITY OF FUZZY CONTROLLERS

Kazuo Tanaka

Department of Mechanical and Control Engineering
University of Electo-Communications
1-5-1 Chofugaoka, Chofu, Tokyo 182 Japan
ktanaka@mce.uec.ac.jp

7.1 INTRODUCTION

Stability is one of the most important concepts of analysis and design of control systems. Stability analysis of fuzzy control systems has been difficult since fuzzy systems are essentially nonlinear systems. Recently, stability analysis techniques [1]-[6] [13]-[16] based on nonlinear stability theory have been reported. Langari and Tomizuka [2, 6] analyzed stability of fuzzy linguistic control systems. Kitamura [3] realized stability analysis of fuzzy control systems by using extended circle criterion. Farinwata [5] and Kato [14] utilized phase plane analysis type of techniques. tanaka and co-authors [1,3,4] applied Lyapunov approach to fuzzy control systems and recently developed LMI-based designs [10, 17, 28-37], where LMI stands for nonlinear matrix inequality.

The purpose of this chapter is to show a methodology for stability analysis and design of fuzzy control systems. The methodology discussed here is simple and natural. The stability analysis is based on Lyapunov stability theory. The design utilizes the concept of the so-called "parallel distributed compensation" (PDC) [4, 10, 17]. The basic principle for stability analysis and design will be discussed. For the advanced techniques "LMI based designs", see [10, 17. 28-37].

Fuzzy controls have been applied to a wide variety of industrial applications. On the other hand, we have experienced difficulties in design of fuzzy controllers, i.e., derivation of fuzzy control rules and determination of membership functions, since we have lacked methodologies for analysis and design of fuzzy control systems. This chapter will present an answer for the problem. A keyword for the answer is a model-based fuzzy control.

Section 7.2 shows the Takagi-Sugeno's fuzzy model (T-S fuzzy model) and stability conditions based on Lyapunov approach. Section 7.3 presents a fuzzy controller design via the PDC. The Takagi-Sugeno fuzzy model is utilized to represent the dynamics of nonlinear systems. The PDC offers a procedure to design fuzzy controllers from the T-S fuzzy systems. The stability analysis of the feedback system is reduced to a problem of finding a common Lyapunov function for a set of Lyapunov inequalities. A convex optimization technique based on linear matrix inequalities (LMIs) is utilized to find a common Lyapunov function.

7.2 STABILITY CONDITIONS BASED ON LYAPUNOV APPROACH

7.2.1 Takagi and Sugeno's fuzzy model

The fuzzy model proposed by Takagi and Sugeno [7] is described by fuzzy IF-THEN rules which represent local linear input-output relations of a nonlinear system. The i-th rules of the T-S fuzzy model are of the following forms:
<Continuous fuzzy system : CFS>

Plant Rule i:

IF $z_1(t)$ is M_{i1} and ... and $z_p(t)$ is M_{ip}, THEN

$$\dot{\mathbf{x}}_i(t) = \mathbf{A}_i\mathbf{x}(t) + \mathbf{B}_i\mathbf{u}(t), \qquad 1 = 1, 2, \ldots, r. \tag{7.1}$$

<Discrete fuzzy system : DFS>

Plant Rule i:

IF $z_1(t)$ is M_{i1} and ... and $z_p(t)$ is M_{ip}, THEN

$$\mathbf{x}_i(t+1) = \mathbf{A}_i\mathbf{x}(t) + \mathbf{B}_i\mathbf{u}(t), \qquad i = 1, 2, \ldots, r. \tag{7.2}$$

M_{ij} is the fuzzy set. CFS and DFS denote the continuous fuzzy system and the discrete fuzzy system, respectively. $\mathbf{x}(t) \in \mathbf{R}^n$ is the state vector, $\mathbf{u}(t) \in \mathbf{R}^m$ is the input vector, $\mathbf{A}_i \in \mathbf{R}^{n \times n}$, $\mathbf{B}_i \in \mathbf{R}^{n \times m}$, and r is the number of IF-THEN rules. $z_1(t), \ldots, z_p(t)$ are the premise variables. It is assumed in this chapter that the premise variables do not depend on the input variables $\mathbf{u}(t)$. It should be emphasized that the stability analysis shown in this chapter can be applied even to the case that the premise variables $z_1(t), \ldots, z_p(t)$ depend on the input variables u(t). This restriction is required only for avoidance of complicated defuzzification process [21]. Each linear consequent equation represented by $\mathbf{A}_i\mathbf{x}(t) + \mathbf{B}_i\mathbf{u}(t)$ is called "subsystem".

Given a pair of $(\mathbf{x}(t), \mathbf{u}(t))$, the final outputs of the fuzzy systems are inferred as follows, where $\mathbf{z}(t) = [z_1(t), z_2(t), \ldots, z_p(t)]$.

<CFS>

$$\begin{aligned} \dot{\mathbf{x}}(t) &= \left(\sum_{i=1}^{r} w_i(\mathbf{z}(t))[\mathbf{A}_i\mathbf{x}(t) + \mathbf{B}_i\mathbf{u}(t)]\right) / \left(\sum_{i=1}^{r} w_i(\mathbf{z}(t))\right) \\ &= \sum_{i=1}^{r} h_i(\mathbf{z}(t))[\mathbf{A}_i\mathbf{x}(t) + \mathbf{B}_i\mathbf{u}(t)]. \end{aligned} \tag{7.3 }$$

<DFS>

$$\begin{aligned} \mathbf{x}(t+1) &= \left(\sum_{i=1}^{r} w_i(\mathbf{z}(t))[\mathbf{A}_i\mathbf{x}(t) + \mathbf{B}_i\mathbf{u}(t)]\right) / \left(\sum_{i=1}^{r} w_i(\mathbf{z}(t))\right) \\ &= \sum_{i=1}^{r} h_i(\mathbf{z}(t))[\mathbf{A}_i\mathbf{x}(t) + \mathbf{B}_i\mathbf{u}(t)]. \end{aligned} \tag{7.4 }$$

In (12.3) and (12.4),

$$w_i(\mathbf{z}(t)) = \prod_{j=1}^{p} M_{ij}(z_j(t)), \quad \left\{ \begin{array}{l} \sum_{i=1}^{r} w_i(\mathbf{z}(t)) > 0 \\ w_i(\mathbf{z}(t)) \geq 0, \quad i = 1, 2, \ldots, r. \end{array} \right. \tag{7.5 }$$

for all t. $M_{ij}(z_j(t))$ is the grade of membership of $z_j(t)$ in M_{ij}. From (12.3), (12.4), and (12.5), we have

$$\left\{ \begin{array}{l} \sum_{i=1}^{r} h_i(\mathbf{z}(t)) = 1 \\ h_i(\mathbf{z}(t)) \geq 0, \quad i = 1, 2, \ldots, r. \end{array} \right. \tag{7.6 }$$

for all t.

Example 7.1 *Assume in the DFS that* $p = n$,

$$z_1(t) = x(t), \quad z_2(t) = x(t-1), \ldots, \; z_n(t) = x(t-n+1).$$

Then, the plant rules can be represented as follows, where

$$\mathbf{x}(t) = [x(t), x(t-1), \ldots, x(t-n+1)]^T.$$

Plant Rule i:

IF $x(t)$ is M_{i1} and ... and $x(t-n+1)$ is M_{in}, THEN

$$\mathbf{x}_i(t+1) = \mathbf{A}_i\mathbf{x}(t) + \mathbf{B}_i\mathbf{u}(t), \qquad i = 1, 2, \ldots, r.$$

7.2.2 Stability conditions

The open-loop systems of (12.3) and (12.4) are defined as follows.

<CFS>

$$\begin{aligned}\dot{\mathbf{x}}(t) &= \left(\sum_{i=1}^{r} w_i(\mathbf{z}(t))\mathbf{A}_i\mathbf{x}(t)\right) / \left(\sum_{i=1}^{r} w_i(\mathbf{z}(t))\right) \\ &= \sum_{i=1}^{r} h_i(\mathbf{z}(t))\mathbf{A}_i\mathbf{x}(t). \end{aligned} \tag{7.7}$$

<DFS>

$$\begin{aligned}\mathbf{x}(t+1) &= \left(\sum_{i=1}^{r} w_i(\mathbf{z}(t))\mathbf{A}_i\mathbf{x}(t)\right) / \left(\sum_{i=1}^{r} w_i(\mathbf{z}(t))\right) \\ &= \sum_{i=1}^{r} h_i(\mathbf{z}(t))\mathbf{A}_i\mathbf{x}(t). \end{aligned} \tag{7.8}$$

Stability conditions for ensuring stability of (12.7) and (12.8) can be derived via Lyapunov approach. The conditions are given in Theorems 7.1 and 7.2.

Theorem 7.1 *(CFS) [15, 20]. The equilibrium of a continuous fuzzy system described by (12.7) is asymptotically stable in the large if there exists a common positive definite matrix* $\mathbf{P}$ *such that*

$$\mathbf{A}_i^T\mathbf{P} + \mathbf{P}\mathbf{A}_i < \mathbf{0} \tag{7.9}$$

for $i = 1, 2, \ldots, r$*, i.e., for all the subsystems.*

Theorem 7.2 *(DFS) [1, 4]. The equilibrium of a discrete fuzzy system described by (12.8) is asymptotically stable in the large if there exists a common positive definite matrix* $\mathbf{P}$ *such that*

$$\mathbf{A}_i^T\mathbf{P}\mathbf{A}_i - \mathbf{P} < \mathbf{0} \tag{7.10}$$

for $i = 1, 2, \ldots, r$*, i.e., for all the subsystems.*

These theorems are reduced to the Lyapunov stability theorems for linear continuous systems and linear discrete systems respectively when r=1. Theorems 7.1 and 7.2 give sufficient conditions for ensuring stability of (7.7) and (7.8), respectively. For the systems (7.7) and (7.8), a question naturally arises is whether the systems are stable if all its subsystems are stable, i.e., all $\mathbf{A}_i$'s are stable. The answer is no in general [4, 10, 17].

Example 7.2 *Consider the following CFS.*

Rule 1:

$$IF\ z(t)\ is\ Small, \quad THEN\ \dot{\mathbf{x}}(t) = \mathbf{A}_1\mathbf{x}(t).$$

Rule 2:

$$IF\ z(t)\ is\ Big, \quad THEN\ \dot{\mathbf{x}}(t) = \mathbf{A}_2\mathbf{x}(t).$$

In the CFS,

$$\mathbf{A}_1 = \begin{bmatrix} -2 & -0.5 \\ 1 & 0 \end{bmatrix}, \qquad \mathbf{A}_2 = \begin{bmatrix} -1 & -3 \\ 1 & 0 \end{bmatrix}.$$

Assume that "Small" and "Big" are arbitrary fuzzy sets satisfying (12.5). If we select

$$\mathbf{P} = \begin{bmatrix} 0.3654 & 0.1778 \\ 0.1778 & 0.9309 \end{bmatrix} > \mathbf{0},$$

then

$$\mathbf{A}_1^T\mathbf{P} + \mathbf{P}\mathbf{A}_1 = \begin{bmatrix} -1.1061 & 0.3926 \\ 0.3926 & -0.1778 \end{bmatrix} < \mathbf{0}$$

and

$$\mathbf{A}_2^T\mathbf{P} + \mathbf{P}\mathbf{A}_2 = \begin{bmatrix} -0.3752 & -0.3432 \\ -0.3432 & -1.0669 \end{bmatrix} < \mathbf{0}.$$

Therefore, the fuzzy system is asymptotically stable in the large.

Example 7.3 *Consider the following DFS.*

Rule 1:

$$IF\ z(t)\ is\ Small, \quad THEN\ \mathbf{x}(t+1) = \mathbf{A}_1\mathbf{x}(t).$$

Rule 2:

$$IF\ z(t)\ is\ Big, \quad THEN\ \mathbf{x}(t+1) = \mathbf{A}_2\mathbf{x}(t).$$

In the DFS,

$$\mathbf{A}_1 = \begin{bmatrix} 1.503 & -0.588 \\ 1 & 0 \end{bmatrix}, \qquad \mathbf{A}_2 = \begin{bmatrix} 1 & -0.361 \\ 1 & 0 \end{bmatrix}.$$

Assume that "Small" and "Big" are arbitrary fuzzy sets satisfying (12.5). If we select

$$\mathbf{P} = \begin{bmatrix} 29.3 & -16.3 \\ -16.3 & 11.1 \end{bmatrix} > \mathbf{0},$$

then

$$\mathbf{A}_1^T\mathbf{P}\mathbf{A}_1 - \mathbf{P} = \begin{bmatrix} -1 & 0 \\ 0 & -1 \end{bmatrix} < \mathbf{0}$$

and

$$\mathbf{A}_2^T \mathbf{P} \mathbf{A}_2 - \mathbf{P} = \begin{bmatrix} -21.5 & 11.6 \\ 11.6 & -7.28 \end{bmatrix} < \mathbf{0}.$$

Therefore, the fuzzy system is asymptotically stable in the large.

Example 7.4 *Assume that*

$$\mathbf{A}_1 = \begin{bmatrix} 1 & -0.5 \\ 1 & 0 \end{bmatrix}, \qquad \mathbf{A}_2 = \begin{bmatrix} -1 & -0.5 \\ 1 & 0 \end{bmatrix}$$

in Example 3. It is shown in [4] that there is no common $\mathbf{P}$ *satisfying (7.10) for the matrices* $\mathbf{A}_1$ *and* $\mathbf{A}_2$*.*

For the case of Example 7.4, we should be able to check the existence of a common $\mathbf{P}$. The following theorem for the DFS was derived to check it.

Theorem 7.3 (4) . *Assume that* $\mathbf{A}_i$ *is a stable matrix for* $i = 1, 2, \ldots, r$. $\mathbf{A}_i\mathbf{A}_j$ *is a stable matrix for* $i, j = 1, 2, \ldots, r$ *if there exists a common positive definite matrix* $\mathbf{P}$ *satisfying (7.10).*

The contraposition of this theorem means that there dose not exist a common $\mathbf{P}$ satisfying (7.10) if one of the $\mathbf{A}_i\mathbf{A}_j$'s is at least an unstable matrix. Hence, the only problem is how to find a common Lyapunov function satisfying (7.9) or (7.10). The construction of a common Lyapunov function will be discussed in Section 7.3.2.

7.3 FUZZY CONTROLLER DESIGN

7.3.1 Parallel distributed compensation

The history of the PDC (parallel distributed compensation) started in a mode l-based design procedure proposed by Kang and Sugeno (e.g. [21]). However, the stability of the control systems was not discussed in the design procedure. The design procedure was slightly improved and the stability of the control systems was analyzed in [4]. The developed design procedure is named "parallel distributed compensation" in [10].

The PDC [4, 10, 17] offers a procedure to design a fuzzy controller from the T-S fuzzy model. To realize the PDC, a controlled object (nonlinear system) should be represented by a T-S fuzzy model. We emphasize that many real systems, e.g., mechanical systems, have been represented by T-S fuzzy models. For instance, see [11], [12], [17]-[19], [24] for more detail of the fuzzy modeling for nonlinear systems.

Each control rule is derived from the corresponding rule of a T-S fuzzy model in the PDC. The designed fuzzy controller shares the same fuzzy sets with the fuzzy model in the premise parts. For the fuzzy models (1) and (2), the following fuzzy controller is designed via the PDC.

Control Rule i:

IF $z_1(t)$ is M_{i1} and ... and $z_p(t)$ is M_{ip}, THEN

$$\mathbf{u}_i(t) = -\mathbf{F}_i\mathbf{x}(t), \qquad r = 1, 2, \ldots, r.$$

Example 7.5 *If the controlled object is represented as the fuzzy plant rules shown in Example 7.1, the control rules can be described as follows:*

Control Rule i*:*

IF $x(t)$ *is* M_{i1} *and ... and* $x(t-n+1)$ *is* M_{ip}*, THEN*

$$\mathbf{u}_i(t) = -\mathbf{F}_i\mathbf{x}(t), \qquad r = 1, 2, \ldots, r.$$

The fuzzy control rules have linear state feedback laws in the consequent parts. The overall fuzzy controller is calculated by

$$\begin{aligned}\mathbf{u}(t) &= -\left(\sum_{i=1}^{r} w_i(\mathbf{z}(t))\mathbf{F}_i\mathbf{x}(t)\right) / \left(\sum_{i=1}^{r} w_i(\mathbf{z}(t))\right) \\ &= -\sum_{i=1}^{r} h_i(\mathbf{z}(t))\mathbf{F}_i\mathbf{x}(t). \qquad (7.11)\end{aligned}$$

By substituting (7.11) into (7.3) and (7.4), we obtain (7.12) and (7.13) respectively, where

$$\mathbf{G}_{ij} = \mathbf{A}_i - \mathbf{B}_i\mathbf{F}_j.$$

<CFS>

$$\begin{aligned}\dot{\mathbf{x}}(t) &= \sum_{i=1}^{r}\sum_{j=1}^{r} h_i(\mathbf{z}(t))h_j(\mathbf{z}(t))[\mathbf{A}_i - \mathbf{B}_i\mathbf{F}_j]\mathbf{x}(t) \\ &= \sum_{i=1}^{r} h_i(\mathbf{z}(t))\mathbf{G}_{ii}\mathbf{x}(t) \\ &\quad + 2\sum_{i<j}^{r} h_i(\mathbf{z}(t))h_j(\mathbf{z}(t))\left(\frac{\mathbf{G}ij + \mathbf{G}ji}{2}\right)\mathbf{x}(t). \qquad (7.12)\end{aligned}$$

<DFS>

$$\mathbf{x}(t+1) = \sum_{i=1}^{r}\sum_{j=1}^{r} h_i(\mathbf{z}(t))h_j(\mathbf{z}(t))[\mathbf{A}_i - \mathbf{B}_i\mathbf{F}_j]\mathbf{x}(t)$$

$$= \sum_{i=1}^{r} h_i(\mathbf{z}(t))\mathbf{G}_{ii}\mathbf{x}(t)$$

$$+2\sum_{i<j}^{r} h_i(\mathbf{z}(t))h_j(\mathbf{z}(t)) \left(\frac{\mathbf{G}ij + \mathbf{G}ji}{2}\right) \mathbf{x}(t). \quad (7.13)$$

7.3.2 Stability conditions

Stability conditions for the CFS and the DFS can be derived by applying Theorems 7.1 and 7.2 to (7.12) and (7.13), respectively.

Theorem 7.4 *(CFS). The equilibrium of a continuous fuzzy control system described by (12.12) is asymptotically stable in the large if there exists a common positive definite matrix* $\mathbf{P}$ *such that*

$$\mathbf{G}_{ii}^T\mathbf{P} + \mathbf{P}\mathbf{G}_{ii} < \mathbf{0}, \quad (7.14)$$

$$\left(\frac{\mathbf{G}ij + \mathbf{G}ji}{2}\right)^T \mathbf{P} + \mathbf{P}\left(\frac{\mathbf{G}ij + \mathbf{G}ji}{2}\right) < \mathbf{0}, \quad i < j, \quad (7.15)$$

for all i *and* j *excepting the pairs* (i, j) *such that* $h_i(\mathbf{z}(t))h_j(\mathbf{z}(t)) = 0$ *for all* t.

The proof of Theorem 4 follows directly from Theorem 1.

Theorem 7.5 *(DFS). The equilibrium of a discrete fuzzy control system described by (7.13) is asymptotically stable in the large if there exists a common positive definite matrix* $\mathbf{P}$ *such that*

$$\mathbf{G}_{ii}^T\mathbf{P}\mathbf{G}_{ii} - \mathbf{P} < \mathbf{0}, \quad (7.16)$$

$$\left(\frac{\mathbf{G}ij + \mathbf{G}ji}{2}\right)^T \mathbf{P}\left(\frac{\mathbf{G}ij + \mathbf{G}ji}{2}\right) - \mathbf{P} < \mathbf{0}, \quad i < j, \quad (7.17)$$

for all i *and* j *excepting the pairs* (i, j) *such that* $h_i(\mathbf{z}(t))h_j(\mathbf{z}(t)) = 0$ *for all* t.

The proof of Theorem 7.5 follows directly from Theorem 2.

The design problem of the fuzzy controller is to determine $\mathbf{F}_j$ $(j = 1, 2, \ldots, r)$ satisfying the conditions of Theorem 7.4 or Theorem 7.5 for a common positive definite matrix $\mathbf{P}$.

Remark 1. To check stability of the fuzzy control system, it has long been considered difficult to find a common positive definite matrix $\mathbf{P}$ satisfying the conditions of Theorem 7.4 or Theorem 7.5. Most of the time, a trail-and-error type of procedure is used [4, 22]. In [23], a procedure to construct a common P is given for second-order fuzzy systems, i.e., the dimension of the state is two. We first stated in [10, 17] that the common P problem for the fuzzy controller design can be solved numerically. To solve the problem numerically, a very important observation is that the stability conditions of Theorems 7.1, 7.2, 7.4, and 7.5 are expressed in linear matrix inequalities (LMIs) [25, 26]. To check stability of Theorem 7.1, we need to find $\mathbf{P}$ satisfying the LMI

$$\mathbf{P} > \mathbf{0}, \qquad \mathbf{A}_i^T\mathbf{P} + \mathbf{P}\mathbf{A}_i < \mathbf{0}, \quad r = 1, 2, \ldots, r,$$

or determination that no such $\mathbf{P}$ exists, where r is the number of if-then rules. This is a convex feasibility problem. Numerically, this feasibility problem can be solved very efficiently by means of the most powerful tools available to date in the mathematical programming literature. For instance, recently developed interior-point methods [27] are extremely efficient in practice.

The following design procedure can be constructed using the PDC and a convex optimization technique based on LMIs.

Design procedure:

(Step 1) Determine $\mathbf{F}_i$ from $\mathbf{A}_i$ and $\mathbf{B}_i$ using the linear control theory.

(Step 2) By using a convex optimization technique based on LMIs, find a common $\mathbf{P}$ satisfying the conditions of Theorem 7.4 (or Theorem 7.5). If there does not exist a common $\mathbf{P}$, then go back to Step 1 and select another $\mathbf{F}_i$.

Remark 2. In the design procedure, the feedback gains $\mathbf{F}_i$ are first determined. Then, a convex optimization technique of LMIs is employed to find a common $\mathbf{P}$. This is still a trial-and-error procedure. However, it is first shown in [17,28] that by using a convex optimization technique we can directly find the feedback gains Fi satisfying the stability conditions of Theorems 7.4 and 7.5 or determination that no such $\mathbf{P}$ exists. In other words, by using a convex optimization technique, we can judge whether there exists the feedback gains $\mathbf{F}_i$ satisfying the stability conditions of Theorems 7.4 and 7.5 or not. This is not a trial-and-error procedure. For more details, see [10,17,28-37].

We will show design examples only for the CFS.

Example 7.6 *Consider the following CFS.*

Plant Rule 1*:*

IF $z(t)$ *is Small, THEN* $\dot{\mathbf{x}}(t) = \mathbf{A}_1\mathbf{x}(t) + \mathbf{b}_1 u(t)$.

Plant Rule 2*:*

IF $z(t)$ is Big, THEN $\dot{\mathbf{x}}(t) = \mathbf{A}_2\mathbf{x}(t) + \mathbf{b}_2 u(t)$.
In the CFS,

$$\begin{aligned}\mathbf{A}_1 &= \begin{bmatrix} 2 & 5 \\ 1 & 0 \end{bmatrix}, \qquad \mathbf{b}_1 = \begin{bmatrix} 1 \\ 0 \end{bmatrix}, \\ \mathbf{A}_2 &= \begin{bmatrix} 1 & -3 \\ 1 & 0 \end{bmatrix}, \qquad \mathbf{b}_2 = \begin{bmatrix} 7 \\ 0 \end{bmatrix}.\end{aligned}$$

"Small" and "Big" are arbitrary fuzzy sets satisfying (7.5). The fuzzy controller can be designed via the PDC.

Control Rule 1*:*

IF $z(t)$ is Small, THEN $u(t) = -\mathbf{f}_1\mathbf{x}(t)$.

Control Rule 2*:*

IF $z(t)$ is Big, THEN $u(t) = -\mathbf{f}_2\mathbf{x}(t)$.

$\mathbf{f}_1$ *and* $\mathbf{f}_2$ *are the feedback gains. Choose the closed-loop eigenvalues to be* $[-2, -2]$, *we have*

$$\mathbf{f}_1 = [6, 9], \qquad \mathbf{f}_2 = [0.7143, 0.1529].$$

Then, we obtain

$$\begin{aligned}\mathbf{G}_{11} &= \mathbf{A}_1 - \mathbf{b}_1\mathbf{f}_1 = \begin{bmatrix} -4 & -4 \\ 1 & 0 \end{bmatrix}, \\ \frac{\mathbf{G}_{12} + \mathbf{G}_{21}}{2} &= \frac{\{\mathbf{A}_1 - \mathbf{b}_1\mathbf{f}_2\} + \{\mathbf{A}_2 - \mathbf{b}_2\mathbf{f}_1\}}{2} = \begin{bmatrix} -19.8571 & -30.5714 \\ 1 & 0 \end{bmatrix}, \\ \mathbf{G}_{22} &= \mathbf{A}_2 - \mathbf{b}_2\mathbf{f}_2 = \begin{bmatrix} -4 & -4 \\ 1 & 0 \end{bmatrix}.\end{aligned}$$

If we select

$$\mathbf{P} = \begin{bmatrix} 0.0269 & 0.0222 \\ 0.0222 & 0.2542 \end{bmatrix} > \mathbf{0},$$

then

$$\begin{aligned}&\mathbf{G}_{11}^T\mathbf{P} + \mathbf{P}\mathbf{G}_{11} = \mathbf{P} = \begin{bmatrix} -0.1708 & 0.0579 \\ 0.0579 & -0.1775 \end{bmatrix} < \mathbf{0}, \\ &\left(\frac{\mathbf{G}_{12} + \mathbf{G}_{21}}{2}\right)^T \mathbf{P} + \mathbf{P}\left(\frac{\mathbf{G}_{12} + \mathbf{G}_{21}}{2}\right) = \begin{bmatrix} -1.0238 & -1.0087 \\ -1.0087 & -1.3568 \end{bmatrix} < \mathbf{0}, \\ &\mathbf{G}_{22}^T\mathbf{P} + \mathbf{P}\mathbf{G}_{22} == \begin{bmatrix} -0.1708 & 0.0579 \\ 0.0579 & -0.1775 \end{bmatrix} < \mathbf{0}.\end{aligned}$$

Therefore, the fuzzy controller guarantees the stability of the control system. The common **P** *is found by a convex optimization technique.*

Remark 3. In Example 7.7, the feedback gains are obtained by locally solving Riccati equations. Note that the designed fuzzy controller guarantees only the local stability. A fuzzy controller design that globally minimizes the upper bound of a quadratic performance function is discussed in [35-37].

Example 7.7 *Consider the following CFS.*

Plant Rule 1:

IF $z_1(t)$ *is Small 1 and* $z_2(t)$ *is Small 2, THEN*

$$\dot{\mathbf{x}}(t) = \mathbf{A}_1\mathbf{x}(t) + \mathbf{B}_1\mathbf{u}(t).$$

Plant Rule 2:

IF $z_1(t)$ *is Small 1 and* $z_2(t)$ *is Big 2, THEN*

$$\dot{\mathbf{x}}(t) = \mathbf{A}_2\mathbf{x}(t) + \mathbf{B}_2\mathbf{u}(t).$$

Plant Rule 3:

IF $z_1(t)$ *is Big 1 and* $z_2(t)$ *is Small 2, THEN*

$$\dot{\mathbf{x}}(t) = \mathbf{A}_3\mathbf{x}(t) + \mathbf{B}_3\mathbf{u}(t).$$

Plant Rule 4:

IF $z_1(t)$ *is Big 1 and* $z_2(t)$ *is Big 2, THEN*

$$\dot{\mathbf{x}}(t) = \mathbf{A}_4\mathbf{x}(t) + \mathbf{B}_4\mathbf{u}(t).$$

In the CFS,

$$\mathbf{A}_1 = \begin{bmatrix} -2.5294 & -4.0766 & 1.6098 & 1.6137 \\ 1 & 0 & 0 & 0 \\ 0 & 1 & 0 & 0 \\ 0 & 0 & 1 & 0 \end{bmatrix},$$

$$\mathbf{B}_1 = \begin{bmatrix} 8.4122 & 0 & 0 & 0 \\ 0 & 3.4515 & 0 & 0 \\ 0 & 0 & 4.6747 & 0 \\ 0 & 0 & 0 & 7.4787 \end{bmatrix},$$

$$\mathbf{A}_2 = \begin{bmatrix} -4.1889 & -4.0188 & 0.2763 & 1.5403 \\ 1 & 0 & 0 & 0 \\ 0 & 1 & 0 & 0 \\ 0 & 0 & 1 & 0 \end{bmatrix},$$

$$\mathbf{B}_2 = \begin{bmatrix} 0.0693 & 0 & 0 & 0 \\ 0 & 3.4507 & 0 & 0 \\ 0 & 0 & 0.6016 & 0 \\ 0 & 0 & 0 & 3.7547 \end{bmatrix},$$

$$\mathbf{A}_3 = \begin{bmatrix} 1.6810 & 0.8008 & 3.8739 & 3.1155 \\ 1 & 0 & 0 & 0 \\ 0 & 1 & 0 & 0 \\ 0 & 0 & 1 & 0 \end{bmatrix},$$

$$\mathbf{B}_3 = \begin{bmatrix} 4.7424 & 0 & 0 & 0 \\ 0 & 0.8277 & 0 & 0 \\ 0 & 0 & 5.8853 & 0 \\ 0 & 0 & 0 & 3.7740 \end{bmatrix},$$

$$\mathbf{A}_4 = \begin{bmatrix} 1.8107 & 3.6929 & 2.3598 & -2.1379 \\ 1 & 0 & 0 & 0 \\ 0 & 1 & 0 & 0 \\ 0 & 0 & 1 & 0 \end{bmatrix},$$

$$\mathbf{B}_4 = \begin{bmatrix} 0.4272 & 0 & 0 & 0 \\ 0 & 6.6247 & 0 & 0 \\ 0 & 0 & 2.9541 & 0 \\ 0 & 0 & 0 & 5.6937 \end{bmatrix}.$$

"Small 1", "Small 2", "Big 1", and "Big 2" are arbitrary fuzzy sets. The fuzzy controller can be designed via the PDC.

Control Rule 1*:*

IF $z_1(t)$ *is Small 1 and* $z_2(t)$ *is Small 2, THEN* $u(t) = -\mathbf{f}_1\mathbf{x}(t)$.

Control Rule 2*:*

IF $z_1(t)$ *is Small 1 and* $z_2(t)$ *is Big 2, THEN* $u(t) = -\mathbf{f}_2\mathbf{x}(t)$. **Control Rule 3***:* *IF* $z_1(t)$ *is Big 1 and* $z_2(t)$ *is Small 2, THEN* $u(t) = -\mathbf{f}_3\mathbf{x}(t)$. **Control Rule 4***:* *IF* $z_1(t)$ *is Big 1 and* $z_2(t)$ *is Big 2, THEN* $u(t) = -\mathbf{f}_4\mathbf{x}(t)$.

$\mathbf{f}_1 \sim \mathbf{f}_4$ *are the feedback gains. Choose the feedback gains to optimize*

$$J = \int_0^\infty \mathbf{x}^T(t)\mathbf{Q}\mathbf{x}(t) + \mathbf{u}^T(t)\mathbf{R}\mathbf{u}(t)dt,$$

where

$$\mathbf{G} = \begin{bmatrix} 1 & 0 & 0 & 0 \\ 0 & 1 & 0 & 0 \\ 0 & 0 & 1 & 0 \\ 0 & 0 & 0 & 1 \end{bmatrix}, \qquad \mathbf{R} = \begin{bmatrix} 1 & 0 & 0 & 0 \\ 0 & 1 & 0 & 0 \\ 0 & 0 & 1 & 0 \\ 0 & 0 & 0 & 1 \end{bmatrix},$$

we have

$$\mathbf{F}_1 = \begin{bmatrix} 0.7362 & -0.0342 & 0.1022 & 0.0697 \\ -0.0140 & 1.0308 & 0.0684 & -0.0095 \\ 0.0568 & 0.0926 & 1.0201 & 0.0613 \\ 0.0620 & -0.0206 & 0.0980 & 1.0090 \end{bmatrix},$$

$$\mathbf{F}_2 = \begin{bmatrix} 0.0081 & -0.0005 & 0.0039 & 0.0015 \\ -0.0262 & 1.2145 & 0.7435 & -0.0082 \\ 0.0339 & 0.1296 & 0.7380 & 0.0520 \\ 0.0816 & -0.0090 & 0.3247 & 1.0316 \end{bmatrix},$$

$$\mathbf{F}_3 = \begin{bmatrix} 1.3697 & 0.8153 & 0.4607 & 0.3833 \\ 0.1423 & 0.7414 & 0.0603 & 0.0334 \\ 0.5717 & 0.4288 & 1.2677 & 0.3716 \\ 0.3026 & 0.1512 & 0.2364 & 1.1034 \end{bmatrix},$$

$$\mathbf{F}_4 = \begin{bmatrix} 0.2853 & 0.1095 & 0.0713 & -0.0637 \\ 1.6987 & 1.6962 & 0.6788 & -0.3977 \\ 0.4929 & 0.3027 & 1.1241 & -0.0636 \\ -0.8494 & -0.3435 & -0.1225 & 1.2125 \end{bmatrix}.$$

If we select

$$\mathbf{P} = \begin{bmatrix} 0.3194 & 0.1053 & 0.1514 & -0.0093 \\ 0.1053 & 0.2707 & 0.0464 & 0.0276 \\ 0.1514 & 0.0464 & 0.4300 & 0.0346 \\ -0.0093 & 0.0276 & 0.03460.2621 & \end{bmatrix} > \mathbf{0},$$

then

$$\mathbf{G}_{ii}^T\mathbf{P} + \mathbf{P}\mathbf{G}_{ii} < \mathbf{0}, \qquad \mathbf{H}_{ij}^T\mathbf{P} + \mathbf{H}_{ij} < \mathbf{0}, \quad i < j$$

for $i, j = 1, 2, 3, 4$, *where*

$$\mathbf{H}_{ij} = (\mathbf{G}_{ij} + \mathbf{G}_{ji})/2.$$

Therefore, the fuzzy controller guarantees the stability of the control system. The common $\mathbf{P}$ *is found by a convex optimization technique.*

7.3.3 Common **B** *matrix case*

This subsection will present the property of the stability in the common **B** case, i.e.,$\mathbf{B}_1 = \mathbf{B}_2 = \cdots = \mathbf{B}_r$. The fuzzy feedback system can be linearized via a fuzzy controller designed by the PDC [10, 17] if it is possible to select the feedback gains $\mathbf{F}_i$ such that $\mathbf{A}_i - \mathbf{B}\mathbf{F}_i = \mathbf{G}$, where $\mathbf{G}$ is a stable matrix. Then, the linearized feedback system is represented as $\dot{\mathbf{x}}(t) = \mathbf{G}\mathbf{x}(t)$ or $\mathbf{x}(t+1) = \mathbf{G}\mathbf{x}(t)$.

In this case, (7.3) and (7.4) are represented as follows.

<CFS>

$$\begin{aligned} \dot{\mathbf{x}}(t) &= \left(\sum_{i=1}^{r} w_i(\mathbf{z}(t))[\mathbf{A}_i\mathbf{x}(t) + \mathbf{B}\mathbf{u}(t)]\right) \Big/ \left(\sum_{i=1}^{r} w_i(\mathbf{z}(t))\right) \\ &= \sum_{i=1}^{r} h_i(\mathbf{z}(t))[\mathbf{A}_i\mathbf{x}(t) + \mathbf{B}\mathbf{u}(t)], \end{aligned} \tag{7.18}$$

<DFS>

$$\begin{aligned} \mathbf{x}(t+1) &= \left(\sum_{i=1}^{r} w_i(\mathbf{z}(t))[\mathbf{A}_i\mathbf{x}(t) + \mathbf{B}\mathbf{u}(t)]\right) \Big/ \left(\sum_{i=1}^{r} w_i(\mathbf{z}(t))\right) \\ &= \sum_{i=1}^{r} h_i(\mathbf{z}(t))[\mathbf{A}_i\mathbf{x}(t) + \mathbf{B}\mathbf{u}(t)]. \end{aligned} \tag{7.19}$$

By substituting (7.11) into (7.18) and (7.19), we obtain (7.20) and (7.21) respectively, where $\mathbf{G} = \mathbf{A}_i - \mathbf{B}\mathbf{F}_i$,

<CFS>

$$\dot{\mathbf{x}}(t) = \sum_{i=1}^{r} h_i(\mathbf{z}(t))\{\mathbf{A}_i - \mathbf{B}\mathbf{F}_i\}\mathbf{x}(t) = \mathbf{G}\mathbf{x}(t), \tag{7.20}$$

<DFS>

$$\mathbf{x}(t+1) = \sum_{i=1}^{r} h_i(\mathbf{z}(t))\{\mathbf{A}_i - \mathbf{B}\mathbf{F}_i\}\mathbf{x}(t) = \mathbf{G}\mathbf{x}(t). \tag{7.21}$$

The feedback system is represented as $\dot{\mathbf{x}}(t) = \mathbf{G}\mathbf{x}(t)$ or $\mathbf{x}(t+1) = \mathbf{G}\mathbf{x}(t)$ if we can select the feedback gains $\mathbf{F}_i$ such that $\mathbf{A}_i - \mathbf{B}\mathbf{F}_i = \mathbf{G}$, where $\mathbf{G}$ is a stable matrix. However, is should be emphasized that we can not always choose $\mathbf{K}_i$ such that $\mathbf{A}_i - \mathbf{B}\mathbf{K}_i = \mathbf{G}$ for all i.

Theorem 7.6 *Consider the common* **B** *case, i.e.,* $\mathbf{B}_1 = \mathbf{B}_2 = \cdots = \mathbf{B}_r$ *in the CFS. The fuzzy feedback system can be linearized via a fuzzy controller* *designed by the PDC if it is possible to select the feedback gains* $\mathbf{F}_i$ *such that* $\mathbf{A}_i - \mathbf{B}\mathbf{F}_i = \mathbf{G}$, *where* $\mathbf{G}$ *is a stable matrix. The linearized feedback system is represented as* $\dot{\mathbf{x}}(t) = \mathbf{G}\mathbf{x}(t)$.

Theorem 7.7 *Consider the common* $\mathbf{B}$ *case, i.e.,* $\mathbf{B}_1 = \mathbf{B}_2 = \cdots = \mathbf{B}_r$ *in the DFS. The fuzzy feedback system can be linearized via a fuzzy controller designed by the PDC if it is possible to select the feedback gains* $\mathbf{F}_i$ *such that* $\mathbf{A}_i - \mathbf{B}\mathbf{F}_i = \mathbf{G}$, *where* $\mathbf{G}$ *is a stable matrix. The linearized feedback system is represented as* $\mathbf{x}(t+1) = \mathbf{G}\mathbf{x}(t)$.

Example 7.8 *Consider the following CFS.*

Plant Rule 1:

IF $z(t)$ *is Small, THEN* $\dot{\mathbf{x}}(t) = \mathbf{A}_1\mathbf{x}(t) + \mathbf{b}u(t)$.

Plant Rule 2:

IF $z(t)$ *is Big, THEN* $\dot{\mathbf{x}}(t) = \mathbf{A}_2\mathbf{x}(t) + \mathbf{b}u(t)$.
In the CFS,

$$\mathbf{A}_1 = \begin{bmatrix} 2 & 5 \\ 1 & 0 \end{bmatrix}, \quad \mathbf{A}_2 = \begin{bmatrix} 1 & -3 \\ 1 & 0 \end{bmatrix}, \quad \mathbf{b} = \begin{bmatrix} 1 \\ 0 \end{bmatrix}.$$

"Small" and "Big" are arbitrary fuzzy sets satisfying (12.5). The fuzzy controller can be designed via the PDC.

Control Rule 1:

IF $z(t)$ *is Small, THEN* $u(t) = -\mathbf{f}_1\mathbf{x}(t)$.

Control Rule 2:

IF $z(t)$ *is Big, THEN* $u(t) = -\mathbf{f}_2\mathbf{x}(t)$.
$\mathbf{f}_1$ *and* $\mathbf{f}_2$ *are the feedback gains. Choose the closed-loop eigenvalues to be* $[-2, -2]$, *we have*

$$\mathbf{f}_1 = [6, 9], \qquad \mathbf{f}_2 = [5, 1].$$

Then, we obtain

$$\mathbf{A}_1 - \mathbf{b}\mathbf{f}_1 = \mathbf{A}_2 - \mathbf{b}\mathbf{f}_2 = \begin{bmatrix} -4 & -4 \\ 1 & 0 \end{bmatrix} = \mathbf{G}.$$

Therefore, the fuzzy controller linearizes and stabilizes the control system.

Recently, new stability conditions are derived in [31, 32] by relaxing the above stability results. On the other hand, robust stability [e.g., 18, 29, 30] and optimal control design [e.g., 38, 39] are discussed. A mixed control problem for robust-optimal fuzzy control is also presented in [35-37]. Furthermore, fuzzy observer designs are discussed in [32-34]. As mentioned above, the advanced works, i.e., LMI based designs, for the stability results discussed here have been presented in [10, 17, 28-37].

References

[1] Tanaka, K. and Sugeno, M. (1990) Stability analysis of fuzzy systems using Lyapunov's direct method, *Proceedings of NAFIPS'90*, 133-136.

[2] Langari, R. and Tomizuka, M. (1990) Analysis and synthesis of fuzzy linguistic control systems, *1990 ASME Winter Annual Meeting*, 35-42.

[3] Kitamura, S. and Kurozumi, T. (1991) Extended circle criterion and stability analysis of fuzzy control systems, *Proceedings of the International Fuzzy Eng. Symp.'91*, **2**, 634-643.

[4] Tanaka, K. and Sugeno, M. Stability analysis and design of fuzzy control systems, *Fuzzy Sets and Systems*, **45**(2), 135-156.

[5] Farinwata S. S. et al. (1993) Stability analysis of the fuzzy logic controller designed by the phase portrait assignment algorithm, *Proceedings of 2nd IEEE International Conference on Fuzzy Systems*, 1377-1382.

[6] Langari, R. and Tomizuka, M. (1990) Stability of fuzzy linguistic control systems, *Proceedings of IEEE Conference on Decision and Control*, 2185-2190.

[7] Takagi, T. and Sugeno, M. (1985) Fuzzy identification of systems and its applications to modeling and control, *IEEE Trans. on SMC*, **15**(1), 116-132.

[8] Tanaka, K. and Sano, M. (1993) Fuzzy stability criterion of a class of nonlinear systems, *Information Sciences*, **71**, 3-26.

[9] Tanaka, K. and Sugeno, M. (1993) Concept of stability margin or fuzzy systems and design of robust fuzzy controllers, *Proceedings of 2nd IEEE International Conference on Fuzzy Systems*, **1**, 29 -34.

[10] Wang, H., Tanaka, K., and Griffin, M. (1995) Parallel distributed compensation of nonlinear systems by Takagi and Sugeno's fuzzy model, *Proceedings of FUZZ-IE EE'95*, 531-538.

[11] Kawamoto, S. et al. (1993) Fuzzy-type Lyapunov function, *Proceedings of Second IEEE International Conference on Fuzzy Systems.*

[12] Tanaka, K. and Sano, M. (1995) Trajectory stabilization of a model car via fuzzy control, *Fuzzy Sets and Systems*, **70**, 155-170.

[13] Singh, S. (1992) Stability analysis of discrete fuzzy control systems, *Proceedings of First IEEE International Conference on Fuzzy Systems*, 527-534.

[14] Katoh, R. et al. (1993) Graphical stability analysis of a fuzzy control system, *Proceedings of IEEE International Conference on IECON'93*, **1**, 248-253.

[15] Chen, C. -L. et al. (1993) Analysis and design of fuzzy control systems, *Fuzzy Sets and Systems*, **57**, 125-140.

[16] Filev, D. (1996) Polytopic TSK fuzzy systems: analysis and synthesis, *Fifth IEEE International Conference on Fuzzy Systems (FUZZ-IEEE'96)*, 687-693.

[17] Wang, H., Tanaka, K., and Griffin, M. (1996) An approach to fuzzy control of nonlinear systems: Stability and Design Issues, *IEEE Transactions on Fuzzy Systems*, **4**(1), 14-23.

[18] Tanaka, K. and Sano, M. (1994) A robust stabilization problem of fuzzy controller systems and its applications to backing up control of a truck-trailer, *IEEE Trans. on Fuzzy Systems*, **2**(2), 119-134.

[19] Tanaka, K. and Yoshioka, K. (1995) Design of fuzzy controller for backer-upper of a five-trailers and truck, *4th IEEE International Conference on Fuzzy Systems*, **3**, 1543-1548.

[20] Tanaka, K. (1994) *Advanced Fuzzy Control*, Kyouritu Publisher(in Japanese).

[21] Sugeno, M. (1998) *Fuzzy Control*, Nikangougyou-shinnbunsha Publisher, (in Japanese).

[22] Tanaka, K. and Sano, M. (1993) Fuzzy stability criterion of a class of nonlinear systems, *Information Sciences*, **71**, 3-26.

[23] Kawamoto, S. et al. (1992) An approach to stability analysis of second order fuzzy systems, *Proceedings of First IEEE International Conference on Fuzzy Systems*, **1**, 1427-1434.

[24] Tanaka, K. and Kosaki, T. (1995) Path tracking control of a vehicle robot with a trailer using parallel distributed compensation, *Proceedings of International Conference of CFSA/IFIS/SOFT'95 on Fuzzy Theory and Applications*, 555- 560.

[25] Boyd, S. et. al. (1994) Linear matrix inequalities in systems and control theory, *SIAM*, Philadelphia.

[26] Boyd, S. et. al. (1994) Control system analysis and design via linear matrix inequalities, *MOVIC Tutorial Workshop Notes.*

[27] Nesterov, Yu. and Nemirovsky, A. (1994) Interior-point polynomial methods in convex programming, *SIAM*, Philadelphia.

[28] Wang, H., Tanaka, K., and Kosaki, K. (1996) An LMI-based stable fuzzy control of nonlinear systems and its application to backing of a truck with multiple trailers, *Fifth IEEE International Conference on Fuzzy Systems (FUZZ-IEEE'96)*, 1433-1438.

[29] Tanaka, K., Ikeda, T., and Wang, H. (1996) Robust stabilization of a class of uncertain nonlinear systems via fuzzy control: quadratic stabilizability,

control theory and linear matrix inequalities, *IEEE Transactions on Fuzzy Systems*, **4**(1), 1-13.

[30] Xia, L. et al. (1995) Assessment on robustness properties of a class of non-linear systems with fuzzy logic controllers, *Proceedings of the International Joint Conference of CFSA/IFIS/SOFT'95 on Fuzzy Theory and Applications*, 271-276.

[31] Tanaka, K., Ikeda, T., and Wang, H. (1996) Design of fuzzy control systems based on relaxed LMI stability conditions, *Proceedings of 35th IEEE Conference on Decision and Control*, 598-603.

[32] Tanaka, K., Ikeda, T., and Wang, H. (1997) Fuzzy regulators and fuzzy observers: relaxed stability conditions and LMI based designs, *IEEE Transactions on Fuzzy Systems*, (to appear).

[33] Tanaka, K., Ikeda, T., and Wang, H. O. (1997) Fuzzy regulators and fuzzy observers: a linear matrix inequality approach, *J. 36th IEEE Conference on Decision and Control*, San Diego.

[34] Tanaka, K., Taniguchi, T., and Wang, H. (1997) Fuzzy controller and observer for backing control of a trailer truck, *Engng. Applic. Artif. Intell.*, **10**(5), 441-452.

[35] Tanaka, K., Taniguchi, T., and Wang, H. O. (1998) Model-based fuzzy control of TORA system: fuzzy observer design via LMIs that represent decay rate, disturbance rejection, robustness, optimality, *J. FUZZ-IEEE*, Alaska (to appear).

[36] Tanaka, K., Nishimura, M., and Wang, H. (1998) Multi-objective fuzzy control of high rise/high speed elevators using LMIs, *1998 American Control Conference*, Philadelphia (to appear).

[37] Tanaka, K., Taniguchi, T., and Wang, H. O. (1998) A mixed control problem of robust fuzzy control and optimal fuzzy control and its solution via linear matrix inequalities, *IEEE Transactions on Fuzzy Systems* (submitted).

[38] Kacprzyk, J. (1992) Fuzzy optimal control revised, *2nd Int. Conf. on Fuzzy and Neural Network*, Lizuka, 17-22.

[39] Filev, D. and Angelov, P. (1992) Fuzzy optimal control, *J. Fuzzy Sets and Systems*, **47**, 151-156.

8 LEARNING AND TUNING OF FUZZY RULES

Hamid R. Berenji

Intelligent Inference Systems Corp.
Computational Sciences Division, MS: 269-2
NASA Ames Research Center
Mountain View, CA 94035
berenji@ptolemy.arc.nasa.gov

8.1 INTRODUCTION

The main idea in integrating fuzzy logic systems with learning and tuning techniques is to use the strength of each one collectively in the resulting adaptive fuzzy system. This fusion produces a major group of techniques required for computing with words and soft computing [1, 2, 3]. In this chapter, we discuss a small set of these techniques for generation of fuzzy rules from data and also, a set of techniques for tuning fuzzy rules. For clarity, we refer to the process of generating rules from data as the learning problem and distinguish it from tuning an already existing set of fuzzy rules. For learning, we touch on unsupervised learning techniques such as fuzzy c-means, fuzzy decision tree systems, fuzzy genetic algorithms, and linear fuzzy rule generation methods. For tuning, we discuss Jang's ANFIS architecture, Berenji-Khedkar's GARIC architecture and its extensions in GARIC-Q. We show that the hybrid techniques capable of learning and tuning fuzzy rules, such as CART-ANFIS, RNN-FLCS, and GARIC-RB, are desirable in development of a number of future intelligent systems.

8.2 LEARNING FUZZY RULES

A vast set of learning techniques have been used for clustering (e.g., Ruspini [4], Krishnapuram, Naasraoui, and Frigui [5] Nakamori and Ryoke [6]) and for generating fuzzy rule bases (e.g., Wang and Mendel [7], Takagi and Sugeno [8], Kosko [9] and Dickerson and Kosko [10]). The general idea here is to generate a set of fuzzy rules that not only best describe the data at hand but also are robust enough to show good generalization capabilities. The obtained rules should perform well under test conditions not included in the training data set.

8.2.1 Fuzzy C-Means Clustering

Fuzzy C-Means Clustering (FCM) of Bezdek [11], also known as ISODATA, smoothens the boundary of clusters generated by the Hard C-means (HCM) approach [12] and allows each data point to belong to a cluster using a membership function. In FCM, a data point can belong to several clusters with the condition

$$\sum_{i=1}^{c} u_{ij} = 1, \forall j = 1, ..., n. \tag{8.1 }$$

where c is the number of clusters and u_{ij} is the degree that data point j belongs to cluster i. The FCM cost function is defined as

$$J(U, c_1, c_2, .., c_c) = \sum_{i=1}^{c} J_i = \sum_{i=1}^{c} \sum_{j}^{n} u_{ij}^m d_{ij}^2 \tag{8.2 }$$

where c_i is the cluster center for fuzzy group i; u_{ij} is between 0 and 1; $d_{ij} = ||c_i - x_j||$ is the Euclidean distance between the ith cluster center and jth data point; and m is a weighting exponent. In every step of FCM, the following two necessary conditions must hold [11]

$$c_i = \frac{\sum_{j=1}^{n} u_{ij}^m x_j}{\sum_{j=1}^{n} u_{ij}^m}, \tag{8.3 }$$

and

$$u_{ij} = \frac{1}{\sum_{k=1}^{c} \left(\frac{d_{ij}}{d_{kj}}\right)^{\frac{2}{(m-1)}}}. \tag{8.4 }$$

In each iteration, FCM initializes the membership matrix U with random values consistent with the constraint in (8.1). Then it calculates c fuzzy cluster centers using (8.3). Next, FCM calculates the cost function according to (8.2) and if this cost is below a certain threshold or it does not significantly improve the observed error, the process stops. Otherwise, a new U matrix is calculated based on (8.4). For more recent extensions of the FCM, see [13], [5], and [14].

8.2.2 Decision tree systems

Decision tree systems partition a data set into mutually exclusive regions based on their input spaces. Decision Trees contain a number of decision nodes and some terminal nodes which can be singletons or functions. If linear functions are used at the terminal nodes, then the decision tree can easily interpret a representation of a non-linear input-output maping with piecewise linear surfaces. Two strong families of decision tree methods have been used. The first family is based on Classification and Regression Trees (CART) of Brieman [15] and the second family includes ID3 proposed by Quinlan [16] and its newer extension in terms of C4 procedure [17]. CART grew mostly from the statistic community inducing binary trees with strong re-sampling techniques for tree pruning and error estimation. ID3 and C4 were developed mostly as machine learning methodologies that induce decision trees which are pruned to reduce the effects of noisy data. Berenji has used C4 to derive a fuzzy controller [18] out of a data set obtained from a control experiment. Later we will discuss CART-ANFIS which is a combined decision tree and adaptive neuro-fuzzy technique proper for rule induction and tuning.

8.2.3 Using Genetic Algorithms to generate Fuzzy Rules

Karr [19, 20] used Genetic Algorithms to generate fuzzy rules with triangular membership functions. Berenji and Malyshev [21] developed a system that generates fuzzy rules using a combined Genetic Algorithms and Reinforcement Learning methodologies. The total number of time steps that a controller can run without experiencing a failure has been used in developing a fitness function. Other criteria such as consideration of the phase-plane generated by a string representation of a controller were used in further improvement of the fitness function.

Sun [22] provides applications of fuzzy genetic algorithms to game playing. For more details, see Chapter 11 of this handbook by Geyer and Shulz.

8.2.4 Berenji-Khedkar's linear fuzzy rules generation method

Given a data set of input-output mappings, Berenji and Khedkar [23] proposed an algorithm that generates a set of fuzzy rules with linear consequents from data using radial basis functions and clustering in product space. The algorithm, now referred to as BK-Clustering, uses output information in conjunction with adding and pruning neurons in order to generate a compact structure and its rough approximation quickly from one pass over the data. This approximation is done by a set of fuzzy rules which use fuzzy preconditions but have crisp outputs similar to the Takagi-Sugeno-Kang (TSK) rules [8]. In particular, a neuron represents a fuzzy rule r:

if s_1 is ξ_{r1} and s_2 is ξ_{r2} .. **then** y_{r1} is $c_{r10} + c_{r11}s_1 + .. + c_{r1n}s_n$ and y_{r2} is $c_{r20} + c_{r21}s_1 + .. + c_{r2n}s_n$

where the conclusion of a rule is a linear function of the input sample $\langle s_1, ..., s_n \rangle$. Each neuron/rule r has a range of influence in the input space, which is governed by a spread parameter σ_r. The fuzzy set against which $\mathbf{s}$ is matched is *not* a cartesian product of fuzzy sets on the real line, as is usually the case in fuzzy theory, but instead is a spherically symmetric shape centered on $(\xi_1, ...\xi_n)$. The degree of membership is given by a Gaussian centered on this point. Thus the input space is partitioned into arbitrarily positioned fuzzy spheres, rather than a fuzzy cartesian partition, which gives greater flexibility. The number of rules is determined based on the data itself. The total mean squared error over all samples in the training set is given by $E = \frac{1}{2}\sum_{\mathbf{s}}\sum_{k=1}^{m}(O_k - d_k)^2$. Since learning is to be one-pass and online, and the net structure is constantly changing, the update-after-each-sample method of carrying out gradient descent is used [23]

$$\frac{\partial E}{\partial y_{jk}} = \sum_s (O_k - d_k)\frac{\partial O_k}{\partial y_{jk}} = \sum_s (O_k - d_k)\rho_j$$

$$\frac{\partial E}{\partial c_{jki}} = \sum_s (O_k - d_k)\rho_j s_i.$$

If updates are done online after each sample, and the learning rate is η, then the update rule for output coefficients is

$$\Delta c_{jki} = -\eta\frac{\partial E}{\partial c_{jki}} = \eta(d_k - O_k)\rho_j s_i.$$

If the sample is located far from all the rules, the knowledge of the net is incomplete. Hence a new cluster or neuron must be added at or near the sample. After the data has been processed, the neurons must be pruned in order to get rid of redundant rules and make the knowledge more compact. This may lead to slightly higher error, but the tradeoff is a reduction in the rulebase size. Rather than doing pruning after each data point is processed, which leads to a high computational load as well as low robustness, pruning is done at the very end. At this stage, all pairs of neurons are made to compete, and a neuron is deleted if:

- its location falls within the radius of influence of another competing neuron, and
- its output hyperplanes are almost the same as the corresponding hyperplanes of the above competing neuron.

If two neurons cover each other's center and have similar outputs, then the two are merged into one neuron. The winner is shifted toward the loser's

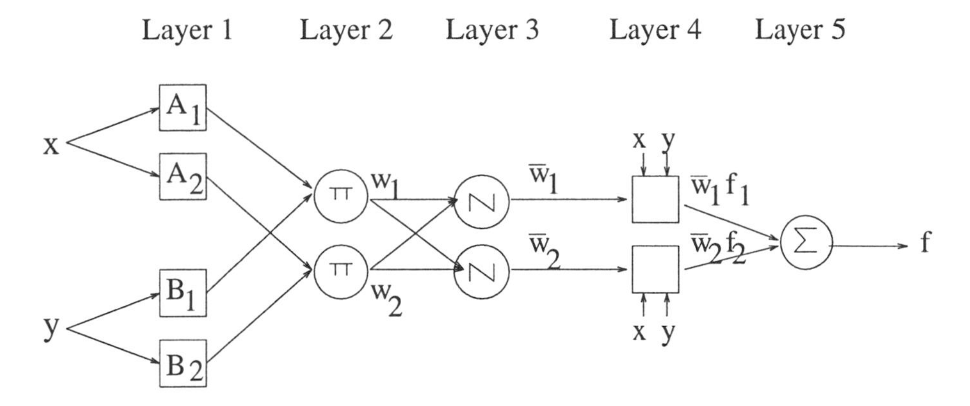

Figure 8.1 The ANFIS architecture for a two-input first-order Sugeno fuzzy model with two rules.

coordinates by an amount proportional to the relative volume of the shells (with radius σ) over which the two neurons have influence. The radius of influence of the winner is increased, if necessary, to include the region within the radius of the loser. For more details, see [23].

8.3 TUNING FUZZY RULES

In this section we assume that a set of fuzzy rules have been obtained either from experts or automatically generated using clustering and other structure identification techniques, such as those described in the last section.

Procyk and Mamdani [24] pioneered the development of self-organizing fuzzy controllers. Chapter 5 of the book by Dubois and Prade [25] reviews some of the earlier works related to learning in fuzzy systems including the work by Wee and Fu [26] and Asai and Kitajima [27, 28] on fuzzy automaton and a reinforcement algorithm, and Saridis and Stephano [29] on control of a prosthetic arm. In this section, we review some of the more recent techniques in tuning fuzzy rules.

8.3.1 ANFIS

Adaptive Neuro-Fuzzy Inference Systems (ANFIS) is a simple but powerful architecture proposed by Jang [30]. It uses first-order Sugeno fuzzy model [8] to represent fuzzy systems in term of a multi-layer neural network. Figure 8.1 illustrates a simple ANFIS architecture for two rules each having two preconditions, two labels for each precondition, a product conjunction operator, and a weighted average combination rule for defuzzification. In particular, the above figure models the following two rules:

Rule 1: If x is A_1 and y is B_1, then $f_1 = p_1x + q_1y + r_1$,
Rule 2: If x is A_2 and y is B_2, then $f_2 = p_2x + q_2y + r_2$,

In layer 2 a product operator shown with π is used that calculates the product of the membership functions of a rule preconditions. Layer 3 calculates normalized firing strength:

$$O_{3,i} = \bar{w}_i = \frac{w_i}{w_1 + w_2}, i = 1, 2. \quad (8.5\ )$$

A node in Layer 4 is adaptive:

$$O_{4,i} = \bar{w}_i f_i = \bar{w}_i(p_i x + q_i y + r_i), \quad (8.6\ )$$

Finally, a node in Layer 5 computes the overall output:

$$overall\ output = O_{5,1} = \sum_i \bar{w}_i f_i = \frac{\sum_i w_i f_i}{\sum_i w_i} \quad (8.7\ )$$

If non-Sugeno type systems, such as Tsukomoto or Mamdani's Max-Min compositions, are used, then ANFIS needs to connect the outputs of layer 2 directly to the nodes in layer 4. This modification makes the ANFIS architecture more similar to the ASN network in GARIC which is described next.

8.3.2 GARIC

The Generalized Approximate Reasoning based Intelligent Control (GARIC) is an architecture for tuning fuzzy rules based on Reinforcement Learning [31]. GARIC determines a control action by using a neural network which implements fuzzy inference. Another neural net will learn to become a good evaluator of the current state and will serve as an internal critic. Both networks will adapt their weights concurrently so as to improve performance. Figure 8.2 shows the GARIC architecture which has three components:

- The Action Selection Network (ASN) maps a state vector into a recommended action F, using fuzzy inference.
- The Action Evaluation Network (AEN) maps a state vector and a failure signal into a scalar score which indicates state goodness. This is also used to produce internal reinforcement $\hat{r}$.
- The Stochastic Action Modifier (SAM) uses both F and $\hat{r}$ to produce an action F' which is applied to the plant.

The ensuing state is fed back into the controller, along with a reinforcement signal. Learning occurs by fine-tuning of the free parameters in the two networks: in the AEN, the weights are adjusted; in the ASN, the parameters describing the fuzzy membership functions change.

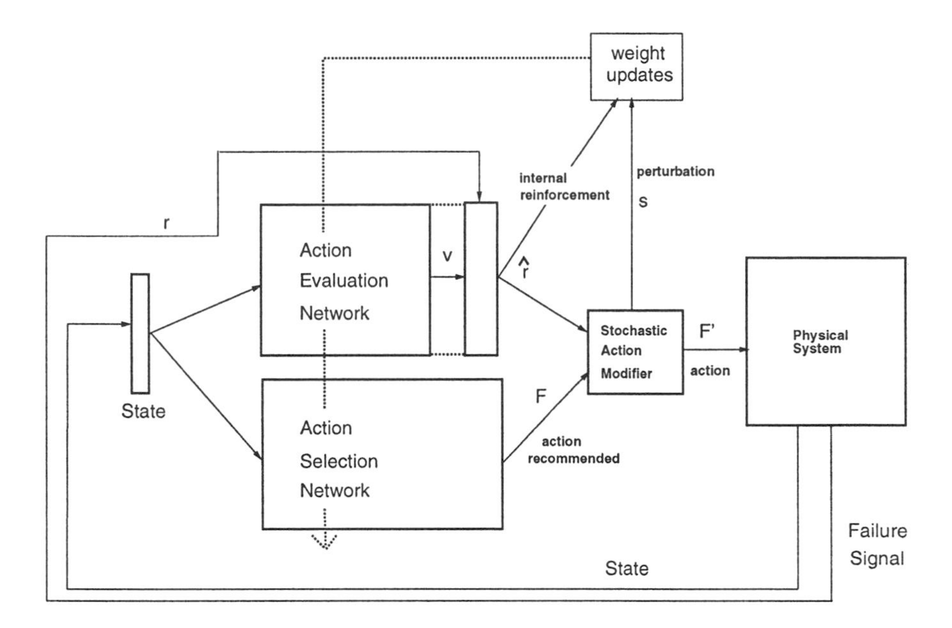

Figure 8.2 The Architecture of GARIC.

The Action Evaluation Network. The AEN plays the role of an adaptive critic element (ACE) [32] and constantly predicts reinforcements associated with different input states. The only information received by the AEN is the state of the physical system in terms of its state variables and whether or not a failure has occurred.

The AEN is a standard two-layer feedforward net with sigmoids everywhere except in the output layer. The input is the state of the plant, and the output is an evaluation of the state (a score), denoted by v. This v-value is suitably discounted and combined with the external failure signal to produce internal reinforcement $\hat{r}$.

This network evaluates the action recommended by the action network as a function of the failure signal and the change in state evaluation based on the state of the system at time $t+1$:

$$\hat{r}[t+1] = \begin{cases} 0 & \text{start state ;} \\ r[t+1] - v[t,t] & \text{failure state;} \\ r[t+1] + \gamma v[t,t+1] - v[t,t] & \text{otherwise} \end{cases} \qquad (8.8)$$

where $0 \leq \gamma \leq 1$ is the *discount rate.* In other words, the change in the value of v plus the value of the external reinforcement constitutes the heuristic or internal reinforcement $\hat{r}$ where the future values of v are discounted more, the further they are from the current state of the system. For example, the value of v generated one time step later is given less weight than the current value of v. This method of estimating reinforcement gives an approximate exponential trace of v, where the series is truncated after two terms.

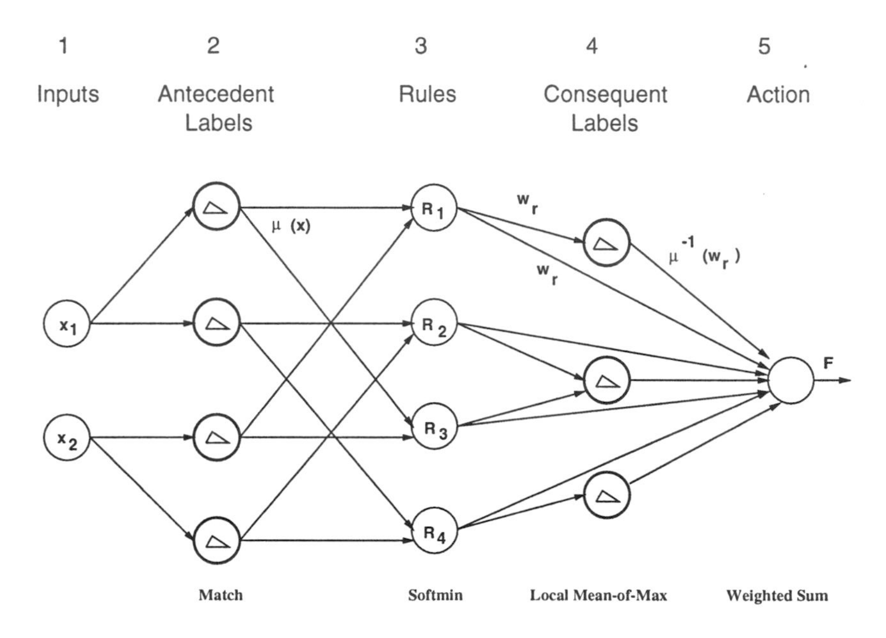

Figure 8.3 The Action Selection Network

Action Selection Network. Given the current state of the plant, this network selects an action. Figure 8.3 shows the 5-layer neural network representation of a fuzzy controller where Layer 1 is the input layer. A node in Layer 2 corresponds to one possible value of one of the linguistic variables in Layer 1. The function is given by

$$\mu_{c_V, s_{VL}, s_{VR}}(x)$$

where V indicates a linguistic value (e.g. *large*), and c, s_L, s_R correspond to the center, left spread and right spread of the fuzzy membership function of label V. c_V serves as a reference point (the mode), and the spreads characterize length scales on either side of the center, thus permitting asymmetry. More parameters may be included if desired. An instance of a smooth membership function is

$$\mu(x) = \frac{1}{1 + |\frac{x-c}{s}|^b}$$

where $s = s_{VL}$ or s_{VR} accordingly as $x < c$ or $x \geq c$ and b controls the curvature. For triangular shapes, this function is given by

$$\mu_{c,s_L,s_R}(x) = \begin{cases} 1 - |x - c|/s_R, & x \in [c, c + s_R] \\ 1 - |x - c|/s_L, & x \in [c - s_L, c) \\ 0 & \text{otherwise} \end{cases} \tag{8.9 }$$

Layer 3 implements the conjunction of all the antecedent conditions in a rule. A node in layer 3 corresponds to a rule in the rule-base. Its inputs come from all nodes in Layer 2 which participate in the *if* part of that rule. The node itself performs the *min* operation, which has been softened to the following continuous, differentiable *softmin* operation:

$$O_{R3} = w_r = \frac{\sum_i \mu_i e^{-k\mu_i}}{\sum_i e^{-k\mu_i}} \tag{8.10 }$$

Here, μ_i is the degree of match between a fuzzy label occurring as one of the antecedents of rule r, and the corresponding input variable. This *softmin* operation gives w_r, the degree of applicability of Rule r. The parameter k controls the hardness of the softmin operation, and as $k \to \infty$, we recover the usual *min* operator. However, for k finite, we get a differentiable function of the inputs, which makes it convenient for calculating gradients during the learning process. The choice of k is not critical.

A **Layer 4** node corresponds to a consequent label. Its inputs come from all rules which use this particular consequent label. For each of the w_r supplied to it, this node computes the corresponding output action as suggested by rule r. This mapping may be written as

$$\mu^{-1}_{c_V,s_{VL},s_{VR}}(w_r)$$

where V indicates a specific consequent label, c, s_L, s_R parameterize the membership function as before, and the inverse is taken to mean a suitable defuzzification procedure applicable to an individual rule. For defuzzification, a modified version of the Mean-of-Maxima, called LMOM is used. For triangular functions, LMOM gives

$$\mu^{-1}_{c_V,s_{VL},s_{VR}}(w_r) = c_V + \frac{1}{2}(s_{VR} - s_{VL})(1 - w_r) \tag{8.11 }$$

For the case $w_r = 0$, the limiting value of $\mu^{-1}(w_r \to 0^+)$ is used (which is $c_V + (s_{VL} + s_{VR})/2$). It is easy to see that the set $\mu^{-1}([0,1])$ is the projection of the median of the triangular membership function on the X-axis. If the membership function is monotonic, then $\mu^{-1}(w_r)$ is just the standard mathematical inverse, with appropriate limiting values.

The unusual feature of a unit in Layer 4 is that it may have multiple outputs carrying different values, since sharing of consequent labels is allowed. For each rule feeding it a degree, it should produce a corresponding output action which is fed to the next layer. However, this nonstandard feature can be eliminated for many classes of membership functions. For triangular functions, such a node needs to output only the value

$$O_{V4} = (c_V + \frac{1}{2}(s_{VR} - s_{VL}))(\sum_r w_r) - \frac{1}{2}(s_{VR} - s_{VL})(\sum_r w_r^2) \qquad (8.12\ )$$

In general, whenever $\mu^{-1}(x)$ is polynomial in x, only one output is sufficient, regardless of the number of inputs. This is true in the ANFIS architecture too. This transformation is possible because of the form of the computation done in the next layer.

Layer 5 will have as many nodes as there are output action variables. Each output node combines the recommendations from all the fuzzy control rules in the rulebase, using the following weighted sum, the weights being the rule strengths:

$$F = \frac{\sum_r w_r \mu^{-1}(w_r)}{\sum_r w_r} \qquad (8.13\ )$$

By taking advantage of the transformation used in layer 4, this may be rewritten as

$$F = \frac{\sum_V O_{V4}}{\sum_R O_{R3}} \qquad (8.14\ )$$

where the inputs come from Layer 3 and Layer 4. The node simply sums up each set of inputs and takes their quotient. This delivers a continuous output variable value which is the action selected by the ASN.

Modifiable weights are present on input links into Layer 2 and 4 only. The other weights are fixed at unity. This means that the gradient descent procedure effectively works on only two layers of weights, rather than all five.

Stochastic Action Modifier. The Stochastic Action Modifier (SAM) uses the values of $\hat{r}$ from the previous time step and the action F recommended by the ASN to stochastically generate an action F' which is a Gaussian random variable with mean F and standard deviation $\sigma(\hat{r}(t-1))$. This $\sigma()$ is some non-negative, monotone decreasing function, e.g. $\exp(-\hat{r})$. The action F' is what is actually applied to the plant. The stochastic perturbation in the suggested action leads to a better exploration of state space and better generalization ability. The magnitude of the deviation $|F' - F|$ is large when $\hat{r}$ is low, and small when the internal reinforcement is high. The result is that a large random step away from the recommendation results when the last action performed is bad, but the controller remains consistent with the fuzzy control rules when the previous action selected is a good one. The actual form of the function $\sigma()$, especially its scale and rate of decrease, should take the units and range of variation of the output variable into account.

The learning mechanisms in GARIC are based on the reward/punishment strategy of reinforcement learning, using the internal reinforcement and a measure of change resulting from the stochastic action modification [31].

8.3.3 Fuzzy Q-Learning and GARIC-Q

Fuzzy Q-Learning extends Watkin's Q-learning method [33] for decision processes in which the goals and/or the constraints, but not necessarily the system under control, are fuzzy in nature. An example of a fuzzy constraint is: "the weight of object A must not be *substantially* heavier than w" where w is a specified weight. Similarly, an example of a fuzzy goal is: "the robot must be in the *vicinity* of door k".

The Q-learning algorithm maintains an estimate $Q(x, a)$ of the value of taking action a in state x. Similarly, in the FQ-Learning algorithm, we maintain an estimate $FQ(x, a)$ for taking action a in state x, where the actions can have fuzzy constraints on them, and continuing with the optimal policy after a new state is reached. The value of a state can be defined as the value of the state's best state-action pair:

$$V(x) = Max_a \; FQ(x, a) \tag{8.15}$$

We define the FQ value as the expected value of the *confluence* [34] the immediate reinforcements plus the discounted value of the next state *and* the constraints on performing action a in state x

$$FQ(x, a) = E\{(r + \gamma V(y)) \wedge \mu_C(x, a)\} \tag{8.16}$$

where $v(x)$ is the value of the current state, $v(y)$ is the value of the state y reached after applying action a at state x, $\wedge$ represents a conjunction or an "and" operator (e.g., the "minimum") and γ is a discount factor. Note that if an action has no constraints then $\mu_C(x, a) = 1$. The FQ values are updated according to:

$$\Delta FQ(x, a) = \beta[(r + \gamma V(y)) \wedge \mu_C(x, a) - FQ(x, a)] \tag{8.17}$$

where β is a learning rate.

Initialize FQ values
Until FQ values converge or max. # of trials do {

1. $x \leftarrow$ current state

2. Select the action using Boltzmann exploration:

$$P(x,a) = \frac{e^{FQ(x,a_i)/T}}{\sum_{i=1}^{n} e^{FQ(x,a_i)/T}} \tag{8.18}$$

where $P(x,a)$ is the probability of selecting action a in state x, T is the temperature parameter which can be decreased over time to decrease exploration, and n is the number of actions possible in state x.

3. Apply action, observe the new state (y) and reward (r)

4. Update
$\Delta FQ(x,a) = \beta[(r + \gamma V(y)) \wedge \mu_C(x,a) - FQ(x,a)]$
where $v(x)$ is the value of the current state, $v(y)$ is the value of the state y reached after applying action a at state x,$\wedge$ represents a conjunction or an "and" operator (e.g., the "minimum") }

Figure 8.4 The FQ-Learning Algorithm

The FQ-Learning method proceeds in the following manner. A state is selected and the FQ values for all the possible actions from that state are compared. The action with the maximum FQ value is selected and gets applied. As the result of applying this action, the system moves to a new state and an immediate reinforcement (if any) is received. The process continues until the FQ values converge to their optimal values or a maximum number of trials is exceeded. The algorithm, as depicted in Figure 8.4, is applicable for discrete incremental dynamic programming. In [35], two applications of this method for a multi-stage decision problem and a typical robotic path planning problem have been shown. However, for continuous problems, such as typical control engineering applications, generalization between similar situations and similar actions is needed in order to apply fuzzy Q-Learning.

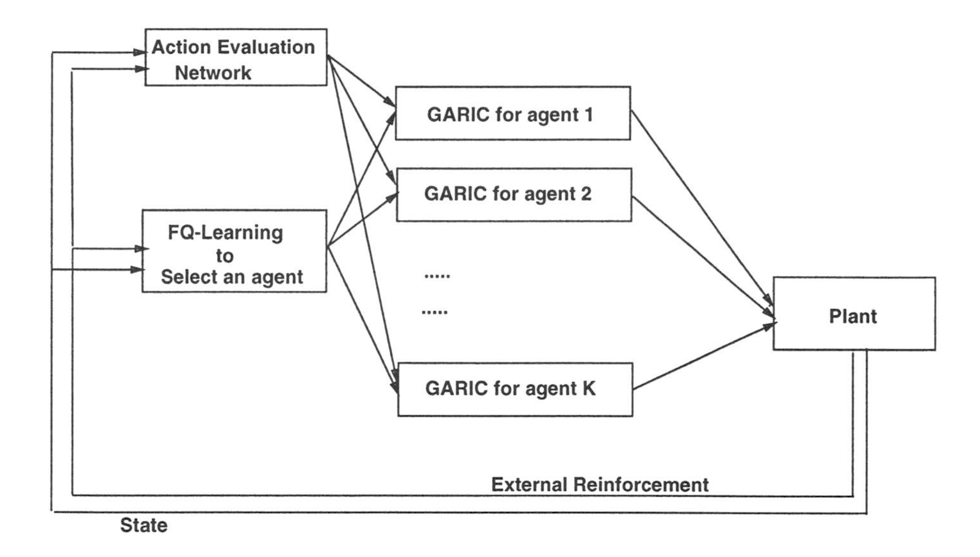

Figure 8.5 The architecture of GARIC-Q

GARIC-Q. GARIC-Q introduces a new method for doing incremental Dynamic Programming [36] using a society of intelligent agents which are controlled at the top level by Fuzzy Q-Learning [35, 37] and at the local level, each agent learns and operates based on GARIC. GARIC-Q improves the speed and applicability of Fuzzy Q-Learning through generalization of input space by using fuzzy rules and bridges the gap between Q-Learning and rule based intelligent systems.

Using Fuzzy Q-Learning, GARIC-Q selects a winner among the GARIC agents and switches the control to that agent at that time step. The agent takes over and:

1. Calculates what action to apply using the current set of rules, within the selected agent, and their fuzzy labels.
2. Using SAM and $\hat{r}(t-1)$ calculates a new action F'
3. Applies the new action and observes the new state
4. Calculates the internal reinforcement $\hat{r}(t)$
5. Updates the weights of the AEN
6. Updates the parameters of the fuzzy labels in ASN

7. Updates the fq values of all the rules used by the agent (to be described next).

The approach that is used in selecting a rule base among the competing rule bases has similarities with the approach first introduced by Glorennec [38]. Assuming that there are K agents and each agent k has R_k rules, then the total number of rules considered by the system is $R = \sum_{k=1}^{K} R_k$. The notation R_{ij} is used to refer to rule number i of agent j. Associated with each rule R_{ij} is a fq_{ij} which represents the fq of rule R_{ij}. The FQ value for an agent k is calculated from:

$$FQ_k = \frac{\sum_{i=1}^{R_k} fq_i * \alpha_i}{\sum_{i=1}^{R_k} \alpha_i} \tag{8.19}$$

where α_i represents the strength of rule i. The FQ values are updated according to:

$$\Delta FQ = \beta(r(t) + \gamma V(y) - V(x)) \tag{8.20}$$

where x is the current state, y is the state reached after applying action a_k at state x, γ is the discount factor, β is a learning rate, $V(x)$ is the value of state x and action a_k selected through a Boltzmann process, and $V(y)$ is the value of the best state-agent pair defined by:

$$V(y) = Max_k FQ(y, a_k) \tag{8.21}$$

where $k = 1$ to K, is the agent number and a_k is its recommended final action. The reinforcement $r(t)$ can take:

$$r(t) = \begin{cases} +1 & if\ within\ the\ success\ region \\ 0 & Viable\ zone \\ -1 & Failure \end{cases} \tag{8.22}$$

Within each agent or rule base k, the reward or punishment is distributed based on the activity of rule i:

$$\rho_i = \frac{\alpha_i}{\sum_{i=1}^{R_k} \alpha_i} \tag{8.23}$$

where α_i is the strength of rule i. The fq values are updated over time for the selected agent j using:

$$\Delta fq_i = \lambda * \rho_i * \Delta FQ \tag{8.24}$$

where λ is a learning rate. Upon each success or failure, the state of the system is returned to an initial state (can be a random state) in the viable zone and learning restarts. The competition between different agents and updating of the fq values continues until the whole process converges to a unique agent (i.e., a unique rule base) or a combination of different agents have been able to control the process for an extended time.

GARIC-Q improves the speed of GARIC and more importantly, provides the facility to design and test different types of agents. These agents may have different number of rules, use different learning strategies on the local level, and have different architectures. GARIC-Q provides the first step toward a true intelligent system where at the lower level, agents can explore the environment and learn from their experience, while at the top level, a super agent can monitor the performance and learn how to select the best agent for each step of the process.

8.4 LEARNING AND TUNING FUZZY RULES

Learning and tuning of fuzzy rules are both required in many applications of fuzzy systems. In the previous two sections, we discussed some of the techniques that are developed to do either learning or tuning. In this section, we discuss a number of methods that can both learn and tune fuzzy rules.

8.4.1 CART-ANFIS

CART-ANFIS [39] is a combination of CART of Brieman [12] and Jang's ANFIS methods. In CART-ANFIS, the CART algorithm is used to identify an appro priate decision tree and to determine relevant inputs. The terminal nodes in this tree may also be characterized by linear equations. Instead of the usual crisp values used in decision trees, CART-ANFIS fuzzifies the values using membership functions. For example, in the rule

If $x < a$ and $y \geq c$ then $z = f_1$

a sigmoidal membership function can be used for c

$$\mu_{y \geq c}(y; \alpha) = sig(y; \alpha, c) = \frac{1}{1 + exp[-\alpha(y - c)]}$$

where α and c are the modifiable parameters of the membership function. Then a modified version of ANFIS to represent the resulting decision tree is developed. ANFIS can then adjust the parameters.

8.4.2 RNN-FLCS

Lin and Lee have proposed RNN-FLCS [40] that learns and tunes a set of fuzzy rules for use in control systems. RNN-FLCS has many similarities to GARIC and in particular, uses a 5-layer neural networks as the action network similar to the ASN, uses a fuzzy predictor similar to the AEN, and learns by incorporating reinforcement learning. However, RNN-FLCS performs structure identification within its learning process and if needed, adds rules in addition to parameter adjustment. The structure identification used in RNN-FLCS is based on an earlier work of Lin and Lee [41]. RNN-FLCS can construct a fuzzy logic control system based on fuzzy information feedback such as high, too high, low, and too low. The system demonstrates hybrid learning and tuning of fuzzy rules.

8.4.3 GARIC-RB

GARIC-RB presents an algorithm which refines an initial set of fuzzy rules which has been developed using radial basis functions through the BK-clustering approach described earlier in Section 8.2.4. It tunes rule parameters based on Fuzzy Reinforcement Learning as used in GARIC.

In GARIC-RB, the conclusion of a rule is a linear function of the input state variables $\langle s_1, ..., s_n \rangle$ and a rule may have multiple outputs. GARIC-RB extends GARIC in a number of important directions including:

1. GARIC uses general, single-axis, fuzzy rules with fuzzy labels in the antecedent and consequent of the rules. GARIC-RB uses Radial basis functions in its antecedents and linear crisp functions in its consequents.

2. GARIC rules can have only a single output while GARIC-RB allows multiple outputs for the rules and the controller.

3. In GARIC-RB, rules with n preconditions are modeled as points in a n-dimensional hyperspace. These points can individually be moved in this space in order to tune the controller.

The ASN is a map from input to output space and the intent of computing the action(s) is to maximize the output of AEN or v, so that the system ends up in a good state and avoids failure. Hence, v is the objective function which needs to be maximized as a function of the parameters of the fuzzy rules, given the state. This can be done by gradient descent, which estimates the derivative $\partial v / \partial p$, for each particular parameter p, and uses the learning rule

$$\Delta p = \eta \frac{\partial v}{\partial p} = \eta \frac{\partial v}{\partial \mathbf{F}} \frac{\partial \mathbf{F}}{\partial p} = \eta \sum_i \frac{\partial v}{\partial F_i} \frac{\partial F_i}{\partial p} \tag{8.25}$$

to adjust the parameter values [23].

The ASN in GARIC-RB is similar to an ANFIS model but uses an Action Evaluation Network (AEN) and reinforcement learning to tune its parameters instead of a supervised learning method as used in ANFIS.

Architectures	**Learning or Tuning**	**Main Technology**
Fuzzy C-Means	Learning	Fuzzy clustering
CART	Learning	Statistical regression trees
ID3,C4	Learning	Decision Trees
Fuzzy Genetic Algorithms	Learning	Genetic algorithms
BK-Clustering	Learning	Radial basis functions, TSK rules
GARIC	Tuning	Reinforcement learning
ANFIS	Tuning	Error back propagation
GARIC-Q	Tuning	Reinforcement learning (Q-Learning)
CART-ANFIS	Learning and tuning	CART, ANFIS
GARIC-RB hline RNN-FLCS	Learning and tuning Learning and tuning	GARIC, BK-Clustering Reinforcement learning

Figure 8.6 Some learning and tuning architectures

8.5 SUMMARY AND CONCLUSION

In this chapter, we discussed some independent techniques for use for learning or tuning fuzzy rules. Figure 8.6 presents a summary. We also emphasized the importance of a combined learning and tuning strategy for many applications of fuzzy rules in intelligent systems. The application domain is the decisive factor in determining the balance between learning and tuning and it should be carefully studied to decide when to learn, when to stop learning and begin to tune, or have a combined approach all the way.

References

[1] Zadeh, L. A. (1996). Fuzzy logic = computing with words, *Transactions on Fuzzy Systems*, 4(2), 103-111.

[2] Zadeh, L. A. (1996). The roles of fuzzy logic and soft computing in the conception, design and development of intelligent systems, *Transactions on Fuzzy Systems*, **4**(2), 103-111.

[3] Zadeh, L. A. (1994). Fuzzy logic, neural networks, and soft computing, *Commun. of ACM*, **37**(3), 77-84.

[4] Ruspini, E. H. (1982). Recent Development in Fuzzy Clustering, *Fuzzy Sets and Possibility Theory*, 133-147.

[5] Bezdek, J. C. and Pal, S. K. (editors), (1992). *Fuzzy Models for Pattern Recognition*, IEEE Press, New York.

[6] Nakamori, Y. and Ryoke, M. (1994). Identification of fuzzy prediction models through hyperellipsoidal clustering, *IEEE Trans. Systems, Man and Cybernetics*, **24**(8), 1153-1173.

[7] Wang, L. X. and Mendel, J. M. (1992). Generating fuzzy rules by learning from examples, *IEEE Trans. Systems, Man and Cybernetics*, **22**(6), 1414-1427.

[8] Takagi, T. and Sugeno, M. (1983). Derivation of fuzzy control rules from human operator's control actions, *IFAC Symposium on Fuzzy Information, Knowledge Representation and Decision Analysis*, Marseille, France, pp. 55-60.

[9] Kosko, B. (1992). *Neural Networks and Fuzzy Systems*, Prentice Hall.

[10] Dickerson, J. A. and Kosko, B. (1996). Fuzzy function approximation with ellipsoidal rules, *IEEE Transactions on Systems, Man, and Cybernetics*, **26**(4), 542-560.

[11] Bezdek, J. C. *Fuzzy Mathematics in Pattern Classification.* Ph. D. Thesis, Cornell University, 1973.

[12] Krishnaiah, P. R. and Kanal, L. N. (1982). Classification, pattern recognition, and reduction of dimensionality, in *Volume 2 of Handbook of Statistics*, Krishnaiah, P. R. and Kanal L. N. (Eds.), North Holland, Amsterdam.

[13] Krishnapuram, R. and Nasraoui, O and Frigui, H. (1992). The Fuzzy C Spherical Shells Algorithm: A New Approach, *IEEE Transactions on Neural Networks*, **3**(5), 663-671.

[14] Bezdek, J. C., Krishnapuram, Raghu, and Pal, Nik (editors) (1992) *Fuzzy Models and algorithms for pattern recognition and image processing*, IEEE Press, New York.

[15] Breman, L., Friedman, J. H., Olshen, R. A., and Stone, C. J. (1984). *Classifications Decision Trees*, Wadsworthm, Inc. Belmont, CA.

[16] Quinlan, J. R. (1986). Induction of Decision Trees, *Machine Learning*, **1**, 81-106.

[17] Quinlan, J. R. (1990). Decision trees and decision making, *IEEE Trans. Systems, Man and Cybernetics*, **20**(2), 339-346.

[18] Berenji, H. R. (1990). Machine learning in fuzzy control, *Int. Conf. on Fuzzy Logic and Neural Networks*, Iizuka, Fukuoka, Japan, pp. 231-241.

[19] Karr, C. L. and Stanley, D. A. (1991). Fuzzy logic and genetic algorithms in time-varying control problems, *Proceedings of the NAFIPS'91*, pp. 285-290.

[20] Karr, C. L. (1991). Applying genetics to fuzzy logic, *AI Expert*, **6**(3), 38-43.

[21] Berenji, H. R. and Malyshev, S. "A fuzzy reinforcement learning system based on genetic algorithm", Technical Report, NASA SBIR phase 1 final report, July, 1994.

[22] Jang, J. -S. R. and Sun, C. -T. (1997) *Neuro-Fuzzy and Soft Computing*, Prentice-Hall, Inc., Upper Saddle River, NJ.

[23] Berenji, H. R. and Khedkar, P. (1993). Clustering in product space for fuzzy inference, *Second IEEE Interntional Conference on Fuzzy Systems*, pp. 1402-1407, San Francisco, CA.

[24] Procyk, C. W. and Mamdani, E. H. (1979) A linguistic self-organizing process controller, *Automatica*, **15**(1), 15-30.

[25] Dubois, D. and Prade, H. (1980) *Fuzzy Sets and Systems: Theory and Applications*, Academic Press, New York.

[26] Wee, W. G. and Fu, K. S. (1969) A formulation of fuzzy automata and its application as a model of learning systems, *IEEE Transactions on Systems, Man, and Cybernetics*, SSC-5, 215-223.

[27] Kiyoji Asai and Seizo Kitajima (1971) Learning Control of Multimodal Systems by Fuzzy Automata, In K.S. Fu, editor, *Pattern Recognition and Machine Learning*, pp. 195-203, Plenum Press, New York.

[28] Kiyoji Asai and Seizo Kitajima (1971) A Method for Optimizing Control of Multimodal Systems using Fuzzy Automata, *Information Sciences*, **3**, 343-353.

[29] Saridis, G. N. and Stephano, H. E. (1977) Fuzzy decision making in prosthetic devices, In M.M Gupta, G.N. Saridis, and B.R. Gaines, editors, *Fuzzy Automata and Decision Making Processes*, pp. 387-402, North Holland, Amsterdam.

[30] Jang, J. S. (1993). Adaptive neural network-based fuzzy inference systems, *IEEE Transactions on Systems, Man, and Cybernetics*, **23**(3).

[31] Berenji, H. R. and Khedkar, P. (1992) Learning and Tuning Fuzzy Logic Controllers Through Reinforcements, *IEEE Transaction on Neural Networks*. **3**(5).

[32] Barto, A. G.,Sutton, R. S., and Anderson, C. W. (1983). Neuronlike adaptive elements that can solve difficult learning control problems, *IEEE Transactions on Systems, Man, and Cybernetics*, **13**, 834-846.

[33] Watkins, C. J. C. H. *Learning with Delayed Rewards*, Ph. D. Thesis, Cambridge University, Psychology Department, 1989.

[34] Bellman, R. and Zadeh, L.A. (1970). Decision-making in a fuzzy environment, *Management Science*, **17**(4), 141-164.

[35] Berenji, H. R. (1994). A new approach for fuzzy Dynamic programming problems, *Third IEEE International Conference on Fuzzy Systems*, pp. 486-491, Orlando, FL.

[36] Bellman, R. (1957). *Dynamic Programming*, Princeton University Press, Princeton.

[37] Berenji, H. R. (1993). Fuzzy reinforcement learning and dynamic programming, *International Joint Conference on Artificial Intelligence, Workshop on Fuzzy Logic Control*, Chambery, France.

[38] Glorennec, P. Y. (1994). Fuzzy Q-learning and dynamical Fuzzy Q-learning. In *FUZZ-IEEE*, pp. 474-479, Orlando, FL.

[39] Jang, J. S. (1994). Structure determination in fuzzy modeling: a fuzzy CART approach, *Third IEEE International Conference on Fuzzy Systems*, pp. 480-485, Orlando, FL.

[40] Lin, C. T. and Lee, C. S. G. (1994). Reinforcement Structure/Parameter Learning for Neural-Network-Based Fuzzy Logic Control Systems, *IEEE Trans. on Fuzzy Systems*, **2**, 46-63.

[41] Lin, C. T. and Lee, C. S. G. (1992). Real-time Supervised structure / parameter learning for fuzzy neural network, *IEEE International conference on Fuzzy Systems*, San Diego, CA. pp. 1283-1290.

9 NEUROFUZZY SYSTEMS

Witold Pedrycz[1], Abraham Kandel[2], Yan-Qing Zhang[3]

[1]Department of Electrical and Computer Engineering
University of Manitoba
Winnipeg, Canada R3T 2N2
pedrycz@ee.umanitoba.ca

[2]Department of Computer Science and Engineering
University of South Florida
Tampa, FL 33620
U.S.A.
kandel@csee.usf.edu

[3]School of Computer and Applied Sciences
Georgia Southwestern State University
Americus, GA 31709
U.S.A.
yqz@canes.gsw.peachnet.edu

9.1. INTRODUCTION

A neurofuzzy system is a hybrid system with integration of fuzzy logic and neural networks, which is capable of performing high-level fuzzy reasoning by using trained fuzzy neural networks which are constructed by learning from sample data. Such a neurofuzzy integration brings high-level fuzzy IF-THEN rules into neural networks, and provides low-level numerical learning mechanisms for fuzzy logic systems. In general, the neurofuzzy system is much more powerful than either neural networks or fuzzy logic systems since it can incorporate the advantages of both, shown in Table 9.1 [86][200].

The current wave of fuzzy neural systems, fuzzy neural controllers or neurofuzzy classifiers, etc. is spectacular. What makes this research so vital and fruitful? The main reason behind this successful fusion of fuzzy sets and neurocomputing is that these technologies are highly complementary. As often emphasized in the literature, fuzzy sets are focused on knowledge representation issues including the way in which various factors of vagueness are taken care of. As primarily normative in their essence, fuzzy sets cannot cope with the prescriptive aspects of phenomena to be modeled (Pedrycz, 1992; Yamakawa, 1989) and accommodate efficacies implied by the underlying data. Hence the superiority of neurocomputing is overwhelming. Neural networks tend to be more efficient when it comes to learning (Rumelhart and McLelland, 1986) and therefore are naturally inclined to address the descriptive factors of the problem at hand. The two technologies are ideally geared into the handling of the evident duality in the perspective - descriptive duality accompanying any problem statement. The agendaof this study is twofold:

-first, we propose a general taxonomy of hybrid neuro fuzzy topologies by studyingvarious temporal and architectural aspects of this symbiosis.

Table 9.1: Comparison Between Neural Networks and Fuzzy Logic Systems

No.	Features	Neural Networks	Fuzzy Logic Systems
1	High-level Knowledge	Implicit representation by weights	Explicit representation by fuzzy rules
2	Model-Free Estimator	Trainable dynamical systems	Structured numerical systems
3	Knowledge Acquisition	From sample data	From experts
4	Uncertain Information	Quantitative	Quantitative and qualitative
5	Uncertain Cognition	Perception	Decision making
6	Reasoning Mechanism	parallel computations	Heuristic search
7	Reasoning Speed	High	Low
8	Fault-tolerance	Very high	Low
9	Adaptive Learning	Adjusting weights	Induction
10	Knowledge Storage	In neurons and links	In fuzzy rule base
11	Natural Language	Implicit	Explicit

-second, our intent is to thoroughly review some representative examples of hybrid structures illustrating the already introduced topology.

Neurofuzzy systems have been around for over 25 years. The history is reviewed decade by decade and the main works on neurofuzzy systems and their applications are shown in Table 9.2.

(1) In 1970s

Lee and Lee were the first to study the concept fuzzy neurons in 1970 [187]. Kandel, Lee and Lee then introduced the theory of fuzzy sets to the conventional McCulloch-Pitts model, and finally analyzed fuzzy neural networks based on the principle of neural networks and the mechanism of fuzzy automata [150,188-190]. However, the development of research in neurofuzzy systems was very slow since (1) there were few researchers who did work on either neural networks or fuzzy logic systems and (2) the researchers didn't find the powerful learning algorithms for neural networks and didn't have the real applications of fuzzy logic systems.

(2) In 1980s

After a relatively difficult period with neural networks and fuzzy logic in the1970's, neural networks and fuzzy logic systems attracted a resurgence of attention from a lot of researchers in a variety of scientific and engineering areas in the 1980's. The first reason for this resurgence was that Hopfield and Tank designed a neural network to solve constraint satisfaction problems such as the "Traveling Salesman Problem", and Rumelhart, Hinton and Williams [235] refined and publicized an effective Backpropagation algorithm for multilayer neural networks which had been first investigated by Werbos[289]. The second reason was that some companies, most in Japan, had successfully made a lot of fuzzy logic products such as fuzzy washing machines, fuzzy air conditioners, and fuzzy subway trains [6,267,268]. With the rapid development of techniques of neural networks and fuzzy logic systems, neurofuzzy systems were attracting more and more interest since they would be more efficient and powerful than either neural networks or fuzzy logic systems. Keller and Hunt studied how to incorporate fuzzy membership functions into the perceptron in 1985[163]. Shiue and Grondin [249] studied fuzzy learning neural-automata in 1987. Takagi and Hayashi [263] analyzed artificial-neural-network driven fuzzy reasoning in 1988. Furuya et al. [72] proposed a neurofuzzy inference system in 1988. Amano et al. [3] used neural nets and fuzzy logic in speech recognition in 1989. Hayashi et al. [93] studied the artificial-neural-network-driven fuzzy control and its application to the inverted pendulum problem in 1989. Kuncicky and Kandel [181] studied a fuzzy neuron model in which the output of one neuron is represented by a fuzzy level of confidence in 1989. Yamakawa and Tomoda discussed the fuzzy neuron and its application to pattern recognition in

1989 [298,299].

Table 9.2: History of Neurofuzzy Systems

Year	Main Works on Neurofuzzy Systems
1970	Fuzzy neurons.
1971-79	Fuzzy neural networks and fuzzy automata.
1980-89	(1) Fuzzy neurons; (2) Pattern recognition; (3) Fuzzy learning neural-automata; (4) A neurofuzzy inference system; (5) The artificial-neural-network-driven fuzzy control.
1990-95	(1) Learning algorithms for fuzzy neural networks; (2) Learning algorithms for neural fuzzy networks; (3) Genetic Algorithms for Neurofuzzy Systems; (4) Consumer Products (neurofuzzy washing machine, neuro-fuzzy fan heater, etc.) (5) Fusion of fuzzy logic, neural networks and chaos; (6) Softcomputing; (7) Applications of Neurofuzzy systems (Fuzzy Control, Fuzzy ARTMAP, Petri Net, Generating Fuzzy Rules, Decision System, Pattern Recognition, Adjusting Membership Functions, Systems Engineering, Medicine, Game theory, Fuzzy Mathematics, Expert systems, etc.).

(3) In 1990s

Important progress with neurofuzzy systems has been made in recent years. Most research papers and applications dealing with neurofuzzy systems appeared in the early 1990's.

In the theoretical realm, many effective learning algorithms for neurofuzzy systems have been developed and a lot of structures of neurofuzzy systems have been proposed. For example, Jang's adaptive-network-based fuzzy inference systems [145], Lin's neural-network-based fuzzy logic control and decision system [197], Wang's several adaptive fuzzy systems [287], the fuzzy ARTMAP by Carpenter et al. [36], the fuzzy Kohonen clustering networks by Bezdek et al. [13], the fuzzy neural network with fuzzy signals and weights by Hayashi et al. [106], the fuzzy neural net with fuzzy inputs and fuzzy targets by Ishibuchi et al. [137], the fuzzy neural network learning fuzzy control rules and membership functions by fuzzy error backpropagation by Nauck and Kruse [217], etc.

In application, neurofuzzy systems have been widely used in control systems, pattern recognition, consumer products, medicine, expert systems, fuzzy mathematics, game theory, etc. The details are in section 6.

In the interdisciplinary aspect, more and more other techniques such as genetic algorithms, chaos, probability, and AI are being applied to improve neurofuzzy systems [33,68,141,155-160,178,221,310-312]. Therefore, the neurofuzzy system, an important research area of softcomputing, will become more and more powerful and efficient over the future.

9.2 SYNERGY OF NEURAL NETWORKS AND FUZZY LOGIC

Since Lee and Lee [187] first defined the fuzzy neuron in 1970, various definitions of fuzzy neurons have arisen. We start off with a brief summary of a generic type of neuron and then embark on a diversity of logic-driven neurons. It will be emphasized that their functional variety helps design system more efficiently by encapsulating domain knowledge.

A biological neuron is a simple processing element that combines signals from 1000 to 10000 other neurons through dendrites. If the combined signal exceeds a threshold, the neuron excites, and generates an output signal to other neurons through the axon which connects to dendrites of the other neurons. Based on the properties of the biological neuron, an artificial neuron can be constructed as in Fig.9.1. First, the aggregation function S combines the input signals x_i $(i = 1, 2, ..., n)$ with respective weights w_i $(i = 1, 2, ..., n)$. Second, the activation function F generates the output signal y of the neuron based on the value of S. Commonly used aggregation function S and activation function F are given below,

$$S = \sum^{n} w_i x_i - \theta \tag{9.2.1}$$

$$y = F(S) = \frac{1}{1 + e^{-\alpha S}} \tag{9.2.2}$$

where θ is the threshold (or bias) while α stands for the activation gain.

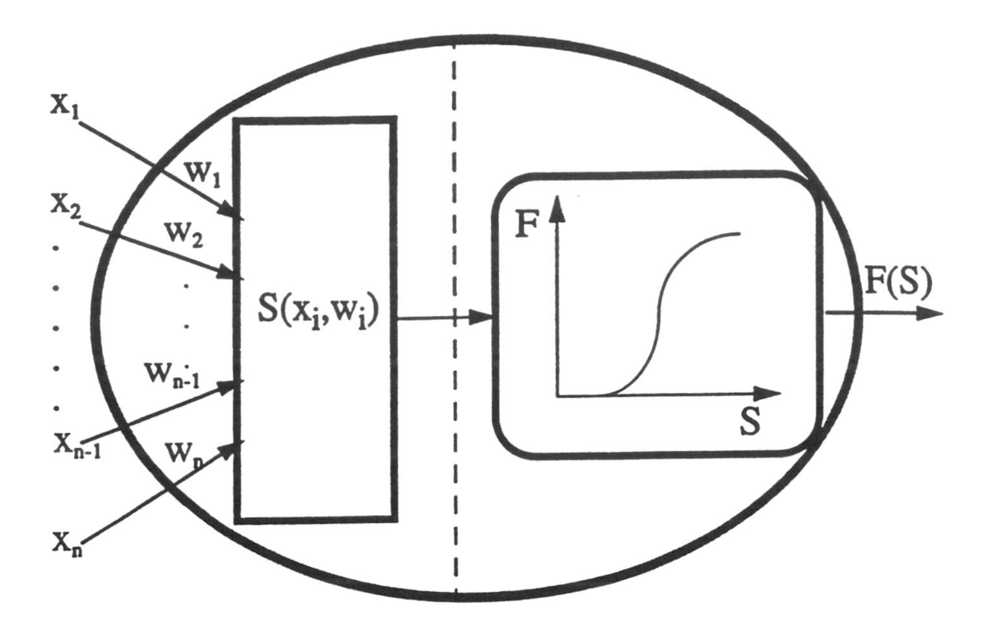

Fig. 9.1 The Model of An Artificial Neuron

An artificial neural network consists of multiple layers each of which includes many artificial neurons. Generally, outputs of an artificial neural network are nonlinear mapping functions of inputs of the artificial neural network. A typical 3-layer artificial neural network is given in Fig. 9.2.

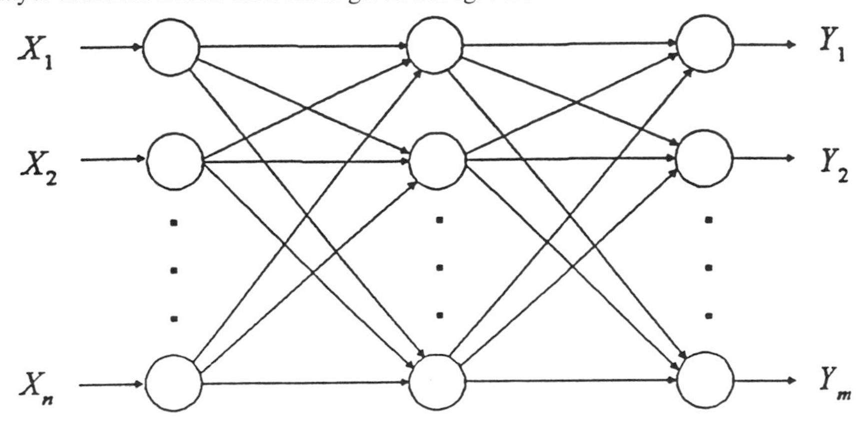

Fig. 9.2 A Three-layer Artificial Neural Network

In the general fuzzy set - neural network tandem, neural networks are providers of useful computational facilities. Fuzzy sets as based on the mechanisms of set theory and logic are chiefly preoccupied with the variety of aspects of knowledge representations and somewhat weaker as far as its processing capabilities are concerned (this claim is valid in the case of all constructs originating from set theory). In particular, set-theoretic operations do not cope directly with repetitive information and cannot reflect this throughout their outcomes - the most evident example arises in terms of highly non-interactive maximum and minimum operations.

There are a number of instances where neural networks are used directly to support or realize computing with fuzzy sets. In general, in most of these cases neural networks are aimed at the straightforward computing utilizing membership values. Similarly, there are various approaches spearheaded along the line of the development of neural networks with the intent of processing fuzzy information. In this case there is not too much direct interaction and influence originating from the theory of fuzzy sets.

9.3 FUZZY SETS IN THE TECHNOLOGY OF NEUROCOMPUTING

The general role of fuzzy sets is to enhance neural networks by incorporating knowledge-oriented mechanisms. Generally speaking, these knowledge - based enhancements of neural networks are threefold:

- preprocessing of training data that could easily lead to the improvement in learning and/or enhanced robustness characteristics of the network
- enhancements of specific training procedures through knowledge-based learning schemes (including learning metarules)
- linguistic interpretation of results produced by neural networks

Each of these areas has specific and highly representative instances - we review them to expose the reader to the very nature of the functional links between fuzzy sets and neural networks.

9.3.1 Fuzzy sets in the preprocessing and enhancements of training data

The function of fuzzy sets in this particular framework is to provide an interface between the data (environment) and the neural network regarded primarily as a processing vehicle. The intent of the interface is to expose the network to the most essential features of the data that need to be captured through the subsequent mechanisms of learning. These features are usually revealed as a part of the underlying domain knowledge. The notion of a cognitive perspective develops a certain learning environment. By selecting a collection of so-called linguistic landmarks (Zadeh, 1979; Pedrycz 1990) one can readily meet several important objectives:

- performing a nonlinear normalization of the training data. Note that by transforming any real data, the pattern $x \in R^n$ becomes converted into the corresponding element of highly dimensional unit hypercube.
- defining a variable (as opposed to fixed) processing resolution of neural networks
- coping with uncertainty in the training data

In the ensuring discussion we elaborate on the nature of these enhancements.

Nonlinear data normalization. For each coordinate (variable) we define "c" linguistic terms - these are denoted in blocks as $A = \{A_1, A_2, \ldots, A_c\}$ for the first coordinate, $B = \{B_1, B_2, \ldots, B_c\}$ for the second, etc. Then the linguistic preprocessing P carries out the mapping of the form

$$P = R^n \rightarrow [0, 1]^{nc}$$

Observe also that this preprocessing is essentially a nonlinear normalization as opposed to the commonly exploited linear transformation defined as

$$\frac{x - x_{min}}{x_{max} - x_{min}}$$

where x_{min} and x_{max} are the bounds of the variable.

The positive effect of data normalization has been often underlined in many studies on neural networks. The normalization is always recommended, especially if the ranges of the individual variables are very distinct, say [0, 0.05] vis a vis $[10^6, 10^8]$. Using such rough (unscaled) data could easily lead to a completely unsuccessful learning. The nonlinear effect of (1) originates due to the nonlinear membership functions of the linguistic terms.

The linguistic preprocessing increases the dimensionality of the problem, however could also decrease the learning effort. The similar speedup effect in training is commonly observed in Radial Basic Function (RBF) neural networks (Chen et al, 1991; Moody and Darken, 1989). The improvement in the performance achieved in this setting stems from the fact that the individual receptive fields modeled by the RBFs identify homogeneous regions in the multidimensional space of input variables. Subsequently, the updates of the connections of the hidden layers are less demanding as a preliminary structure has been already established and the learning is oriented towards less radical changes of the connections and, practically, embarks on some calibration of the receptive fields. By modifying the form of the RBFs some regions exhibiting a significant variability of the approximated function are made smaller so that a single linear unit can easily adjust to that. Similarly, the regions over which the approximated function does not change drastically can be made quite large by adapting radial basis functions of lower resolution - is could eventually lead to the concept of multiresolution - like neural networks.

Variable processing resolution. By defining the linguistic terms (modeling landmarks) and specifying their distribution along the universe of discourse we can orient (focus) the main learning effort of the network. To clarify this idea, let us refer to Fig. 9.3.

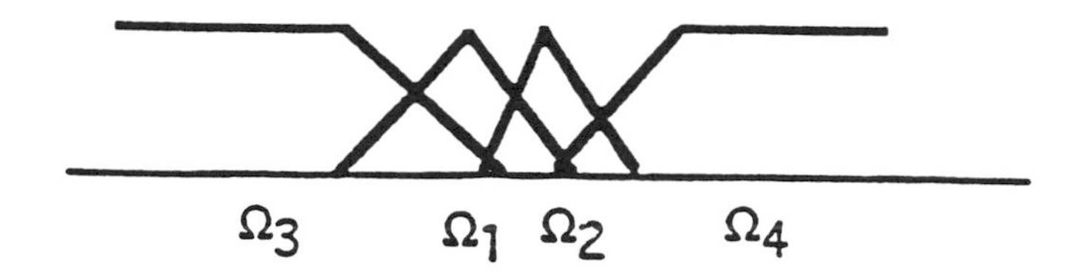

Fig.9.3 Fuzzy partition completed through a series of linguistic terms

The partition of the variable through A assigns a high level of information granularity to some regions (say Ω_1 and Ω_2) and sensitizes the learning mechanism accordingly. On the other hand, the data falling under --- are regarded internally (at the level they are perceived by the networks) as equivalent (by having the same numeric representation in the unit hypercube).

Uncertainty representation. The factor of uncertainty or imprecision can be quantified by exploiting some uncertainty measures as commonly exploited in the theory of fuzzy sets. The underlying rationale is to equip the internal format of information available to the network by some indicators describing how uncertain the given datum is. Considering possibility and necessity measures this quantification is straightforward: once $Poss(X, A_k) \neq Nec(X, A_k)$ then X is regarded uncertain (the notion of uncertainty is also context-sensitive and depends on A_k) (Dubois and Prade, 1988). for numerical data one always arrives at the equality of the two measures that clearly points at the certainty of X. In general, the higher the gap between the possibility and necessity measures, $Poss(X, A_k) = Nec(X, A_k) + \delta$, the higher uncertainty level associated with X. The uncertainty gap attains its maximum for $\delta = 1$. One can also consider the compatibility measure instead of the two used above - this provides us with more flexibility and discriminatory power yet becomes computationally demanding.

The way of treating the linguistic term makes a real difference between the architecture outlined above and RBF neural networks. The latter ones do not have any provisions to deal with and quantify uncertainty. The forms of the membership function (RBFs) is very much a secondary issue. In general, one can expect that the fuzzy sets used therein can exhibit a variety of forms (triangular, Gaussian, etc.) while RBFs are usually more homogeneous (e.g., all Gaussian). Furthermore, there are no specific restrictions on the number of RBFs used as well as their distribution within the universe of discourse. For fuzzy sets one restricts this number to a maximum of 9 terms (more exactly, 7 ± 2); additionally we should make sure that the fuzzy sets are kept distinct and thus retain their semantic identity.

9.3.2 Metalearning and fuzzy sets

Even though guided by detailed gradient - based formulas, the learning of neural networks can be enhanced by making use of some domain knowledge acquired via intense experimentation (learning). By running a series of a mixture of successful and unsuccessful learning sessions one can gain a qualitative knowledge on how an efficient learning scenario should look like.

In particular, some essential qualitative associations can be established by linking the performance of the learning process and the parameters of the scheme being utilized. Two detailed examples follow:

1. The highly acclaimed Backpropagation (BP) scheme used in the training multilevel neural networks is based upon the gradient of the performance index (objective index) Q. The basic update formula read as

$$w_{ij} = w_{ij} - \alpha \frac{\partial Q}{\partial w_{ij}}$$

where w_{ij} denotes a connection (weight) between the two neurons (i and j) and α is regarded as a learning rate, $\alpha \in (0, 1)$. Similarly $\frac{\partial Q}{\partial w_{ij}}$ describes a gradient of

Q expressed with respect to w_{ij}. It is obvious that higher values of α result in more profound changes (updates) of the connection. Higher values of α could result in faster learning that comes at the expense of its stability (oscillations and overshoots in the values of Q). After a few learning sessions one can easily reveal some qualitative relationships that could be conveniently encapsulated in the form of "if - then" rules (Silva and Almeida, 1990).

if changes of Q (ΔQ) then changes in $\Delta\alpha$

A list of pertinent rules is shown in Table 9.3.

These rules are fairly monotonic (yet not symmetric) and fully comply with our intuitive observations. In general, any increase in Q calls for some decrease of α; when Q decrease, the increases in α are made more conservative. The linguistic terms in the corresponding rules are defined in the space (universe) of changes of Q (antecedents) and a certain subset of [0,1] (conclusions). Similarly, the BP learning scheme can be augmented by taking into account a so-called momentum term; the primary intent of this expansion is to avoid eventual oscillations or reduce their amplitude in the values of the minimized performance index. This makes the learning more stable yet adds one extra adjustable learning parameter in the update rule.

Table 9.3: BP - oriented learning rules

ΔQ	$\Delta\alpha$
NB	PB
NM	PM
NS	PS
Z	Z
PS	NM
PM	NB
PB	NB

The learning metarules rules can also be formulated at the level of some parameters of the networks. The essence of the following approach is to modify activation functions of the neurons in the network. Consider the sigmoid nonlinearity (that is commonly encountered in many neural architectures).

$$y = \frac{1}{1 + \exp(-\gamma u)}$$

We assume that the steepness factor of the sigmoid function (γ) is modifiable. As the changes of the connections are evidently affected by this component (γ), we can easily set up the following metarules:

- if performance index then γ

Summing up, two design issues should be underlined:

- the considered approach is fully experimental. The role of fuzzy sets is to represent and properly summarize the available domain knowledge.
- while the control protocol (rules) seems to be universal, the universes of discourse should be modified according to the current application (problem). In other words, the basic linguistic terms occurring therein need to be adjusted (calibrated). The realization of this phase calls for some additional computational effort and could somewhat offset the benefits originating from the availability of the domain knowledge.

9.3.3 Fuzzy clustering in revealing relationships within data

In the second approach the domain knowledge about learning is acquired through some preprocessing data of the training before running the learning scheme. This is the case in the construction known as a fuzzy perceptron (Keller and Hunt, 1985). In its original setting a single-layer perceptron is composed of a series of linear processing units equipped with the threshold elements. The basic perceptron-based scheme of learning is straightforward. Let us start with a two-category multidimensional classification problem. If the considered patterns are linearly separable then there exists a linear discriminant function f such that

$$f(X, W) = W^T X > 0 \quad \text{if } X \text{ in class 1}$$

$$f(X, W) = W^T X < 0 \quad \text{if } X \text{ in class 2}$$

where $X, W \in R^{n+1}$.

After multiplying the class 2 patterns by -1 we obtain the system of inequalities.

$$f(X_k, W) > 0$$

k = 1, 2 ... N. In other words, $f(X_k, W) = 0$ defines a hyperplane partitioning the patterns; all class 1 patterns are located at the same side of this hyperplane. Assuming that X_ks are *linearly* separable. The perception algorithm (see below) guarantees that the discriminating hyperplane (vector W) is found in a *finite* number of steps.

do for all vectors X_k, =1,2,...,N

if $W^T X_k \leq 0$ then update the weights (connections)

$W = W + cX_k$

$c > 0$

end;

the loop is repeated until no updates of w occur.

The crux of the preprocessing phase as introduced in Keller and Hunt (1985) is to carry out clustering of data and determine the prototypes of the clusters as well as the class membership of the individual patterns. The ensuing membership values are used to monitor the changes. Let u_{1k} and u_{2k} be the membership grades of the k-th pattern. Definitely, $u_{1k} + u_{2k} = 1$. The outline of the learning algorithm is the same as before. The only difference is that the updates of the weights are governed by the expression

$$W = W + cX|u_{1k} + u_{2k}|^p$$

$p > 1$. These modifications depend very much on the belongingness of the current pattern in the class. If $u_{1k} = u_{2k} = 0.5$ then the correction term is equal zero and no update occurs. On the other hand if $u_{1k} = 1$ then the updates of the weights are the same as these encountered in the original perceptron.

In comparison to the previous approaches, the methods stemming from this category require some extra computing (preprocessing) but relieve us from the calibration of the linguistic terms (fuzzy sets) standing in the rules.

9.3.4 A linguistic interpretation of computing with neural networks

This style of the usage of fuzzy sets is advantageous in some topologies of neural networks, especially those having a substantial number of outputs and whose learning is carried out in unsupervised form. Fuzzy sets are aimed at the interpretation of results produced by such architectures and facilitates processes of data mining.

To illustrate the idea, we confine ourselves to self-organizing maps. These maps allow us to organize multidimensional patterns in such a way that their vicinity (neighborhood) in the original space is retained when the pattern are distributed in low dimensional space - in this way the map attempts to preserve the main topological properties of the data set. Quite often, the maps are considered in the form of the two-dimensional arrays of regularly distributed processing elements. The mechanism of self-organization is established via competitive learning; the unit that is the "closest" to the actual pattern is given an opportunity to modify its connections and follow the pattern. These modifications are also allowed to affect the neurons situated in the nearest neighborhood of the winning neuron (node) of the map.

Once training has been completed the map can locate any multidimensional pattern on the map by identifying the most active processing unit. Subsequently, the linguistic labels are essential components of data mining by placing the activities of the network in a certain linguistic context.

This concept is visualized in Fig 9.4. Let us consider that for each variable we have specified a particular linguistic term (context) defined as a fuzzy set in the corresponding space, namely A_1 and A_2 ... and A_{n1} for X_1, B_1 , B_2 ... and B_{n2} for X_2, etc.

When exposed to an input pattern the map responds with the activation levels computed at each node in the grid. The logical context leads to an extra two - dimensional grid whose elements are activated based on the corresponding activation levels of the nodes located at the lower layer as well as the level of the contexts assumed for the individual variables. These combinations are of the AND form - the upper grid is constructed as a series of the AND neurons, cf. Section 9.5.1. The activation region obtained in this way indicates how much the linguistic description (descriptors)

$$A_i \text{ and } B_j \text{ and } C_k \text{ and} \ldots$$

"covers" (activates) the data space. The higher the activation level of the region (higher values of F), the more visible the imposed linguistic pattern within the data set. Fig. 9.5 summarizes some possible patterns of activation - note a diversity the size of these regions, their intensity as well as compactness.

By performing analysis of this type for several linguistic data descriptors one can develop a collection of such descriptors that cover the entire data space, see Fig. 9.6.

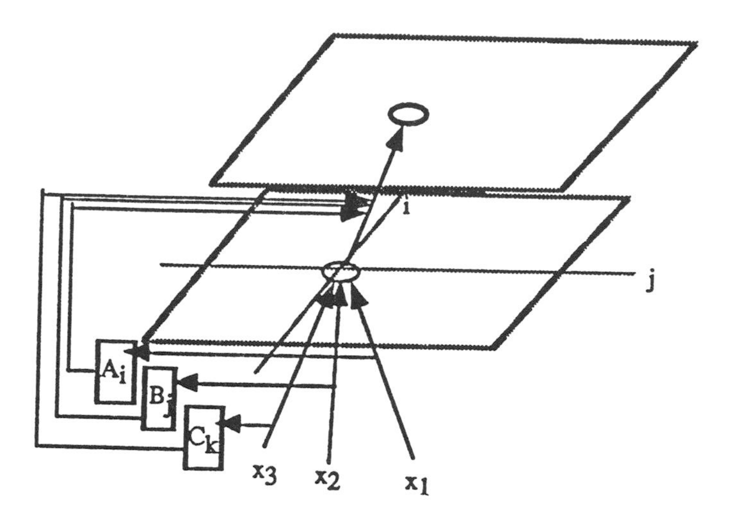

Fig. 9.4 Self-organizing map and its linguistic interpretation

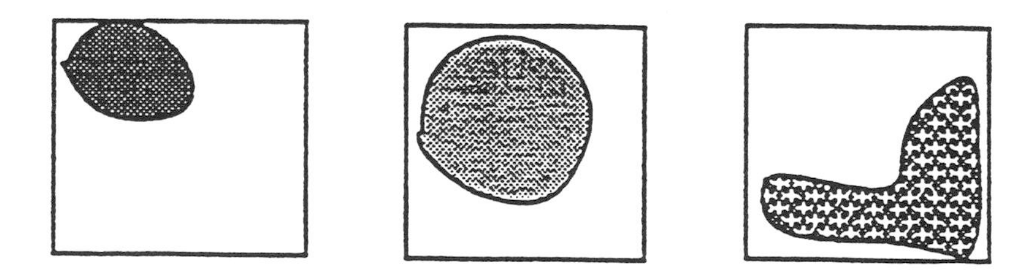

Fig. 9.5 Activation regions in the self organizing map induced by linguistic descriptors

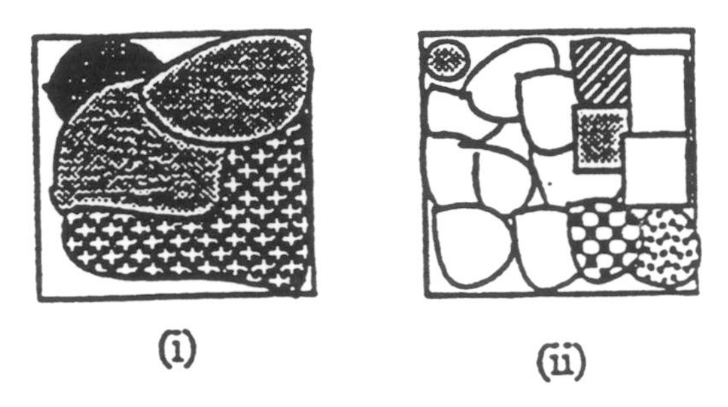

Fig. 9.6 Collection of descriptors induced by various contexts (i) coarse descriptors with overlap (ii) fine descriptors with limited overlap

We may eventually require that this collection should cover the entire map to a high extent, meaning that

$$\underset{\alpha>0}{\exists} \quad \underset{i,\, j=\ 1,2,\,\dots,\, n}{\forall} \quad \underset{c\in C}{\bigcup} N(i,\, j,\, c) \geq \alpha$$

where $N(i, j, c)$ is the response of the neuron located at the (i, j) and considered (placed) in context c from a certain family of contexts C. Some other criteria could be also anticipated; e.g., one may request that the linguistic descriptions are well-separated, meaning that their activation regions in the map are kept almost disjoint.

9.4 HYBRID FUZZY NEURAL COMPUTING STRUCTURES

When it comes to the combination of the technologies of fuzzy sets and neural networks, we can distinguish two key facets one should take into account:

- architectural
- temporal

Fig. 9.7 emphasizes two main features that are essential in establishing any vital relationship between fuzzy and neural computation. These are exemplified in the sense of the plasticity and explicit knowledge representation of the resulting neuro - fuzzy structure.

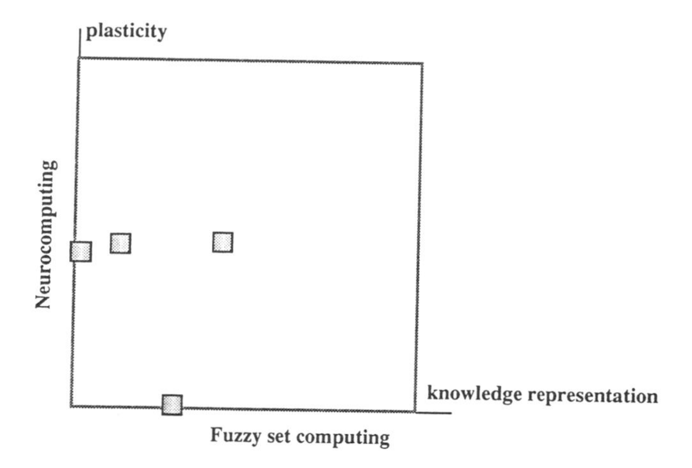

Fig. 9.7 Synergy between fuzzy sets and neural networks

The strength of the interaction itself can vary from the level at which the technology of fuzzy sets and neurocomputing are combined and co - exist to the highest one where there exists a genuine fusion between them.

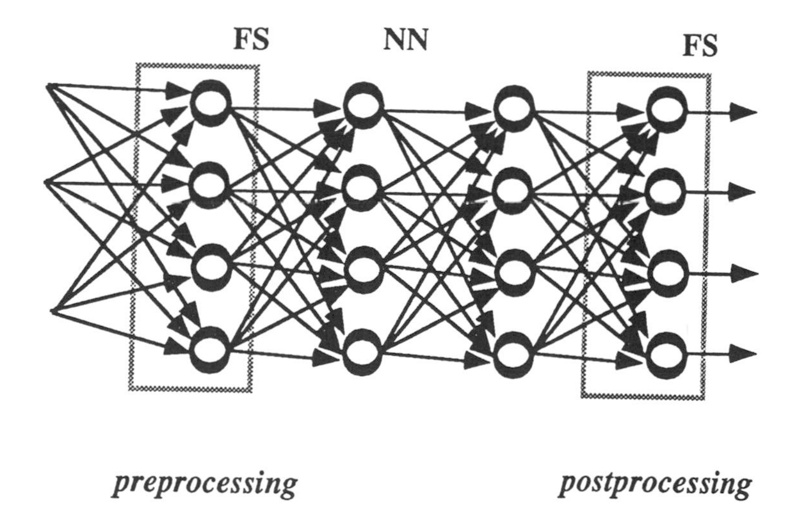

Fig. 9.8 Architecture synergy of fuzzy sets and neural networks

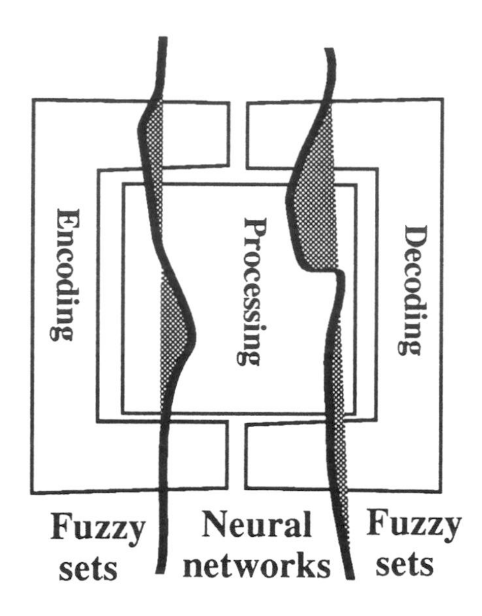

Fig. 9.9 Architectural links between fuzzy sets and neural networks considered in the setting of fuzzy modeling

The essence of the architectural interaction of fuzzy sets and neural networks is shown in Fig. 9.8. To a significant extent the form of interaction is similar to that encountered in fuzzy modeling (Pedcryz, 1995), cf. Fig. 9.9.

Fuzzy sets are more visible at the input and the output layers of any multilayer structure of the network. The role of these layers is much more oriented toward the capturing the semantics of data rather than focusing on pure numeric processing.

The temporal aspects of interaction arise when dealing with the various levels of intensity of learning. Again the updates of the connections are much more vigorous at the hidden layers - we conclude that their plasticity is higher than the others situated close to the input and output layers (Fig. 9.10).

9.5 FUZZY NEUROCOMPUTING - A FUSION OF FUZZY AND NEURAL TECHNOLOGY

In this section we discuss a certain category of hybrid processing where the

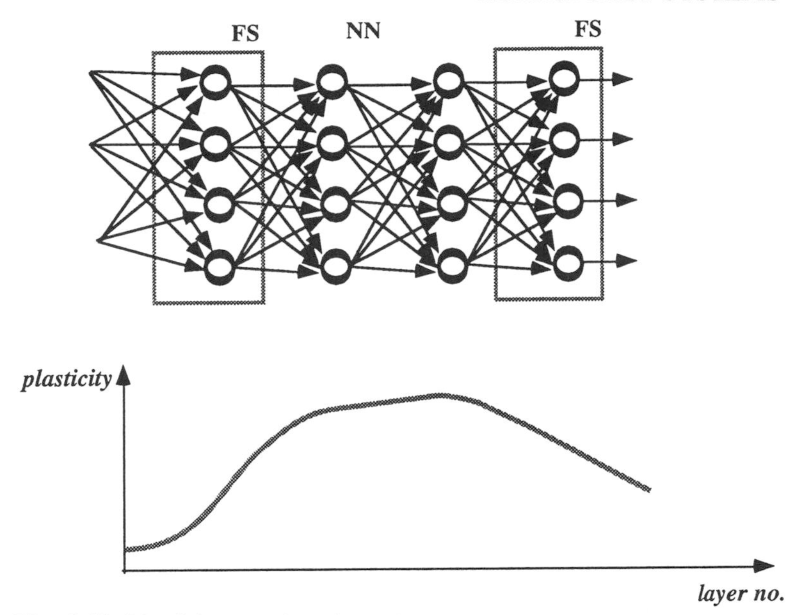

Fig. 9.10 Plasticity as a function of position of the layer in the network

neurons combine a series of features that essential to neural networks and symbolic processing. In fact, this is one of the approaches among these reported in the literature (Lin and Lee, 1994; Ishibuschi et al, 1992; Lee, 1975; Requena, 1992).

9.5.1 Basic types of logic neurons

The general topology of the processing units is outlined in Fig. 9.11. The main rationale behind this choice (Pedrycz, 1993; 1995; Pedrycz and Rocha, 1993) is that the resulting neural networks effortlessly combine learning capabilities with the mechanism of knowledge representation in its *explicit* manner). The neurons are split into two main categories, namely aggregative and referential neurons. The second group utilizes the aggregative neurons as a part of its processing structure.

OR neuron The n-input processing unit is governed by the expression

$$y = OR(x;w)$$

namely

$$y = S^{n}_{i=1}(w_i T x_i)$$

The inputs of the neuron are described by x and w denote its connections. One look at the neuron as a fuzzy relational equation with the s-t composition.

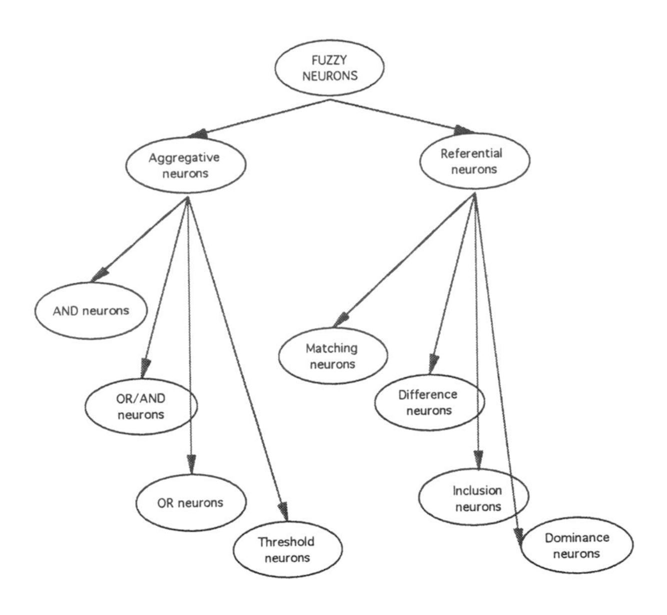

Fig. 9.11 Taxonomy of logic neurons

AND neuron This neuron is described in the following form

$$y = AND(x;w)$$

or equivalently

$$y = T^{n}_{i=1}(w_i S x_i)$$

For comparative purposes the table below outlines the links between the fuzzy neurons and fuzzy relational equations - it is assumed that the cardinality of the output space Y is equal 1, card(Y)=1.

In particular, if all the connections of the OR neuron are set to 1, we derive

$$y = S^{n}_{i=1}(x_i)$$

that is also referred to as an optimistic neuron. Similarly, for the weights equal 0, the AND neuron realizes a pessimistic aggregation (pessimistic neuron)

$$y = T^{n}_{i=1}(x_i).$$

Table 9.4.

fuzzy neuron	fuzzy relational equation
AND	t-s composition
OR	s-t composition
EQUALITY	equality equation
DIFFERENCE	difference equation
DOMINANCE	adjoint equation
INCLUSION	adjoint equation
OR/AND	convex combination

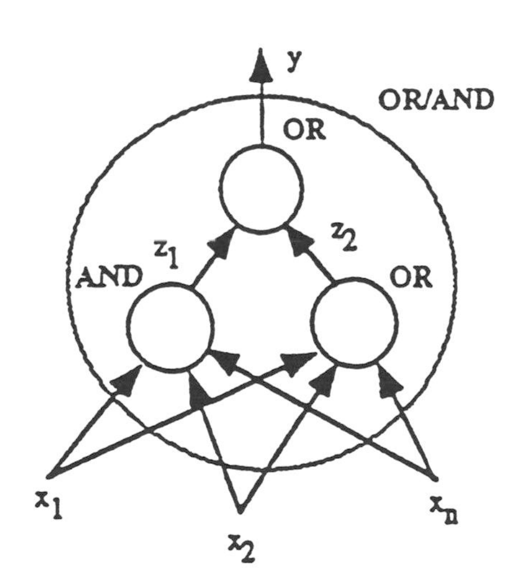

Fig. 9.12 Architecture of an OR/AND neuron

9.5.2 OR/AND neurons

As a straightforward extension of the two aggregative neurons discussed so far, we introduce a neuron with intermediate logical characteristics. The OR/AND neuron is constructed by bundling several AND and OR neurons into a single two-layer structure as shown in Fig. 9.12.

The main motivation behind combining several neurons and considering them as a single computational entity (which, in fact, constitutes a small fuzzy neural network), lies in an ability of this neuron to synthesize intermediate logical characteristics. These characteristics can be located anywhere in-between the "pure" OR and AND characteristics generated by the previous neurons. The influence coming from the OR (AND) part of the neuron can be properly balanced by selecting suitable values of the connections v_1 and v_2 during the learning of this neuron. Especially, if $v_1 = 1$ and $v_2 = 0$, the OR/AND neuron operates as a pure AND neuron. In the second extreme situation (for which $v_1 = 0$ and $v_2 = 1$), the structure functions as a pure OR neuron. The notation

$$y = OR/AND(x;w, v)$$

clearly specifies the nature of the intermediate characteristics produced by the neuron. The relevant detailed formulas describing this architecture read as,

$$y = OR([z_1, z_2];v)$$

$$z_1 = AND(x;w_1)$$

and

$$z_2 = OR(x;w_2)$$

with $v = [v_1, v_2]$, $w_i = [w_{i1}, w_{i2}, ..., w_{in}]$, $i = 1, 2$, being the connections of the corresponding neurons. We can encapsulate the above expressions into a single formula by writing down

$$y = OR/AND(x;connections)$$

where the connections summarize all the connections of the network.

9.5.3 Referential logic-based neurons

In comparison to the AND, OR and OR/AND neurons realizing operations of the aggregative character, the class of neurons discussed now is useful in realizing reference computations. The main idea behind this structure is that the input signals are not directly aggregated as this has been done in the aggregative neuron but rather than that they are first processed in a referential form using the given point of reference. The form of referential computing include such operations as matching, inclusion, difference, and dominance. In general, one can distinguish between the reference neuron in the disjunctive and conjunctive form

$$y = OR(REFx;(reference - point), w)$$

(a disjunctive form of aggregation)
or

$$y = AND(REFx;(reference - point), w)$$

(a conjunctive form of aggregation). The term REF(.) stands for the reference operation carried out with respect to the provided point of reference.

Depending on the reference operation, the functional behavior of the neuron is described accordingly (all the formulas below pertain to the disjunctive form of aggregation),
(i) MATCH neuron:

$$y = MATCH(x;r, w)$$

or equivalently

$$y = S_{i=1}^{n}[w_i T(x_i \approx r_i)]$$

where $r \in [0, 1]^n$ stands for a reference point defined in the unit hypercube.

To emphasize the referential character of this processing carried out by the neuron one can rewrite ( ) as

$$y = OR(x \approx r;w)$$

see also Fig. 9.13.

The use of the OR neuron indicates an "optimistic" (disjunctive) character of the final aggregation. The pessimistic form of this aggregation is produced by using the AND operation.

(ii) difference neuron. The neuron combines degrees to which x is different from the given reference point $g = [g_1, g_2, \ldots, g_n]$. The output is interpreted as a global level of difference observed between the input -- and this reference point,

$$y = DIFFER(x;w, g)$$

i.e.,

$$y = S_{i=1}^{n}[w_i T(x_i \approx |g_i\rangle]$$

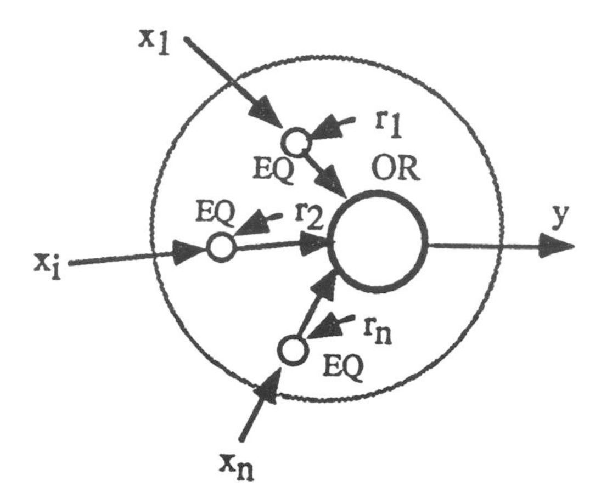

Fig. 9.13 Superposition of referential and aggregative computation in a referential neuron

where the difference operator is defined as a complement of the already used equality index,

$$(a \approx |b\rangle = 1 - (a \approx b)$$

As before, the referential character of processing is emphasized by noting that

$$DIFFER(x;w, g) = OR(x \approx |g;w\rangle$$

(iii) the inclusion neuron summarizes the degrees of inclusion to which x is included in the reference point f.

$$y = INCL(x;w, f)$$

$$y = S_{i=1}^{n}[w_i T(x_i \rightarrow f_i)]$$

To model complex situations, the referential neurons can be encapsulated into a form of a neural network. An example is a tolerance neuron which consists of DOMINANCE and INCLUSION neurons placed in the hidden layer and a single AND neuron in the output layer, Fig. 9.14.

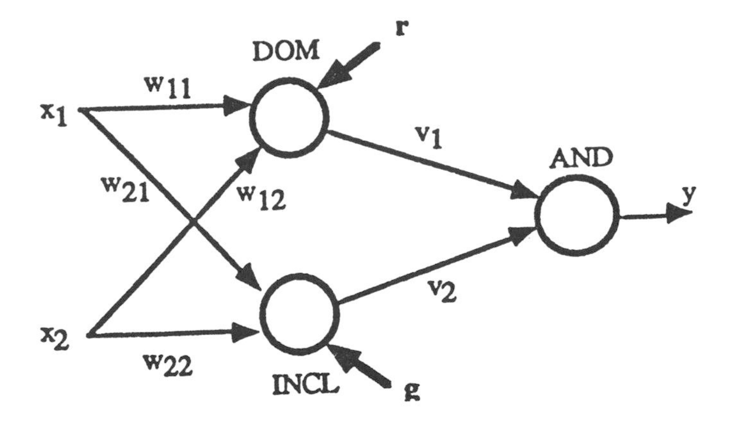

Fig. 9.14 Tolerance neuron as a combination of DOMINANCE and INCLUSION neuron

The above neuron generates a tolerance region as shown in Fig. 9.15.

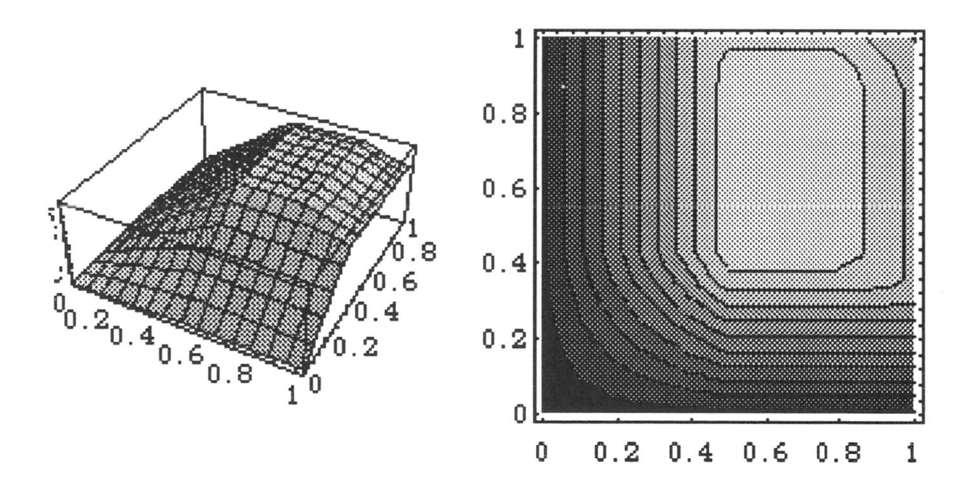

Fig. 9.15 2D and 3D characteristics of a tolerance neuron

AND neuron: min operator,

INCL and DOM neuron: $a \rightarrow b = min(1, b/a)$, $a, b \in [0, 1]$

$$w_{ij} = 0.05, v_i = 0.0$$

reference points: INCL neuron: r=[0.8,0.9], DOM neuron: g=[0.5,0.4]

For sake of completeness, let us report on some other types of neurons - in some cases they are very reduced constructs e.g. in the sense of the number of inputs they accommodate.

Fuzzification neuron: Given a membership function A, this unit accepts a real input(x) and produces an output that is a membership value of A at this value, A(x).

Defuzzification Neuron: A defuzzification neuron, described in Fig. 9.16, can generate the final crisp value y based on the inputs x_i $(i = 1, 2, \ldots, n)$ and the weights w_i^k $(i = 1, 2, \ldots, n$ and $k = 1, 2, \ldots, m)$. Here, the weights w_i^k are the parameters of the output membership functions. For example, a typical defuzzification scheme is described in (9.5.1) and the common output membership function with the adjustable parameters w_i^1 and w_i^2 $(i = 1, 2, \ldots, n)$ is given in (9.5.2).

$$D(x_1, x_2, \ldots, x_n) = \frac{\sum_{i=1}^{n} w_i^1 w_i^2 x_i}{\sum_{i=1}^{n} w_i^2 x_i} \tag{9.5.1}$$

Where w_i^1 and w_i^2 $(i = 1, 2, \ldots, n)$ are the centers and the widths of the output membership functions (9.6.2, 9.6.3), respectively.

$$\mu_{Y_i}(y) = e^{\left[-\left(\frac{y - w_i^1}{w_i^2}\right)^2\right]} \tag{9.5.2}$$

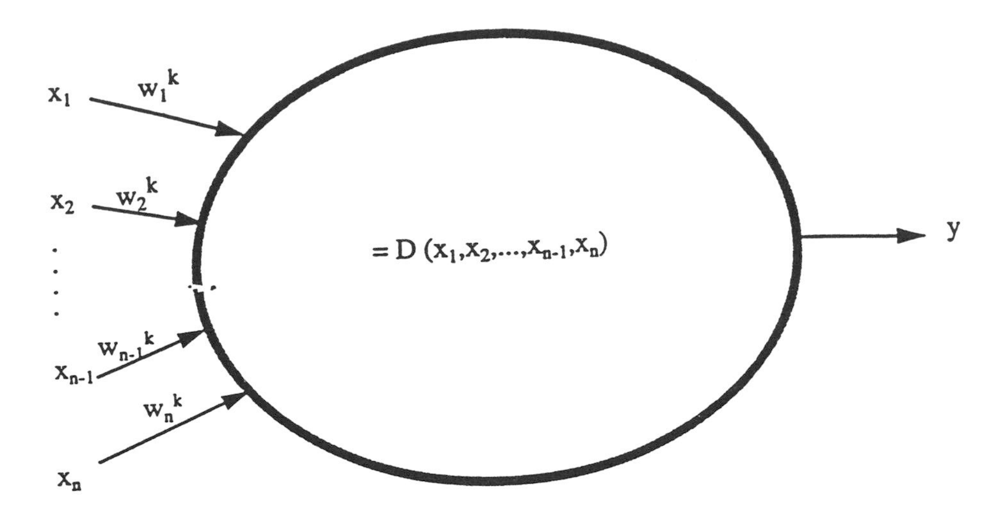

Fig. 9.16 The Model of Defuzzification Neuron

Compensatory Fuzzy Neuron: It is somewhat similar to the OR/AND neuron already discussed. It is composed of the pessimistic and optimistic neurons whose outputs are summarized using a two-input unit with the characteristics

$$y = z_1^{1-\gamma} z_2^{\gamma}$$

where $\gamma (\gamma \in [0, 1])$ stands for the degree of compensation. Note that for $\gamma = 0$ $y = z_1$ meaning that the optimistic part is present. The second boundary condition, $\gamma = 1$, implies the pessimistic characteristics of the entire neuron. The value of γ set to 0.5 yields the highest compensation level, $y = \sqrt{z_1 z_2}$.

9.5.4 Approximation of logical relationships - development of the logical processor

An important class of fuzzy neural networks concerns an approximation of mappings between the unit hypercubes (namely, from $[0, 1]^n$ to $[0, 1]^m$ or

$[0, 1]$ for $m = 1$). These mappings are realized in a logic-based format. To fully comprehend the fundamental idea behind this architecture, let us remind some very simple yet powerful concepts from the realm of two-valued systems. The well known Shannon's theorem states that any Boolean function $\{0, 1\}^n \rightarrow \{0, 1\}$ can be uniquely represented as a logical sum (union) of minterms (a so-called SOM representation) or, equivalently, a product of some maxterms (known as a POM representation). By the minterm we mean an AND combination of all the input variables of this function; they could appear either in a direct or complemented (negated) from. Similarly, the maxterms consists of the variables that now occur in their OR combination. A complete list of minterms and maxterms for Boolean functions of two variables consists of the expressions

minterms: $\bar{x}_1 AND \bar{x}_2$, $x_1 AND \bar{x}_2$, $\bar{x}_1 AND x_2$, $x_1 AND x_2$

maxterms: $\bar{x}_1 OR \bar{x}_2$, $x_1 OR \bar{x}_2$, $\bar{x}_1 OR x_2$, $x_1 OR x_2$

From a functional point of view, the minterms can be identified with the AND neurons while the OR neurons can be used to produced the corresponding maxterms. It is also noticeable that the connections of these neurons are restricted to the two-valued set $\{0,1\}$ therefore making these neurons two-valued selectors.

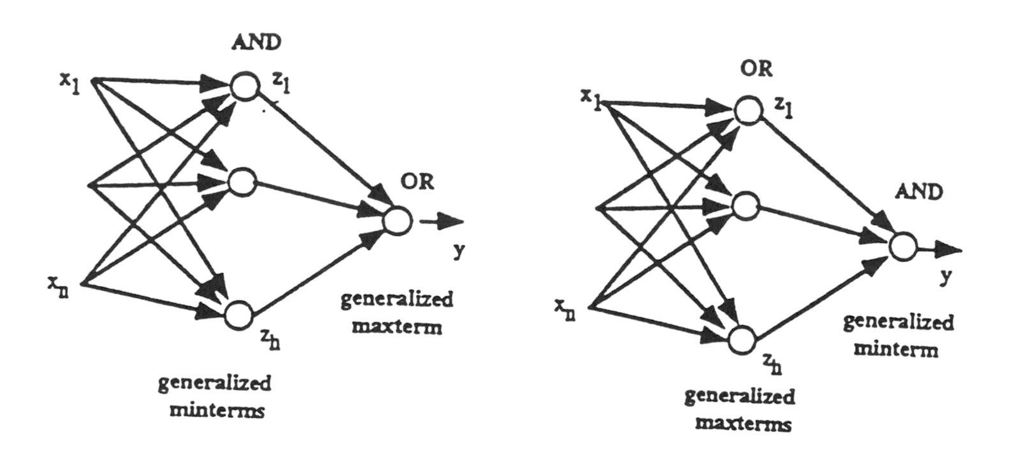

Fig. 9.17 SOM and POM versions of a logic processor

Taking into account the fundamental representation of the Boolean functions, two complementary (dual) architectures are envisioned. In the first case, the network includes a single hidden layer that is constructed with the aid of the AND neurons and the output layer consisting of the OR neurons (SOM version of the network). The dual type of the network is of the POM type in which the hidden layer consists of some OR neurons while the output layer is formed by the AND neurons. The generalization of these networks for the continuous case of the input - output variables will be called a logic processor. Analogously to the topologies of the networks outlined so far in the Boolean situation, we will be interested in the two versions of the logic processor (LP) as illustrated in Fig. 9.17.

Depending on the values of "m" we will be referring either to a scalar or vector version of the logic processor. Its scalar version, m=1, could be reviewed as a generic LP architecture.

Two points are worth making here that contrast between the logic processors in their continuous and two-valued versions:

(i) the logic processor *represents* or *approximates* data. For the Boolean data, assuming that all the input combinations are different, we are talking about a representation of the corresponding Boolean function. In this case the POM and SOM versions of the logic processors for the same Boolean function are equivalent.

(ii) the logic processor used for the continuous data approximates a certain unknown fuzzy function. The equivalence of the POM and SOM types of the obtained LPs is not guaranteed at all.

When necessary, we will be using a concise notation $LP(x;w, v)$ to describe the network with the connections w and v standing between the successive layers. The detailed formulas are given below

SOM version

$$z_i = AND(x;v_i)$$

$$y = OR(z;w)$$

$i = 1, 2, \ldots, h$.

POM version

$$z_i = OR(x;v_i)$$

$$y = AND(z;w)$$

$i = 1, 2, \ldots, h$.

where $v_1, v_2, \ldots, v_h$ are the connections between the input and hidden layer whereas w summarizes the connections between the hidden and output layer.

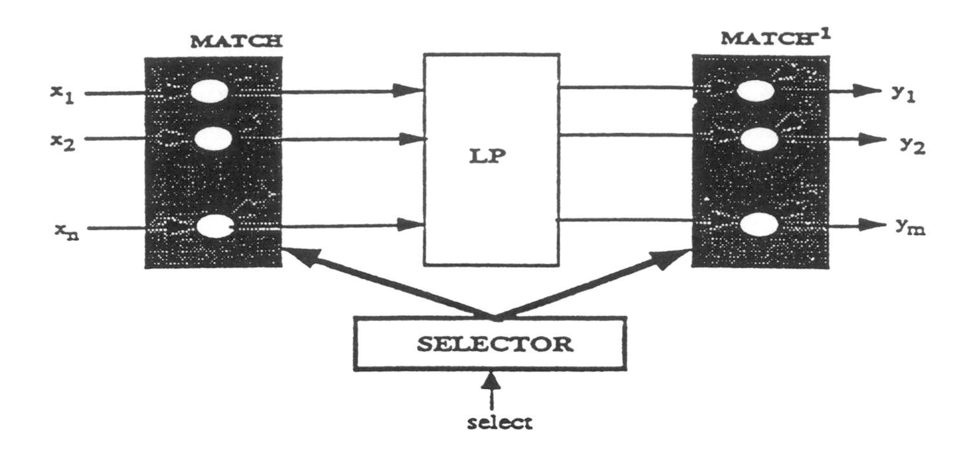

Fig. 9.18 General Architecture of Logic Processor and Analogical Processor

The logic Processor can be augmented by auxiliary processing layers. In particular, for the input layer consisting of the EQUALITY neurons we obtain a structure called Analogical Processor. Both the Logic Processor and Analog Processor can be encapsulated into a single structure using an extra selection variable, Fig. 9.18.

9.5.5 Learning

As usual, a taxonomy of learning mechanisms embraces parametric versus nonparametric learning and supervised versus unsupervised learning. The following discussion will be concerned with the supervised mode of parametric learning. In general, all learning tasks can be partitioned into two groups depending whether they pertain to parametric or structural meaning. Most of the existing schemes of learning are preoccupied by the parametric learning whose aim is to optimize the parameters of the fuzzy neural network. On the other hand, the structural learning being definitely more demanding, is devoted to the optimization of the structure of the network. The structural learning can be accomplished in many different ways, e.g., by changing the number of the layers, adding, replacing, and deleting the individual neurons. The techniques of falling under this category require nonparametric methods such as those coming from Generic Programming and Genetic Algorithms.

The predominant idea of parametric learning, no matter how this is implemented, can be posed accordingly. For a given collection of the input-output pairs of data $(x_1, t_1), \ldots, (x_N, t_N)$, modify the parameters of the network (both the connections as well as eventual reference points) to minimize the predefined performance index Q. The general scheme of learning can be qualitatively described as

$$\Delta - connections = -\zeta \frac{\partial Q}{\partial connections}$$

where ζ denotes a learning rate. Quite often the above formula is augmented by a so-called momentum term whose role is to "filter" out high frequencies in the changes of the connections and assure smooth modifications of the connections. The update of the connections is governed by the expression

$$\Delta - connections(iter + 1) = -\zeta \frac{\partial Q}{\partial connections}(\beta \Delta - connections(iter))$$

where the actual increments of the connections (at time "iter+1") depend also on the previous increments at the previous iteration: -- denotes a momentum rate.

In the sequel, the parameters of the network are adjusted following these increments,

$$\text{new\_connections} = \text{connections} + \Delta\text{\_connections}$$

The relevant details of the learning scheme can be fully specified once the topology of the network as well as some other details regarding the form of triangular norms have been made available.

While most of these detailed computations are fairly standard, the calculations of the derivatives for the maximum and minimum operations deserve a special attention. The problem has been initially addressed in Pedrycz (1995). Briefly speaking, the main issue originates from a piecewise character of these operations. Thus from a formal point of view, the derivative $\frac{\partial max(a, x)}{\partial x}$ and $\frac{\partial min(a, x)}{\partial x}$ can be defined for all x's but $x = a$. This produces the formulas,

$$\frac{\partial min(a, x)}{\partial x} = \begin{cases} 1, if x < a \\ 0, if x > a \end{cases}$$

Similarly

$$\frac{\partial max(a, x)}{\partial x} = \begin{cases} 1, if x > a \\ 0, if x < a \end{cases}$$

Note that both of them do not include the case $x = a$. One can argue that the probability of such a pointwise event $\{x = a\}$ is zero and therefore an impact it might have on the learning algorithm is practically negligible. One can eventually slightly modify these definitions by admitting at this critical point the values of the derivatives equal to 1. The main learning problem is rather associated with a Boolean (two-valued) character of these derivatives rather than their detailed and specific formulations. The potential, and essentially quite pragmatic aspect of the derivatives defined above is that the learning algorithm could eventually terminate not finding any local minimum. This is primarily caused by an accidental zeroing all the derivatives that might occur for some configuration of the connections and the learning data. To avoid this highly undesirable phenomenon, several paths have been pursued:

(i) the above derivatives can be viewed as a two-valued predicates (returning either 1 or 0). One can look at the above derivative as a Boolean predicate "equal to" that returns 1 (true) if and only if both the arguments are equal. This predicate can be relaxed by its multivalued version of the term "included in" that yields

$$\frac{\partial min(a, x)}{\partial x} = INCL(x, a)$$

and allows for a smooth transition between a full inclusion and complete dominance.

For example, the Lukasiewicz implication induces a linear character of the derivative

$$\frac{\partial min(a, x)}{\partial x} = \begin{cases} 1, if x \leq a \\ 1 - x + a, if x < a \end{cases}$$

(ii) The modification proposed in [227] is quite similar to that explained in (i) but now the derivative is defined as a sigmoid-like function.

(iii) the maximum and minimum can be replaced by their smooth albeit still good approximations of the original relationships. Feldkamp et al (1992) have considered a parametric version of them being of the type

$$min_{\delta}(x, a) = \frac{1}{2}\left[x + a - \sqrt{(x-a)^2 + \delta^2} + \delta^2\right]$$

$$max_{\delta}(x, a) = \frac{1}{2}\left[x + a + \sqrt{(x-a)^2 + \delta^2} - \delta^2\right]$$

where δ assumes values close to zero, say $\delta = 0.02$.

9.6 CONSTRUCTING HYBRID NEUROFUZZY SYSTEMS

A neural network has a massively parallel and distributed structure which is composed of many simple processing elements (i.e., artificial neurons) with nonlinear mapping functions. The neurons in a neural network can communicate with each other through the links (i.e.,weights) between the neurons. Similarly, a fuzzy neural network consists of many simple fuzzy neurons which perform some kinds of fuzzy operations and process fuzzy information. In the narrow sense, the neural network made by only logic-oriented fuzzy neurons is the standard fuzzy neural network which is called fuzzy neural network since the fuzzy logic operations are directly applied to the conventional non-fuzzy neural networks. In general, the neural network made by some logic-oriented fuzzy neurons and the control-oriented fuzzy neurons is called neural fuzzy network since the learning mechanism and essences of artificial neurons of the conventional neural networks are used to perform the conventional fuzzy logic control system. However, a neurofuzzy system can be constructed on either fuzzy neural networks or neural fuzzy networks.

9.6.1 Architectures

The fuzzy neural network has 3 types: (1) The first type of fuzzy neural network (FNN1) has real number inputs but fuzzy set weights; (2) The second type of fuzzy neural network (FNN2) has fuzzy inputs and real weights; (3) The third type of fuzzy neural network (FNN3) has both fuzzy inputs and fuzzy weights [29]. The typical structure of a fuzzy neural network is shown in Fig.9.19.

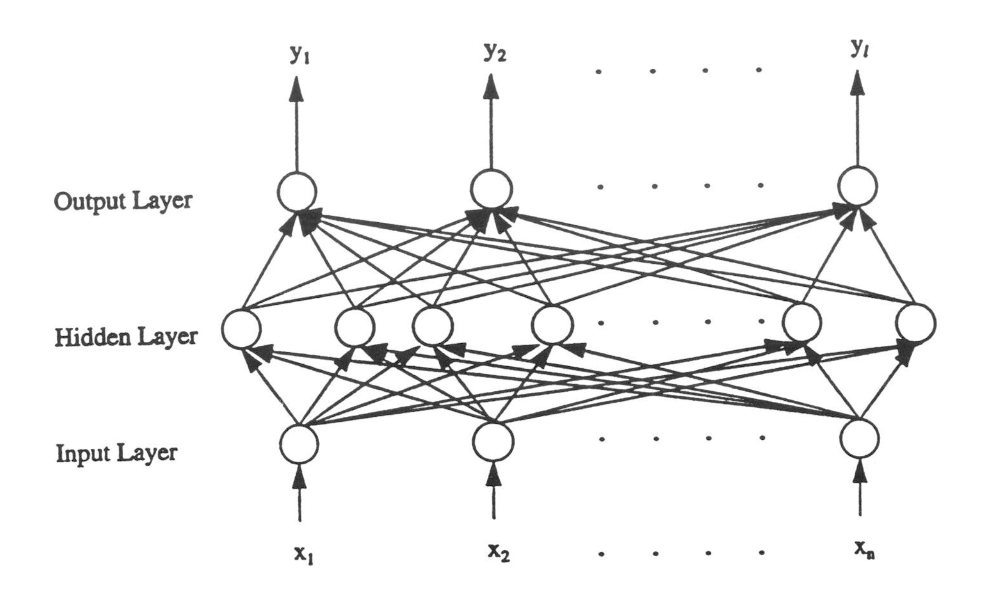

Fig. 9.19 Architecture of A Fuzzy Neural Network

According to the mechanism of fuzzy logic control systems, the neural fuzzy network usually has 5 functional layers: (1) Layer 1 is the input layer. (2) Layer 2 is the fuzzification layer; (3) Layer 3 is the fuzzy reasoning layer which may consist of AND layer and OR layer [244]; (4) Layer 4 is the defuzzification layer; (5) Layer 5 is the output layer. The architecture of a neural fuzzy network is described in Fig. 9.20. Usually, the neural fuzzy network maps crisp inputs x_i $(i = 1, 2, \ldots, n)$ to crisp output y_j $(j = 1, 2, \ldots, m)$. A neural fuzzy network is constructed layer by layer according to linguistic variables, fuzzy IF-THEN rules, the fuzzy reasoning method and the defuzzification scheme of a fuzzy logic control system.

The compensatory neural fuzzy network usually has 6 functional layers: (1) Layer 1 is the input layer; (2) Layer 2 is the fuzzification layer; (3) Layer 3 is the pessimistic-optimistic-operation layer; (4) Layer 4 is the compensatory operation layer; (5) Layer 5 is the defuzzification layer; (6) Layer 6 is the output layer. The architecture of a compensatory neural fuzzy network is generally represented in Fig. 9.21. A compensatory neural fuzzy network is constructed layer by layer according to linguistic variables, fuzzy IF-THEN rules, the pessimistic and optimistic operations, the fuzzy reasoning method and the defuzzification scheme of a fuzzy logic control system.

Many learning algorithms for fuzzy neural networks have been developed. Yamakawa et al. [299-301] proposed learning algorithms for fuzzy neurons. Reguena and Delgado [233] applied a Boltzmann machine algorithm to train the weights in an FNN2. Ishibuchi et al. [130,131] presented a learning algorithm for an FNN2. Hayashi et al. [98,105] developed a fuzzy delta rule for learning the fuzzy weights of an FNN3. Furthermore, some researchers applied the genetic learning algorithms to improve fuzzy neural nets [33,68,141,178]. The other learning algorithms of fuzzy neural networks are given in [29,35,36,66,86,217,310,312,etc.].

9.6.2 Learning Algorithms for Neural Fuzzy Networks

Given the n-dimensional input data vectors x^p ($x^p = \left(x_1^p, x_2^p, \ldots, x_n^p\right)$) and the l-dimensional output data vectors y^p ($y^p = \left(y_1^p, y_2^p, \ldots, y_l^p\right)$) for p=1,2,...,N. We need to design a supervised training algorithm to optimally adjust the centers and widths of both input and output membership functions for the neurofuzzy system.

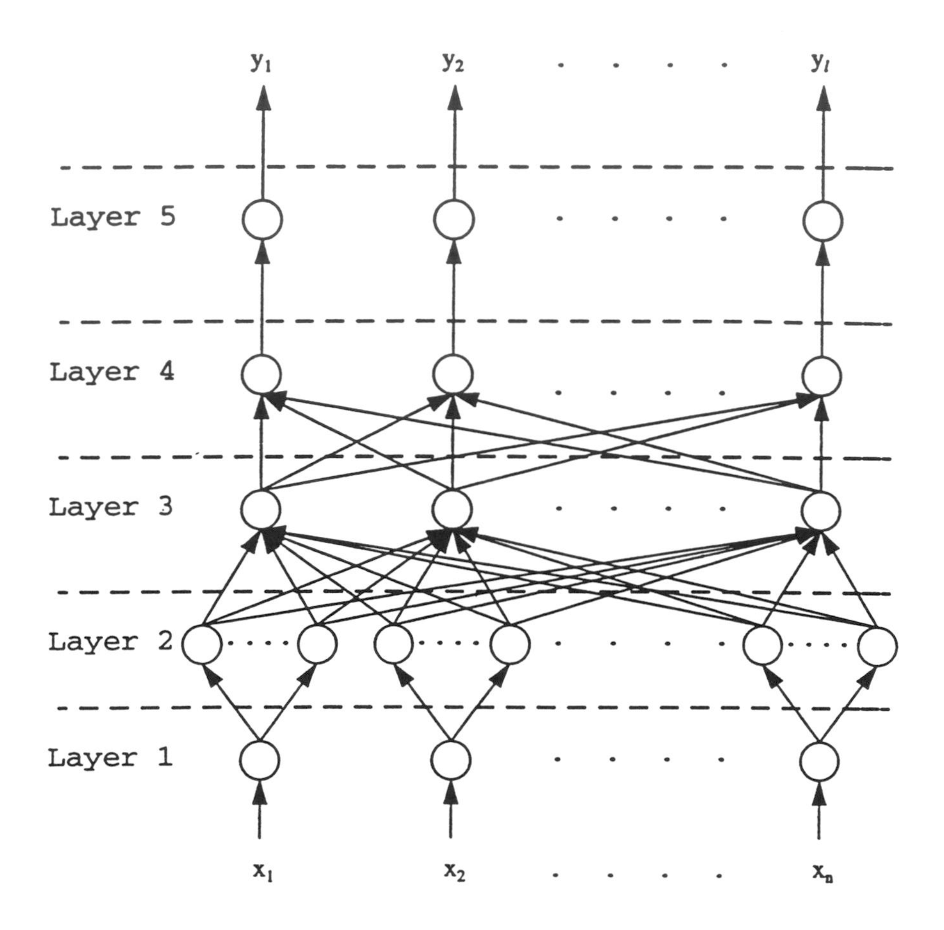

Fig. 9.20 Architecture of A Neural Fuzzy Network

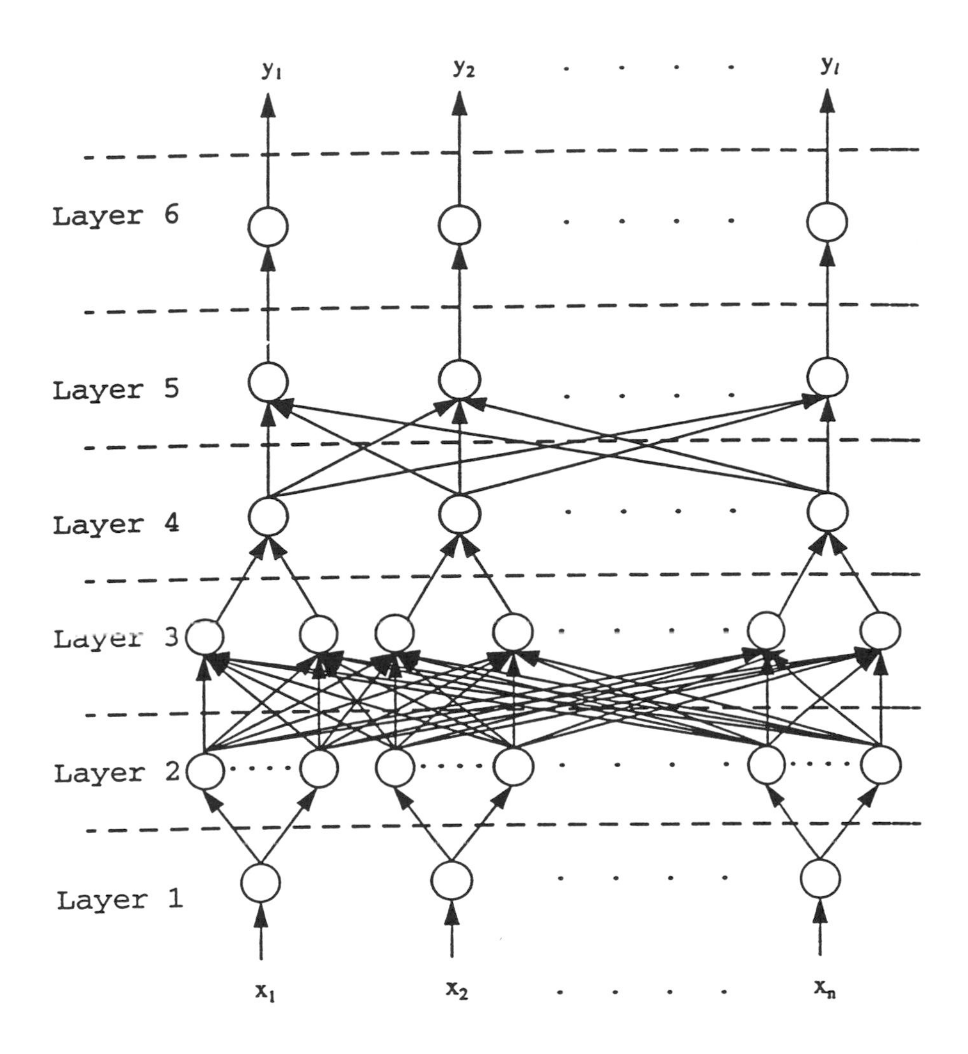

Fig. 9.21 Architecture of A Compensatory Neural Fuzzy Network

The m fuzzy IF-THEN rules of the n-input-l-output neurofuzzy system are described below,

$$\mathrm{FR}^{(k)}: \text{ IF } x_1 \text{ is } A^k_{1j} \text{ and ...... and } x_n \text{ is } A^k_{nj} \text{ THEN } y_j \text{ is } B^k_j \tag{9.6.1}$$

where A^k_{ij} and B^k_j are fuzzy sets in $U_i \subset R$ and $V_j \subset R$, respectively, and $x_i \in U_i$ and $y_j \in V_j$ are linguistic variables for $i = 1, 2, \ldots, n$, $j = 1, 2, \ldots, l$ and $k = 1, 2, \ldots, m$.

The fuzzy membership functions of A^k_{ij} and B^k_j are defined by (9.6.2) and (9.6.3), respectively.

$$\mu_{A^k_{ij}}(x_i) = \exp\left(-\left(\frac{x_i - a^k_{ij}}{\sigma^k_{ij}}\right)^2\right) \tag{9.6.2}$$

$$\mu_{B^k_i}(y_j) = \exp\left(-\left(\frac{y_j - b^k_j}{\delta^k_i}\right)^2\right) \tag{9.6.3}$$

The product-operation fuzzy implication is given by (9.6.4),

$$\mu_{A \to B}(x, y) = \mu_A(x)\mu_B(y) \tag{9.6.4}$$

The defuzzifier using the singleton fuzzifier is defined by (9.6.5).

$$f_j(x^p) = \frac{\sum_{k=1}^{m} b^k_j \delta^k_j z^k_j}{\sum_{k=1}^{m} \delta^k_j z^k_j} \tag{9.6.5}$$

Where

$$z_j^k = \prod_{i=1}^{n} \exp\left(-\left(\frac{x_i^p - a_{ij}^k}{\sigma_{ij}^k}\right)^2\right) \tag{9.6.6}$$

The objective function is given by (9.6.7).

$$E^p = \frac{1}{2}\left[f_j(x^p) - y_j^p\right]^2 \tag{9.6.7}$$

(1) Train the centers of output membership function

We have

$$\frac{\partial E^p}{\partial b_j^k} = \frac{\left[f_j(x^p) - y_j^p\right]\delta_j^k z_j^k}{\sum_{k=1}^{m} \delta_j^k z_j^k} \tag{9.6.8}$$

Then we have

$$b_j^k(t+1) = b_j^k(t) - \eta \left.\frac{\partial E^p}{\partial b_j^k}\right|_t \tag{9.6.9}$$

(2) Train the widths of output membership functions

We have

$$\frac{\partial E^p}{\partial \delta_j^k} = \frac{\left[f_j(x^p) - y_j^p\right]\left[b_j^k - f_j(x^p)\right]\delta_j^k z_j^k}{\sum_{k=1}^{m} \delta_j^k z_j^k} \tag{9.6.10}$$

Then we obtain the training algorithm:

$$\delta_j^k(t+1) = \delta_j^k(t) - \eta \left. \frac{\partial E^p}{\partial \delta_j^k} \right|_t \tag{9.6.11}$$

(3) Train the centers of input membership functions
We have

$$\frac{\partial E^p}{\partial a_{ij}^k} = \frac{2\left[f_j(x^p) - y_j^p\right]\left[b_j^k - f_j(x^p)\right]\left[x_i^p - a_{ij}^k\right]\delta_j^k z_j^k}{\left(\sigma_{ij}^k\right)^2 \sum_{k=1}^{m} \delta_j^k z_j^k} \tag{9.6.12}$$

Then we obtain the training algorithm:

$$a_{ij}^k(t+1) = a_{ij}^k(t) - \eta \left. \frac{\partial E^p}{\partial a_{ij}^k} \right|_t \tag{9.6.13}$$

(4) Train the centers of input membership functions
We have

$$\frac{\partial E^p}{\partial \sigma_{ij}^k} = \frac{2\left[f_j(x^p) - y_j^p\right]\left[b_j^k - f_j(x^p)\right]\left[x_i^p - a_{ij}^k\right]^2 \delta_j^k z_j^k}{\left(\sigma_{ij}^k\right)^3 \sum_{k=1}^{m} \delta_j^k z_j^k} \tag{9.6.14}$$

Then we obtain the training algorithm:

$$\sigma_{ij}^k(t+1) = \sigma_{ij}^k(t) - \eta \left. \frac{\partial E^p}{\partial \sigma_{ij}^k} \right|_t \tag{9.6.15}$$

Where, η is the learning rate and $t = 0, 1, 2, \ldots$.

9.6.3 Learning Algorithms for Compensatory Neural Fuzzy Networks

Here, we introduce the compensatory learning algorithm (see details in [312]) for the compensatory neural fuzzy network shown in Fig. 9.21. We also use the defuzzification scheme (9.5.1). In (9.5.1), the compensatory, pessimistic and optimistic operations of the respective fuzzy neurons are defined by (9.6.16), (9.6.17) and (9.6.18), respectively.

$$z_j^k = \left(u_j^k\right)^{1-\gamma_j}\left(v_j^k\right)^{\gamma_j} \tag{9.6.16}$$

Where the compensatory degrees $\gamma_j = \dfrac{c_j^2}{c_i^2 + d_i^2}$.

$$u_j^k = \prod_{i=1}^{n} \mu_{A_{ij}^k}\left(x_i^p\right) \tag{9.6.17}$$

$$v_j^k = \frac{1}{n}\sum_{i=1}^{n} \mu_{A_{ij}^k}\left(x_i^p\right) \tag{9.6.18}$$

$$\mu_{A_{ij}^k}\left(x_i^p\right) = \exp\left(-\left(\frac{x_i^p - a_{ij}^k}{\sigma_{ij}^k}\right)^2\right) \tag{9.6.19}$$

(1) Train the centers of output membership function

$$b_j^k(t+1) = b_j^k(t) - \eta\alpha(t) \tag{9.6.20}$$

(2) Train the widths of output membership functions

$$\delta_j^k(t+1) = \delta_j^k(t) - \eta\alpha(t)\beta(t) \tag{9.6.21}$$

(3) Train the centers of input membership functions

$$a_{ij}^{k}(t+1) = a_{ij}^{k}(t) - \frac{2\eta\alpha(t)\beta(t)\lambda(t)\left[x_i^p - a_{ij}^k(t)\right]}{\left(\sigma_{ij}^k(t)\right)^2} \tag{9.6.22}$$

(4) Train the centers of input membership functions

$$\sigma_{ij}^{k}(t+1) = \sigma_{ij}^{k}(t) - \frac{2\eta\alpha(t)\beta(t)\lambda(t)\left[x_i^p - a_{ij}^k(t)\right]^2}{\left(\sigma_{ij}^k(t)\right)^3} \tag{9.6.23}$$

(5) Train the compensatory degree

$$\frac{\partial E^p}{\partial \gamma_j} = \alpha(t)\beta(t)\ln\left[\frac{\sum_{i=1}^{n}\mu_{A_{ij}^k}\left(x_i^p\right)}{\prod_{i=1}^{n}\mu_{A_{ij}^k}\left(x_i^p\right)}\right], \tag{9.6.24}$$

$$c_j(t+1) = c_j(t) - \eta\left(\frac{2c_j(t)d_j^2(t)}{\left[c_j^2(t)+d_j^2(t)\right]^2}\right)\frac{\partial E^p}{\partial \gamma_j}\bigg|_t, \tag{9.6.25}$$

$$d_j(t+1) = d_j(t) + \eta\left(\frac{2c_j^2(t)d_j(t)}{\left[c_j^2(t)+d_j^2(t)\right]^2}\right)\frac{\partial E^p}{\partial \gamma_j}\bigg|_t, \tag{9.6.26}$$

$$\gamma_j(t+1) = \frac{c_j^2(t+1)}{c_i^2(t+1)+d_i^2(t+1)}, \tag{9.6.27}$$

where

$$\alpha(t) = \left.\frac{\left[f_j(x^p) - y_j^p\right]\delta_j^k z_j^k}{\sum_{k=1}^{m} \delta_j^k z_j^k}\right|_t \tag{9.6.28}$$

$$\beta(t) = \left.\left[b_j^k - f_j(x^p)\right]\right|_t \tag{9.6.29}$$

$$\lambda(t) = 1 - \gamma_j + \left.\frac{\gamma_j \mu_{A_{ij}^k}\left(x_i^p\right)}{\sum_{i=1}^{n} \mu_{A_{ij}^k}\left(x_i^p\right)}\right|_t \tag{9.6.30}$$

In all above formulas, η is the learning rate and $t = 0, 1, 2, \ldots$.

9.6.4 Designing Neurofuzzy Systems

We use a typical example to show how to systematically design a neurofuzzy system by using high-level constructed knowledge such as fuzzy IF-THEN rules and the low-level learning algorithm. We take the simple XOR problem as an example.

In order to analyze performances of neurofuzzy systems, we use 0.2 and 0.8 to represent False and True respectively for the XOR Problem in Table 9.5.

Table 9.5: XOR Problem

x1	x2	y
0.2	0.2	0.2
0.2	0.8	0.8
0.8	0.2	0.8
0.8	0.8	0.2

The procedure of designing the neurofuzzy system for XOR problem is given step by step so as to clearly show how to design the neurofuzzy system.

Step 1: Building Initial Fuzzy Rule Base

Based on the property of XOR problem in Table 9.5, we can easily make the fuzzy rules FR1, FR2, FR3 and FR4 such that

FR1: IF x_1 is *small* and x_2 is *small* THEN y is *small*

FR2: IF x_1 is *small* and x_2 is *large* THEN y is *large*

FR3: IF x_1 is *large* and x_2 is *small* THEN y is *large*

FR4: IF x_1 is *large* and x_2 is *large* THEN y is *small*

Where the x_1, x_2 and y are the linguistic variables. The fuzzy linguistic values *small* and *large* of x_1, x_2 and y are subjectively defined by (9.6.31) and (9.6.32) respectively.

$$\mu_{small}(z) = \exp\left(-\left(\frac{z-0.25}{0.5}\right)^2\right) \tag{9.6.31}$$

$$\mu_{large}(z) = \exp\left(-\left(\frac{z-0.75}{0.5}\right)^2\right) \tag{9.6.32}$$

Where $z \in \{x_1,x_2,y\}$.

At this point, the subjective parameters 0.25, 0.5 and 0.75 are not optimal for the inputs (x_1, x_2) and the output y. But these subjective parameters from the human brain can be used to initially set the neurofuzzy system at the relatively good state which probably approximates the optimal state. Therefore, the subjective fuzzy rules may speed up the convergence of the learning algorithm by using the high-level knowledge. In this sense, the adaptive learning algorithm of neurofuzzy systems intelligently incorporating fuzzy rules is much more powerful than that of conventional neural networks blindly selecting random initial weights.

Step 2: Building an Initial Neurofuzzy System

The *4* fuzzy IF-THEN rules of the *2*-input-*1*-output neurofuzzy system are described below for k=1,2,3,4,

$$\mathrm{FR}^{(k)}: \text{IF} \quad x_1 \ \text{is} \ A_1^k \ \text{and} \ x_2 \ \text{is} \ A_2^k \ \text{THE} \ y \ \text{is} \ B^k \tag{9.6.33}$$

The fuzzy membership functions of A_1^k, A_2^k and B^k are defined by (9.6.34) for i=1,2 and (9.6.35), respectively.

$$\mu_{A_i^k}(x_i) = \exp\left(-\left(\frac{x_i - a_i^k}{\sigma_i^k}\right)^2\right) \tag{9.6.34}$$

$$\mu_{B^k}(y) = \exp\left(-\left(\frac{y - b^k}{\delta^k}\right)^2\right) \tag{9.6.35}$$

According to the fuzzy rule base (9.6.33) and the structure of the neural fuzzy network in Fig. 9.21, we show the initial neurofuzzy system in Fig. 9.22. The neurons in Layer 1 directly map inputs x_1 and x_2 to the outputs with the same values of x_1 and x_2. Then the neurons in Layer 2 transfer the inputs to the respective outputs by the membership functions (9.6.34) and (9.6.35) with initial forms of (9.6.31) and (9.6.32). The neurons in Layer 3 perform the fuzzy reasoning by using the functions (9.6.6). Finally, the neurons in Layer 4 realize the defuzzification using the function (9.6.5). Here, suppose the neuron in Layer 5 simply maps the output y' of the neuron of Layer 4 to the output y of Layer 5 with $y = y'$.

Step 3: Training The Neurofuzzy System

Here, we use the fuzzy BP learning algorithm to train the neurofuzzy system based on the initial parameters in step 1 and the sample data in Table 9.5. The satisfactory results given in Table 9.6 and Fig. 9.23-9.25 have shown that the neurofuzzy system can very quickly learn XOR logic based on the initial knowledge from human experts, and the conventional neural networks with 2, 4 and 7 hidden neurons need much more training epochs to learn XOR logic than the neurofuzzy system.

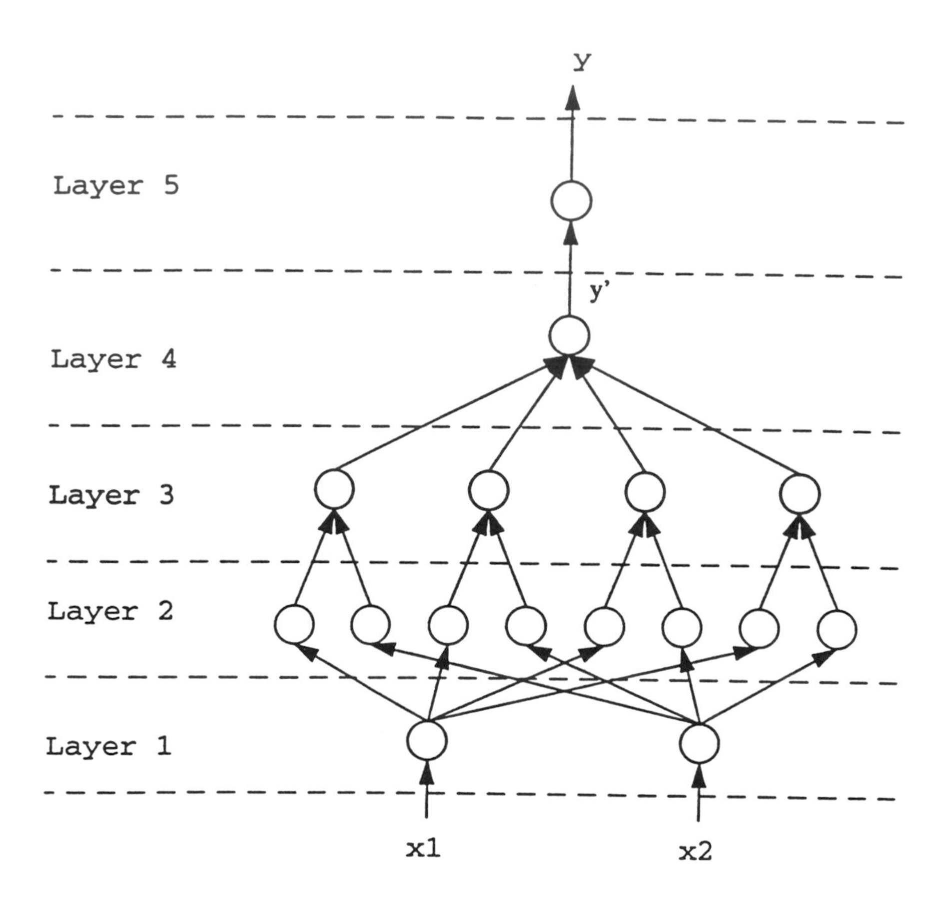

Fig. 9.22 The Initial Neurofuzzy System

Step 4: Constructing the Trained Neurofuzzy System

After training the neurofuzzy system with the error E<0.000001, we get the trained parameters of the fuzzy membership functions shown in Table 9.7. At this point, the trained neurofuzzy system has been constructed by using the new fuzzymembership functions of x_1, x_2 and y. Through analyzing the data in Table 9.7, we can find that the original 4 linguistic values of x_1, 4 ones of x_2 and 4 ones of y can be reduced to 2 linguistic values for each of them. Therefore, the trained neurofuzzy system shown in Fig. 9.26 is finally constructed by the trained fuzzy rule base (TFR1,TFR2,TFR3,TFR4) as follows,

TFR1: IF x_1 is *small* and x_2 is *small* THEN y is *small*

TFR2: IF x_1 is *small* and x_2 is *large* THEN y is *large*

TFR3: IF x_1 is *large* and x_2 is *small* THEN y is *large*

TFR4: IF x_1 is *large* and x_2 is *large* THEN y is *small*

The trained fuzzy linguistic values *small* and *large* of x_1, x_2 and *small* and *large* of y are defined by (9.6.36), (9.6.37) and (9.6.38), (9.6.39), respectively.

$$\mu_{small}(x_i) = \exp\left(-\left(\frac{x_i - 0.17}{0.41}\right)^2\right) \tag{9.6.36}$$

$$\mu_{large}(x_i) = \exp\left(-\left(\frac{x_i - 0.825}{0.41}\right)^2\right) \tag{9.6.37}$$

$$\mu_{small}(y) = \exp\left(-\left(\frac{y - 0.06}{0.5}\right)^2\right) \tag{9.6.38}$$

$$\mu_{large}(y) = \exp\left(-\left(\frac{y - 0.935}{0.5}\right)^2\right) \tag{9.6.39}$$

Table 9.6: Training Epochs of BP algorithm and Fuzzy BP algorithm for XOR Problem.

Error ($E = \frac{1}{2}\sum_{p=1}^{4} E^p$)	0.01	0.001	0.0001	0.00001	0.000001
BP (2-2-1)	352	411	477	546	617
BP (2-4-1)	385	445	510	579	649
BP (2-7-1)	282	338	400	465	532
Fuzzy BP	4	4	6	7	8

Table 9.7: The Trained Centers and Widths of Membership Functions of x_1, x_2 and y

k	Fuzzy Rules	a_1^k	σ_1^k	a_2^k	σ_2^k	b^k	δ^k
1	TFR(k)	0.17	0.40	0.16	0.40	0.05	0.50
2	TFR(k)	0.17	0.41	0.82	0.41	0.93	0.50
3	TFR(k)	0.83	0.41	0.17	0.41	0.94	0.50
4	TFR(k)	0.83	0.42	0.82	0.42	0.07	0.50

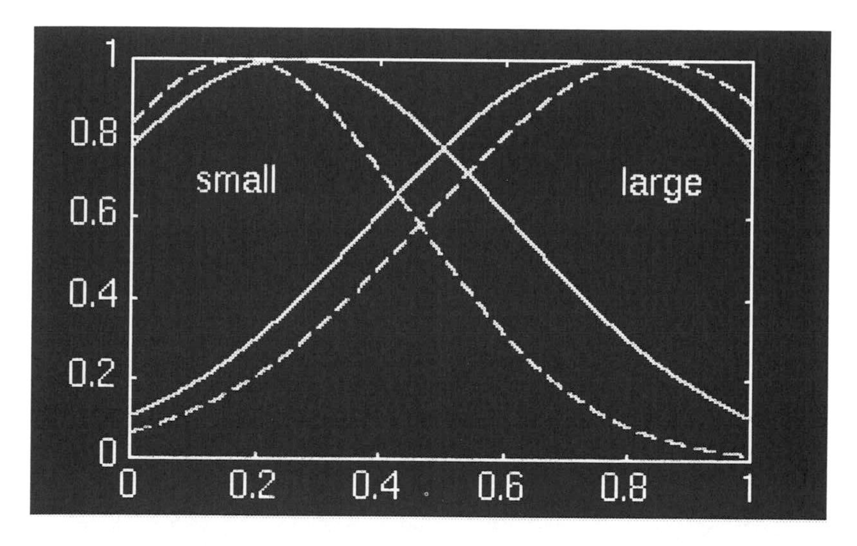

Fig. 9.23 Membership Functions of x_1 Before Training (solid) and After Training (dashed)

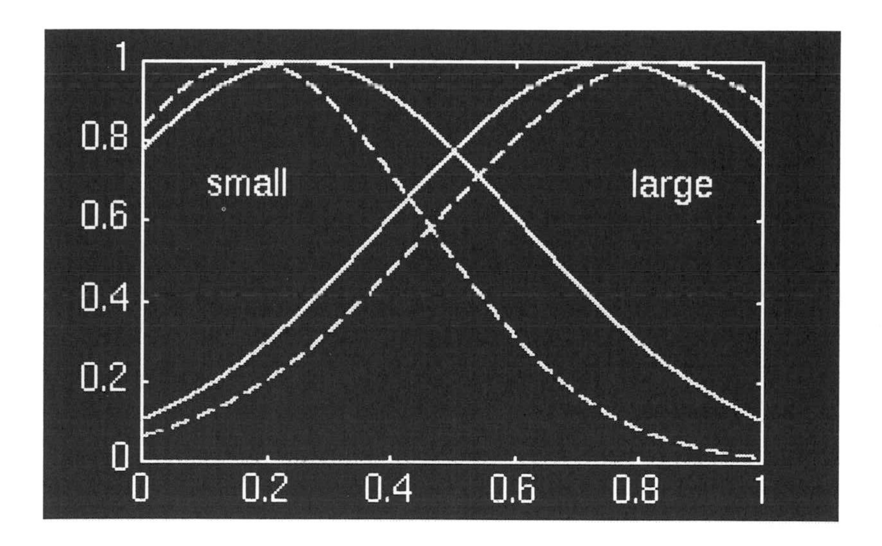

Fig. 9.24 Membership Functions of x_2 Before Training (solid) and After Training (dashed)

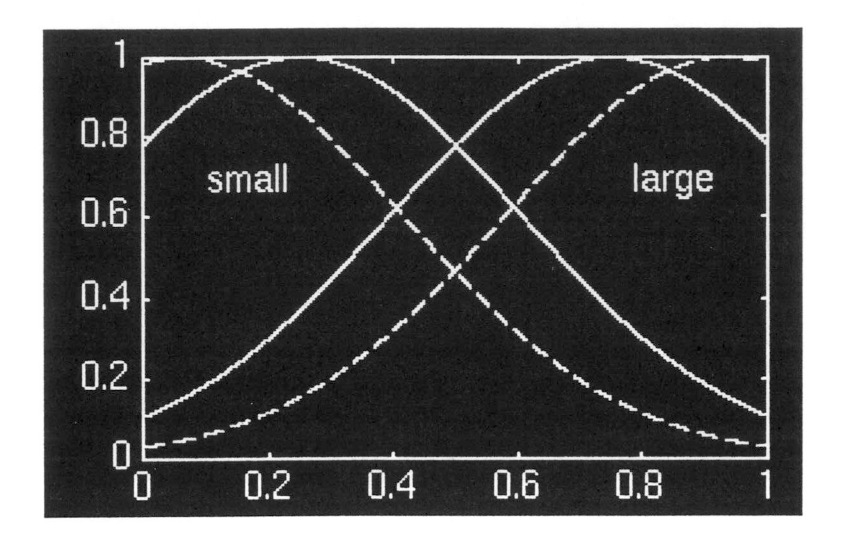

Fig.9.25 Membership Functions of y Before Training (solid) and After Training (dashed)

The following example is simple. A complex neurofuzzy system may be designed in a similar way. That is, first, design the fuzzy rule base (this step can be omitted since the neurofuzzy system can be trained by random initial fuzzy rules); second, make the neural fuzzy network; third, train it, and finally construct the trained neurofuzzy system based on the trained parameters of the membership functions. The details may be found in many references listed at the end of this chapter.

9.7 SUMMARY

This chapter is a tutorial review on history, basic concepts, main principles, applications and bibliography of neurofuzzy systems. Since the research of neurofuzzy systems is still in its infancy and is a very competitive area of interdisciplinary science and engineering, we hope that neurofuzzy systems will play a more and more important role in various applications through the incorporation of more powerful techniques such as genetic algorithms, probability, chaos, AI, and so on.

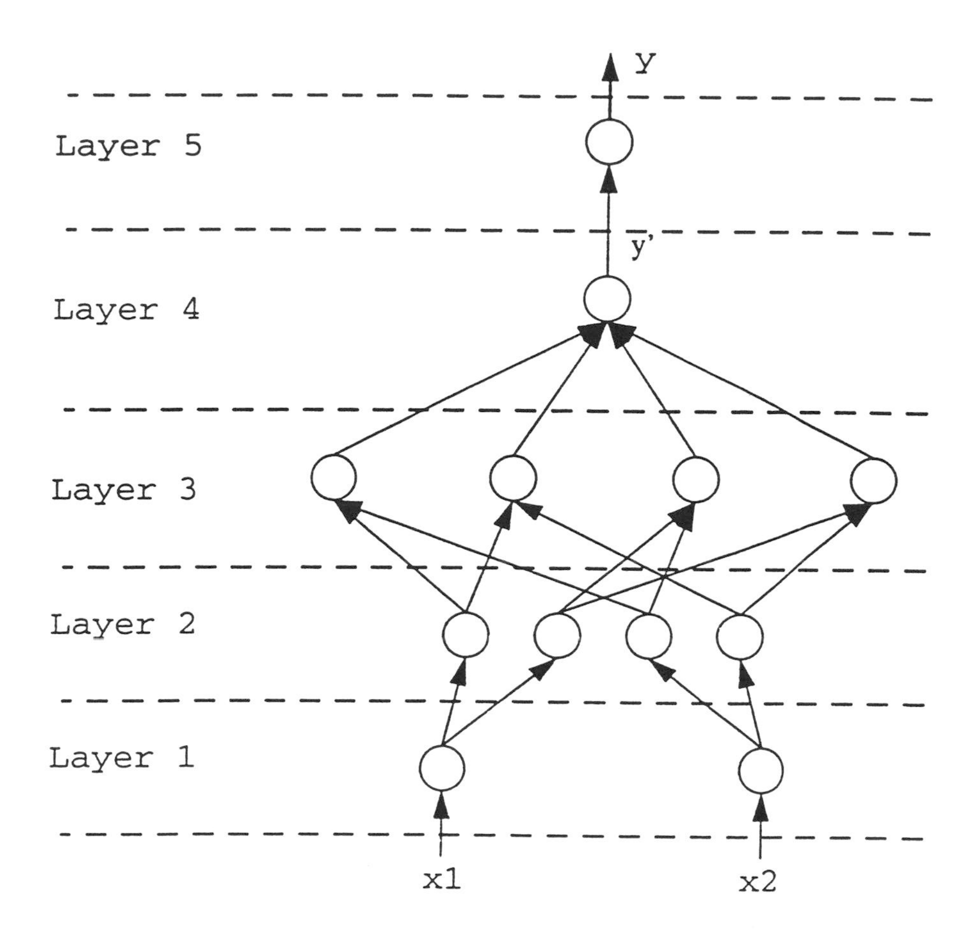

Fig. 9.26 The Trained Neurofuzzy System

Fuzzy sets and neurocomputing are two supplementary technologies. The two-way integration is possible and beneficial. The knowledge-based facilities are well handled by the technology of fuzzy sets while the learning activities are chiefly addressed by neural networks. Interestingly, there are a number of new constructs combining the ideas stemming from fuzzy sets and neural networks. We have investigated various levels of symbiosis and proposed a consistent classification of the systems emerging as an outcome of the symbiosis of these two technologies.

References

(Note: cited references are marked by * at their reference numbers' superscripts.)

[1] Adeli H. and Hung S.-L.,"*Machine learning - Neural Networks, Genetic Algorithms, and Fuzzy Systems*," John Wiley & Sons, Inc., 1995

[2] Ahson S.I.,"*Petri net models of fuzzy neural networks*," IEEE Trans. on SMC, Vol.25, No.6, pp.926-932, June,1995.

[3] *Amano A.,et al.,"*On the use of neural networks and fuzzy logic in speech recognition*," Proc. Int. Joint Conf. on Neural Networks, **Vol.1**, pp.1301-1306, 1989.

[4] Andlinger P. and Reichl E.R.,"*Fuzzy-Neunet: A non standard neural network*," In (prieto, Goos & Hartmanis, 1991), pp.173-180.

[5] Archer N.P. and Wang S.,"*Fuzzy set representation of neural network classification boundaries*," IEEE Trans. on SMC, **Vol.21**, No.4, pp.735-742, July/ Aug.,1991.

[6] *Asakawa K. and Takagi H.,"*Neural networks in Japan*," Communications of the ACM, **Vol.37**, No.3,pp.106-112,March 1994.

[7] Bastian A.,"*Handling the nonlinearity of a fuzzy logic controller at the transition between rules*," Fuzzy Sets and Systems, **71**(1995) 369-387.

[8] Barto A.G.,"*Reinforcement learning and adaptive critic methods*," in (White & Sofge, 1992), pp.469-491, Van Nostrand Reinhold.

[9] Berenji H.R.,"*Neural networks and fuzzy logic in intelligent control*," Proc. 5th IEEE Int. Symp. on Intelligent Control, **Vol.2**, pp.916-920, 1990.

[10] Berenji H.R.,"*A simple direct adaptive fuzzy controller derived from its neural equivalent*," Proc. IEEE Int. Conf. on Fuzzy Systems, pp.345-350, 1993.

[11] Berenji H.R.,"*Neural networks for fuzzy logic inference*," Proc. 2nd IEEE Int. Conf. on Fuzzy Systems,**Vol. II**, pp.1395, 1993.

[12] Berenji H.R.,Malkani A. and Copeland C.,"*Tether control using fuzzy reinforcement learning*," Proc. of FUZZ-IEEE/IFES'95, **Vol. III**, pp.1315-1322,1995.

[13] *Bezdek J.C., Tsao E.C.-K., and Pal N.R., "*Fuzzy Kohonen Clustering Networks*,"IEEE Intl. Conf. on Fuzzy Systems, pp.1035-1043,1992.

[14] Blanco A., Delgado M. and Requena I.,"*Solving Fuzzy Relational Equations by max-min neural networks*," Proc. of Third IEEE Int'l Conf. on Fuzzy System, **Vol.3**, pp1737-1742,1994.

[15] Bothe H.H. and Wieden E.A., "*A NeuroFuzzy approach for Modeling Lips*

Movements," Proc. of Third IEEE Int'l Conf. on Fuzzy System, **Vol.1**, pp.234-237,1994.

[16] Bremner F.,et al.,"*Statistical analysis of fuzzy-set data from neuronal networks*," Behav. Res. Methods. Comput., **21**, pp.209-212, 1989.

[17] Bridgett N.A.,Brandt J. and Harris C.J.,"*A neurofuzzy route to breast cancer diagnosis and treatment*,"Proc. of FUZZ-IEEE/IFES'95, **Vol. II**, pp.641-648,1995.

[18] Brown M.,et al.,"*Intelligent self-organizing controllers for autonomous guided vehicles: comparative aspects of fuzzy logic and neural nets*," Int. Conf. on Control, **Vol.1**, pp.134-139,1991.

[19] Brown M. and Harris C.J.,"*Neurofuzzy adaptive modelling and control*," Prentice-Hall, 1994.

[20] Brown M., An P.E. and Harris C.J.,"*On the condition of adaptive neurofuzzy models*," Proc. of FUZZ-IEEE/IFES'95, **Vol. II**, pp.663-670,1995.

[21] Buckley J.J.,et al.,"*On the equivalence of neural nets and fuzzy expert system*," Proc. IEEE IJCNN, **Vol.II**, pp.691-695,1992.

[22] Buckley J. J. and Hayashi Y. , "*Fuzzy neural nets and applications*," Fuzzy Systems and AI, **1**:11-14,1992.

[23] Buckley J. J. and Hayashi Y., "*Are regular fuzzy neural nets universal approximators?*," Proc. Int'l Joint Conf. on Neural Networks, **Vol.1**, pp.721-724, 1993.

[24]Buckley J. J. and Hayashi Y., "*Can fuzzy neural nets approximate continuous fuzzy functions?*," Fuzzy Set and Systems, **Vol.60**, pp.-,1993.

[25] Buckley J.J.,et al.,"*On the equivalence of neural nets and fuzzy expert system*," Fuzzy Sets and Systems, **53**(1993) 129-134.

[26] Buckley J.J. and Czogala E.,"*Fuzzy models, fuzzy controllers and neural nets*," Arch. Theoret. Appl. Comput. Sci. **5**(1993) 149-165.

[27] Buckley J.J. and Hayashi Y.,"*Numerical relationships between neural networks, continuous functions, and fuzzy systems*," Fuzzy Sets and Systems, **60**(1993) 1-8.

[28] Buckley J.J. and Hayashi Y.,"*Hybrid neural nets can be fuzzy controllers and fuzzy expert systems*," Fuzzy Sets and Systems, **60**(1993) 135-142.

[29] *Buckley J.J. and Hayashi Y.,"*Fuzzy neural networks*," in: R.R. Yager and L.A. Zadeh, eds., Fuzzy Sets, Neural Networks and Soft computing (Van Nostrand Reinhold, 1994) 233-249.

[30] Buckley J.J. and Hayashi Y.,"*Fuzzy neural networks: a survey*," Fuzzy Sets and Systems, **66**(1994) 1-13.

[31] Buckley J. J. and Hayashi Y.,"*Hybrid Fuzzy Neural Nets Are Universal Approximators*," Proc. of Third IEEE Int'l Conf. on Fuzzy System, **Vol.1**, pp.238-243,1994.

[32] Buckley J. J. and Hayashi Y.,"*Neural nets for fuzzy systems*," Fuzzy Sets and

Systems, **71**(1995) 265-276.
[33] *Buckley J. J. and Krishnamraju P., Reilly K., Hayashi Y.,"*Genetic LearningAlgorithms for Fuzzy Neural Nets,*" Proc. of Third IEEE Int'l Conf. on Fuzzy System, **Vol.3**, pp1969-1974, 1994.
[34] Carpenter G. A. , Grossberg S., and Rosen D.B., "*Fuzzy ART: An Adaptive Reasonance Algorithm for Rapid, Stable Classification of Anolog Patterns,*" Proc. of IJCNN, **Vol.II**, pp.411-420, 1991.
[35] *Carpenter G. A. , Grossberg S.,et al, "*Fuzzy ARTMAP: A Neural Network Architecture for Incremental Supervised Learning of Analog Multi dimensional Maps,*"IEEE Trans.Neural Networks, **Vol.3**, No.5, pp.698-713,1992.
[36] *Carpenter G.A., Grossberg S.,et al.,"*Fuzzy ARTMAP: a neural network architecture for incremental supervised learning of analog multidimensional maps,*" IEEE Trans. on Neural Networks, **Vol.3**, No.5, pp.698-713,Sept. 1992.
[37] Chaiberge M. and Reyneri L.M.,"*Cinta: a neuri-fuzzy real-time controller for low-power embedded systems,*" IEEE Micro, pp.40-47, June 1995.
[38] Chang R.-I and Hsiao P.-Y.,"*A Fuzzy-Neuro Approach for Timing-Driven System Partitioning in VLSI Multi-Chip Design,*" Proc. of Third IEEE Int'l Conf. on Fuzzy System, **Vol.1**, pp.302-307,1994.
[39] Chen C.-L. and Chen W.-C.,"*Fuzzy controller design by using neural network techniques,*" IEEE Trans. on Fuzzy Systems, **Vol.2**, No.3, pp.235-244, Aug. 1994.
[40] Chen H., Mitzumoto M. and Ling Y.-F,"*Automatic control of sewerage pumpstation by using fuzzy controls and neural networks,*" Proc. 2nd IEEE Int. Conf. on Fuzzy Systems, pp.877-882,1993.
[41] Chen S., Cowan C.F.N. and Grant P.M.,"*Orthogonal least squares learning algorithm for radial basis function networks,*" IEEE Trans. on Neural Networks, **Vol. 2**, pp.302-309, 1991.
[42] Chen T.,"*Fuzzy neural network application in medicine,*"Proc. of FUZZ-IEEE/ IFES'95, **Vol. II**, pp.627-634,1995.
[43] Cho S.,"*On-line Handwriting recognition with a neuro-fuzzy method,*"Proc. of FUZZ-IEEE/IFES'95, **Vol. II**, pp.1131-1136,1995.
[44] Cohen M.E. and Hudson D.L.,"*Approaches to the handling of fuzzy input data in neural networks,*" Proc. FUZZ-IEEE'92, pp.93-100,1992.
[45] Culliere T., Titli A. and Corrieu J.M.,"*Neuro-fuzzy modeling of nonlinear systems for control purposes,*" Proc. of FUZZ-IEEE/IFES'95, **Vol. IV**, pp.2009-2016,1995.
[46] Dagli C.H.,et al.,(Eds.)"*Proc. of the Artificial Neural Networks in Eng. Conf.,*" St. Louis, ASME.
[47] Dahanayake B.W. and Upton A.R.M.,"P*aralysis free fast learning: smart neural nets,*" Proc. of FUZZ-IEEE/IFES'95, **Vol. III**, pp.1527-1534,1995.
[48] D'Alche-Buc F.,et al.,"*Learning fuzzy control rules with a fuzzy neural*

network," Artificial Neural Networks, **2**, pp.715-719, 1992.
[49] Da Rocha A.,"*Neural fuzzy point processes*," Fuzzy Sets and Systems, **5**, pp.127-140, 1981.
[50] Davis J.P., Warms T.M. and Winters W.R.,"*A neural net implementation of the fuzzy c-means clustering algorithm*," Proc. Int. Joint Conf. Neural Networks (IJCNN'91), **Vol II**,1991.
[51] Diamond J.,et al.,"*A fuzzy cognitive system: examination of a referential neural architecture*," Proc. Int. Joint Conf. on Neural Networks, **Vol.2**, pp.617-622,1990.
[52] Diekgerdes H.,"*NEFCON-I: Entwurf und Implementierung einer Entwicklungsumgebung fur Neuronale Fuzzy-Regler*," Diplomarbeit, Technische Universitat Braunschweig, 1993.
[53] Dote Y.,"*Fuzzy and neuro controllers*," Proc. WANN'91, Aubern Univ., Feb. 1991.
[54] Dote Y.,et al.,"*Neural fuzzy transmission control for automobile*," Proc.IEEE Round Table Discussion Neural Network, Fuzzy, Vehicle Applications, 1991.
[55] Dote Y.,"*Neuro fuzzy robust controllers for drive systems*," Proc. IEEE int. Symp. on Industrial Electronics, pp.229-242, 1993.
[56] *Dubois D. and Prade H.,"*Possibility Theory - An Approach to Computerized Processing of Uncertainty*," Plenum Press, New York, 1988.
[57] Eklund P. and Klawonn,"*Neural fuzzy logic programming*," IEEE Trans. on Neural Networks, **Vol.3**, No.5, pp.815-818,1992.
[58] Eppler W."*Implementation of fuzzy oroduction systems with neural networks*," In (Eckmiller,et al.,1991), pp.249-252, Elsevier Science (North Holland).
[59] Eppler W.,"*Prestructuring of neural networks with fuzzy rules*," in Neuro-Nimes'90. 3rd Int. Workshop on Neural Networks and Their Applications, pp.227-241, 1990.
[60] Esogbue, A.O.,"*A fuzzy adaptive controller using reinforcement learning neural networks*," Proc. IEEE Int. Conf. on Fuzzy Systems 1993, pp.178-183, 1993.
[61] *Feldkamp L.A.,et al.,"*Architecture and training of a hybrid neural-fuzzy system*," Proc. 2nd Int'l Conf. on Fuzzy Logic and Neural Networks (IIZKA'92), pp.131-134,1992.
[62] Feldkamp L. and Puskorius G.,"*Trainable fuzzy and neural-fuzzy systems for Idle-speed control*," Proc. 2nd IEEE Int. Conf. on Fuzzy Systems, pp.45-51,1993.
[63] Fernando V.-V. and Angel R.-V.,"*Using building blocks to design analog neuro-fuzzy controllers*," IEEE MICRO, **Vol.15**, No.4, pp.49-57, Aug. 1995.
[64] Foerster S.,"*Zu kombinationen neuronaler netze und fuzzy systems*," Diplomarbeit, Technische Universitat Braunschweig, 1993.
[65] Freisleben B. and Kunkelmann T,"*Combining fuzzy logic and neural networks to control an autonomous vehicle*," Proc. 2nd IEEE Int. Conf. on Fuzzy Systems,**Vol. I**,pp.321-326,1993.

[66] *Fu L.-M.,"*Backpropagation in neural networks with fuzzy conjunction units*," Proc. Int. Joint Conf. on Neural Networks 1990, **Vol.1**, pp.613-618,1990.
[67] Fukuda T.,et al.,"*Multi-sensor integration system based on fuzzy inference and neural network*," J. Inform. Sci. **71**(1,2)(1993) 27-41.
[68] *Fukuda T., et al.,"*Structure optimization of fuzzy neural network by genetic algorithm*," Fifth IFSA World Congress, pp.964-967, 1993.
[69] Furukawa M. and Yamakawa T.,"*The advanced method to optimize cross-detecting lines for a fuzzy neuron*," Proc. 5th IFSA Congr., pp.857-860,1993.
[70] Furukawa M. and Yamakawa T.,"*The design algorithms of membership functions for a fuzzy neuron*," Fuzzy Sets and Systems, **71**(1995) 329-344.
[71] Furuhashi, et al.,"*An adaptive fuzzy controller using fuzzy neural networks*,"Fifth IFSA World Congress, pp.769-772, 1993.
[72] *Furuya T.,et al.,"*NFS: Neuro fuzzy inference system*," in Int. Workshop on Fuzzy System Applications (IIZUKA'88), pp.219-230, 1988.
[73] Glorennec P.-Y.,"*Associating a neural network and fuzzy rules for dynamic process control*," in Neuro-Nimes'90. 3rd Int. Workshop on Neural Networks and Their Applications, pp.211-225, 1990.
[74] Gomide F. and Rocha A.,"*Neurofuzzy controllers*," Fifth IFSA world congress, 1993.
[75] Gonzalez A. and Perez R.,"*Structural learning of fuzzy rules from noised examples*," Proc. of FUZZ-IEEE/IFES'95, **Vol. III**, pp.1323-1330,1995.
[76] Goode P.V. and Chow M.-Y. ,"*A Hybrid Fuzzy/Neural System Used to Extract Heuristic Knowledge from a Fault Detection Problem*," Proc. of Third IEEE Int'l Conf. on Fuzzy System, **Vol.3**, pp.1731-1736,1994.
[77] Gupta M.M.,"*Fuzzy neural network approach to control system*," Proc. IEEE 1990 Americam Control Conf., **Vol.3**, pp.3019-3022,1990.
[78] Gupta M.M. and Knopf G.K.,"*Fuzzy neural network approach to control systems*," Proc. of First Int. Symp. on Uncertainty Modeling and Analysis, pp.483-488,1990.
[79] Gupta M.M. and Gorzalczany,"*Fuzzy neuro-computational technique and its application to modelling and control*," Proc. IEEE IJCNN, **Vol.2**, pp.1454-1457,1991.
[80] Gupta M.M. and Qi J.,"*On fuzzy neuron models*," Int. Joint Conf. on Neural Networks, pp.431-456, 1991.
[81] Gupta M.M. and Qi J.,"*Fusion of fuzzy logic and neural networks with applications to decision and control problem*," Proc. NAFIPS'91, pp.327-328,1991.
[82] Gupta M.M. and Qi J.,"*Connections (AND, OR, NOT) and T-operators in fuzzy reasoning*," in Conditional Logic in Expert Systems, pp.211-233,1991.
[83] Gupta M.M.,"*Fuzzy neural computing systems*," Proc. 2nd Int. Conf. on Fuzzy Logic and Neural Networks, pp.17-22,1992.

[84] Gupta M.M.,"*Fuzzy logic and neural networks*," Proc. 2nd Int. Conf. on Fuzzy Logic and Neural Networks, pp.257-160,1992.
[85] Gupta M.M. and Knopf G.K.,"*Dynamic neural network for fuzzy inference*," SPIE's Conf. on Applications of Fuzzy Logic Technology,1993.
[86] *Gupta, M.M. and Rao D.H.,"*On the principles of fuzzy neural networks*," Fuzzy Sets and Systems, **Vol.61**, pp.1-18, 1994.
[87] Halgamuge S. and Glesner M.,"*The fuzzy neural controller FuNe II with a new adaptive defuzzification strategy based on CBAD distributions*," European Congress on Fuzzy and Intelligent Technologies, Aachen,1993.
[88] Han G.,et al.,"*Fuzzy Lapart Supervised Learning Through Inferencing for Stable Category Recognition*," Proc. of Third IEEE Int'l Conf. on Fuzzy System, **Vol.1**, pp.46-51,1994.
[89] Hataoka N.,et al.,"*Large vocabulary speed recognition using neural-fuzzy and concept networks*,"Proc. Int.Conf. on Acoustics, Speech and Signal Processing, pp.513-516,1990.
[90] Hauptmann W. and Heesche K.,"*A prototype of an integrated fuzzy-neural system*," Proc. of the EUFIT'94, 1994.
[91] Hauptmann W. and Heesche K.,"*A neural net topology for bidirectional fuzzy-neuro transformation*," Proc. of FUZZ-IEEE/IFES'95, **Vol. III**, pp.1511-1518,1995.
[92] Hayashi H.,et al.,"*Neuro fuzzy transmission control for automobile*," in (Dagli C.H.,et al.,1991), pp.283-288.
[93] *Hayashi I.,et al.,"*Artificial-neural-network-driven fuzzy control and its application to the learning of inverted pendulum system*," Proc. 3rd IFSA Congress, pp.610-613,1989.
[94] Hayashi I.,et al.,"*Construction of fuzzy inference rules by neural network driven fuzzy reasoning and neural network driven fuzzy reasoning with learning functions*," Int. J. Approximate Reasoning, 6, pp.241-266, 1992.
[95] Hayashi K.,et al.,"*Neuro fuzzy transmission control for automobile with variable loads*," IEEE Trans. on Control Sytems Technology, **Vol.3**, No.1, March 1995.
[96] Hayashi Y. and Imura A.,"*Fuzzy neural expert system with automated extraction of fuzzy if-then rules from a trained neural network*," Proc. 1st Int. Symp. on Uncertainty Modeling and Analysis, pp.489-494,1990.
[97] Hayashi Y., Buckley J.J. and Czogala E.,"*Systems engineering applications of fuzzy neural networks*," Proc. Int. Joint Conf. Neural Networks (IJCNN'92), **Vol. II**, pp.412-418, 1992.
[98] *Hayashi Y.,et al.,"*Direct fuzzification of neural network and fuzzified delta rule*," Proc. 2nd Int'l Conf. on Fuzzy Logic and Neural Networks (IIZUKA'92), pp.73-76, 1992.
[99] Hayashi Y., Buckley J.J. and Czogala E.,"*Fuzzy neural network with fuzzy*

signals and weights," Proc. Int. Joint Conf. on Neural Networks , **Vol II**, pp.696-701,1992.
[100] Hayashi Y., Buckley J.J. and Czogala E.,"*Systems engineering applications of fuzzy neural networks*," J. Systems Eng. **2**(1992) 232-236.
[101] Hayashi Y., Buckley J.J. and Czogala E.,"*Fuzzy neural controller*," Proc. IEEE Int. Conf. on Fuzzy Systems, pp.197-202,1992.
[102] Hayashi Y., Buckley J.J. and Czogala E.,"*Fuzzy expert systems versus neural networks*," Proc. Int. Joint Conf. on Neural Networks , **Vol II**, pp.720-726,1992.
[103] Hayashi Y., Buckley J.J. and Czogala E.,"*Approximation between fuzzy expert systems and neural networks*," Proc. 2nd Int. Conf. on Fuzzy Logic and neural networks, **Vol. I**, pp.135-139, 1992.
[104] Hayashi Y.,"*A neural expert system using fuzzy teaching input*," Proc. IEEE Int. Conf. on Fuzzy Systems, pp.485-491,1992.
[105] *Hayashi Y. and Buckley J.J.,"*Fuzzy max-min neural controller*," unpublished manuscript.
[106] *Hayashi Y., Buckley J.J. and Czogala E., "*Fuzzy Neural Network with Fuzzy Signals and Weights*,"International J. of Intelligent Systems, **Vol.8**, pp.527-537, 1993.
[107] Hayashi Y. and Buckley J.J.,"*Fuzzy controller and fuzzy expert systems as hybrid neural nets*," Proc. 5th IFSA World Congr., **Vol. I**, pp.66-69,1993.
[108] Hayashi Y. and Buckley J.J.,"*Direct fuzzification of neural networks*," Proc. 1st Asian Fuzzy Systems Symposium, pp.560-567, 1993. pp.721-724,1993.
[109] Hayashi Y. and Buckley J.J.,"*Approximations between fuzzy expert systems and neural networks*," Int. J. Approx. Reasoning 10(1994) 63-73.
[110] Hellendoorn H. and Baudrexl R.,"*Fuzzy-neural traffic control and forecasting*," Proc. of FUZZ-IEEE/IFES'95, **Vol. IV**, pp.2187-2194,1995.
[111] Higgins C.M. and Goodman R.M.,"*Learning fuzzy rule-based neural networks for function approximation*," Proc. IEEE IJCNN, **Vol.I**, pp.251-256,1992.
[112] Hirota K. and Pedrycz,"OR/AND neuron in modeling of fuzzy sets connectives," IEEE Trans. on Fuzzy Systems, **Vol.2**, No.2, pp.151-161, 1994.
[113] Hitachi,"*Neuro and fuzzy logic automatic washing machine and fuzzy logic frier*," Hitachi News Rel., No.91-024 (Feb. 26, 1991). Hitachi,1991. (In Japanese)
[114] Holtzman J.,"*Coping with broadband traffic uncertainties: statistical uncertainty, fuzziness, neural networks*," Proc. IEEE Global Telecommunications Conf. and Exhibition 1990-'Communications: Connecting the Future',**Vol.1**, pp.7-11,1990.
[115] Hollatz J.,"*Neuro-fuzzy in legal reasoning*," Proc. of FUZZ-IEEE/IFES'95, **Vol. II**, pp.655-662, 1995.
[116] Hong S., Kim S. and Lee J.,"*The minimum cost path finding algorithm using a hopfield type neural network*," Proc. of FUZZ-IEEE/IFES'95,**Vol.IV**,pp.1719-

1726,1995.
[117] Horikawa S.I., et al.,"*A fuzzy controller using a neural network and its capability to learn expert's control rules*," Proc. Int'l Conf. on Fuzzy Logic and Neural Networks, IZUKA'90, pp.103-106, 1990.
[118] Horikawa S.,et al,"*Composition methods of fuzzy neural networks*," IEEE/ IECON Conf. Proc., pp.1253-1258,1990.
[119] Horikawa S.,et al.,"*A study on fuzzy modeling using fuzzy neural networks*," Proc.IFES, pp.562-573,1991.
[120] Horikawa S.I., et al.,"*On fuzzy modeling using fuzzy neural networks with the back propagation algorithm*," IEEE Trans. on Neural Networks, **Vol.3**,No.5,pp.801-806, 1992.
[121] Hsieh C., Su M. and Chienn S.,"*Use of a self-learning neuro-fuzzy system for syllabic labeling of continuous speech*," Proc. of FUZZ-IEEE/IFES'95, **Vol. IV**, pp.1727-1734,1995.
[122] Hsu Y., Lin Y. and Sun T.,"*Engineering design optimization as a fuzzy control process*," Proc. of FUZZ-IEEE/IFES'95, **Vol. IV**, pp.2001-2008,1995.
[123] Ichihashi H.,"*Fuzzy systems-optimization by neuro-like fuzzy model*," Japanese-European Symp. on Fuzzy Systems, 1992.
[124] Ichihashi H. and Turksen I.B.,"*A neuro-fuzzy approach to data analysis of pairwise comparisons*," Int.J.Approximate Reasoning, **Vol.9**, pp.227-248,1993.
[125] Ichihashi H. and Turksen I.B.,"*Neuro-fuzzy data analysis and its future directions*," Proc. of FUZZ-IEEE/IFES'95, **Vol. IV**, pp.1919-1926,1995.
[126] Ichihashi H., et al.,"*Neuro-fuzzy finite elememt method and a posteriori error estimation*," Proc. of FUZZ-IEEE/IFES'95, **Vol. IV**, pp.2335-2342,1995.
[127] Ichihashi H.,Harada N. and Nagasaka K.,"*Selection of the optimum number of hidden layers in neuro-fuzzy GMDH*," Proc. of FUZZ-IEEE/IFES'95, **Vol. III**, pp.1519-1526,1995.
[128] Imasaki N.,et al.,"*A fuzzy neural network and its application to elevator group control*," Proc. Int. Conf. on Fuzzy Logic and Neural Networks (IIZUKA'90), pp.119-122,1990.
[129] Imasaki N.,Kubo S. and Nakai S.,"*Elevator group control system tuned by a fuzzy neural network applied method*," Proc. of FUZZ-IEEE/IFES'95, **Vol.IV**, pp.1735-1740,1995.
[130] *Ishibuchi H.,et al.,"*An architecture of neural networks for input vectors of fuzzy numbers*," Proc. IEEE Int'l Conf. on Fuzzy Systems (FUZZ-IEEE'92), pp.1293-1300, 1992.
[131] *Ishibuchi H.,et al.,"*Interpolation of fuzzy if-then rules by neural networks*," Proc. 2nd Int. Conf. on Fuzzy Logic and Neural Networks (IIZUKA'92), pp.337-340,1992.
[132] *Ishibuchi H.,et al.,"*Learning of neural networks from fuzzy inputs and fuzzy*

targets," Proc. Int'l Joint Conf. on Neural Networks, **Vol.3**, pp.447-452,1992.
[133] Ishibuchi H.,et al.,"*Implementation of fuzzy if-then rules by fuzzy neural networks with fuzzy weights*," Proc. 1st European Congress on Fuzzy and Intelligent Technologies, **Vol.1**, pp.209-215, 1993.
[134] Ishibuchi H.,et al.,"*Fuzzy neural networks with fuzzy weights and biases*," Proc. IEEE Int'l Conf. on Neural Networks, **Vol.3**, pp.1650-1655,1993.
[135] Ishibuchi H., Tanaka H. and Okada H., "*Fuzzy Neural Networks with Fuzzy' Weights and Fuzzy Biases*,"Proc. of 1993 IEEE International Conf. on Neural Network,"pp.1650-1655,1993.
[136] Ishibuchi H., Fujioka R. and Tanaka H., "*Neural Networks That Learn From Fuzzy If-Then Rules*,"IEEE Trans. on Fuzzy Systems, **Vol.1**, pp.85-97, 1993.
[137] *Ishibuchi H., Kwon K. and Tanaka H., "*Learning of Fuzzy Neural Networks from Fuzzy Inputs and Fuzzy Tragets*," Proc. of Fifth IFSA World Congress, vol.I, pp.147-150,1993.
[138] Ishibuchi H., et al.,"*Neural networks that learn from fuzzy if-then rules*," IEEE Trans. on Fuzzy Syetems, **Vol.1**, pp.85-97, 1993.
[139] Ishibuchi H., Kwon K. and Tanaka H., "*A Fuzzy Neural Network with Trapezoid Fuzzy Weights*," Proc. of Third IEEE Int'l Conf. on Fuzzy System, **Vol.1**, pp.228-233,1994.
[140] Ishibuchi H.,Kwon K. and Tanaka H.,"*A learning algorithm of fuzzy neural networks with triangular fuzzy weights*," Fuzzy Sets and Systems, **71**(1995) 277-294.
[141] *Ishigami H.,et al.,"*Structure optimization of fuzzy neural network by genetic algorithm*," Fuzzy Sets and Systems, **71**(1995) 257-264.
[142] Iwata T.,et al.,"*Fuzzy control using neural network techniques*," Proc. Int. Joint Conf. on Neural Networks 1990, **Vol.3**, pp.365-370,1990.
[143] Jang J.-S. R.,"*Fuzzy modeling using generalized neural networks and Kalman filter algorithm*," Proc. Ninth National Conf. on Artificial Intelligence (AAAI-91), pp.762-767,1991.
[144] Jang J.-S. R.,"*Rule extraction using generalized neural networks*," Proc. 4th IFSA Congress, pp.82-86, (in the Volume for Artificial Intelligence), July 1991.
[145] *Jang J.-S. R.,"*ANFIS: Adaptive-Network-Based Fuzzy Inference Systems*," IEEE Trans. SMC, **Vol.23**, No.3, pp.665-685, May/June 1993.
[146] Jang J.-S.R.,"*Self-learning fuzzy controllers based on temporal back propagation*," IEEE Trans. on Neural Networks, **Vol.3**, pp.714-723, 1992.
[147] Jin Y.,et al.,"*Neural network based fuzzy identification and its application to modeling and control of complex systems*," IEEE Trans. on SMC, **Vol.25**, No.6, pp.990-997, June 1995.
[148] Kamimura R., Yager R.R., and Shohachiro,"*Representing Acquired Knowledge of Neural Networks by Fuzzy Sets:Control of Internal Information of Neural Networks by Entropy Minimization*," Proc. of Third IEEE Int'l Conf. on

Fuzzy System, **Vol.1**, pp.59-63.,1994.
[149] Kamimura R. and Nakanishi S.,"*Learning by a-Divergence*," Proc. of FUZZ-IEEE/IFES'95, **Vol. III**, pp.1535-1542,1995.
[150] *Kandel A. and Lee S.C.,"*Fuzzy switching and automata: theory and applications*," Crane Russak, New York, 1979.
[151] Kandel A. (Ed.),"*Fuzzy expert systems*," Addison-Wesley, 1987.
[152] Kandel A. and Langholz G.,"*Hybrid architectures for intelligent systems*," CRC Press, 1992.
[153] Kandel A., Zhang Y.-Q. and Miller T.,"*Fuzzy neural decision system for fuzzy moves*," the 3rd World Congress on Expert Systems, 1996.
[154] Katayama R.,et al.,"*Self generating radial basis function as neuro-fuzzy model and its application to nonlinear prediction of chaotic time series*," Proc. IEEE Int. Conf. on Fuzzy Systems, pp.407-414,1993.
[155] *Katayama R.,et al.,"*Neuro, fuzzy and chaos technology and its applications to consumer electronics*," Joint Japanese-European Symp. on Fuzzy Systems, Doc. 21, 1992.
[156] *Katayama R.,et al.,"*Self generating radial basis function as neuro-fuzzy model and its application to nonlinear prediction of chaotic time series*," Proc. 2nd IEEE Int. Conf. on Fuzzy Systems, pp.407-414,1993.
[157] *Katayama R.,et al.,"*Comparative analysis on learning ability of self generating algorithm for neuro fuzzy model with various radial basis functions*," Proc. 5th IFSA Congr., pp.139-142,1993.
[158] *Katayama R.,et al.,"*Developing tools and methods for applications incorporating neuro, fuzzy and chaos technology*," Comput. Indust. Engrg. **24**(1993) 579-592.
[159] *Katayama R.,et al.,"*Dimension analysis of chaotic time series using self generating neuro-fuzzy model*," Proc. 5th IFSA Congr., pp.857-860,1993.
[160] *Katayama R.,et al.,"*Embedding dimension estimation of chaotic time series using self-generating radial basis function network*," Fuzzy Sets and Systems, **71**(1995) 311-328.
[161] Kawamura A.,et al.,"*A prototype of neuro-fuzzy cooperation system*," Proc. FUZZ-IEEE'92, pp.1275-1282, 1992.
[162] Kayama M, Sugita Y. and Morooka Y. and Saito Y.,"*Adjusting neural networks for accurate control model tuning*,"Proc. of FUZZ-IEEE/IFES'95, **Vol.IV**, pp.1995-2000,1995.
[163] *Keller J.M. and Hunt D.J.,"*Incorporating fuzzy membership functions into the perceptron algorithm*," IEEE Trans. on Pattern Anal. Mach. Intell., pp.693-699,1985.
[164] Keller J.M.,"*Experiments on neural network architectures for fuzzy logic*," in (Lea & Villareal, 1991), pp.201-216.

[165] Keller J.M. and Tahani H.,"*Implementation of conjunctive and disjunctive fuzzy logic rules with neural networks*," Int. J. Approx. Reasoning **6**(1992) 221-240.
[166] Keller J.M. and Tahani H.,"*Backpropagation neural networks for fuzzy logic*," Inform. Sci. **62**(1992) 205-221.
[167] Keller J.M., Yager R.R., and Tahani H.,"*Neural Network Implementation of Fuzzy Logic*," Fuzzy Sets and Systems, **45**:1-12, 1992.
[168] Keller J.M. and Krishnapuram R.,"*Evidence aggregation networks for fuzzy logic inference*," IEEE Trans. on Neural Networks, **Vol.3**, No.5, pp.761-769,1992.
[169] Khan E. and Venkatapuram P.,"*NeuFuz: Neural Network Based Fuzzy Logic Design Algorithms*,"FUZZ-IEEE'93 Proceedings, **Vol.1**, pp.647-654, 1993.
[170] Khan E.,"*Neural Network Based Algorithms for Rule Evaluation & Defuzzification in Fuzzy Logic Design*," WCNN'93 Proceedings, vol.2, pp.31-38, 1993.
[171] Khan E.,"*NeuFuz:An Intelligent Combination of Fuzzy Logic with Neural Nets*,"IJCNN'93 Proceedings, **Vol.3**, pp.2945-2950, 1993.
[172] Kiguchi K. and Fukuda T.,"*Fuzzy neural controller for robot manipulator force control*,"Proc. of FUZZ-IEEE/IFES'95, **Vol.II**, pp.869-874,1995.
[173] Kim S.-W. and Lee J.-J,"*Design of a fuzzy controller with fuzzy sliding surface*," Fuzzy Sets and Systems, **71**(1995) 359-368.
[174] Kiszka J.B. and Gupta M.M.,"*Fuzzy logic neural network*," BUSEFAL **4**(1990) 104-109.
[175] Kong S.-G. and Kosko B.,"*Comparison of fuzzy and neural truck backerupper control systems*," Proc. Int. Joint Conf. on Neural Networks 1990, **Vol.3**, pp.349-358,1990.
[176] Kosko B.,"*Adaptive inference in fuzzy knowledge networks*," Proc. IEEE 1st Int. Conf. Neural Networks, pp.261-268, July,1987.
[177] Kosko B.,"*Neural Networks and Fuzzy Systems - a dynamical systems approach to machine intelligence*," Prentice-Hall, 1992.
[178] *Krishnamraju P.V.,et al.,"*Genetic learning algorithms for fuzzy neural networks*," Proc. IEEE Int. Conf. on Fuzzy Systems,**Vol. III**, pp.1969-1974,1994.
[179] Krishnapuram R. and Lee J.,"*Fuzzy-set-based hierarchical networks for information fusion in computer vision*," Neural Networks, **5**(2):335-350, 1992.
[180] Kuo Y.-H.,et al,"*A fuzzy neural network model and itd hardware implementation*," IEEE Trans. on Fuzzy Systems, **Vol.1**, No.3, pp.171-183, Aug. 1993.
[181] Kuncicky D.C. and Kandel A.,"*A fuzzy interpretation of neural networks*," Proc. of the Third IFSA Congress, pp.113-116,1989.
[182] Kwan H.K. and Cai Y.,"*A fuzzy neural network and its application to pattern recognition*," IEEE Trans. on Fuzzy Systems, **Vol.2**, No.3, pp.185-193, Aug. 1994.
[183] Langari R. and Wang L.,"*Fuzzy models, modular networks, and hybrid*

learning," Proc. of FUZZ-IEEE/IFES'95, **Vol. III**, pp.1291-1298,1995.
[184] Lea R.N. and Villareal J. (Eds.),"*Proc. of 2nd Joint Technology Workshop on Neural Networks and Fuzzy Logic*," Lyndon B. Johnson Space Center, Houston, Texas. NASA.
[185] Lee C.-C.,"*A self-learning rule-based controller employing approximate reasoning and neural network concepts*," Int.J. Intell. Syst., 5, 1990.
[186] Lee K.-C.,et al.,"*An adaptive fuzzy current controller with neural network for field-oriented controlled induction machine*," Proc. 2nd Int. Conf. on Fuzzy Logic and Neural Networks (IIZUKA'92), pp.449-452,1992.
[187] *Lee S.C. and Lee E.T.,"*Fuzzy neurons and automata*," Proc. 4th Princeton Conf. on Information Science and Systems, pp.381-385, 1970.
[188] *Lee S.C.,"*Fuzzy sets and neural networks*," US-Japan Seminar on Fuzzy Sets and their Application, Berkeley, CA, 1974.
[189] *Lee S.C. and Lee E.T.,"*Fuzzy sets and neural networks*," J.Cybernet. **4**, pp.83-103, 1974.
[190] *Lee S.C. and Lee E.T.,"*Fuzzy neural networks*," Math Biosci. **23**, 151-177, 1975.
[191] Li C.-C. and Wu C.-H J.,"*Generating Fuzzy Rules for A Neural Fuzzy Classifier*," Proc. of Third IEEE Int'l Conf. on Fuzzy System, **Vol.3**, pp.1719-1724,1994.
[192] Li W.,"*Optimization of A Fuzzy Controller Using Neural Networks*," Proc. of Third IEEE Int'l Conf. on Fuzzy System, **Vol.1**, pp.223-227,1994.
[193] Li W. and Tan Q.,"*An efficient self-organizing fuzzy controller using neural networks*," Proc. of 1994 IEEE Int.Conf. on SMC, pp.1797-1802,1994.
[194] Li W.,"*Neuro fuzzy systems for intelligent robot navigation and control under uncertainty*," Proc. of FUZZ-IEEE/IFES'95, **Vol. IV**, pp.1747-1754,1995.
[195] Lin C.-J. and Jou C.-C,"*A fuzzy-neural hybrid system for identifying nonlinear dynamic systems*," Proc. Int. Joint Conf. on Neural Networks, **Vol. III**, pp.625-630,1992.
[196] Lin C.-J. and Lin C.-T.,"*Reinforcement learning for ART-based fuzzy adaptive learning control networks*," Proc. of FUZZ-IEEE/IFES'95, **Vol. III**, pp.1299-1306,1995.
[197] *Lin C.-T. and Lee C.S.G., "*Neural-Network-Based Fuzzy Logic Control and Decision System*,"IEEE Trans. on Computers-Special Issue on Artificial Neural Networks, **40**(12):1320-1336, Dec. 1991.
[198] Lin C.-T. and Lee C.S.G.,"*Real-time supervised structure-parameter learning for fuzzy neural network*," Proc.IEEE Internat. Conf. on Fuzy System, pp.1283-1290, 1992.
[199] Lin C.-T. and Lee C.S.G.,"*Reinforcement structure-parameter learning for neural-network-based fuzzy logic control systems*," Proc. IEEE Internat. Conf. on

Fuzzy Systems, pp.88-93,1993.
[200] *Lin C.-T.,"*Neural fuzzy control sytems with structure and parameter learning*," World scientific, 1994.
[201] Lin C.-T., "A neural fuzzy control system with structure and parameter learning," Fuzzy Sets and Systems 70(1995) 183-212.
[202] Lin Yinghua and A.G. George III, "*A New Fuzzy Approach for Setting the Initial Weights in a Neural Network*," Proc. of Third IEEE Int'l Conf. on Fuzzy System, **Vol.1**, pp.40-45.,1994.
[203] Linkens D.A. and Nie J.,"*Fuzzified RBF network-based learning control: structure and self-construction*," Proc. IEEE Int. Conf. on Neural Networks, pp.1016-1021,1993.
[204] Makkonen A. and Koivo H.N.,"*Autotuner for fuzzy controller- relay based approach*," Proc. of FUZZ-IEEE/IFES'95, Vol. IV, pp.1987-1994,1995.
[205] Miyazaki A.,et al., "*Fuzzy Regression Analysis by Fuzzy Neural Networks and Its Application*," Proc. of Third IEEE Int'l Conf. on Fuzzy System, **Vol.1**, pp.52-58.,1994.
[206] *Moody J. and Darken C.,"*Fast learning networks of locally-tuned processing units*," **Vol.1**, pp.281-294, 1989.
[207] Morita T.,et al.,"*Photo-copier image density control using neural network and fuzzy theory*," Proc. 2nd Int. Workshop on Industrial Fuzzy Control and Intelligent Systems, pp.10-16,1992.
[208] Morita T.,et al.,"*Fan heater using fuzzy control and neural networks*," Sanyo Tech. Rev. **23**, 3 (Mar. 1991), 93-100. (In Japanese)
[209] Nakamura K.,et al.,"*Fuzzy network production system*," Proc. 2nd Int. Conf. Fuzzy Logic Neural Networks (IIZUKA'92), pp.127-130,1992.
[210] Narazaki H. and Ralescu A.L.,"*A synthesis method for multi-layered neural network using fuzzy sets*," IJCAI-91:Workshop on Fuzzy Logic in Artificial Intelligence, pp.54-66,1991.
[211] Nauck D.,"*NEFCON-I: eine simulationsumgebung fur neuronal fuzzy-regler*," in 1. GI-Workshop Fuzzy-Systeme'93 Braunschweig, 1993.
[212] Nauck D.,"*Modellierung neuronaler fuzzy-regler*," Ph.D. thesis, Technische Universitat Braunschweig.
[213] Nauck D., Klawonn F. and Kruse R.,"*Fuzzy sets, fuzzy controllers and neural networks*," Wissenschaftliche Zeitschrift der Humboldt-Universitat zu Berlin, R. Medizin, **41**(4), 99-120, 1992.
[214] Nauck D., Klawonn F. and Kruse R."*Combining neural networks and fuzzy controllers*," Fuzzy Logic in Artificial Intelligence (FLAI93), pp.35-46 Berlin.Springer-Verlag,1993.
[215] Nauck D. and Kruse R.,"*Interpreting changes in the fuzzy sets of a self-adaptive neural fuzzy controller*," Proc. 2nd Int. Workshop on Industrial

Applications of Fuzzy Control and Intelligent Systems (IFIS'92), pp.146-152, 1992.
[216] Nauck D. and Kruse R.,"*A neural fuzzy controller learning by fuzzy error backpropagation*," Proc. of NAFIPS, **Vol.2**, pp.388-397,1992.
[217] *Nauck D. and Kruse R.,"*A fuzzy neural network learning fuzzy control rules and membership functions by fuzzy error backpropagation*," Proc. IEEE Internt. Conf. on Neural Networks, pp.1022-1027,1993.
[218] Nauck D. and Kruse R.,"*NEFCON-I: an x-window based simulator for neural fuzzy controllers*," Proc. IEEE Int.Conf. Neural Networks 1994.
[219] Nie J. and Linkens D.,"*Fuzzy-neural control - principles, algorithms and applications*," Prentice Hall, 1995.
[220] Nikkei Electronics,"*New trend in consumer electronics: Combining neural networks and fuzzy logic*," Nikkei Elec. **528** (1991), 165-169. (In Japanese)
[221] *Nobre F.S.M.,"*Genetic-neuro-fuzzy systems: a promising fusion*," Proc. of FUZZ-IEEE/IFES'95, **Vol. I**, pp.259-266,1995.
[222] Nomura H.,et al.,"*A learning method of fuzzy inference rules by descent method*," Proc. IEEE Int. Conf. on Fuzzy Systems 1992, pp.203-210.
[223] O'Hagan M.,"*A fuzzy neuron based upon maximum entropy ordered weighted averaging*," in Uncertainty in Kowledge Bases, Lecture Notes in Computer Science, Vol.521. B. Bouchon-Meunier, R.R. Yager and L.A. Zadeh eds. New York: Springer-Verlag, pp.598-609,1991.
[224] Okada H.,et al.,"*Initializing multilayer neural networks with fuzzy logic*," Proc. Int. Conf. on Neural Networks, pp.239-244, 1992.
[225] Pal S.K. and Mitra S.,"*Multilayer perceptron, fuzzy sets and classification*," IEEE Trans. on Neural Networks, **Vol.3**, No.5, pp.683-697,Sept. 1992.
[226] *Pedrycz W."*Fuzzy sets framework for development of a perception perspective*," Fuzzy Sets and Systems, **37**, pp.123-137, 1990.
[227] *Pedrycz W.,"*Selected issues of frame of knowledge representation realized by means of linguistic labels*," Int. J. of Intelligent Systems, **Vol. 7**, pp.155-170, 1992.
[228] *Pedrycz W. and Rocha A.,"*Fuzzy-set based models of neurons and knowledge-based neurons*," IEEE Trans. on Fuzzy Systems, **Vol.1**, No.4, pp.254-266, 1993.
[229] *Pedrycz W.,"*Fuzzy neural networks and neurocomputations*," Fuzzy Sets and Systems, **56**, pp.1-28, 1993.
[230] *Pedrycz W.,"*Fuzzy sets engineering*," CRC Press, 1995.
[231] *Pedrycz W., Poskar C.H. and Czezowski P.J.,"*A reconfigurable fuzzy neural network with in-site learning*," IEEE MICRO, **Vol.15**, No.4, pp.19-30, Aug. 1995.
[232] Rao V.B. and Rao H.V.,"*C++ Neural networks and fuzzy logic*," MIS, Inc.,1993.
[233] *Reguena I. and Delgado M.,"*R-FN: a model of fuzzy neuron*," Proc. 2nd Int.

Conf. Fuzzy Logic Neural Networks (IIZUKA'92), pp.793-796,1992.
[234] Rose B.K.,"*Expert system, fuzzy logic, and neural network applications in power electronics and motion control*," Proc. IEEE, **Vol.82**, No.8, pp.1303-1323,Aug. 1994.
[235] *Rumelhart D.E., McClelland, and the PDP Research Group,"*Parallel distributed processing: explorations in the microstructure of cognition*," Vol.1:Foundations. Cambridge,MA: MIT Press, 1986.
[236] Sanchez E.,"*Fuzzy connectionist expert systems*," Proc. 1st Int. Conf. on Fuzzy Logic and Neural Networks (IIZUKA'90), pp.31-35, 1990.
[237] Sanyo,"*Neuro and fuzzy logic automatic washing machine ASW-50v2*," Sanyo News Rel. (Feb. 29,1991). Sanyo, 1991. (In Janpanese)
[238] Sarkodie-Gyan T. and Lam C.W.,"*The paradigm of neuro-fuzzy in autonomous vehicles*," IMC, Belfast, 1994.
[239] Sarkodie T. and Willumeit H.-P.,"*Neuro-fuzzy based autonomous vehicles*," Proc. of FUZZ-IEEE/IFES'95, **Vol. II**, pp.855-862,1995.
[240] Sekine S.,"*Two-degree-of-freedom fuzzy neural network control system and its application to vehicle control*," 5th IFSA'93 Cong.Proc., pp.1121-1124,1993.
[241] Sekine S.,Imasaki N. and Endo T.,"*Application of fuzzy neural network controlto automatic train operation and tuning of its control rules*,"Proc. of FUZZ-IEEE/IFES'95, **Vol. IV**, pp.1741-1747,1995.
[242] Shann J.J. and Fu H.C.,"*A fuzzy neural network for knowledge learning*," Proc. 5th IFSA Congr.,1993.
[243] Shann J.J. and Fu H.C.,"*Backpropagation learning for acquiring fine knowledge of fuzzy neural networks*," Proc. WCNN'93, 1993.
[244] Shann J.J. and Fu H.C.,"*A fuzzy neural network for rule acquiring on fuzzy control systems*," Fuzzy Sets and Systems, **71**(1995) 345-358.
[245] Sharaf A. and Lie T.,"*A Hybrid Neuro-Fuzzy Power System Stabilizer*," Proc. of Third IEEE Int'l Conf. on Fuzzy System, **Vol.3**, pp.1608-1613,1994.
[246] Sharpe R.N.,et al.,"*A methodology using fuzzy logic to optimize feedforward artificial neural network configuration*," IEEE Trans. SMC, **Vol.24**, No.5, pp.760-768, May 1994.
[247] Shimojima K.,et al.,"*Multi-sensor integration system utilizing fuzzy inference and neural network*," J.Robotics and Mechatronics **4**(5) (1992) 416-421.
[248] Shimojima K.,et al.,"*Self-tuning fuzzy modeling with adaptive membership function, rules, and hierarchical structure based on genetic algorithm*," Fuzzy Sets and Systems, **71**(1995) 295-310.
[249] *Silva F.M. and Almeida L.B.,"*Acceleration Techniques for the Back-Propagation Algorithm*," Lecture Notes in Computer Science, Springer-Verlag, **Vol.412**, pp.110-119, 1990.
[250] Simoes M.G.,"*Using fuzzy neural networks (FNN) in modeling fuzzy*

estimators on power electronic waveforms," Univ. of Tenn. Internal Rep., Dec.1992.
[251] Simpson P.,"*Fuzzy min-max classification with neural networks*," J. Knowledge Eng., **4**, 1-9, 1991.
[252] Simpson P.,"*Fuzzy min-max neural networks for function approximation*," Proc. IEEE Int. Conf. on Neural Networks, **Vol. III**, pp.1967-1972,1993.
[253] Simpson P.,"*Fuzzy min-max Neural Networks - Part 1: Clustering*," IEEE Trans. on Fuzzy Systems, **1**, 1, 32-45, 1993.
[254] Simpson P.,"*Fuzzy min-max Neural Networks - Part 2: Clustering*," IEEE Trans. on Fuzzy Systems, **1**, 1, 32-45, 1993.
[255] Song K. and Sheen L.,"*Fuzzy-neuro control design for obstacle avoidance of a mobile robot*," Proc. of FUZZ-IEEE/IFES'95, **Vol. I**, pp.71-76,1995.
[256] Steinmuller H. and Wick O.,"*Fuzzy and neurofuzzy applications in European washing machines*," EUFIT'93- 1st European Congress on Fuzzy and Intelligent Technologies in Aachen, pp.1031-1035, 1993.
[257] Suh I.H. and Kim T.W.,"*Fuzzy membership function based neural networks with applications to the visual servoing of robot manipulators*," IEEE Trans. on Fuzzy Systems, **Vol.2**, No.3,pp.203-220, Aug. 1994.
[258] Sun C.-T. and Jang J.S.,"*A neuro-fuzzy classifier and its applications*," Proc. IEEE Internat. Conf. on Fuzzy Systems, pp.94-98,1993.
[259] Srinivasa N. and Ziegert J. C.,"*An Application of Fuzzy ARTMAP Neural Network to Real-Time Learning and Prediction of Time-Variant Machine Tool Error Maps*," Proc. of Third IEEE Int'l Conf. on Fuzzy Systcm, **Vol.3**, pp.1725-1730,1994.
[260] Stephen T.W.,"*Neural network and fuzzy logic applications in C/C++*," -- ,1994.
[261] Sulzberger S.M.,et al.,"*FUN: optimization of fuzzy rule based systems using neural networks*," Proc. IEEE Int. Conf. on Neural Networks 1993,pp.312-316.
[262] Swiniarski R.,et al.,"*Neural-fuzzy variable structure control systems*," Advances in Modeling & Simulation, **25**, 51-64, 1991.
[263] Takagi H. and Hayashi I.,"*Artificial-neural-network driven fuzzy reasoning*," Proc. IIZUKA'88, pp.183-184, 1988.
[264]Takagi H.,"*Fusion technology of fuzzy theory and neural networks-survey and future directions*," First Int'l Conf. on Fuzzy Logic and Neural Networks, pp.13-26,1990.
[265] Takagi H. and Hayashi I.,"*Neural Networks-Driven Fuzzy Reasoning*," Int. Jour. of Approximate Reasoning," pp.191-212, 1991.
[266] Takagi H., et al.,"*Neural networks designed on approximate reasoning architecture and their applications*," IEEE Trans. on Neural Networks, **Vol.3**, No.5, Sept. 1992.
[267] *Takagi H.,"*Application of neural networks and fuzzy logic to consumer*

products," Proc. Int. Conf. on Industrial Fuzzy Electronics, Control Instrumentation, and Automation (IECON'92), **Vol.3**, pp.1629-1639,1992.,
[268] *Takagi H.,"*Cooperative system of neural networks and fuzzy logic and its application to consumer products*," Industrial Applications of Fuzzy Control and Intelligent Systems, Van Nostrand Reinhold, New York,1994.
[269] Takahashi H. and Minami H.,"*Subjective evaluation modeling using fuzzy logic and a neural network*," Proc. 3rd IFSA Congress, pp.520-523,1989.
[270] Tang Z.,et al.,"*An adaptive ULR fuzzy controller through reinforcement learning*," Proc. of FUZZ-IEEE/IFES'95, **Vol. III**, pp.1307-1314,1995.
[271] Tokunaga M.,et al.,"*Learning mechanism and an application of FFS-network reasoning system*," Proc. 2nd Int. Conf. Fuzzy Logic Neural Networks (IIZUKA'92), pp.123-126,1992.
[272] Tokunaga M. and Ichihashi H.,"*Backer-Upper control of a trailer truck by neuro-fuzzy optimal control*," Proc. of 8th Fuzzy System Symposium, pp.49-52,1992.
[273] Trebi-Ollennu A.,Stacey B.A. and White B.A.,"*A direct adaptive fuzzy SMC, a case study of ROV*," Proc. of FUZZ-IEEE/IFES'95, Vol. IV,pp.1979-1986,1995.
[274] Uebele U., et al.,"*A neural-network-based fuzzy classifier*," IEEE Trans. on SC, **Vol.25**, No.2, pp.353-361,Feb. 1995.
[275] Uehara K. and Fujise M.,"*Learning of fuzzy-inference criteria with artificial neural network*," Proc. IIZUKA'90, **Vol.1**, pp.193-198, 1990.
[276] Ullah Z. and Khan N.,"*Development of a fast charger for secondary batteries by using nuefuz neural-fuzzy technology*," Computer Design Fuzzy Logic '94 Conf. in San Diego, pp.79-84, 1994.
[277] Ullah Z.,Neely W.S. and Khan E.,"*A microcontroller implementation of a neufuztm control system*," Proc. of FUZZ-IEEE/IFES'95, **Vol. IV**, pp.1785-1792,1995.
[278] Unal F.A. and Kham E.,"*A Fuzzy Finite State Machine Implementation Based on a Neural Fuzzy System*," Proc. of Third IEEE Int'l Conf. on Fuzzy System, **Vol.3**, pp1749-1754,1994.
[279] Vachkov G. and Nikolov M.,"*Successive fuzzy rule based tuning of industrial control systems*," Proc. of FUZZ-IEEE/IFES'95, **Vol. IV**, pp.2173-2178,1995.
[280] Vasilakos A.V. and Zikidis K.C.,"*A.S.A.F.ES, 2: a novel, neuro-fuzzy architecture for fuzzy computing, based on functional reasoning*,"Proc. of FUZZ-IEEE/IFES'95, **Vol. II**, pp.671-678, 1995.
[281] Von Altrock C.,"*NeuroFuzzy technologies*," Computer Design Magazine 6/94, pp.82-83,1994.
[282] Von Altrock C.,"*Enhanced fuzzy systems using data analysis techniques and neural networks*," Computer Design Fuzzy Logic Conf., San Diego, 1994.
[283] Wakami N.,"*Engineering application of fuzzy systems- fuzzy control and*

neural networks: applications for (Matsushita) home appliances," Joint Janpanese-European Symp. on Fuzzy Systems in Berlin, Vol. 8, 1994.
[284] Wang C.-H., et al.,"*Fuzzy B-spline membership function (BMF) and its applications in fuzzy-neural control,*" IEEE Trans. on SMC, **Vol.25**, No.5, pp.841-851, May 1994.
[285] Wang D. and Chai T.,"*A self-tuning fuzzy controller via fuzzy inference network,*" Proc. of FUZZ-IEEE/IFES'95, **Vol. IV**, pp.2201-2206,1995.
[286] Wang L.-X. and Mendel J.M.,"Backpropagation fuzzy system as nonlinear dynamic system identifiers," --,pp.1409-1418,1992.
[287] *Wang L.-X.,"*Adaptive Fuzzy Systems and Control,*" Prentice-Hall, Englewood Cliffs, N.J., 1994.
[288] Wang P.-Z. and Loe K.-F,"*Between mind and computer - fuzzy science and engineering,*" World Scientific, 1993.
[289] *Werbos P.J.,"*Beyond regression: new tools for prediction and analysis in the behavioral sciences,*" Ph.D. thesis, Harvard Univ., Nov., 1974.
[290] Werbos P.J.,"*Neurocontrol and fuzzy logic: connections and designs,*" Int. J. Approx. Reasoning **6**(1992) 185-219.
[291] White D.A. and Sofge D.A.(Eds.),"*Handbook of Intelligent Control. Neural, Fuzzy, and Adaptive Approaches,*" Van Nostrand Reinhold, New York, 1992.
[292] Wohlke G., "*A Neuro-Fuzzy Based System Architecture for the Intelligent Control of Multi-Finger Robot Hands,*" Proc. of Third IEEE Int'l Conf. on Fuzzy System, **Vol.1**, pp.64-69, 1994.
[293] Wong F.,et al.,"*A stock selection strategy using fuzzy neural networks,*" Computer Science in Economics and Management, **4**, 77-89, 1991.
[294] Wu C.J., Sung A.H. and Soliman H.S.,"*An ART Network with Fuzzy Control for Image Data Compression,*" Proc. of Third IEEE Int'l Conf. on Fuzzy System, **Vol.3**, pp1743-1748, 1994.
[295] Yager R.R.,"*On the interface of fuzzy sets and neural networks,*" Int. Workshop on Fuzzy System Applications (IIZUKA'88), pp.215-216,1988.
[296] Yager R.R,"*Implementing fuzzy logic controllers using a neural network framework,*" Fuzzy Sets and Systems, **Vol.48**, pp.53-64,1992.
[297] Yager R.R. and Zadeh L.A.,"*Fuzzy sets, neural networks, and soft computing,*" Van Nostrand Reinhold, 1994.
[298] *Yamakawa T.,"*A membership function circuit and its applications to a singleton-consequent fuzzy logic controller and a fuzzy neuron,*" 5th Fuzzy Syst. Symp., pp.13-18, 1989.
[299] *Yamakawa T. and Tomoda S.,"*A fuzzy neuron and its application to pattern recognition,*" Proc.Third Int. Fuzzy System Associat. Congress, pp.30-38, 1989
[300] *Yamakawa T.,"*Pattern recognition hardware system employing a fuzzy neuron,*" Proc. Int. Conf. Fuzzy Logic, pp.934-938, 1990.

[301] *Yamakawa T.,et al.,"*A neo fuzzy neuron and its application to fuzzy system identification and prediction of the system behavior*," Proc. 2nd Int'l Conf. on Fuzzy Logic and Neural Networks (IIZKA'92), pp.477-483,1992.
[302] Yamakawa T. and Furukawa M.,"*A design algorithm of membership functions for a fuzzy neuron using example-based learning*," Proc. FUZZ-IEEE'92, pp.943-948,1992.
[303] Yamakawa T.,"*A fuzzy inference engine in nonlinear analog mode and its application to a fuzzy logic control*," IEEE Trans. on Neural Networks **4**(3), pp.496-522, May 1993.
[304] Yao C. and Kuo Y.,"*A fuzzy neural network model with three-layered structure*," Proc. of FUZZ-IEEE/IFES'95, **Vol. III**, pp. 1503-1511,1995.
[305] Yeh Z. and Chen H.,"*A self-organizing decentralized fuzzy neural net controller*," Proc. of FUZZ-IEEE/IFES'95, **Vol. IV**, pp.2179-2186,1995.
[306] Yen J., Wang H. and Daugherty W.C.,"*Design issues of a reinforcement-based self-learning fuzzy controller for petrochemical process control*," Proc. Workshop of NAFIPS (NAFIPS92), pp.135-142,1992.
[307] *Zadeh L.A.,"*Fuzzy Logic, Neural Networks, and Soft Computing*," Communications of the ACM, **Vol.37**, No. 3, pp.77-84, March 1994.
[308] Zhang D., Kamel M. and Elmasry M.I.,"*Fuzzy Clustering Neural Network (FCNN) Using Fuzzy Competitive Learning*," WCNN'93, Portland II, pp.22-25, 1993.
[309] Zhang D., Kamel M. and Elmasry M.I.,"*Mapping Fuzzy Clustering Neural Networks onto Systolic Arrays*," Proc. of Third IEEE Int'l Conf. on Fuzzy System, **Vol.1**, pp.218-222,1994.
[310] *Zhang Y.-Q. and Kandel A.,"*Genetic-Guided Compensatory Neurofuzzy Systems*," IPMU Congress'96, **Vol.I**, pp.181-185, 1996.
[311] *Zhang Y.-Q., Kandel A. and Friedman M.,"*Hybrid decision-making system for fuzzy moves*," Proc. of FUZZ-IEEE/IFES'95, **Vol. II**, pp.621-626,1995.
[312] *Zhang Y.-Q. and Kandel A., "*Compensatory Neurofuzzy Systems with Fast Learning Algorithms*," IEEE Trans. on Neural Networks 9 (1), pp.83-105, January 1998.
[313] Zimmermann H.-J. and Zysno P.,"*Latent connective in human decision making*," Fuzzy Sets and Systems, **Vol.4**, pp.37-51, 1980.
[314] Zadeh L.A.,"*Fuzzy sets and information granularity*," in: M.M. Gupta, R.K. Ragade, R.R. Yager, eds., Advances in Fuzzy Set Theory and Applications, North Holland, Amsterdam, pp.3-18, 1979.

10 NEURAL NETWORKS AND FUZZY LOGIC

Nadipuram (Ram) R. Prasad

The Klipsch Scool of Electrical and Computer Engineering
New Mexico State University
Las Cruces, NM 88003-8001
rprasad@nmsu.edu

10.1 INTRODUCTION

In Chapter 9 (this volume), Pedrycz, Kandel and Zhang present an extensive survey of Neurofuzzy systems. The authors give a clear exposition of the complementary technologies offered by fuzzy logic and neural networks. The basic functions of a "fuzzy neuron" are developed and learning algorithms for neuro-fuzzy systems are presented. In this chapter, we briefly discuss the synergistic issues in the integration of these technologies. The development and results of an integrated system are presented.

While conventional approaches to control system development have been very successful, the benefits of such approaches in the sense of "intelligent" control cannot be evaluated. That is to say that the control actions derived cannot be described in a qualitative manner for humans to understand. The cause and effect relationships underlying the process input/output behavior are masked. In addition, there is no direct means for generalizing the application of classical control systems for large parameter variations in the process itself. The development of adaptive capabilities present a formidable task for complex systems. Another issue might be the scaling-up of process variables that result from prototypical bench-top design to the actual commercialization (i.e., large-scale implementation).

Fuzzy logic provides a means for the qualitative description of process variables which subsequently allows the development of qualitative associations between such variables. This is very useful if one has to clearly understand the input-output behavior of the model/process so that logical control actions can always be assured. In other words, the control actions needed to steer the process towards the specified goals will always be the correct actions. As a result, one may conclude that there is a sense of "robustness" that is inherently guaranteed. Of course, robustness includes many factors among which the degree of control reliability, control optimality, (i.e., the best control actions), and system stability are the key issues. Stability of fuzzy logic-based systems is an issue of current research. The degree of reliability, however, may be argued to be very high in fuzzy logic-based approaches since the rules are intended to cover all possible input/output conditions. A shortcoming of fuzzy logic-based systems is that there is an inherent lack of "memory".

Neural networks have found a place in pattern recognition applications. They can be trained and are capable of classifying input variables into several classes. Where there is substantial statistical deviation between the training patterns and test patterns misclassifications among the test patterns can occur. At this point it would be wise to retrain the neural network. Generally, the training set is supposed to capture the expected variations in the process variables. The number of training patterns is a critical issue. There has been significant research in determining the effects of sample size on classifier performance [1-3]. Research has shown that classifier performance may decrease with increased sample sizes. This is known as "peaking phenomenon" where the classifier performance degrades with increased sample size. The notion of optimal size does not exist and hence is a matter of trial and error. A neural network however, is capable of generalizing (i.e., the outcomes can reflect a very broad range of variability compared to the outcomes for which it was originally trained for). Of course, with its capability to classify, the inherent property of neural networks is that it has memory.

Based upon the aforementioned properties of fuzzy logic and neural networks one may conclude that an integration of the two could provide a reasonable means for developing "intelligent" or "smart" systems. Integrated systems can be expected to *learn* and *adapt*. They learn new associations, new patterns, and new functional dependencies. The emergence of such integrated systems have given rise to two basic types of architectures, namely, Neural-Fuzzy Systems and Fuzzy-Neural Systems.

In *neural-fuzzy systems*, neural networks provide fuzzy systems with automatic tuning methods without altering their functionality. The neural network serves in augmenting numerical processing of fuzzy sets, such as the elicitation of membership functions and realization of mappings between fuzzy sets that are utilized as fuzzy rules. Since neural-fuzzy systems preserve the basic structure of fuzzy logic based systems (i.e., fuzzification, the use of fuzzy rules, and defuzzification), they are most commonly used for control applications.

In *fuzzy-neural networks*, the basic properties and architecture of neural networks are preserved while the elements are fuzzified. For example, a crisp neuron can become fuzzy and the response of the neuron to its lower layer activation signal can be of a fuzzy relation type rather than a sigmoid type. Since the basic architecture is preserved, fuzzy-neural systems are mostly used for pattern recognition applications.

A very comprehensive treatment of the synergism in neuro-fuzzy systems is provided in [4]. The architectures and learning rules of adaptive networks are discussed in great detail in [5].

In the literature, fuzzy-neural networks have been classified into three types. The first type of network (FNN1) has crisp inputs to the neural network but the weights between neurons are fuzzy sets. The second type of network (FNN2) has fuzzy inputs but the weights between neurons are crisp. Finally, the third type of network (FNN3) has both fuzzy inputs as well as fuzzy weights. In the previous chapter, the authors provide a very comprehensive listing of references on fuzzy-neural networks. Of particular interest in this chapter is the work performed by Ishibuchi, et al. [6,7,8,9]. In the following sections, the development and comparison of fuzzy-neural controllers of the FNN2 and FNN3 types for a liquid level control problem are presented.

10.2 LIQUID LEVEL CONTROL PROBLEM

Figure 10.1 illustrates the basic scheme for liquid level control in tank. In this figure, valves 1 and 2 control the input and output flow rates of the tank, respectively. These flow rates are measured using two separate flow-rate sensors. A pressure sensor is mounted at the bottom of the tank to measure the head of liquid in the tank. Typically the relationship would be nonlinear as illustrated in Figure 10.2. However, this characteristic would be implicitly accounted for in the linguistic description of fluid head and the manner in which it can be associated with flow characteristics.

Referring to Figure 10.1, we can describe the operation of the liquid level control in the following manner. It is desired that the liquid level be maintained at a desired head. To accomplish this task, we require that the error between the desired head and the actual head equal zero, and the differential flow rate (i.e., the difference between the input and output flow rates) also be equal to zero. At this point, we note that there is no restriction on the flow rates at the input or output valves. This means that a complete shut down on both valves can occur immediately when the error in head reaches zero. This case may not

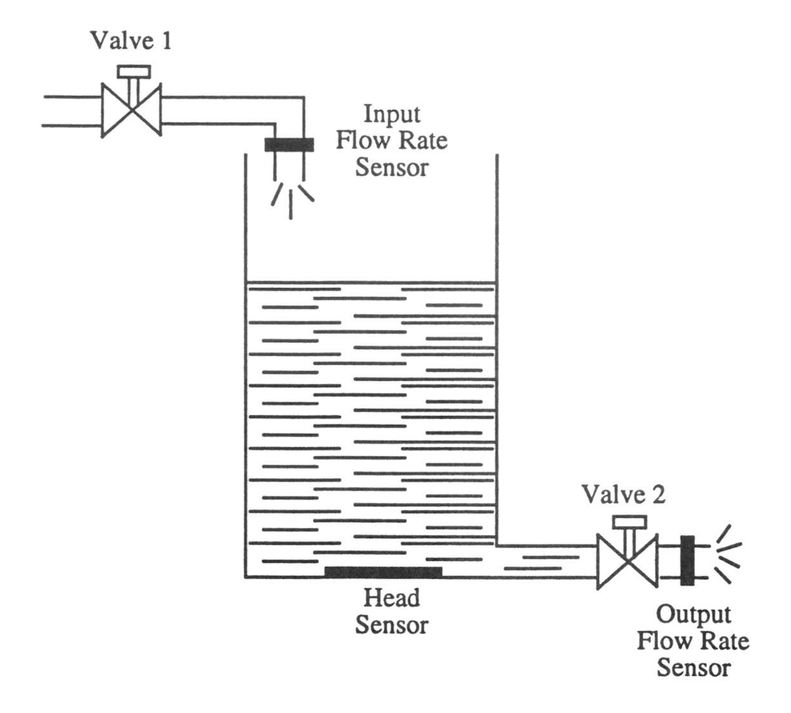

Figure 10.1 Liquid Level Control Problem.

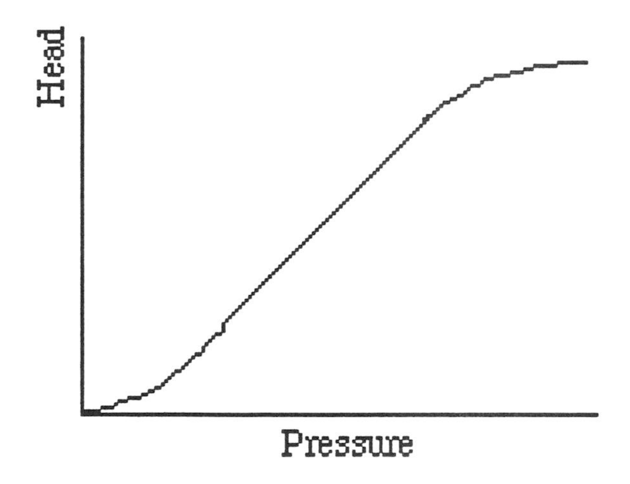

Figure 10.2 Nonlinear Characteristic of the Pressure Sensor.

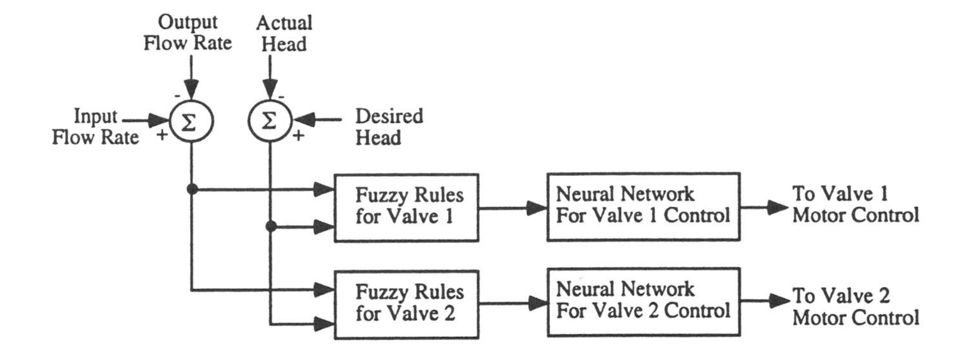

Figure 10.3 Fuzzy-Neural Architecture.

be of general interest. Such a condition could be avoided if a reference is set on the output valve to maintain a desired flow rate. We will consider both cases and develop the appropriate control strategies.

Figure 10.3 illustrates an overview of the fuzzy-neural system architecture that may be employed. In this scheme the fundamental goal is to employ the neural network to recognize the patterns of flow rates and the liquid level in the tank, and actuate the motor control to obtain appropriate valve control actions.

There are several steps needed before such a control scheme can be implemented. These include:

- The development of a set of fuzzy rules for valves 1 and 2.
- Training the neural network.
- Simulation of the control scheme.

We will address each of the above systematically.

10.3 FUZZY RULE DEVELOPMENT

The first step in fuzzy rule development is to identify the input and output variables and define appropriate subsets in terms of membership functions. The input variables are the differential flow rate and difference in head. The outputs of the fuzzy rules are the changes required in valve 10.1 and valve 2 positions. These are depicted in Figure 10.4.

Based upon the linguistic set definitions of the input/output variables, the fuzzy associative memory for the appropriate valve control actions may then be derived as shown in Figure 10.5.

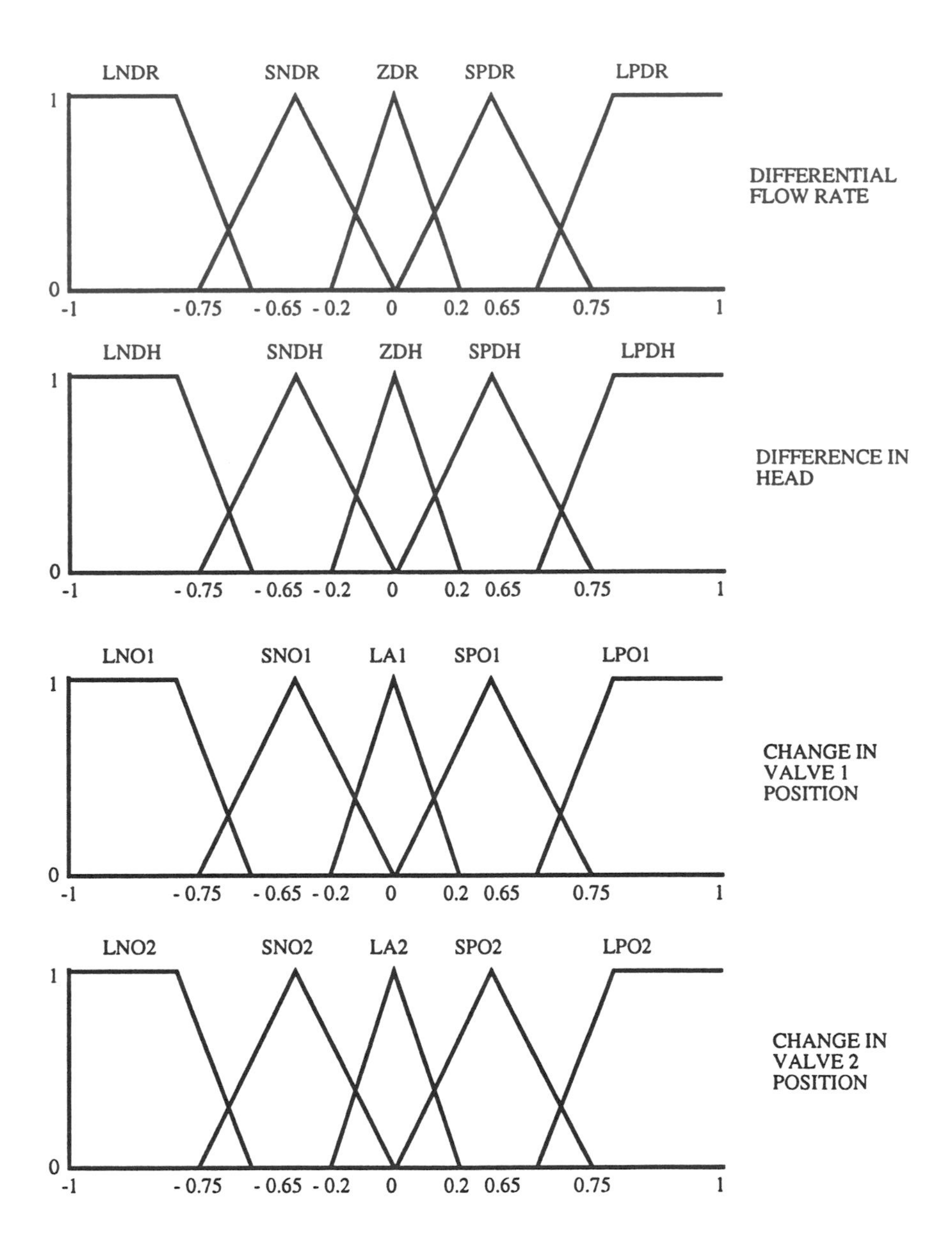

Figure 10.4 Set and Subset Definitions of Input and Output Variables.

Fuzzy Associative Memory For Valve 1 And Valve 2

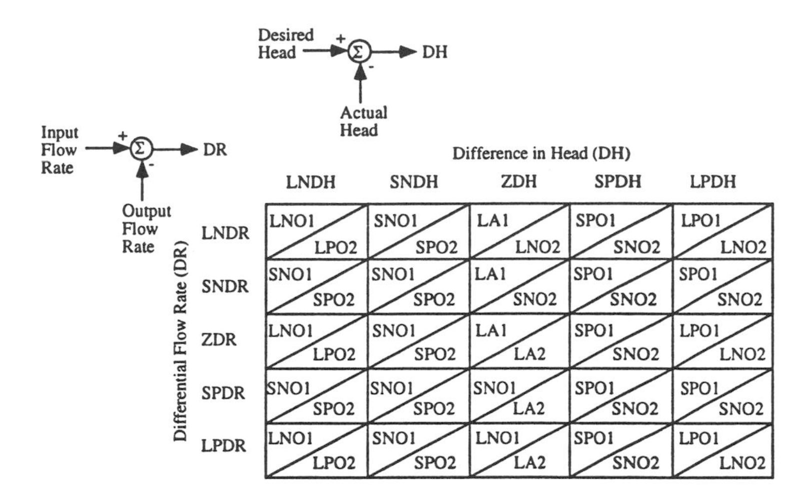

Figure 10.5 FAM for the Liquid Level Controller.

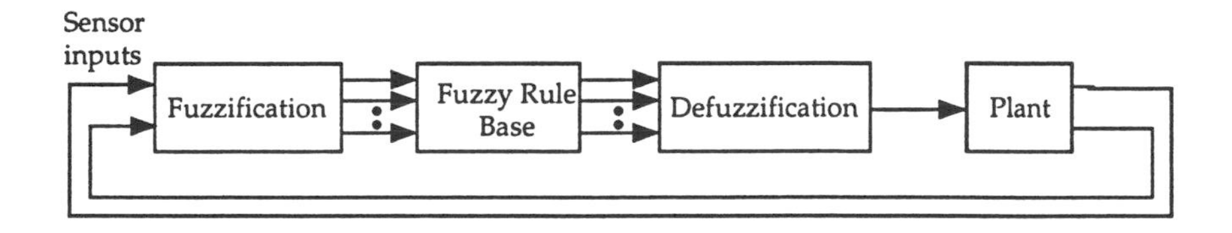

Figure 10.6 Classical Fuzzy Logic-Based Control.

In conventional fuzzy logic-based systems, the rule base is obtained by formulating the If-Then rules from the fuzzy associative memory and with an appropriate choice of a defuzzification method a system controller can be implemented. This is illustrated conceptually in Figure 10.6. The fundamental problem with contollers of the type shown in Figure 10.6 is that they lack the ability to learn and adapt. In this context, emerging technologies in neuro-fuzzy systems play a significant role in the development of integrated systems that exhibit human-like control and inference mechanisms.

In the following discussion, we provide a detailed analysis of the liquid level control problem with the goal of obtaining a "complete" set of pseudo-rules.

10.3.1 Case 1: No Restrictions on Flow Rates

Assuming that the difference in head is zero, zero differential flow rates occur when the input and output flow rates are the same. Hence an infinite number of conditions exist that satisfy this differential flow rate including zero flow rate when both valves are in the fully closed position. The control actions listed in the fuzzy associative memory of Figure 10.5 will result in the above conditions if no reference is placed on the input/output valves.

10.3.2 Case 2: Input/Output Flow Rate Control

If there are restrictions placed on flow rates in terms of desired flow rates we need to investigate all possible control actions that one could place on the valve operation. This would require one set of rules for controlling the output flow rate and another set of rules to control the input flow rate.

10.3.3 Pseudo-Rules

Suppose that a constant output flow rate is desired. Note that for a fixed valve position, the flow rate can vary depending upon the head of liquid in the tank. Hence, we can obtain a set of pseudo-rules that can aid in the controller development. These rules may be listed as follows for the output flow valve:

Rule 1: **If** the head is higher than desired *and*
the output flow rate is higher than desired
Then turn the output valve in the fully closed direction.

Rule 2: **If** the head is higher than desired *and*
the output flow rate is lower than desired
Then turn the output valve in the fully open direction.

Rule 3: **If** the head is lower than desired *and*
the output flow rate is higher than desired
Then turn the output valve in the fully closed direction.

Rule 4: **If** the head is lower than desired *and*
the output flow rate is lower than desired
Then turn the output valve in the fully open direction.

The above rules, especially Rule 4, suggest the implementation of valve control for the input flow rate adjustment also. Hence the following set of pseudo-rules may be written for the input flow conditions.

Rule I: **If** the head is higher than desired **Then** turn the input valve in the fully closed direction.

Rule II: **If** the head is lower than desired **Then** turn the input valve in the fully open direction.

Note that Rules I and II can be merged into a new set of Rules N1 - N4 to form a multi-input-multi-output system. The set of pseudo-rules therefore can be restated in compact form as follows:

Rule N1: **If** the head is higher than desired *and*
the output flow rate is higher than desired
Then turn the output valve in the fully closed direction *and*
turn the input valve in the fully closed direction.

Rule N2: **If** the head is higher than desired *and*
the output flow rate is lower than desired
Then turn the output valve in the fully open direction *and*
turn the input valve in the fully closed direction.

Rule N3: **If** the head is lower than desired *and*
the output flow rate is higher than desired
Then turn the output valve in the fully closed direction *and*
turn the input valve in the fully open direction.

Rule N4: **If** the head is lower than desired *and*
the output flow rate is lower than desired
Then turn the output valve in the fully open direction *and*
turn the input valve in the fully open direction.

In addition to the above set of rules, which take into account the abnormal conditions resulting in decreased/increased head, we recognize that rules are needed to maintain the valve positions at their respective openings during equilibrium conditions. Such is the case when the output flow rate and the head are at their desired values. Failure to include these rules can result in an oscillatory response and hence a marginally stable system. These rules are as follows:

Rule N5: **If** the head is higher than desired *and*
the output flow rate is at the desired value
Then leave the output valve at its current position
and turn the input valve in the fully closed direction.

Rule N5: **If** the head is lower than desired *and*
the output flow rate is at the desired value
Then leave the output valve at its current position
and turn the input valve in the fully open direction.

Rule N5: **If** the head is at the desired value *and* the output flow rate is at the desired value **Then** leave the output valve at its current position *and* leave the input valve at its current position.

The pseudo-rules, namely Rule N1 - Rule N7, form the basis for developing an integrated neurofuzzy system.

10.4 INTEGRATED SYSTEM ARCHITECTURES

In this section we present two fuzzy-neural architectures, namely the FNN2 and FNN3 integrated systems. Recall that in the FNN2 architecture the input to the neural network are fuzzy sets while the weights of the neural network are crisp. Whereas, in the FNN3 architecture both the input as well as the weights are fuzzy sets. In order to train the neural network, we need appropriate targets. In the FNN2 type architecture it would suffice to have bipolar target function to allow the classification of the input pattern in terms of whether to move the valves in the open direction, close direction, or to leave the valve in its current position. In contrast, for the FNN3 type architecture it is necessary to include the target as fuzzy sets to allow the network to provide a degree of opening/closing of the input/output valves. This is necessary for the training algorithm described in Section 10.5 of this chapter.

Table 10.1 provides an input/output truth table using both fuzzy and crisp bipolar training targets.

Table 10.1 Input/Output Truth Table

Head	**Output Flow Rate**	**Output Valve**	**Input Valve**
High (1)	High (1)	Open More (1)	Close More (-1)
High (1)	Low (-1)	Open More (1)	Close More (-1)
Low (-1)	High (1)	Close More (-1)	Open More (1)
Low (-1)	Low (-1)	Close More (-1)	Open More (1)
High (1)	Normal (1)	Leave Alone(0)	Close More (-1)
Low (-1)	Normal (-1)	Leave Alone(0)	Open More (1)
Normal (-1)	Normal (-1)	Leave Alone(0)	Leave Alone(0)

Note that a bipolar activation function is used to provide the capability to leave the valves alone at their respective openings in the event that the head and output flow rate is maintained at the desired values. This "leave alone" condition is generated when the activation function produces a zero output.

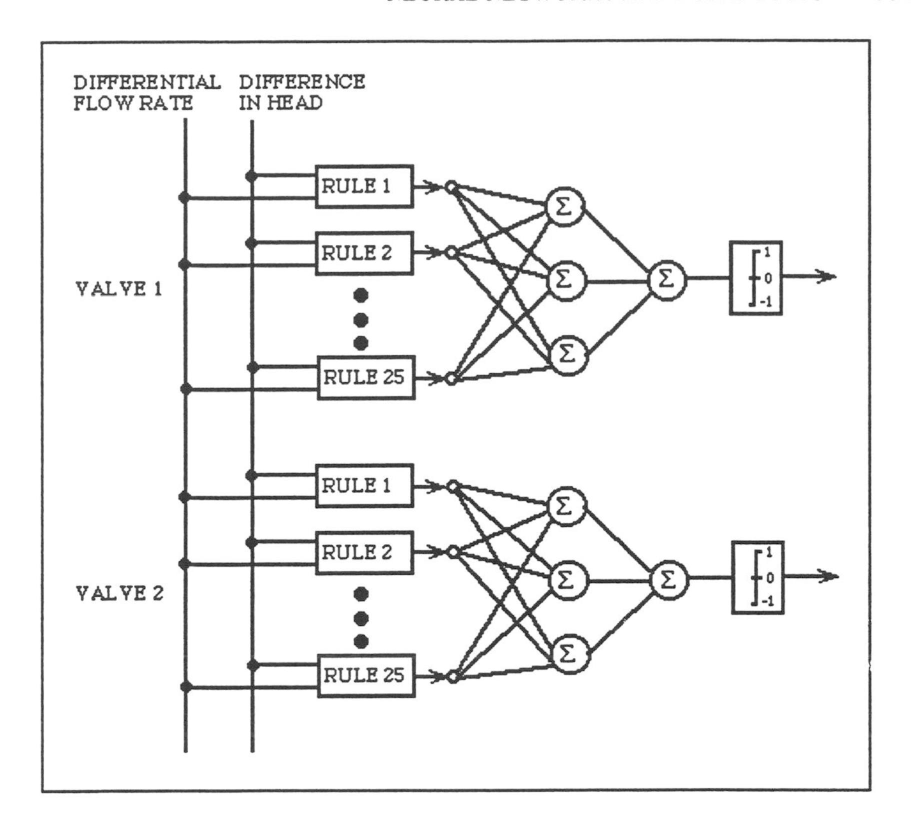

Figure 10.7 FNN2 Valve Control Scheme.

The fuzzy output targets are defined in terms of triangular sets. They can be either symmetrical or nonsymmetrical.

10.4.1 FNN2 Architecture

In this architecture, the input vector to the neural network is a fuzzy set representing the fuzzification of the sensor measurements. Figure 10.7 illustrates the schematic of the FNN2 fuzzy-neural system.

The set associations in the fuzzy rules provide the desired training vectors for the neural network. Figure 10.8 illustrates the control surface generated using the max-min compositional form on the fuzzy rule base.

Figure 10.9 illustrates the prototypical membership functions for Valve 1 and the corresponding set of training vectors. In this example, P1 - P35 represent a set of 35 training vectors used to train the neural network. They represent a

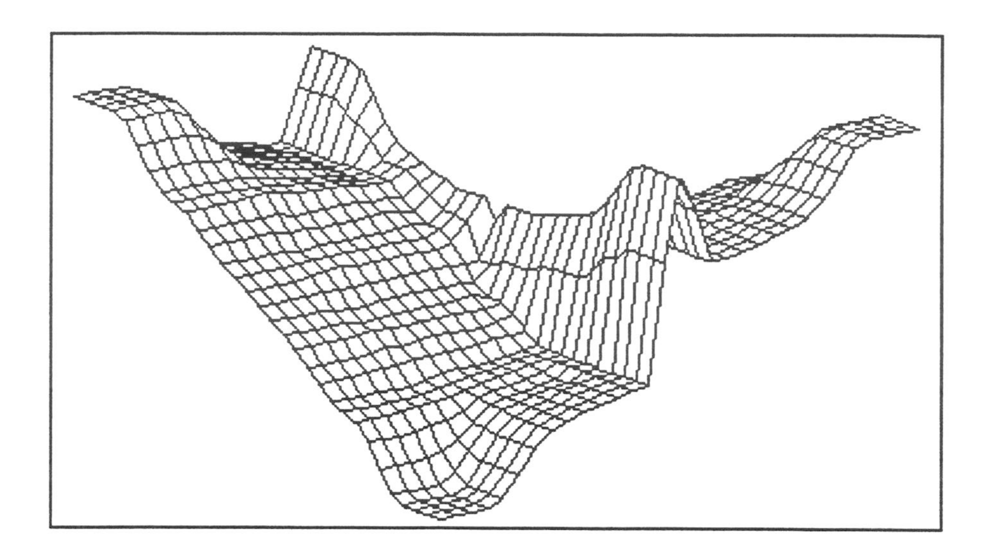

Figure 10.8 Fuzzy Control Surface for FNN2 Architecture.

sliding range of values indicative of the fully open to fully closed positions of each valve. Training was performed using a fuzzy back propagation algorithm, discussed in [10] with an error tolerance of 10^{-6}.

10.5 FNN3 TRAINING ALGORITHM

In general, the training of neural networks are based upon numerical training data. However, in many situations that require learning of expert knowledge, the training data may only be available in qualitative form such as, "IF x_1 is Large AND x_2 is Small, THEN y is Large". In this section, an algorithm for learning expert system knowledge in the form of fuzzy IF-THEN rules is presented.

Suppose the following K fuzzy IF-THEN rules are specified:

$$\text{IF } x_1 \text{ is } A_{p1} \text{ AND } \cdots \text{ AND } x_n \text{ is } A_{pn}, \text{ THEN } y \text{ is } C_p,$$

where $p = 1, 2, \ldots, K$, and A_{pi} and C_p are fuzzy numbers. These fuzzy rules represent fuzzy input-output pairs (A_p, C_p), $p = 1, 2, \ldots, K$, where $A_p = (A_{p1}, A_{p2}, \ldots, A_{pn})$ is a fuzzy vector. (A_p, C_p), $p = 1, 2, \ldots, K$, therefore represents a general form of training data which can be fuzzy, numerical, or a combination of both.

The method for training a fuzzy-neural system with fuzzy weights and fuzzy biases was proposed by Ishibuch et al. [8]. The training algorithm is used

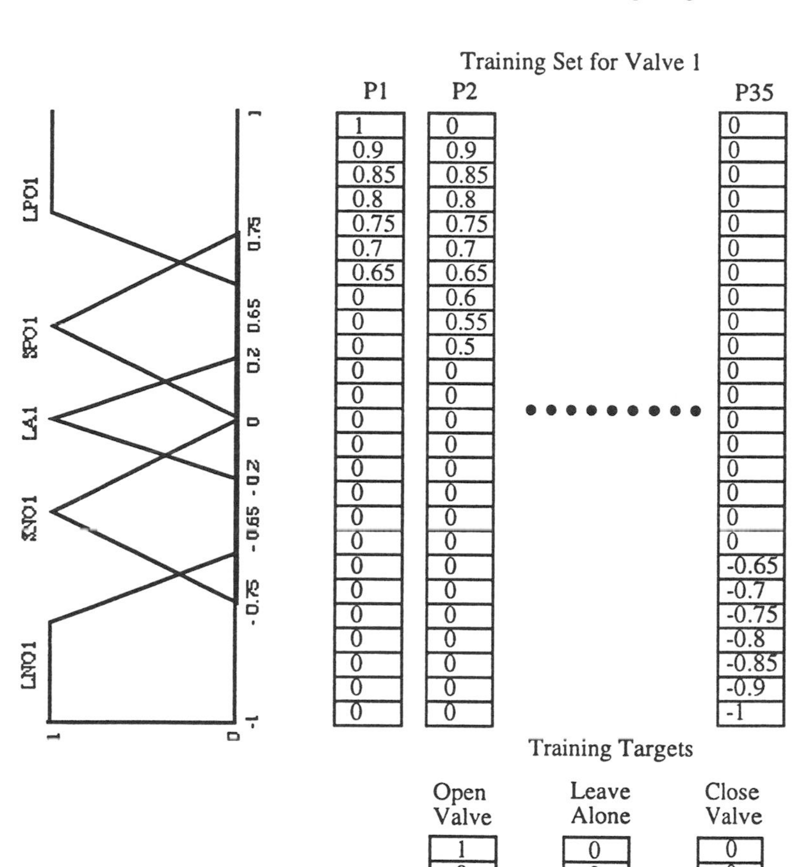

P1	P2	P35
1	0	0
0.9	0.9	0
0.85	0.85	0
0.8	0.8	0
0.75	0.75	0
0.7	0.7	0
0.65	0.65	0
0	0.6	0
0	0.55	0
0	0.5	0
0	0	0
0	0	0
0	0	0
0	0	0
0	0	0
0	0	0
0	0	0
0	0	0
0	0	-0.65
0	0	-0.7
0	0	-0.75
0	0	-0.8
0	0	-0.85
0	0	-0.9
0	0	-1

Open Valve	Leave Alone	Close Valve
1	0	0
0	0	0
0	0	-1

Figure 10.9 Training Set for FNN2 Architecture.

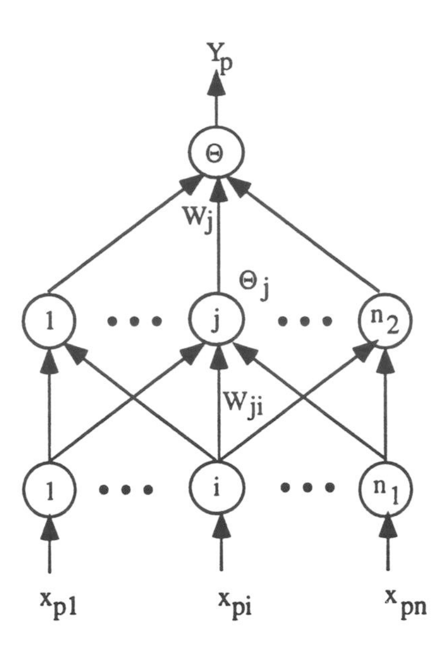

Figure 10.10 FNN3 Architecture.

for fuzzy sets in which symmetric triangular fuzzy numbers are used for fuzzy weights and fuzzy biases. The proposed architecture is illustrated in Figure 10.10.

Referring to Figure 10.10, the fuzzy weights and fuzzy biases are defined as:

$$W_j = \left(w_j^L, w_j^C, w_j^U\right), \qquad W_{ji} = \left(w_{ji}^L, w_{ji}^C, w_{ji}^U\right), \tag{10.1}$$

$$\Theta = (\theta^L, \theta^C, \theta^U), \qquad \Theta_j = (\theta_j^L, \theta_j^C, \theta_j^U), \tag{10.2}$$

where a triangular fuzzy number is denoted by the lower limit (L), the center (C), and the upper limit (U). The following relations hold because of the use of symmetric triangular fuzzy numbers, namely:

$$w_j^C = \left(w_j^L + w_j^U\right)/2, \qquad W_{ji}^C = \left(w_{ji}^L + w_{ji}^U\right)/2, \tag{10.3}$$

$$\theta^C = (\theta^L + \theta^U)/e, \qquad \Theta_j^C = (\theta_j^L + \theta_j^U)/2, \tag{10.4}$$

and the h-level sets of fuzzy weights and biases can be calculated as:

$$[W_j]_h^L = w_j^L(1 - h/2) + w_j^U h/2, \quad [W_j]_h^U = w_j^U(1 - h/2) + w_j^L h/2, \tag{10.5}$$

$$[W_{ji}]_h^L = w_{ji}^L(1 - h/2) + w_{ji}^U h/2, \quad [W_{ji}]_h^U = w_{ji}^U(1 - h/2) + w_{ji}^L h/2. \tag{10.6}$$

For an n-dimensional vector of real inputs $x_p = (x_{p1}, x_{p2}, ..., x_{pn})^T$ the input-output relations at each node of the network in Figure 10.10 can be expressed in terms of an h-level set as follows:

Input nodes:

$$y_{pi} = x_{pi}, \qquad i = 1, 2, \ldots, n. \tag{10.7}$$

Hidden nodes:

$$[Y_{pj}]_h = [f(\mathrm{Net}_{pj})]_h = f([\mathrm{Net}_{pj}]_h), \quad j = 1, 2, \ldots, n_2, \tag{10.8}$$

$$[\mathrm{Net}_{pj}]_h = \sum_{i=1}^{n} [W_{ji}]_h y_{pi} + [\Theta_j]_h, \quad j = 1, 2, \ldots, n_2. \tag{10.9}$$

Output nodes:

$$[Y_p]_h = [f(\mathrm{Net}_p)]_h = f([\mathrm{Net}_p]_h), \tag{10.10}$$

$$[\mathrm{Net}_p]_h = \sum_{j=1}^{n_2} [W_j]_h [Y_{pj}]_h + [\Theta]_h. \tag{10.11}$$

We assume now that K crisp-input and fuzzy-output pairs (x_p, T_p), $p = 1, 2, \ldots, K$, are given. The target outputs T_p can be symmetric or nonsymmetric triangular sets specified as $T_p = (t_p^L, t_p^C, t_p^U)$. The cost functions for the h-level sets can be defined as:

$$E_{ph} = ([T_p]_h^L - [Y_p]_h^L)^2/2 + ([T_p]_h^U - [Y_p]_h^U)^2/2 \tag{10.12}$$

and

$$E_p = \sum_h E_{ph}, \tag{10.13}$$

where, $[Y_p]_h = [[Y_p]_h^L, [Y_p]_h^U]$ and $[T_p]_h = [[T_p]_h^L, [T_p]_h^U]$ are the intervals. The learning rules which constitute the incremental changes for the fuzzy weights W_j and fuzzy biases W_{ji} can then be written as:

$$\Delta w_j^L(t+1) = -\eta h(\partial E_{ph}/\partial w_j^L) + \alpha \Delta w_j^L(t), \tag{10.14}$$

$$\Delta w_j^U(t+1) = -\eta h(\partial E_{ph}/\partial w_j^U) + \alpha \Delta w_j^U(t), \tag{10.15}$$

$$\Delta w_{ji}^L(t+1) = -\eta h(\partial E_{ph}/\partial w_{ji}^L) + \alpha \Delta w_{ji}^L(t), \tag{10.16}$$

$$\Delta w_{ji}^U(t+1) = -\eta h(\partial E_{ph}/\partial w_{ji}^U) + \alpha \Delta w_{ji}^U(t), \tag{10.17}$$

where, η and α are the learning rate and momentum values, and the partial derivatives are obtained using the cost function. Note that the use of a fuzzy back-propagation algorithm results in variations in the learning and momentum values and are based upon the rate of error convergence.

In order to avoid situations when the lower limits of the fuzzy weights are higher than the upper limits, the new fuzzy weights obtained are as follows:

$$w_j^L = \min\left\{w_j^L(t+1), w_j^U(t+1)\right\}, \quad (10.18)$$
$$w_j^U = \max\left\{w_j^L(t+1), w_j^U(t+1)\right\}, \quad (10.19)$$
$$w_{ji}^L = \min\left\{w_{ji}^L(t+1), w_{ji}^U(t+1)\right\}, \quad (10.20)$$
$$w_{ji}^U = \max\left\{w_{ji}^L(t+1), w_{ji}^U(t+1)\right\}. \quad (10.21)$$

The centers of the triangular functions, namely, w_j^C and w_{ji}^C, are then determined according to (10.3). Further details of the algorithm including example calculations and an extension using a fuzzy cost function are discussed in [5].

In [13], another approach is proposed by Hayashi et al. in which the back-propagation algorithm is directly fuzzified based on a fuzzy-valued cost function. This has the advantage of producing a better approximation of the target outputs and for any new inputs to the system.

10.5.1 FNN2 Simulation Results

The mathematical model of the tank was represented by a simple first-order lag with initial conditions chosen such that the input/output flow rates were equal and the difference in head of fluid (i.e., desired-actual) was zero. For a new set-point on the output valve, indicating an increased output flow requirement, both the input and output valves open as expected. This is illustrated in Figure 10.11 in which Valve 2 opening is shown. The corresponding neural network output is shown in Figure 10.12 where the "Open Command" signal is generated for the increased flow requirement.

Upon reaching a steady-state condition, a new and completely arbitrary output flow rate was set. While the neural network does recognize the decreased flow requirement, the valve motion indicates a non-convergent result. This condition is somewhat as expected because of the inability of the neural network to generalize over "all" operating conditions. Infact, the response is chaotic.

10.5.2 FNN3 Architecture

In this architecture, the neural network is composed of five functional layers with the first layer corresponding to the input layer. The second layer is the fuzzification layer where the inputs from the sensors are fuzzified in terms of their linguistic variables. The third layer is the fuzzy reasoning layer which consists of fuzzy neurons that perform logical AND/OR operations. The fourth layer is the defuzzification layer. Finally, the fifth layer is the output layer. Using the learning algorithm suggested in [8] and described in Section 10.4.2

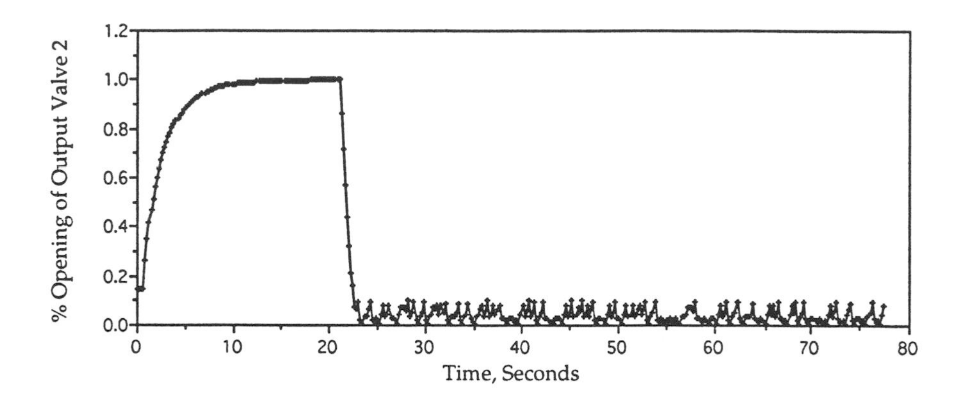

Figure 10.11 FNN2 Simulation of Valve 2 Opening for Variation in Output Flow Rate.

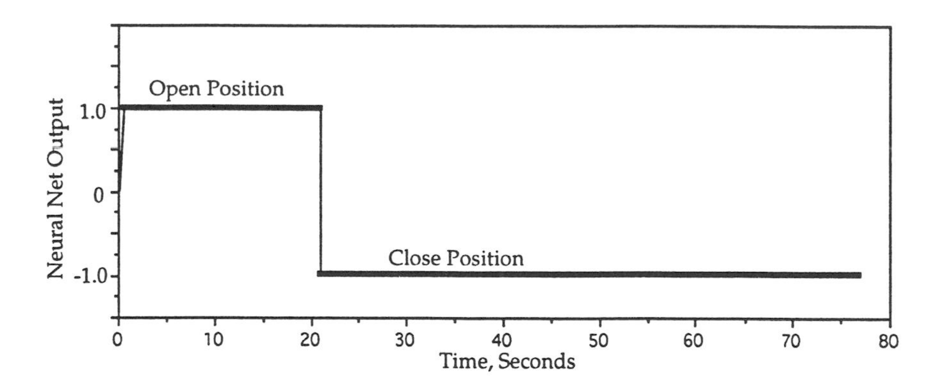

Figure 10.12 Neural Network Output for Variation in Output Flow Rate.

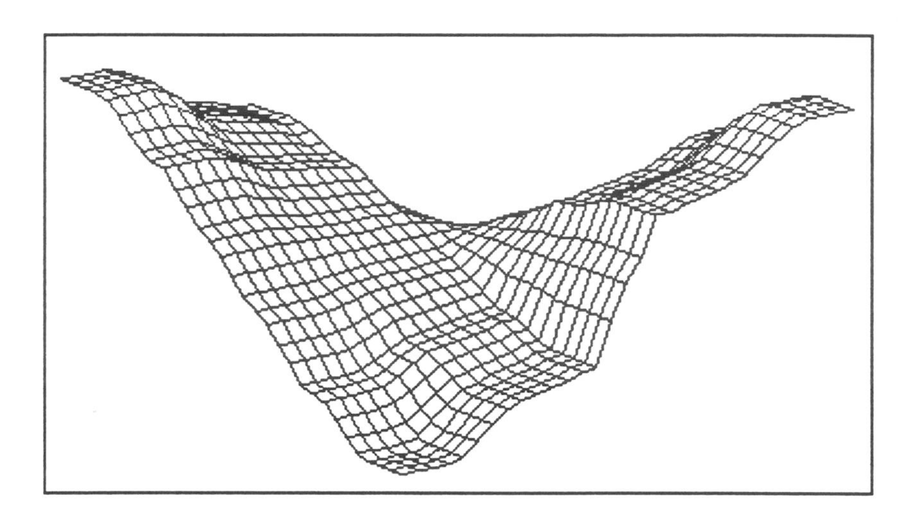

Figure 10.13 Control Surface Using the FNN3 Architecture.

(this chapter), the resulting control surface is llustrated in Figure 10.13. Of particular significance is the smoothing obtained at the point of equilibrium (refer to Figure 10.6 for FNN2).

Network training is based upon the use of fuzzy targets listed in Table 10.1. The edges of the control surface are not affected because the range of the input/output fuzzy sets was defined only over the interval that includes only the triangular region. A convergence tolerance of 10^{-3} was used in conjunction with a fuzzy back-propagation algorithm. A crisp cost function, given by Eqs. (10.12) and (10.13), was used as the error criteria. Lower error tolerances did not show any significant improvement in the overall smoothing of the control surface.

Figures 10.14 and 10.15 illustrate the simulation results using FNN3 architecture. The operating conditions simulated for this architecture are the same as for the FNN2 type system. The results appear more robust and indicate the adaptive capability of the system.

In the FNN3 type architecture, we observe that the behavior of the fuzzy-neural network follows the time-constant of the system thereby yielding a degree of valve opening. This is an important characteristic from the viewpoint of overall stability.

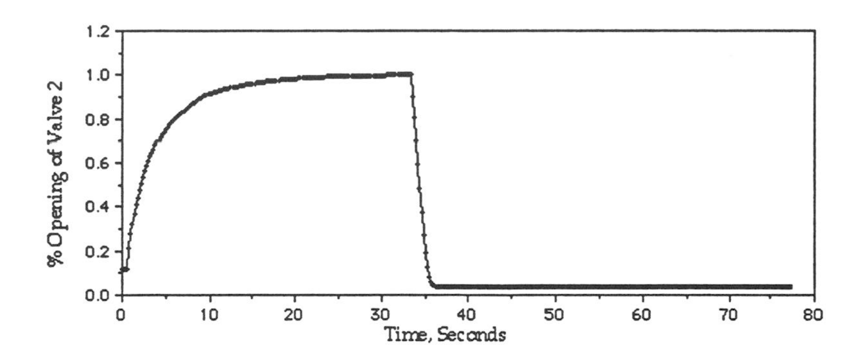

Figure 10.14 FNN3 Simulation of Valve 2 Opening for Variation in Output Flow Rate.

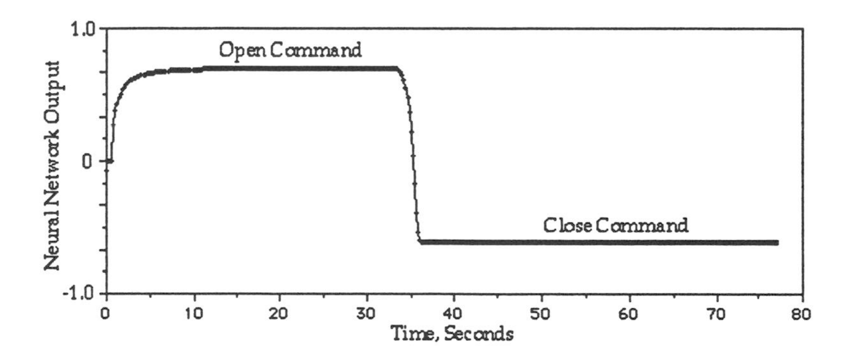

Figure 10.15 FNN3 Output for Variations in Ouput Flow Rate.

10.6 CONCLUSIONS

In this chapter, we discussed the synergistic issues between neural networks and fuzzy logic-based systems. A simple example was chosen to demonstrate the possible behavior of two types of integrated systems. Uncharacteristic behavior was shown to result in systems that are "less" trained and do not have the ability to learn and adapt.

The use of a crisp cost function produced acceptable results in terms of training the fuzzy-neural architecture model. While no comparisons were made between these results and that which can be obtained using a fuzzy cost function, we may speculate that the results are comparable. From numerical standpoint, the number of iterations required in the training phase may be substantially less. However, the significant results may not be affected.

Throughout this chapter, we have assumed that the structure of the fuzzy-neural system is fixed and that the parameter identification problem is solved through hybrid learning. However, to make the approach more complete, the structure identification [11, 12] is equally important for the successful application of fuzzy-neural systems.

References

[1] Fukunaga, K. and Hayes, R. R. (1989). Effects of sample size in classifier design, *IEEE Trans. Pattern Anal. Machine Intell.*,**11**(8), 873-885.

[2] Fukunaga, K. and Hayes, R. R. (1989). Estimation of classifier performance, *IEEE Trans. Pattern Anal. Machine Intell.*,**11**(10), 1087-1101.

[3] Jain, A. K. and Chandrashekaran, B. (1985). Dimensionality and sample size considerations in pattern recognition practice, *Handbook of Statistics*,**2**, P. R. Krishnaiah and L. N. Kanal (Eds.), North Holland, Amsterdam, Netherlands, pp.835-855.

[4] Lin, C. T. and Lee, C. S. G. (1996). *Neural Fuzzy Systems - A Neuro-Fuzzy Synergism to Intelligent Systems*, Prentice-Hall, Englewood, NJ.

[5] Jang, J.-S. R., Sun, C. T., and Mizutani, E. (1997). *Neuro-Fuzzy and Soft Computing - A Computational Approach to Learning and Machine Intelligence*, Prentice-Hall, Englewood, NJ.

[6] Ishibuchi, H., Fujioka, R., and Tanaka, H. (1993). Neural networks that learn from fuzzy if-then rules, *IEEE Trans. on Fuzzy Systems*,**1**, 85-97.

[7] Ishibuchi, H., Kwon, K., and Tanaka, H. (1992). An architecture of neural networks for input vectors of fuzzy numbers, *Proc. IEEE Int. Conf. on Fuzzy Systems (FUZZ-IEEE'92)*, pp. 1293-1300.

[8] Ishibuchi, H., Kwon, K., and Tanaka, H. (1995). A learning algorithm of fuzzy neural networks with triangular fuzzy weights, *Fuzzy Sets and Systems*, **71**, 277-294.

[9] Ishibuchi, H., Kwon, K., and Tanaka, H. (1993). Implementation of fuzzy if-then rules by fuzzy neural networks with fuzzy weights, *Proceedings First European Congress on Fuzzy and Intelligent Technologies*,**1**, pp. 209-215.

[10] Haykin, S. (1994). *Neural Networks - A Comprehensive Foundation*, Macmillan College Publishing Company.

[11] Sugeno, M. and Kang, G. T. (1988). Structure identification of fuzzy model, *Fuzzy Sets and Systems*, **28**, 15-33.

[12] Sun, C. T. (1994). Rulebase structure identification in an adaptive network based fuzzy inference system, *IEEE Trans. on Fuzzy Systems*, **2**(1), 64-73.

[13] Hayashi, Y., Buckley, J. J., and Czogala, E. (1992) Fuzzy neural network with fuzzy signals and weights, *Proc. Int. Joint Conf. Neural Networks*, Vol. II, pp. 696-701, Baltimore, MD.

11 FUZZY GENETIC ALGORITHMS

Andreas Geyer-Schulz

Department of Applied Computer Science,
Institute of Information Processing and Information Economics,
Vienna University of Economics and Business Administration,
Augasse 2–6, A-1090 Vienna, Austria.
geyers@wu-wien.ac.at

11.1 INTRODUCTION

In this chapter we present a tutorial on fuzzy genetic algorithms applied to control problems. The unifying theme of this chapter is the use of fuzzy genetic algorithms to systematically breed better and better control strategies with simulation models of real-world systems and thus to overcome the limitations of classic analytical and numerical optimization methods. As introduction we present the MIT beer distribution game as an example of a complex dynamic decision-problem. In the following section we introduce simple genetic algorithms at a leisurely pace and show why we use a genetic algorithm for control problems. Next, we proceed to fuzzy genetic algorithms. We restrict our exploration of the design space of genetic algorithms to simple genetic algorithms over fuzzy rule-languages. In section 11.3 we show how such rule-languages can be coded with fixed or variable length strings over arbitrary alphabets. Finally, in section 11.4 we extend the representation of the fuzzy rule-language to genetic programming.

At the dawn of the sensor revolution [129] we shall characterize control of complex dynamic systems in terms of intelligent agents acting in some environments. The agent metaphor will provide a unifying framework and a classification of systems according to their complexity. For our purposes an *agent* will be anything which perceives its environment through sensors and acts upon that environment through effectors [128]. To be precise: *"For each possible percept sequence, an ideal rational agent should do whatever action is expected to maximize its performance measure, on the basis of the evidence provided by the percept sequence and whatever built-in knowledge the agent has"* [128], p. 33.

In this section we give a survey on different types of agents and environments and we describe one specific environment, namely the MIT beer distribution game in sufficient detail to be useful as a tutorial [138]. The rest of the chapter is devoted to explain how a genetic algorithm, more specifically a fuzzy genetic algorithm, combined with a simulation of the environment yields a powerful learning method for automatically "breeding" the agent's program (or control strategy).

What is rational for the agent shown in figure 11.1 depends on the performance measure that defines the agent's success, the percept sequence, the knowledge of the agent about the environment and the actions that the agent can perform.

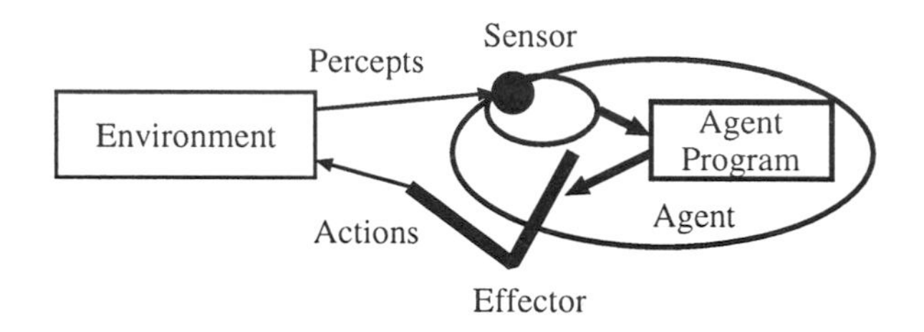

Figure 11.1 Agent Interacting with Its Environment

The notion of agents is a tool for analyzing systems. In principle, we may describe an agent by a mapping from percept sequences to actions. However, in most situations writing down such a list of percept sequence/action pairs which leads to a successful agent is infeasible and we have to come up with a better way to design the agent program. We suggest genetic algorithms which breed the agent programs by conducting controlled experiments in a simulated environment. We replace H. A. Simon's slogan "Understanding by Simulating" from his 1969 lectures on the sciences of the artificial [135] by the more radical "Evolution by Simulation". Simulations compress time and space and – combined with genetic algorithms – they allow controlled experiments and lead to a rapid accumulation of experience. Before we let a genetic algorithm evolve an agent program, we must have a clear picture of the percepts and actions, of the performance measure which the agent should achieve and of the environment it will be operating in [128]. In table 11.1 we show agents of different flavors. We strongly urge the reader to identify and analyze his application domain in terms of percepts, actions, goals and environment.

Table 11.1 Examples of Agents

Agent	Percepts	Actions	Goals	Environment
Inverted pendulum controler	Angle and angular velocity	Adjust direction, velocity of finger	Maintain pendulum in equilibrium	Robot hand balancing a pole
Refinery controler	Temperature and pressure measurements	Open, close valves, adjust temperature	Maximize purity, safety, yield	Refinery
Inventory manager	Inventory, supply line level	Order items	Minimize inventory cost	Inventory system
Product manager	Balance, cost accounting, market data	Adapt production capacity, prices, marketing expenditure	Maximize profit	Market economy with competitors

Agents and environments are always interconnected in the same way: actions are done by the agent on the environment which in turn provides percepts to the agent [128], p. 45. Depending on the internal structure of the agent we find the following basic agent types:

Table-driven agents use a percept sequence/action table in memory to find the next action. They are implemented by a (large) lookup table.

Simple reflex agents are based on condition-action rules and implemented with an appropriate production system. They are stateless devices which do not have memory of past world states.

Agents with memory have internal state which is used to keep track of past states of the world.

Agents with goals are agents which in addition to state information have a kind of goal information which describes desirable situations. Agents of this kind take future events into consideration.

Utility-based agents base their decision on classic axiomatic utility-theory in order to act rationally [12].

Agents may act in a wide variety of environments. Different environment types will require appropriate agent types to deal with them efficiently. The

Table 11.2 Examples of Environments and Their Characteristics

Environment	Accessible	Deterministic	Episodic	Static	Discrete	No adversary
Robot balancing a pole	yes	yes	no	no	no	yes
Refinery	no	no	no	no	no	yes
Inventory	yes	no	no	yes	yes	yes
Market Economy	no	no	no	no	no	no

following list of attributes of environments provides the main criteria for a rough classification of environments:

Accessible/Inaccessible. If an agent's sensors give it access to the complete state of the environment needed to choose an action, the environment is accessible to the agent. Such environments are convenient, since the agent is freed from the task of keeping track of the changes in the environment.

Deterministic/Nondeterministic. An environment is deterministic for an agent if the next state of the environment is completely determined by the current state of the environment and the action of the agent. In an accessible and deterministic environment the agent need not deal with uncertainty.

Episodic/Nonepisodic. An episodic environment means that subsequent episodes do not depend on what actions occurred in previous episodes. Such environments do not require the agent to plan ahead.

Static/Dynamic. An environment which does not change while the agent is thinking is static. In a static environment the agent need not worry about the passage of time while he is thinking, nor does he have to observe the world while he is thinking. In static environments the time it takes to compute a good strategy does not matter.

Discrete/Continuous. If the number of distinct percepts and actions is limited the environment is discrete, otherwise it is continuous.

With/Without rational adversaries. If an environment does not contain other rationally thinking, adversary agents, the agent need not worry about strategic, game theoretic aspects of the environment (which makes life a lot easier). Most engineering environments are without rational adversaries, whereas most social and economic systems get their complexity from the interactions of (more or less) rational agents.

In table 11.2 we have applied this classification to the environments of our agents from table 11.1. As we might expect, the "easiest" environments are accessible, deterministic, epsisodic, static, without adversaries, and with a small number of discrete percepts and actions. In reality, however, most environments seem at a first glance to belong to the opposite category. In modeling an environment we should always strive for simplifying the non-essential aspects of the environment. Usually, a good deal can be learned by starting with quite idealized environments and by moving to more realistic environments step by step. Often, simplified environments are cast into the form of management games. As an example we present below a description of the MIT beer distribution game [138].

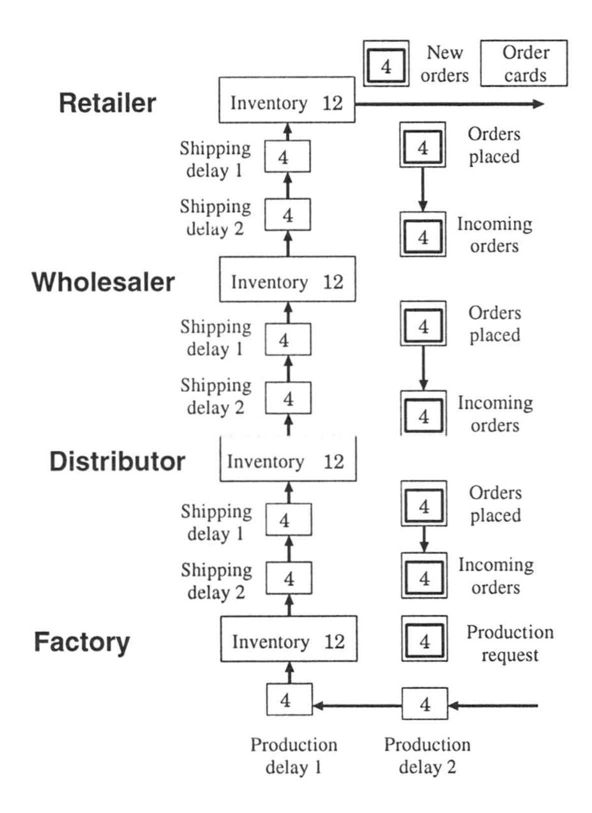

Figure 11.2 The MIT Beer Distribution Game

In the MIT beer distribution game a team of four players manages a simulated industrial production and distribution system with the aim of minimizing total inventory cost of the system during 50 weeks. The game is played on the board shown in figure 11.2 which depicts the production and distribution system of a brewery [138]. Orders for beer and cases of beer are represented by order slips and coins which are manipulated by the players. Each brewery has a four link distribution chain with retailer, wholesaler, distributor, and factory. Each link is managed by one person. The game is played for 50 rounds, each round corresponds to a week. Each week, the retailer receives a customer order and ships the requested beer out of the inventory. The retailer orders beer from the wholesaler who ships the beer out of his inventory. The wholesaler in turn orders and receives beer from the distributor who orders and receives beer from the factory which brews beer. Between each sector shipping delays and order receiving delays exist. Inventory holding costs are 0.5 monetary units per case per week, backlog costs are 1.0 monetary units per case per week. Costs are recorded at each sector of the brewery. Information availability is restricted throughout the game. Communication between players is not allowed within a game. So retailers are the only players who discover customer demand. All others learn only what their customers order with a delay of one week [138]. During an actual play the order cards and order slips are face down at all times.

In each simulated week each of the players has to carry out the following five steps [138]:

1. **Receive inventory and advance shipping delays.** The contents of the field *Shipping delay 1* is moved to the field *Inventory.* The contents of *Shipping delay 2* is moved to *Shipping delay 1.* The factory advances the production delays.
2. **Fill orders.** Retailers take the top card of the *Order cards*, all others examine the order slip in the field *Incoming orders.* The number of beer cases to deliver is made up of the incoming order plus the backlog of the previous week. Orders are always processed as far as the inventory permits. Unprocessed orders add to the backlog.
3. **Record inventory or backlog on the record sheet.**
4. **Advance the order cards.** Order cards are moved from the field *Orders placed* to the field *Incoming orders* of the next link in the chain. The factory 'produces' the contents of the field *Production request* and introduces the appropriate number of coins to the field *Production Delay 2.*
5. **Place orders.** Each player decides what the order should be, records the order on the record sheet and on an order card which he places face down to his *Orders placed* field. The factory places its 'orders' in the field *Production request.*

The game board in figure 11.2 shows the initial conditions in Sterman's experiments. Figure 11.3 shows the customer orders. Customer demand starts at four cases per week. In week 5 customer demand increases to eight cases and remains constant for the rest of the game.

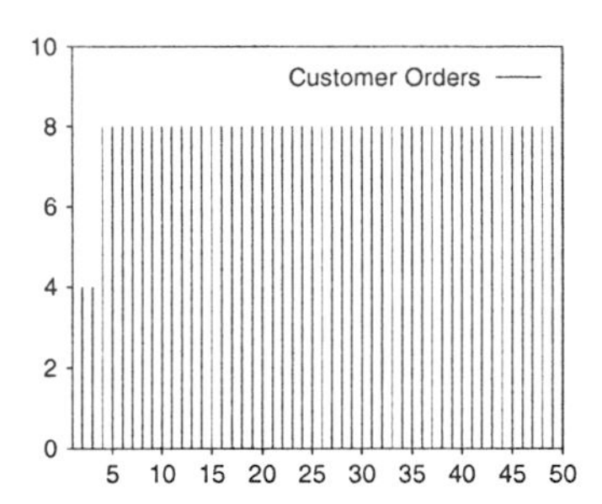

Figure 11.3 Customer Orders

Although this environment looks deceptively simple, it does not belong to the easiest environment class. The MIT beer distribution game is not accessible, deterministic, not episodic, static, discrete, and without adversaries. Clearly, this classification is debatable: It is only deterministic for finding a good reference strategy for J. Sterman's experiment. It is clearly non-deterministic, if we want to find a good strategy for arbitrary order streams. As formulated, it is non-accessible, because backlogs are not recorded on the game board. However, the attribute which makes the game really tough is that it is not episodic: The game has a rich feedback and lag structure. Next, let us analyze the MIT beer distribution game in terms of percepts, actions, goals and the environment.

Percepts. Straightforward modeling of the team of four agents requires that they receive the same information as a human player: they see the whole game board presented in figure 11.2 and receive an order slip in each round of the game. A closer inspection of figure 11.2 suggests drastic simplifications in modeling the percepts of the four agents. Each agent needs only three percepts which matter in choosing an action: one for his (local) supply line, one for his inventory, and one for the order slip received. Note, that the supply line SL_t e.g. of the retailer is the sum of the contents of the fields *Shipping Delay 1*, *Shipping Delay 2*, and *Incoming orders* (of the wholesaler) on the game board shown in figure 11.2. Compared to the number of fields on the game board this is a considerable reduction in the number of percepts. However, this implies that the agent will only employ locally rational decision-rules. Furthermore, a second look on the game board in figure 11.2 reveals that the stock management systems of the first three levels have the same feedback structure. We may use this symmetry to represent the first three agents with the same agent program.

Actions. The only action available to each agent is to order (respectively produce) n cases of beer in each game period.

Goals. The goal of the agents is to minimize team inventory cost. For the deterministic case (the customer orders and initial conditions of J. Sterman's experiment (figures 11.3 and 11.2) this means

$$\min \sum_{i=1}^{50} \sum_{j=1}^{4} (0.5 S_{i,j} + 1.0 B_{i,j}) \tag{11.1 }$$

where $S_{i,j}$ is the stock on inventory and $B_{i,j}$ the backlog in week i for level j. However, for a class of stochastic order processes or a set of scenarios of mixed order streams and initial conditions with no prior information on the probability of the scenarios, we might, for example, wish to minimize expected team inventory cost:

$$\min \frac{1}{N} \sum_{n=1}^{N} \sum_{i=1}^{50} \sum_{j=1}^{4} (0.5 S_{n,i,j} + 1.0 B_{n,i,j}) \tag{11.2 }$$

with $S_{n,i,j}$ and $B_{n,i,j}$ denoting stock and backlog for scenario n in week i for level j. However, other goals may be rational too. E.g. in a tight cash-flow situation we may seek to minimize the maximal inventory cost in the scenario. For an introduction to decision-making under uncertainty, see e.g. [12].

Environment. Each level of the distribution chain can be modeled as an instance of the generic stock management system shown in figure 11.4.

The stock S_t is the sum of the differences of the acquisition rate A_i and the loss rate L_i over $i = t_0, \ldots, t$ added to the initial stock S_{t_0}.

$$S_t = \sum_{i=t_0}^{t} (A_i - L_i) + S_{t_0} \tag{11.3 }$$

The supply line SL_t at time t is simply the sum of orders placed O_i minus those which have been delivered A_i over $i = t_0, \ldots, t$ added to the initial value of the supply line SL_{t_0}.

$$SL_t = \sum_{i=t_0}^{t} (O_i - A_i) + SL_{t_0} \tag{11.4 }$$

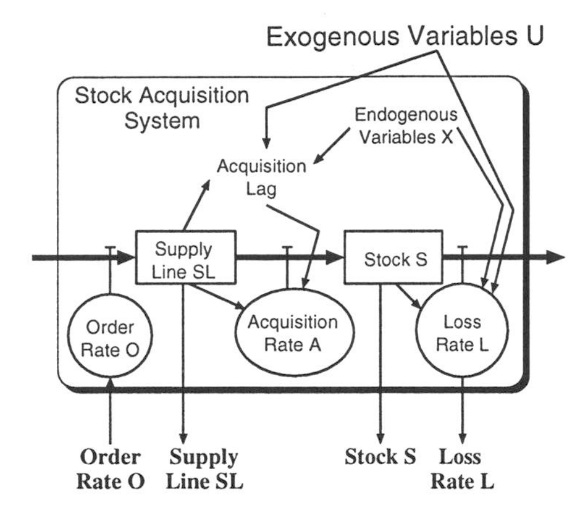

Figure 11.4 The Structure of a Stock Management System

Now we have set the stage for discussing different ways of writing the agent's program or control strategy. We have to bite the bullet and breed the agent's program.

11.2 WHAT IS A GENETIC ALGORITHM?

Genetic algorithms simulate natural evolution processes in order to solve problems by breeding solutions. They are programs for a simple artificial evolution process. As natural evolution acts on populations of individuals, so do genetic algorithms with sets of strings.

```
procedure Simple_GA(int size, real crossoverrate, real mutationrate);
    array_of_members population[size];
begin     INITIALIZE each population member;
    EVALUATE each population member;
    repeat begin
        reproduce each population;
        CROSSOVER each (population, crossoverrate);
        MUTATE each (population, mutationrate) end
    until (not terminated);
    print_result end
```

Figure 11.5 Pseudocode for a Genetic Algorithm

In a simple genetic algorithm strings have a similar role like chromosomes. They encode a set of solutions (parameters, actions, rule, programs) which are evaluated by a fitness function. In our example, a string encodes the parameters of the ordering heuristics in a distribution chain. We compute the fitness value of such a string by decoding and running the simulation of the game with the parameters. The fitness of the string is the total inventory cost of the chain in a game. A simple genetic algorithm generates a new population of strings with three simple genetic operators, namely reproduction, crossover, and mutation which taken together act as a probabilistic state transition function. Figure 11.5 shows the pseudocode of a simple genetic algorithm.

In the reproduction phase of a genetic algorithm strings are replicated (copied) according to the outcome of a Darwinian selection process. Only the best survive. Reproduction is a biased sampling method, biased towards the fittest strings. A straightforward method to implement such a biased method is reproduction proportional to fitness as analyzed in [60] and [61]. The effect of reproduction is to assign an exponentially increasing number of trials on successful individuals and thus to exploit the information in the population.

For example, we have four strings, a, b, c, and d in our population with fitness values of 90, 4, 4, and 3 respectively. The probability of survival with reproduction proportional to fitness is $P_a = 0.9$, $P_b = 0.04$, $P_c = 0.04$, and $P_d = 0.03$. As figure 11.6 shows, it is like a roulette wheel with slot sizes proportional to fitness. In our experiments we have used this simple reproduction method.

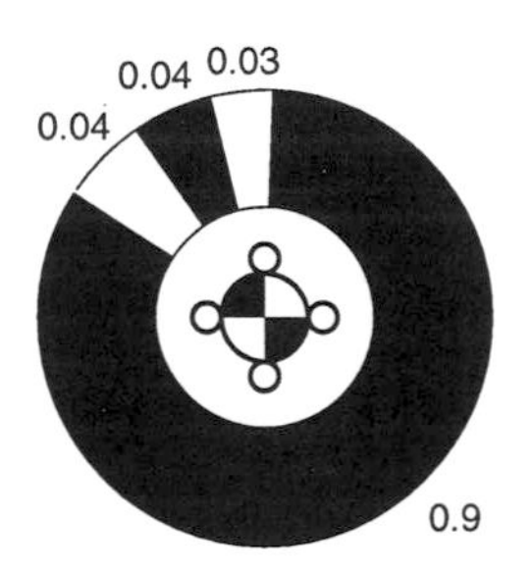

Figure 11.6 Reproduction

However, because of the well known shortcomings of reproduction proportional to fitness (premature convergence and loss of selective pressure towards the end of the run) other selection methods (for example linear rank selection [43] or tournament selection [41], [5], [8], [17]) can be used for reproduction. Linear rank selection does not lead to premature convergence because of a few individuals with above average fitness and it maintains constant selective pressure throughout the run of a genetic algorithm. It is based on a linear function of the fitness rank of an individual in the population. For parallel or distributed implementations of genetic algorithms the reproduction mechanism of choice is tournament selection because of its low communication overhead. The basic idea is to take a small sample of individuals from the population, and the individual with the best fitness in the group is chosen for reproduction.

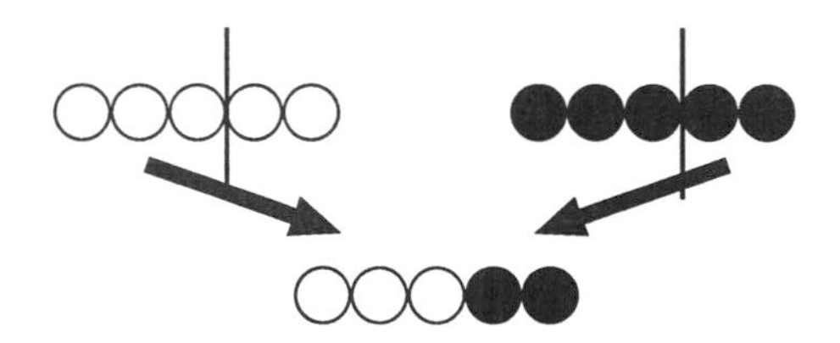

Figure 11.7 Crossover

In the next phase the crossover operator is applied to a pair of strings with a certain probability. The crossover operator creates a new string by recombining two parent strings as shown in figure 11.7. All bits up to a randomly selected position are from the first parent, the rest is from the second. In the canonical simple genetic algorithm both children are created. The effect of the crossover operator is that a "neighborhood" of the parents is explored without losing genetic material and that short, above-average "building-blocks" are combined on one string. Because of its efficiency in exploring the "neighborhood" of strings the crossover operator has often been termed the workhorse of the genetic algorithm. In the version of the crossover operator used in our experiment only one child is produced. For an overview of other variants see e.g. [29].

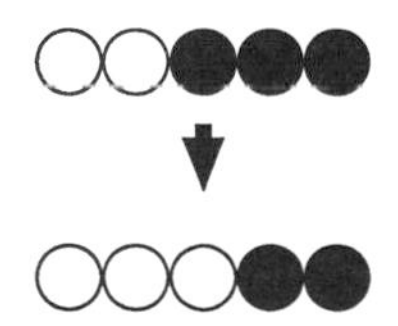

Figure 11.8 Mutation

Finally, as in natural selection, the mutation operator is applied at a very low rate. In our experiments, we use a one-bit mutation operator which changes a randomly selected bit of the string (see figure 11.8). Mutation reintroduces lost genetic material into the population, so that no allele permanently disappears from the population. Its second effect is that it ensures finding the global optimum as the number of trials tends to infinity.

The behavior of genetic algorithms has been the topic of many theoretical investigations and experimental studies of which we summarize the most important below:

Markov Theory. A simple genetic algorithm, of course, is itself a complex, nonlinear, dynamical system. To understand both its transient and asymptotic behavior simple genetic algorithms have been modeled as Markov-chains by several researchers (e.g. [35], [148], [119], [149], [127], [141], [15], [150]).

In these models populations correspond to a point on a surface and the progression from one generation to the next forms a path leading toward a fixed-point of the infinite population model. In the infinite population models populations move toward the local fixed-point in whose basin of attraction they started, they are not global optimizers [150]. However, when in the finite population model the mutation rate is greater than zero, a finite-population genetic algorithm is described as an ergodic Markov-chain. It visits every state infinitely often and escapes every local basin after some time [150]. When adding elitism as absorption mechanism asymptotic global convergence can be shown by elementary means [26]. In addition, short term behavior of a simple genetic algorithm is determined by the local basin in which the initial population is located, whereas long-term behavior is determined by the local fixed-point with the largest basin of attraction [150].

Statistical Decision Theory. Holland characterized adaptive systems in nature by the tension between exploration that is the search for new, useful adaptations and exploitation that is the use and propagation of these adaptations. For its survival in an uncertain environment any adaptive system must find an optimal balance between exploration and exploitation. The system must continue experimenting with new adaptations, because otherwise it would overadapt and become too rigid to cope with new situations, while at the same time it has to incorporate continually and use past experience to improve its future performance. Statistical decision-theory is a natural theoretical framework for comparing such competing adaptive systems in terms of their loss functions. In [60] and [61] Holland succeeded in analyzing the loss function of the best sequential sampling plan which is an optimal plan for the k-armed bandit problem and the loss function of a genetic algorithm. He proved that the genetic algorithm outperforms the optimal k-armed bandit strategy and thus all sequential sampling strategies.

Holland's analysis is based on the notion of schemas. A schema is a binary string with 0 or more wildcard characters (a pattern or a template) which serves to identify sets of binary strings which match the schema. For example, the schema $*1*00$ is matched by each string in the set $\{01000, 01100, 11000, 11100\}$. The observed fitness of a schema in the population is defined as the mean of all instances of the schema in the population.

Holland expressed the loss function of a genetic algorithm with the help of the schema difference equation which gives a lower bound on the expected number of schemas reproduced under crossover and mutation. Because of its importance, the schema difference equation has been nicknamed the "Schema Theorem" or the "Fundamental Theorem" of genetic algorithms. It is based on an analysis of the effects of reproduction, crossover and mutation on the probability of survival of a schema. It has been interpreted to imply that short, low-order schemas whose fitness is above average will receive an exponentially increasing number of trials over time. Reproduction focuses search on subsets of the search space with above average fitness, crossover combines high-fitness "building blocks" (schemas) on the same string, and mutation makes sure that genetic diversity is never lost.

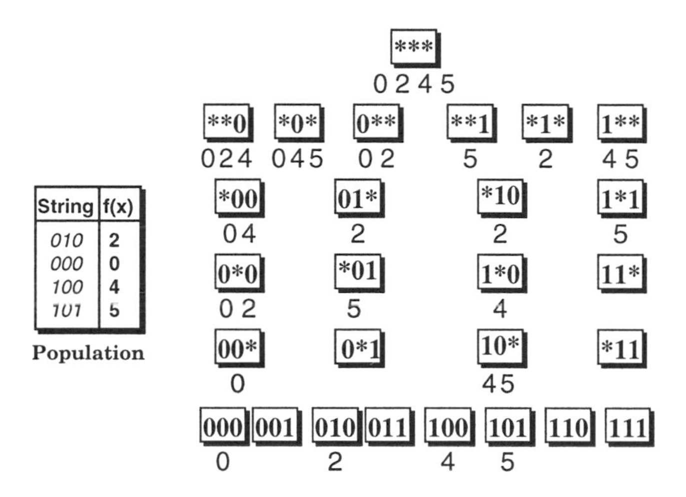

Figure 11.9 Exploitation of Information

The second basic discovery of John H. Holland is the notion of implicit parallelism: While explicitely evaluating the fitness of n strings in a population, the genetic algorithm implicitely processes the observed fitness of schemas with instances in the population. Each individual is thus exploited as a trial for 2^l schemas. Figure 11.9 shows a small example. The 4 individuals in the population are exploited for the implicit computation of the fitness of 20 schemas.

However, the validity of the schema analysis sketched above depends crucially on the Glivenko-Cantelli theorem [125], p. 300 which establishes the convergence of sample moments to distribution moments as the number of trials in the sample increases. Therefore, the major limitation of schema analysis is the requirement that the observed fitness of a schema converges to the "real" fitness of a schema. This does not hold for a small number of trials per schema, for long, high-order schemas, and over several reproduction steps because of selection bias.

Statistical Mechanics. The statistical mechanics approach to the analysis of the dynamics of genetic algorithms pioneered by [122] and [133] aims to describe laws of genetic algorithms behavior in terms of macroscopic statistics like "mean fitness in the population" in analogy with the description of physical systems by quantities like pressure and temperature. Although this approach is still in its infancy and depends heavily on the application domain the approach can be used to determine parameter values for a genetic algorithm which produce desired ratios of minimization speed versus convergence [116].

Hard Problems. What functions are difficult for a genetic algorithm? Investigations of this question lead us to representing functions as Walsh functions and to the analysis of Walsh polynomials. Deceptive functions which have low-order schemas which lead the genetic algorithm away from the global optimum are one class of hard functions. An analysis of this class and a way to construct such functions is given in [39] and [40]. Another class of difficult functions whose Walsh polynomials have all non-zero terms of the same order and whose optimal solution receives positive contributions from each term in the polynomial are called after their inventor Tanese functions. They are investigated in detail in [22].

Experimental Studies. For the practitioner of genetic algorithms knowledge of good and robust parameters for a genetic algorithm is of high importance. De Jong's parameters for a genetic algorithm for function optimization (population size $50 - 100$, crossover rate 0.60 and a mutation rate of 0.001) which he found by hand optimization for his thesis in 1975 have become part of the lore of the genetic algorithm community [131]. His results have been checked by a large brute force experiment by [131]. Grefenstette introduced meta-level genetic algorithms for parameter tuning [46].

Back to our example. For learning a good control strategy for the MIT beer distribution game two basic ingredients are still missing: The fitness or evaluation function and a coding for the agent program. The first is – at least for finding a reference strategy (the deterministic case) – easy. We settle for function 11.1 as goal. For the agent program we strive in our first example for a classic operations research style formulation for comparison purposes. The classic approach to writing an agent program is to express the relationship between percepts and actions as a function of a certain type and structure and to find the parameters of this function which optimize the agent's goals with a genetic algorithm. The parameter values are binary-coded and catenated to a fixed-length binary string. In the simplest case these strings correspond to the points on a regular grid over a suitable region of the parameter space. The decision which function type to choose usually requires a good deal of know-how about the application domain. In addition, deciding what is a suitable region of the parameter space and what is a suitable number of grid-points requires at least some feeling for the numerical behavior of the problem.

For the determination of a suitable function type and structure for the MIT beer distribution game we heavily rely on previous work by [147] and [138] which in the spirit of bounded rationality [13] assume that managers, when they are unable to optimize, employ at least *locally rational* decision rules. Moreover, [147] report that managers often use simple adaptive models combined with anchoring and adjustment heuristics for complex inventory decisions. In [138] we find an ordering heuristic which assumes that managers choose orders in such a way that expected losses from the stock are replaced, that the difference between desired and actual stock level is reduced, and that a reasonable number of unfilled orders are in the supply line (see figure 11.10).

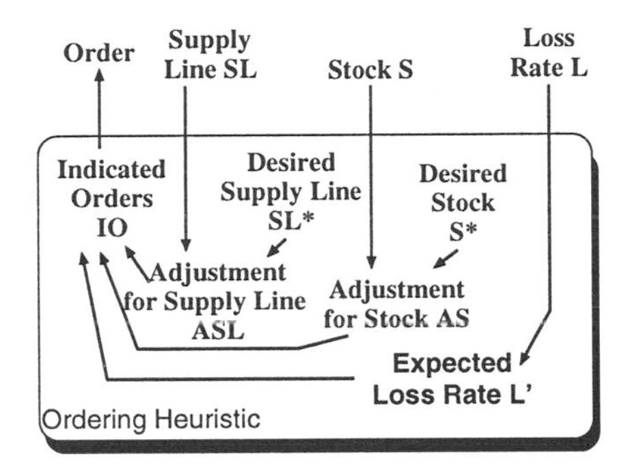

Figure 11.10 The Structure of the Ordering Heuristic

First, observe that in real life orders must be nonnegative:

$$O_t = \max(0, IO_t) \tag{11.5 }$$

where O_t is the order rate and IO_t the indicated order rate in t. The indicated order rate IO_t is based on an anchoring and adjustment heuristic [147] with the expected loss rate L'_t playing the role of a known reference point (the anchor) and with adjustments for the differences between actual and desired stock (AS_t) and the differences between actual and desired supply line (ASL_t):

$$IO_t = L'_t + AS_t + ASL_t \tag{11.6 }$$

In his formalization of the ordering heuristic, Sterman used adaptive expectations for modeling the expected loss rate L'_t:

$$L'_t = \theta \cdot L_{t-1} + (1-\theta) \cdot L'_{t-1} \quad (11.7)$$

with $0 \leq \theta \leq 1$. For the adjustments of stock and supply line, desired stock S^*_t and desired supply line SL^*_t are assumed to be constant. α_S and α_{SL} denote the weights for adjustment of stock and adjustment of supply line, respectively.

$$AS_t = \alpha_S(S^*_t - S_t) \quad (11.8)$$

$$ASL_t = \alpha_{SL}(SL^*_t - SL_t) \quad (11.9)$$

In order to reduce the number of parameters we substitute $\beta = \alpha_{SL}/\alpha_S$ and $S' = S^* + \beta SL^*$ into equations 11.5 to 11.9 and we obtain the following ordering heuristic for each link in the distribution chain:

$$O_t = \max[(0, L'_t + \alpha_S(S' - S_t - \beta SL_t)] \quad (11.10)$$

$$L'_t = \theta L_{t-1} + (1-\theta)L'_{t-1} \quad (11.11)$$

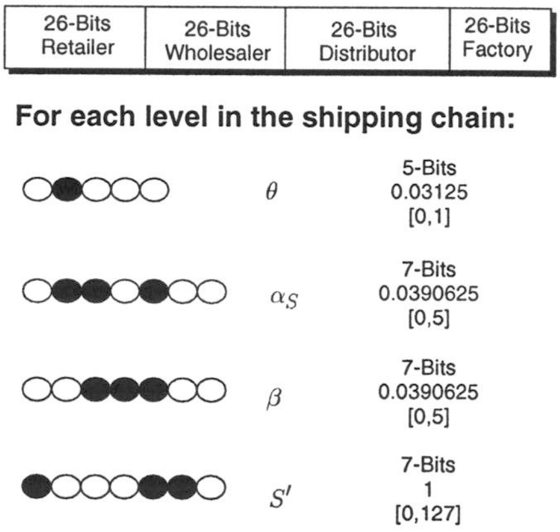

26-Bits Retailer	26-Bits Wholesaler	26-Bits Distributor	26-Bits Factory

For each level in the shipping chain:

	Parameter	Length / Precision / Range
○●○○○	θ	5-Bits 0.03125 [0,1]
○●●○●○○	α_S	7-Bits 0.0390625 [0,5]
○○●●●○○	β	7-Bits 0.0390625 [0,5]
●○○○●●○	S'	7-Bits 1 [0,127]

Figure 11.11 Coding the Beer Distribution Game

As shown in figure 11.11 the chromosome which encodes a parameter set for the ordering heuristics of each link in the distribution chain contains 104 bits. This means 26 bits per link. Length, precision, and range of each parameter shown in figure 11.11. In determining precision and range of each parameter we have once again taken advantage of the work of [138]. However, we have chosen a finer grid than Sterman, because the genetic algorithm allowed us to cope with a larger search space. In this coding we have not taken advantage of the symmetry in the feedback structure on the first three levels of the distribution chain which we have discussed in the previous section. By exploiting this symmetry we could reduce string length from 104 bits to 52 bits and thus search space size from 2^{104} to 2^{56}.

In table 11.3 we show the result of a run of a simple genetic algorithm with elitism with a population of 1000 strings for 1000 generations (a million evaluations). Elitism ensures that the best individual in a population is always reproduced. The best solution we have obtained in this run occurred in generation 532 for the first time. This solution could not be improved during the rest of the run. What is remarkable about the solution is that it relies exclusively on adapting optimally the expected losses from inventory with the simple adaptive model shown in equation 11.11 . The anchor and adjustment heuristics of the second term of equation 11.10 are not used in this solution.

Table 11.3 The Parameters Found with the GA(1000,532) with Inventory Costs 181.5 (GA-Parameters: P(Crossover) = 0.70, P(Mutation) = 0.01, Reproduction Proportional to Fitness, Elite Selection)

Level	θ	α_S	β	S'
Retailer	1.000000	0.000000	0.511811	23
Wholesaler	0.967742	0.000000	2.874016	70
Distributor	0.935484	0.000000	1.574803	36
Factory	0.483871	0.000000	1.535433	45

Well, at least we have got a result. But a lot of questions remain:

1. Is the MIT beer distribution game a difficult problem? Or more general, are control problems difficult to solve?
2. Is the ordering heuristic we have found with the genetic algorithm a good heuristic? Or more general, are genetically bred control strategies good strategies?
3. How is the performance of the genetic algorithm compared to human decision-makers?

The first question is answered for us by Sterman: "The complexity of the system – it is a 23rd order nonlinear equation – renders calculation of the optimal behavior intractable" [138], p. 328. And for dynamical systems in general, a lucid description of the state of the art is given by [57] which is still valid today: "But to achieve any kind of general picture in dimensions higher than two, one must confront limit sets which can be extremely complicated, even for structurally stable systems. It can happen that a compact region contains infinitely many periodic solutions with periods approaching infinity. Poincaré was dismayed by his discovery that this could happen even in the Newtonian three-body problem, and expressed dispair of comprehending such a phenomenon" [57], p. 320f. Despite much progress in recent years in understanding the global behavior of fairly general types of dynamical systems, "we are far from a clear picture of the subject and many interesting problems are unsolved" [57], p. 321.

For the second problem we compare the genetically bred control strategy with a control strategy which has been found by Sterman by complete enumeration over a grid over the plausible parameter space. For the computation of this strategy which, as claimed by J. Sterman, produces minimal inventory cost, Sterman exploited the symmetry in the feedback structure of the game to reduce the parameter space. The "optimal" parameters discovered are $\theta = 0$, $\alpha_S = 1$, $\beta = \frac{\alpha_{SL}}{\alpha_S} = 1$, and $S' = S^* + \beta SL^* = 28$ (20 for the factory) with total inventory cost of 210 [138], p. 329. It is remarkable that Sterman's strategy entirely relies on optimizing the anchoring and adjustment heuristics. The forecasting part implies constant customer orders. The genetically bred control strategy shown in table 11.3 which relies only on the forecasting part of the heuristic constitutes a 15 percent improvement over the previously known best solution (see table 11.6 and figure 11.12). This result is quite instructive, because it shows that even a full enumeration over some fixed grid on the plausible parameter space does not guarantee an optimal solution.

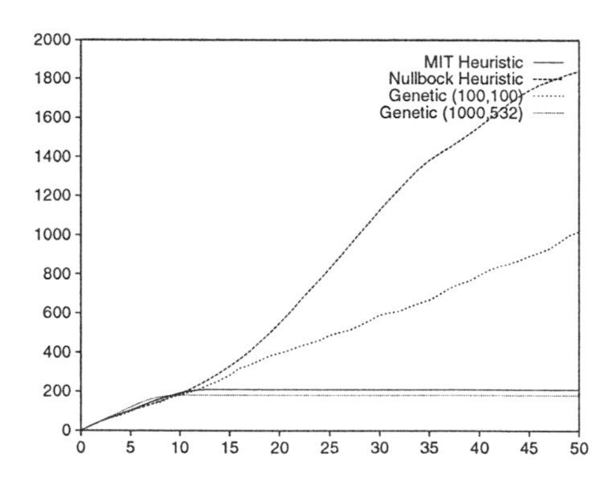

Figure 11.12 Comparing Inventory Cost

But can we expect similar performance of the genetic algorithm for other control problems too? Fortunately, the answer is yes. [15] analyzed the behavior of genetic algorithms in economic systems. Economic systems are an example of systems with rational adversaries which are characterized by state-dependent fitness functions. We call the fitness function of an agent state-dependent if it depends on the control strategies of the other agents in the population. [15] constructed a Markov process for genetic algorithms with state-dependent fitness functions which can be approximated by a system of difference equations which allows analysis of the trajectory of the system and a characterization of locally asymptotically stable states which correspond to economic equilibria. After some transient period the system will be in a uniform state for almost all the time [15], p. 48. This implies that even with rational adversaries the genetic algorithm will find a control strategy which leads to an economic equilibrium. In another game theoretic setting, namely the iterated prisoners dilemma, a genetic algorithm evolved a cooperative strategy which bet Anatol Rapaport's Tit-for-Tat strategy, the long-time champion in this game [120]. Moreover, in our approach to dynamical systems we convert the problem of finding a good control strategy into a function optimization problem which is usually characterized by many local optima. Experiments in optimizing multimodal functions show that genetic algorithms are excellent for such environments [16].

Now let us turn to the third question. For the MIT beer distribution game we have reference data for human decision-makers from J. Sterman's experiments over a four year period described in [138]. In table 11.6 we compare the reference strategies with the average actual costs in the 11 trials without recording error [138]. As a result, we observe that actual behavior is far from optimal. The average team cost is ten times greater than the reference strategies and this difference is highly significant. Figure 11.12 shows the difference between the reference strategies' cost and the actual cost of the Nullbock team.

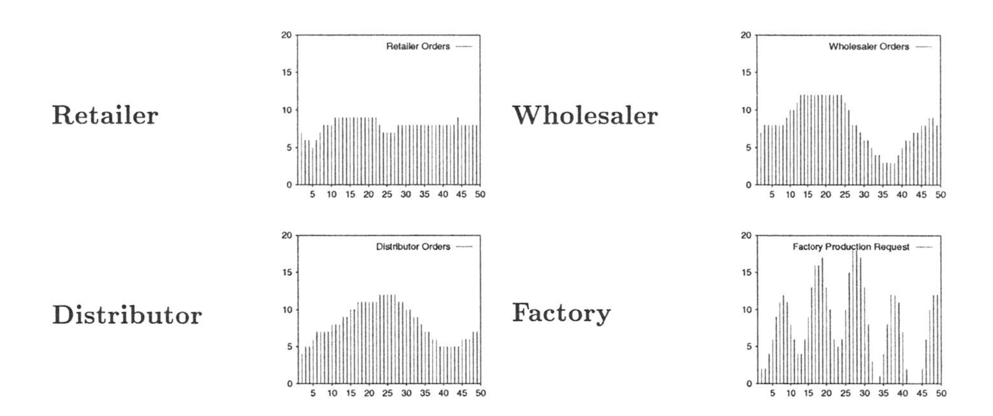

Figure 11.13 Orders of the Nullbock Team

In figure 11.13 we show as a typical example of the actual behavior the orders of the Nullbock team which has about average performance. Analyzing the behavior of the 11 teams in the sample, Sterman identified the following patterns in team behavior [138]:

Oscillation. All trials are characterized by instability and oscillations. Orders and inventory have large amplitude fluctuations.

Amplification. The amplitude and variance of orders increase along the distribution chain.

Phase lag. The peak of the order rate tends to occur later as we move from the retailer to the factory.

The suboptimal behavior of subjects in the beer distribution game can be attributed to risk and uncertainty for retailers [147], bounded rationality [136] in general, and as Sterman has shown, to misperceptions of feedback [138]. In the beer game misperceptions of feed back are the main reason for the oscillations of orders and inventory which are endogenously produced by the interaction of the players' heuristics with the feedback structure of the system. The customer orders are – except for a one time jump – constant. However, it is revealing that most subjects attribute the dynamics of the system to external events [138]. Figure 11.12 indicates a tremendous potential for improvement by genetically breeding superior decision rules.

For finding the solution shown in table 11.3 the genetic algorithm played the game one million times. However, as figure 11.12 shows, we can obtain quite good solutions with less evaluations. With 10000 evaluations we obtained a solution with inventory costs of 1018.50. This indicates that even with quite restricted computation time competitive decision rules (with managers) can be found with a genetic algorithm.

To summarize, genetic algorithms are attractive, because they

1. are robust,
2. are easy to use,
3. may find a global optimal solution with high probability,
4. are fast for satisfying (=good) solutions,
5. can be used in parallel and distributed (=networked) environments,
6. are suitable for nonlinear, multimodal, and multiobjective problems,
7. are suitable for mixed variable problems which are common in real-world problems [16].

11.3 FUZZY GENETIC ALGORITHMS

In this chapter we adapt a narrow definition of a fuzzy genetic algorithm: a fuzzy genetic algorithm is a genetic algorithm which breeds fuzzy logic controllers as agent programs. In contrast to other contributions in this handbook we find it in the context of genetic algorithms productive to emphasize the language aspect of fuzzy logic control: we speak of evolving agent programs in fuzzy rule-languages and we are concerned with the classic computer science topic of specifying languages for fuzzy logic controllers and with coding programs in these languages for genetic algorithms. The components of a language system for a fuzzy language controller are shown in figure 11.14, we assume that the reader is familiar with them.

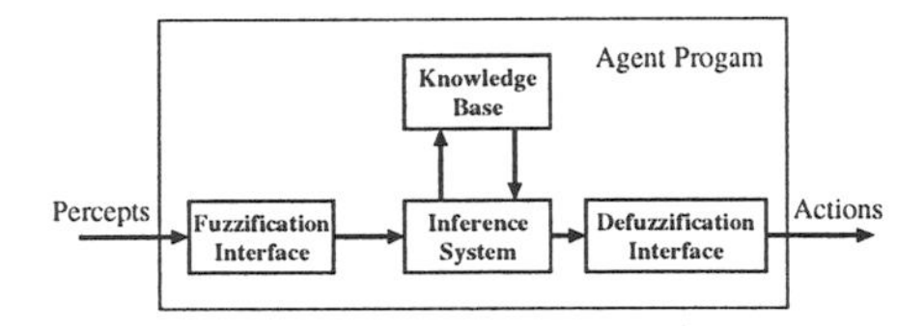

Figure 11.14 The Components of a Fuzzy Logic Controller

In this section we progress from rigid table-driven agents to simple reflex agents which are more flexible. And in coding an agent program we distinguish between coding its syntax that is e.g. the labels of fixed memebership functions or its semantics that is e.g. the parameters which determine the shape of membership functions. In the second category we survey several limited approaches to code variable structures as strings, before we present an elegant and general solution to this problem in section 11.4: with special purpose genetic operators and with alignment and mating restrictions for chromosomes, by a sequence of genetic algorithms which yield partial solutions, and as variants of messy genetic algorithms.

However, flexibility is achieved at a cost: increased search spaces, more involved, less general genetic machinery, more genetic algorithm parameters, and a loss of elegance and simplicity. And still, even the most flexible approaches of this kind breed simple reflex agents, additional language elements can only be added with pains – if at all.

11.3.1 Table-driven agents: Coding the syntax

The easiest way of representing an agent program is as a lookup table for percept sequence action combinations. Each percept sequence action combination results in a fuzzy rule of the form e.g. *if velocity is NM and position is PL the force is ZE* which corresponds to an entry in a fuzzy decision table. The agent program is represented as a fuzzy decision table with l^d entries, where l is the number of adjectives (labels) in the fuzzy partition over the domain of a percept and d is the number of percepts. Each entry represents the label of a fuzzy membership function of the action. What the genetic algorithm should learn, is the content of the fuzzy decision table. The form of the membership functions which define the meaning of an adjective is predefined and fixed.

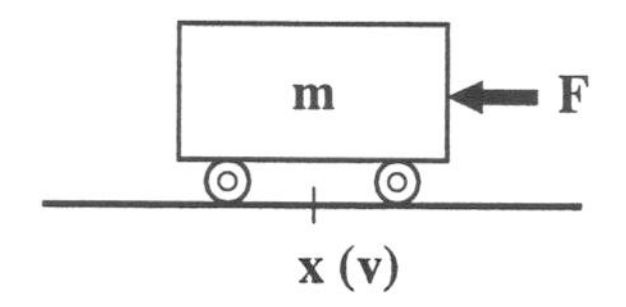

Figure 11.15 The Cart Centering Problem

Let us, for example, consider the cart centering problem [144] shown in figure 11.15: a cart of mass m which moves from an initial position x with speed $v = \dot{x}$ on a frictionless, one-dimensional track should be stopped at a certain position ($x = 0$ and $v = 0$) in the shortest possible time by repeated application of a force F which is set by the controller. The goal is to evolve a fuzzy decision table for the controller. How do we solve this problem with a genetic algorithm? **Write a simulation of the environment.** Like most engineering problems the cart-centering problem is a continuous system which can be described by a set of differential equations. However, for a computer simulation we usually reformulate the problem as a set of difference equations which should closely approximate the continuous problem. For the cart-centering problem the kinematic difference equations are:

$$x(t+\tau) = x(t) + \tau v(t) \tag{11.12}$$

$$v(t+\tau) = v(t) + \tau \frac{F(T)}{m} \tag{11.13}$$

Clearly, x and $v = \dot{x}$ describe the state of the system and in order to experiment with a simulation of this system we have to specify an initial state x_0, v_0, the time step τ of the simulation, the mass m and an interval for the force F which the controller can apply. Choice of initial conditions, system constants, and intervals for action variables usually poses no problem, because they can be directly obtained from the real system under consideration. However, for more complex systems the following (in general unsolved) problem restricts the usefulness of such simulations: As the simulation continues over many time-steps, the discretization errors accumulate. If we choose τ too large, the discretization error may become intolerably large, whereas choosing τ too small increases the computational cost considerably, but without entirely eliminating the error accumulation. For applications a reasonable procedure is to choose τ in such a way that the controller can compute an action in each time step and to check if the error is still acceptable for this application. In addition to the error tolerance, cost/speed trade-offs have to be taken into account. A low enough price for a relatively slow controller may still open a new market for intelligent control, because, as we have shown in the previous section, most human operators are usually very poor controllers. We recommend some experimentation with the simulation of the dynamic system under consideration to become familiar with the (numerical) behavior of the system. For the cart centering problem a reasonable choice for τ is 0.2 seconds (see e.g. [144]).

Identify percepts and actions. In the cart-centering problem, this is – as in most technical problems – easy: the system state corresponds to the percepts: position x and velocity $v = \dot{x}$. Note, that the number of required sensors can be further reduced by approximating v by δx. The only action of the controller is to choose the force F. For all percepts and actions we restrict the domain to an interval which is reasonable for the system under consideration and with this information we define the semantic of a linguistic variable for location, velocity and force with n adjectives as a fuzzy partition over the interval of interest (see figure 11.16). We choose a triangular membership function as shape. What is important is that the membership functions in the fuzzy partition have considerable overlap, so that there are "no holes" in the interval of interest. For a first approximation, the shape of the membership function is of secondary importance. A reasonable number of partitions (adjectives) ranges from 5 to 23.

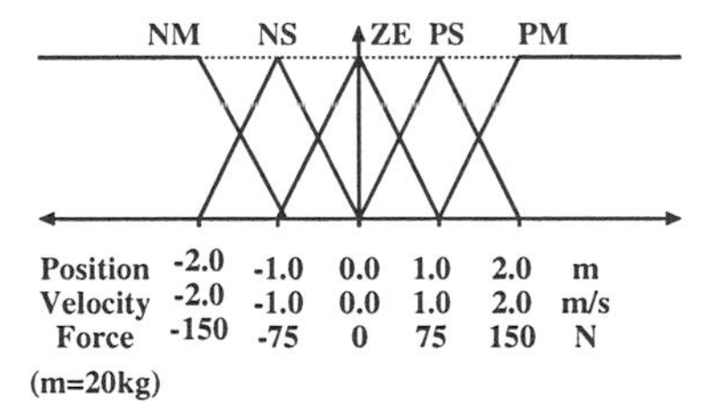

Figure 11.16 The Semantic of Adjectives (from [144] and [63])

However, we can gain a little bit flexibility and the chance to eliminate redundant rules from the decision table by introducing an additional action label U (Unnecessary) with the semantic that rules with action U are pruned (deleted) before the evaluation of the fuzzy controller starts. The agent program for the cart centering problem is represented as fuzzy decision table in figure 11.17.

Position x	**Velocity v**				
	NM	**NS**	**ZE**	**PS**	**PM**
NM	PM	PM	PM	U	U
NS	PM	PM	PM	U	U
ZE	PM	PM	U	NM	NM
PS	PM	U	NS	U	NM
PM	PM	NM	NM	NM	NM

Figure 11.17 A Fuzzy Decision Table (from [144])

If the number of adjectives of a linguistic variable is a power of 2, we can easily code a fuzzy decision table as a binary string by catenating each of the binary coded entries in the decision table in row-major order (see [78]). For the scale $\{U, NL, NM, NS, ZE, PS, PM, PL\}$ we need 3 bits per entry, thus for the fuzzy decision table shown in figure 11.17 a binary string with 75 bits. No modification of the standard simple genetic algorithm is necessary.

If the number of adjectives does not correspond to a power of 2, we have two options: use binary coding with the change that we decode different bit-patterns with the same labels or to use integer coding with a modified mutation operator. The disadvantage of the first variant is that the coding introduces sampling biases which may reduce the performance of the genetic algorithm. The second variant has the disadvantage of a mutation operator which has been tailored specifically to the application: if we mutate an integer position, we add or subtract 1 except for a position with a value of 0 (U) where any other value may be chosen [144].

Formulate the goals of the controller as fitness function. A straightforward formulation of the fitness function is $\max 500 - \frac{1}{N}\sum_{i=1}^{n} T_i$, where T_i is the number of simulation steps needed to center the cart near the goal $(\max(|\ x\ |, |\ y\ |) < 0.5)$ for the i-th initial condition. If the goal has not been reached after 500 time steps, the simulation times out and $T = 500$ [144]. In choosing the training set of initial conditions a reasonable procedure is to choose a small number of equally spaced initial conditions. Again, we have to strike a compromise between computational effort and reliability of the controller. For the record, [144] used 25 initial conditions as training set.

Modelling the fitness function is an important tool to incorporate several goals and/or constraints into the optimization process. As a first example, consider the necessity to evolve a robust controller with as few rules as possible. To achieve this goal, we slightly modify the fitness function:

$$\max w_1 \left(500 - \frac{1}{N} \sum_{i=1}^{N} T_i \right) + w_2 \left(\sum_{i=1}^{25} (l_i = \text{'U'}) \right) \tag{11.14}$$

where w_1, w_2 are weights for minimizing the time to stop the cart at position x and for minimizing the number of rules the controller uses for this purpose. For the example the first goal is more important, so $w_1 > w_2$. The weight of the first term should be an order of magnitude larger than the weight of the second term, so try $w_1 = 10$ and $w_2 = 1$.

Test the controller. If an optimal control strategy is known, compare the optimal strategy with the fuzzy controller for a test set of simulation runs over a randomly selected sample of initial conditions from the intervals of interest. For the example, the average time of the fuzzy controller (3.28 s) is slightly above the average time of the optimal bang-bang controller (2.86 s) [144]. If no optimal control strategy is known, compare with the control strategies of competitors over a sample of initial conditions. If empirical data about the frequency of initial conditions exist, weight the initial conditions with their observed relative frequency in the fitness function. If the controller is mission-critical, apply worst case assumptions and search the initial conditions for which the controller has the worst performance. Is the performance still acceptable?

This approach is suitable for low-dimensional systems with simple dynamics.

11.3.2 Table-driven agents: Coding the semantic

In this section the agent program is still represented as a lookup table for percept sequence action combinations. However, rules are now represented by the parameters (and the types) of the membership functions in their condition and – for a Mamdani-type controller – action parts. For a Sugeno-type controller, the action part consists of the parameters of the function in the action part (e.g. w_1, w_2, w_3 for the rule `IF` X_1 `is` A `and` X_2 *tt* B `THEN` $y = w_1 X_1 + w_2 X_2 + w_3$). A chromosome is a sequence of the binary coded parameters of the membership functions, and for Sugeno-type controllers the binary coded parameters in the action parts. Coding a full decision table for a m-input one-output system of Mamdani-type with membership functions of the same type with k parameters of d bit precision requires $d \cdot k \cdot (m \cdot n + n^m)$ bits with the first summand giving the number of parameters in the condition part and the second the number of parameters in the action part. At an additional cost of t bits per parameter we can represent 2^t different types of k-parametric membership functions. The 2-parametric families include e.g. triangular, Gaussian, sinusoidal, and exponential membership functions.

However, a direct coding of the $m+1$ memberhip functions per rule pays off for sparse rule-sets up to a maximal size $s < \frac{m \cdot n + n^m}{m+1}$ rules. We catenate the $s_{max} \cdot k(m+1)$ parameters with binary coded size parameter which determines how many rules are used in the evaluation of the chromosome.

As application for this kind of agent program we consider the cart-pole centering [77] which is the standard problem in fuzzy genetic algorithms. Figure 11.18 shows a cart with mass m_c which may freely move along a frictionless, one-dimensional track while an inverted pendulum with mass m_l, length l, angle θ and angular velocity $\dot{\theta}$ is free to rotate only in the vertical plane of the cart and the track. The controller can apply a multivalued force F at discrete time intervals horicontally in either direction to the center of the mass of the cart. The objective of the controller is to center the cart at position x_0 while balancing the pole in a vertical position and remaining in $[x_{min}, x_{max}]$.

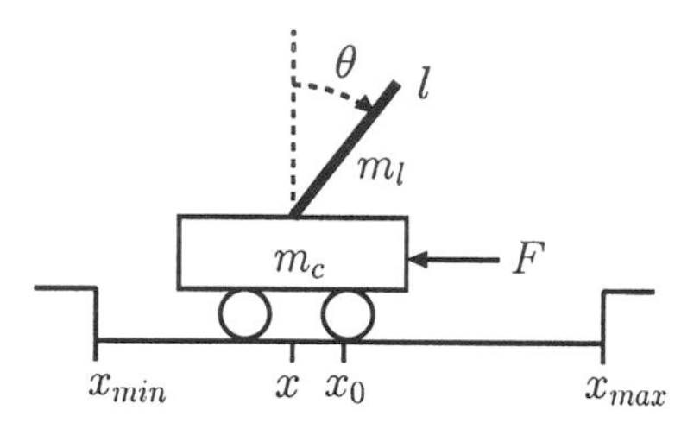

Figure 11.18 Cart Pole Centering Problem (from [77])

The simulation of the environment. We can describe the kinematic equations by the following set of second order differential equations [59], p. 85:

$$\ddot{\theta} = \frac{g \sin\theta + \cos\theta \left(\frac{i - F - m_l l \dot{\theta}^2}{m_c + m_l} \right)}{l \left(\frac{4}{3} - \frac{m_l \cos^2\theta}{m_c + m_l} \right)} \quad (11.15)$$

$$\ddot{x} = \frac{F + m_l l (\dot{\theta}^2 \sin\theta - \ddot{\theta} \cos\theta)}{m_c + m_l} \quad (11.16)$$

The system is inherently unstable and requires fast control, so a reasonable choice for $\tau = 0.02$ seconds [77].

Percepts and actions. System state is described by the position x and the velocity $\dot{x}$ of the cart and the angle θ of the pole with the vertical and the angular velocity $\dot{\theta}$ of the pole. Since we can approximate $\dot{x}$ by δx and $\dot{\theta}$ by $\delta\theta$ for small τ the agent needs a position sensor for the cart and an angle sensor for the pole.

Coding. Actually, we even have more flexibility in coding the rules. Suppose for for the four state variables x, $\dot{x}$, θ, $\dot{\theta}$ we define a fuzzy partition with three adjectives and for the action variable F a fuzzy partition with seven adjectives as shown in figure 11.19.

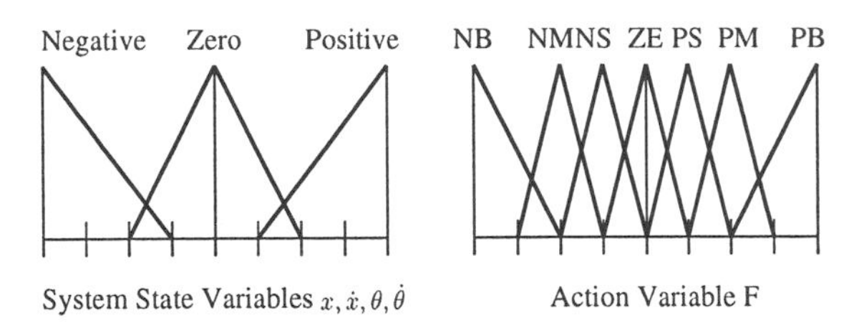

Figure 11.19 State and Action Variables

For the membership functions on the boundary (NEGATIVE, POSITIVE, NB, PB) we need only one parameter, for the rest we need two parameters per label. We first catenate the binary coded parameters for all membership functions in the five fuzzy partitions and then the labels of the actions as 3-bit patterns for the 3^4 actions in the fuzzy decision table. For coding the parameters of the membership functions with 8 bit precision we need 224 bits and for coding the action part of the chromosome we need 243 bits. The whole chromosome thus is 467 bits long.

Consider as alternative a sparse coding with asymmetric triangular membership functions and a Sugeno-type controller with 3 parameters for the function in the action part. For this kind of membership function we need 3 parameters a piece. A rule is thus defined by 15 parameters, at 8 bit precision we need 60 bits per rule. For a sparse decision table with a maximum of 8 rules we need strings with 483 bits.

The coding and the assumptions on the necessary precision of the parameters directly influence the search space size of the genetic algorithm. Therefore, we strongly urge the reader to consider variants and assumptions which reduce search space size and to compute the search space sizes for each coding variant. For example, suppose in the first alternative we are contend to optimize the parameters of the membership functions of the state variables at 5 bit precision: we now need 80 bits for the membership function part. Next, we exploit the symmetry inherent in the problem ...

Goals. We take the opportunity to show the reader a more elaborate example of how multiple objectives and constraints can be handled by means of reward and penalty functions. For additional information on this topic we refer the reader to [115] and [59]. What are the goals of the controller in the cart-pole centering problem?

1. Balancing the pole implies $\theta = 0$ and $\dot{\theta} = 0$. Obviously, we want to minimize the deviation from the equilibrium over the run of a simulation: $\min \sum_{t=0}^{T} \left(\theta(t)^2 + \dot{\theta}(t)^2\right)$. If this goal contributes at most w_1 scores to the fitness, we get $G_1 = \frac{w_1}{1+\sum_{t=0}^{T}\left(\theta(t)^2+\dot{\theta}(t)^2\right)}$.
2. Centering the cart implies $x = 0$ and $\dot{x} = 0$ for the center $x_0 = 0$. We have to minimize the deviations from the center of the track and we get: $G_2 = \frac{w_2}{1+\sum_{t=0}^{T}(x(t)^2+\dot{x}(t)^2)}$.
3. We can roughly classify a controller by the way a simulation ends into the categories failure, time out, and success. Failures imply constraint violations, time-outs mean that the termination condition could not be satisfied, and success that the goal could be reached. Clearly, in the class of successful controllers we prefer those with the minimal time to reach the goal, in the class of failures those with the longest time to failure.
   (a) The pole shoud not fall implies $\mid \theta \mid < 90$. If this constraint is violated, we set the flag $c_1 = 1$. We get $G_3 = \frac{w_3}{1+c_1 \cdot (t_{max}-T)}$.
   (b) Never crashing into the barriers at the end of the track implies $x_{min} < x(t) < x_{max}$. We set $c_2 = (x(t) < x_{min}) \vee (x(t) > x_{max})$. We get $G_4 = \frac{w_4}{1+c_2 \cdot (t_{max}-T)}$.
   (c) We set the flag c_4, if the simulation times out ($T = t_{max}$). The controller is rewarded with $G_5 = c_4 \cdot w_5$.
   (d) Minimizing the time until the pole is balanced and the cart centered requires $c_5 = \max(\mid x(t) \mid, \mid \dot{x}(t) \mid, \mid \theta(t) \mid, \mid \dot{\theta}(t) \mid) < \epsilon$ and we get $G_6 = w_5 + \frac{w_6}{1+T}$.

The fitness function for a single trial (one initial condition) is $G = G_1 + G_2 + G_3 + G_4 + G_5 + G_6$. Further improvement can be expected, if we add fuzzy constraints to the fitness function. The reason for this is that the genetic algorithm reacts more favorable to smooth control surfaces [59]. However, numerically determining robust scores for each partial goal usually restricts fitness function modeling to a few reward and penalty functions.

The combination of a simple genetic algorithm with a fuzzy logic controller has been pioneered by Karr in [79] and [78]. He proposes to encode all triangular membership functions of a rule base on a fixed length binary string. Details of this work and experimental results on the cart-pole-centering problem are reported in [77] and [76]. Karr applied his method to the control of autonomous spacecraft rendezvous ([70] and [71]), fuzzy process control in [72], time-varying control problems in [79], control of an exothermic chemical reaction ([73],[74], and [75]), fuzzy control of pH ([66] and [67]), adaptive process control ([80], [68], and 82] in [159]), and real time process control in [81].

11.3.3 Table-driven agents: Combinations

In [63] the approach of [144] is improved by determining the rules and the base lengths of the triangular membership functions. [63] use an integer coded chromosome. The number of labels determines the number of integers needed for the coding. For a two-input controller with 5 labels the decision table is a 5×5 rule base which needs the first 25 positions on the chromosome. For the labels of each input variable 5 positions are reserved for the coding of the base length. This modification adds 10 additional positions to the chromosome. It is obvious that only 5 different base lengths can be coded with this approach. [63] report improvements in response time of more than thirty percent as the result of controllers bred with limited membership function tuning.

In [139] a combination of self organizing feature maps and genetic algorithms is used to automate the design of fuzzy control systems. In the first step of this algorithm the parameters of gaussian membership functions (center and width) are extracted from the observations by self organizing feature maps. The self organizing feature maps compute n vectors of cluster centers in a d-dimensional feature space whose the first l dimensions correspond to input variables and whose last $m = d - l$ dimensions correspond to output variables. The width vector of the cluster is computed by taking first the absolute value of the difference of the cluster center vector and the nearest observation vector and then dividing by an arbitrary overlap parameter. The ith element of the cluster center vector c_j and the ith element of the cluster width vector w_j are then taken to be the parameters of the gaussian membership function $A_{i,j}$ of a fuzzy variable X_i. representing the ith dimension of the feature space. For each cluster j a rule is defined, where the condition part is the conjunction of the l input variables and the action part the conjunction of the m output variables:

$$\textbf{if } X_1 \textbf{ is } A_{1,j} \textbf{ and } \ldots \textbf{and } X_l \textbf{ is } A_{l,j} \textbf{ then}$$
$$Y_1 \textbf{ is } A_{l+1,j} \textbf{ and } \ldots \textbf{and } Y_l \textbf{ is } A_{l+m,j}$$

In the second step the n rules found for each cluster are encoded as described in [77], only with gaussian membership functions instead of triangular membership functions, and a genetic algorithms is applied to further improve the performance of the fuzzy rule system derived above. [139] apply this combination of algorithms to a classification problem and to a process simulation problem with good results.

11.3.4 Simple reflex agents: Variants of messy genetic algorithms

In [37] messy genetic algorithms which process variable-length character strings are introduced. For additional information see [36] and [38]. Messy genetic algorithms use position independent coding and allow for underspecification and overspecification. Consider for example the string 01101. In the messy genetic algorithm, such a string may be coded as (51)(10)(21)(31)(40) or as (21)(51)(40)(31)(10) with the first position identifying the gene and the second position identifying its meaning. The mutation operator works only on the second position, crossover is replaced by cut and splice operators: For example, if we cut the above chromosomes after the second gene, we obtain the following lists of pairs (51)(10), (21)(31)(40), (21)(51), and (40)(31)(10). The splice operator catenates two arbitrary lists and the result of may be any of the following combinations:

(51)(10)(21)(31)(40), (51)(10)(21)(51), (51)(10)(40)(31)(10),
(21)(51)(40)(31)(10), (21)(51)(51)(10), (21)(51)(21)(31)(40),
(21)(31)(40)(51)(10), (21)(31)(40)(21)(51), (21)(31)(40)(40)(31)(10),
(40)(31)(10)(51)(10), (40)(31)(10)(21)(31)(40), (40)(31)(10)(21)(51).

The following cases can happen:

Overspecification, that is having more than one version of the same gene in the string, is simply handled by positional precedence. The first version of a gene encountered on the string is expressed and used in the evaluation of the string.

Underspecification, that means some genes are missing on the string, is harder to handle. [37] suggest the use of *competitive templates* to fill in the missing information. A *competitive template* is a locally optimal string found by a hill-climbing method.

What are the advantages of this coding? Obviously, we can represent variable-length strings, and it is claimed that good building-blocks have may propagate faster, because the disruptive effects of crossover can be reduced by an appropriate permutation of the genes.

We show, how this coding scheme has been used for coding fuzzy rule-bases by [59]. Hoffmann defines a fuzzy controller as $x_3 = f(x_1, x_2)$ so that we can easily define the percepts x_1, x_2 and the action variable x_3. A gene may be now a pair or e.g. a triple where the first position identifies the variable and the rest codes either the label or represents the membership function. For the first rule specified in the decision table shown in figure 11.17 we get (35)(11)(21) which can be read as IF $v = NM$ AND $x = NM$ THEN $F = PM$.

Next, we have to specify, how overspecified rules are interpreted. We have two choices. First, invoke the dominance rule as suggested in [37] and to use only the first gene for a variable or as suggested in [59] to take the union of the two membership functions. For example, the string (35)(11)(21)(12) is interpreted as IF ($v = NM$ OR $v = NS$) AND $x = NM$ THEN $F = PM$.

What should be done, if a string is underspecified? If no action variable is present, the easiest approach is to generate a gene for the action variable at random. For example, the string (11)(21) may be rewritten to (11)(21)(33) or to (11)(21)(34). We leave the decoding as an exercise to the reader. And what happens, if no input variable is present? Again, we can either discard this rule as meaningless or complete the rule at random.

Nesting of lists allows coding of rule-bases. For example ((11)(21)(34)) ((15)(31)) represents the rules IF $v = NM$ AND $x = NM$ THEN $F = PS$; IF $v = PM$ THEN $F = NM$. Cut and splice work on the rule level and on the rule base level. In addition, [59] provided weights for the rules which allow hierarchically structured fuzzy controllers.

Other variants of variable coding have been proposed in the literature. For example, a context dependent coding method with variable-precision representation of fuzzy numbers and junk code is presented in [108] and [109]. In table 11.4 we have collected a few references to control applications of fuzzy genetic algorithms.

11.4 FUZZY GENETIC PROGRAMMING

In this section we consider agent programs with the syntax of a context-free language and the semantic of a Mamdani-type fuzzy controller with centroid defuzzification. A natural representation of such a fuzzy control strategy is the derivation tree of the context-free expression. We present simple genetic algorithms over k-bounded context-free languages which are a generalization of Koza's genetic programming approach [34]. Next, we summarize several approaches to find good training sets for learning a fuzzy control strategy in non-deterministic environments. Finally, we give a survey of the state of the art in genetic programming theory and its application to fuzzy control.

11.4.1 Context-Free Fuzzy Rule Languages

By $L(G)$ we mean the language L generated by grammar G, this is the set of sentences (words) generated by G. By a *k-bounded* language $L(G)$ we mean the set of words generated by G with at most k derivation steps.

A *context-free grammar* G is a 4-tuple $G = (V_{NT}, V_T, P, S)$, where V_{NT} is a finite set of nonterminal symbols, V_T is a finite set of terminal symbols disjoint from V_{NT}, P is a finite subset of $V_{NT} \times (V_{NT} \cup V_T)^*$ called the production rules or productions of the grammar and S is a distinguished symbol in V_{NT} called the start symbol of G [1]. We denote the empty word by ϵ. A *sentential form* of G is defined recursively: S is a sentential form and if xyz is a sentential form and $y \rightarrow u$ is in P, then xuz is a sentential form too. A *sentence* or a *word* w of $L(G)$ is a sentential form without nonterminal symbols. Clearly, the *programs* in genetic programming are words or sentences of a context-free language. For "real" grammars (for example figure 11.20) we use a version of Backus Naur

Test environments	
Cart centering	[144], [63]
Cart pole centering	[77], [142], [64], [87], [86], [54], [156], [11], [42], [85], [105], [113], [114], [132], [134]
Double inverted pendulum	[111]
Walking of a biped robot	[114]
Chaotic bouncing ball	[83]
Mobile robots	
Goal finding	[58], [59]
Collision Avoidance	[59]
Truck backing	[63]
Corridor tracker	[109], [110], [65]
Multi point turner	[109], [110]
Parallel parker	[109], [110]
Industrial applications	
Flight control	[130]
Twin tank liquid level regulation	[111]
Power plant control	[114]
Battery charger	[140]
Wastewater treatment plant	[21]
Temperature control of Martian oxygen production system	[85]
Tool feed rate in turning	[143]
Magnetic levitation	[51]
Genetic Algorithm	
Dynamic adaptation of GA parameters	[106], [107], [52]

Table 11.4 Control Applications of Fuzzy Genetic Algorithms

```
S := <frb> ;
<frb> := <rb> <rb> <rb> <rb> ;
<rb> :=  <rule> | '(' <rule> 'OR' <rb> ')' |
         '(' <rule> 'OR' <rb> ')' ;
<verbexpr> := <adjective> |
              '(' <adverb> <verbexpr> ')' |
             '(' <adverb> <verbexpr> ')' ;
<adjective> := 'LOW' | 'MEDIUM' | 'HIGH' ;
<adverb> := 'VERY' | 'AROUND' | 'UPPER' |
            'UPPER' | 'LOWER' | 'ABOVE' | 'BELOW' ;
<rule> := '(' <verbexpr> 'IF' <condition> ')' ;
<condition> := '(' <variable> 'IS' <verbexpr> ')' |
               '(' <condition> 'AND' <condition> ')' ;
<variable> := 'L' | 'EL' | 'ST' | 'SL' | 'EL1' ;
```

Figure 11.20 The Grammar for Team fu8 (BNF8)

Form (BNF) [117]: Symbols from V_T are delimited by ", for example "D1", symbols from V_{NT} are delimited by < and >, for example <fe>. *Derives* is denoted by :=, *or* by |, and *catenation* is denoted by juxtaposition of symbols. Whenever we want to represent *catenation* explicitly, we denote *catenation* as $\star$, for example aS corresponds to $a \star S$ with explicit representation of the catenation operation. For simplifying automatic processing of grammars in Backus-Naur form we use a semicolon ; as production rule separator and we include the specification of the start symbol as the *first* derivation.

The syntax of the context-free fuzzy rule language shown in figure 11.20 is directly derived from [121] and [151]. A complete definition of its semantic can be found in [29]. The syntax of the fuzzy rule language used in one of the experiments described below is shown in figure 11.20. The semantic of the adjectives LOW, MEDIUM, and HIGH is defined by the membership functions shown in figure 11.21. The membership function is approximated with 23 equidistant support points over the universe of discourse. In the experiments with the MIT beer distribution game the range of the base set of the variables L, EL, and $EL1$ was $[0, 25]$ and for SL and ST the range was $[0, 50]$.

The fuzzy rule language shown in figure 11.20 has several features which distinguish it from the fuzzy logic controllers of the previous section. The first is the use of adverbs (hedges or modifiers) for generating "derived" adjectives with a modified membership function. In the early days of fuzzy control this

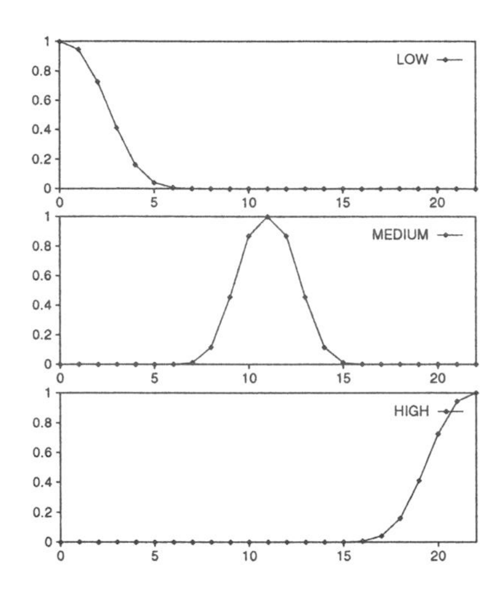

Figure 11.21 The Semantic of the Adjectives LOW, MEDIUM, and HIGH

construction was mainly motivated by the aim of describing human operator control strategies [62]. We reintroduce it into fuzzy control, because with this construction we can approximate any membership function by a combination of derived adjectives sufficiently close with purely syntactic means. In fact, the construction is similar to the approximation of a real number by a rational series (see [20]) and constitutes an efficient and general solution to the problem of simultaneously optimizing the structure of a fuzzy logic controller and the tuning of its membership functions.

In the experiments we have included three types of adverbs, namely adverbs based on concentration and dilation operations (VERY and AROUND), adverbs based on a shift of the membership function up or down in relation to the base set (UPPER and LOWER), and adverbs which set all elements to the maximum reached when scanning from the end of the scale (from the low end: ABOVE, from the high end: BELOW). Evaluation of the fuzzy production rules is implemented as in [151], for the defuzzification of the control actions the center of gravity method is used.

The second feature of the language shown in figure 11.20 is redundancy in the production rules. The motivation for specifying redundant productions is the incorporation of a priori information about the structure of the fuzzy logic controller. Redundant grammar rules have a strong influence on the way the genetic algorithm samples programs.

The third feature of the language shown in figure 11.20 is that four fuzzy controllers are evolved simultaneously. This is specified in the first production rule in figure 11.20. Knowledge about the modules of a large fuzzy controller can be incorporated by simply adding or changing a production rule of the grammar.

What is remarkable about this is that the specification of a context-free grammar allows considerable flexibility in the specification of a fuzzy controller **without** any need for changes in the genetic machinery. As long as the grammar remains context-free, **no changes** are required to the genetic algorithm presented in the next section.

To demonstrate the ease and flexibility of grammar based representations we give a last example. Suppose, we want to find out how much a priori knowledge we really need in order to evolve good fuzzy control strategies for the MIT beer distribution game. Or just suppose, we did not find Kahneman and Tversky's work [147] on adaptive ordering heuristics. Can we do without it? In other words, can we discover the preprocessing necessary for state-space reduction that we have analyzed at length in section 11.1? Although we cannot answer this question, we show how the fuzzy rule language can be extended to include the language elements necessary for specifying preprocessing computations. We replace the first production in figure 11.20 by:

⟨frb⟩ := ⟨preprocess⟩ ⟨rb⟩ ⟨rb⟩ ⟨rb⟩ ⟨rb⟩ ;

We add the following rules for preprocessing:

⟨preprocess⟩ := ' L = FUZZIFY ' ⟨expr⟩ ';ST = FUZZIFY ' ⟨expr⟩ '; S = FUZZIFY ' ⟨expr⟩ ';' ;
⟨expr⟩ := ⟨state variable⟩ | ⟨expr⟩ ⟨op⟩ ⟨expr⟩ ;
⟨op⟩ := '+' | '-' | '×' | '÷' ;
⟨state variable⟩ := all variables in the game board.

11.4.2 Simple Genetic Algorithms over k-Bounded Context-Free Languages

In this section we extend simple genetic algorithms over fixed-length strings from a finite alphabet to simple genetic algorithms over k-bounded context-free languages. Algebraically, the basic idea is a mapping from strings to complete derivation trees and from catenating symbols to tree substitution (loosely speaking). In figure 11.5 we show the well-known pseudo-code of a simple genetic algorithm. The procedures which have to be changed to accomodate the above mapping (INITIALIZE, EVALUATE, CROSSOVER, and MUTATE) are shown in small caps in figure 11.5. Note, that the reproduction mechanism is not influenced by the change in representation. However, before we discuss the changes, we summarize the necessary results about context-free languages:

A *derivation tree D* is a labeled-ordered tree for a context–free grammar $G = (V_{NT}, V_T, P, X)$ with the following properties: X labels the root of D. For all subtrees $D_1, \ldots, D_k$ of the sons of the root X with the root of the subtree D_i labeled X_i, $X \to X_1 \ldots X_k$ is a production of P. If X_i is a nonterminal symbol, D_i is a derivation tree, if X_i is a terminal symbol, D_i is the single node X_i. If the empty word ϵ is the root of D_1, the only subtree of D, then $X \to \epsilon$ is a production in P [1], p. 139.

The *frontier* of a derivation tree is the string obtained by concatenating the leaves of the derivation tree (in order from the left) [1], p. 140.

$\Longrightarrow$ denotes the relation *derives*, $\overset{k}{\Longrightarrow}$ denotes the k-fold product of the relation $\Longrightarrow$ and $\overset{*}{\Longrightarrow}$ denotes the k-fold product of the relation $\Longrightarrow$ for an arbitrary but finite k.

Theorem 11.1 *Suppose $G = (V_{NT}, V_T, P, X)$ is a context-free grammar. Then $X \overset{*}{\Longrightarrow} \alpha$ if and only if there is a derivation tree with the sentential form α as frontier [1], p. 143.*

Proof See [1], p. 141. ∎

From theorem 11.1 the following consequences are obvious: For each word w of $L(G)$ there exists at least one derivation tree with frontier w. We can retrieve a word w from its derivation tree by extracting the frontier of the derivation tree.

A *complete derivation tree* is a derivation tree whose frontier is a word w of $L(G)$. All leaves of a complete derivation tree are terminal symbols and all interior nodes of a complete derivation tree are nonterminal symbols. Three examples of complete derivation trees for the grammar shown in figure 11.20 are presented in figure 11.22.

However, the reason for using complete derivation trees to represent words of $L(G)$ is, that although no string manipulation operation which is closed over words of $L(G)$ is known to us, *a tree-manipulation operation closed over complete derivation trees exists.*

Definition 11.1 *Suppose, D is a complete derivation tree for a context-free grammar $G = (V_{NT}, V_T, P, X)$ which contains at least one $S \in V_{NT}$ with a set of subtrees $D_1, \ldots, D_n$ whose roots are labeled with symbols from $V_{NT} \cup V_T$. I is a complete derivation tree for the context-free grammar $G' = (V_{NT}, V_T, P, S)$ whose root is labeled with S. A subtree substitution is feasible if I replaces a subtree D_i of D which is labeled with S.*

In other words, a subtree substitution is feasible if I replaces a subtree D_i of D whose root has the same label. Check this for the example shown in figure 11.22.

Theorem 11.2 *Feasible subtree substitutions are closed over complete derivation trees and thus over $L(G)$.*

Proof Trivial. The operation is possible, because by assumption D contains at least one S. (However, if we want to do without this assumption, the operation is undefined if D does not contain at least one S and we complete the operation as a right or left identity.) We only have to check that the result of the feasible subtree substitution operation is still a complete derivation tree. Obviously, all parts of D which have not been changed by the subtree substitution are still complete. The new subtree I is complete by its definition. Because the roots of I and of the subtree D_i which should be replaced by I have the same label and because I and D_i are complete derivation trees, no substitution which is not defined by a production is introduced into the result by the replacement operation. Therefore, the resulting derivation tree is complete ■

For complete derivation trees feasible subtree substitution has the same role as catenation for strings.

The algorithms for INITIALIZE, EVALUATE, CROSSOVER and MUTATE are based on the following operations on complete derivation trees: tree traversal, decoding a word, extracting the n-th subtree, substituting the n-th subtree and generating a random complete derivation tree. The change of EVALUATE is trivial, simply plug in the new algorithm for decoding a word (depth-first, left-to-right tree traversal combined with catenating the terminals).

Random generation of complete derivation trees implements INITIALIZE and serves as building block of MUTATE. In Figure 11.23 we present the pseudo-code for the word generating function. Optimal minimax algorithms for CHOOSE_production and CHOOSE_PARTITION for the case of no a priori information about the optimal solution are given in [10].

In addition, MUTATE consists of a stochastic choice function CHOOSE for determining the subtree to replace and the function for substituting the n-th subtree. The pseudocode for MUTATE is shown in figure 11.24.

CROSSOVER employs CHOOSE twice: to determine the subtree to be substituted in the first parent and to determine the subtree to be extracted from the second parent. In addition, substituting the n-th subtree and extracting the n-th subtree are necessary for the subtree swap. An example of the crossover operation is shown in figure 11.22. The pseudocode for CROSSOVER is shown in figure 11.25.

Again, performance of the genetic programming algorithms depends on selecting a "good" probability distribution over the subtrees in the derivation tree. The heuristic of CHOOSE for CROSSOVER and MUTATE is: The probability of choosing a subtree is proportional to the square of the size of the subtree. Simulation results of several heuristics can be found in [32]. However, the anal-

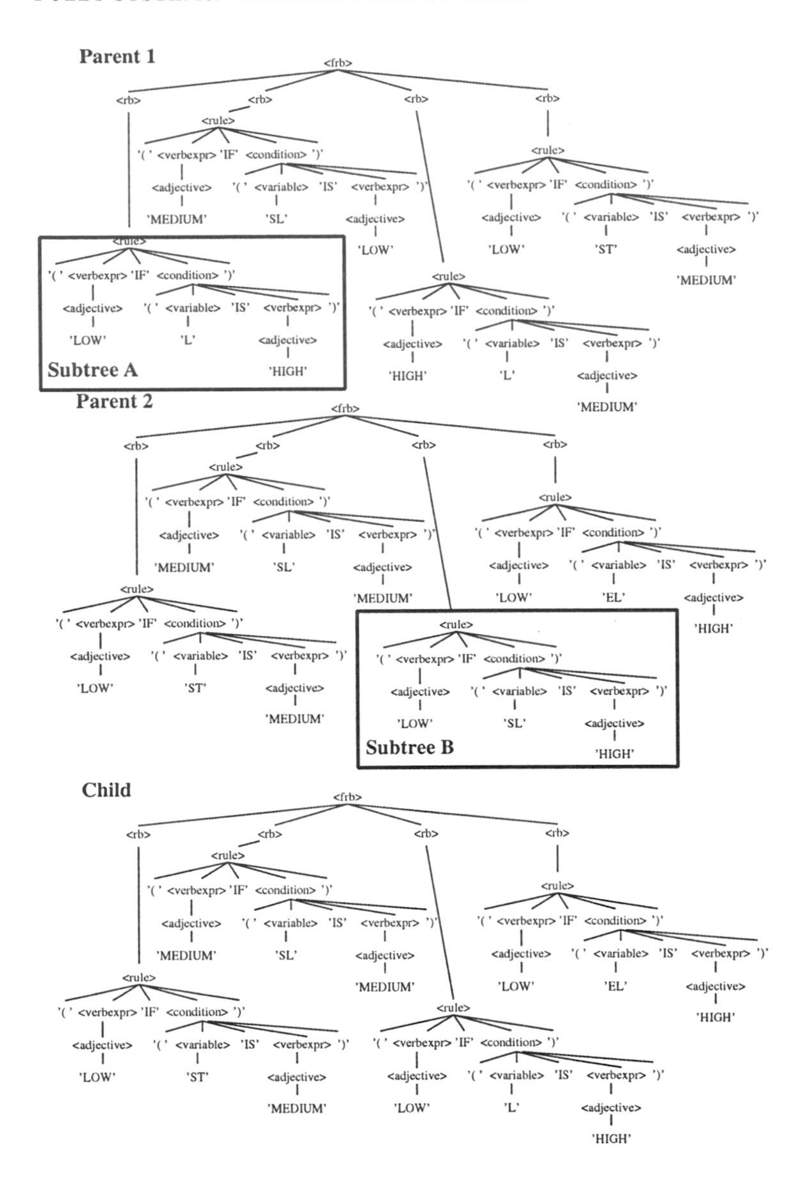

Figure 11.22 Crossover: Feasible Subtree Substitution of Subtree B by Subtree A

ysis of the choice of a probability distribution over the nodes of a complete derivation tree is left for further research.

```
tree function INITIALIZE(symbol root, int d);
 int production; list_of_symbol production_symbols;
 list_of_int partition; list_of_trees subtree_list;
begin
 if terminal(root) then begin
  return(new_tree(root, new_list))end
 else begin
  production := CHOOSE_PRODUCTION(access_ST(root), d-1);
  production_symbols := access_PT(production);
  partition := CHOOSE_PARTITION(production_symbols, d-1);
  subtree_list := new_list;
  while not empty(production_symbols) do begin
   subtree_list := add_list(subtree_list, INITIALIZE(
     head(production_symbols),head(partition)));
   production_symbols := tail(production_symbols);
   partition := tail(partition); end
 return(new_tree(root, subtree_list)) end
end
```

Figure 11.23 Pseudocode for INITIALIZE .

```
procedure MUTATE
begin
    CHOOSE insertion_node in tree;
    d=k-sizeof(tree)+sizeof(subtree);
    INITIALIZE(insertion_node, d);
    insert_subtree in tree
end
```

Figure 11.24 Pseudocode for MUTATE.

Finally, we apply this algorithm to the MIT beer distribution game with reproduction proportional to fitness, elitist selection strategy, and initialization without duplicates and report the result of two experiments. Both experiments were conducted with a population size of 100 for 100 generations and in both experiments retailer, wholesaler, and distributor were assumed to have the same ordering heuristic. However, the depth of the derivation trees was restricted to 10 in the first and 12 in the second experiment. Mutation and crossover rates were 0.025 and 0.35 in the first and 0.05 and 0.70 in the second experiment. Although the fuzzy rule language was the same in both experiments, different (but equivalent) grammars were used as word generating device. In the first

```
procedure CROSSOVER
begin
    CHOOSE insertion_node in tree_1;
    CHOOSE extraction_node with the same label
        as the insertion_node in tree_2;
    if (extraction_node exists) begin
        extract_subtree from tree_2;
        insert_subtree in tree_1 end
    else return tree_1
end
```

Figure 11.25 **Pseudocode for CROSSOVER.**

experiment (labeled `fu6` in figure 11.26) total team inventory cost was 692.5, whereas in the second experiment (labeled `fu8` in figure 11.26) a team inventory cost of 330.5 could be reached.

Table 11.5 Rule Base for fu8

Retailer, Wholesaler, Distributor:

```
Expected Loss Rule Base:
((LOW IF (ST IS (UPPER MEDIUM))) OR
((LOW IF (EL1 IS HIGH)) OR
((MEDIUM IF (ST IS (UPPER MEDIUM))) OR
(LOW IF (EL1 IS HIGH)))))
Order Rule Base:
(LOW IF (EL1 IS HIGH))
```

Factory:

```
Expected Loss Rule Base:
(LOW IF (EL1 IS HIGH))
Order Rule Base:
(((UPPER (AROUND (LOWER (AROUND HIGH)))) IF (EL IS LOW)) OR
(LOW IF (EL1 IS HIGH)))
```

In table 11.5 we show the rule base which evolved as the best ordering heuristic in the second experiment. Nevertheless, compared with the average team's performance in Sterman's experiment, the ordering heuristic of the `fu8` team is a significant improvement. Figure 11.26 shows that both fuzzy ordering heuristics are improvements compared to average team performance. However, even the better fuzzy ordering heuristic is almost 60 percent above the MIT benchmark heuristic and more than 80 percent above the best known solution found by genetic programming.

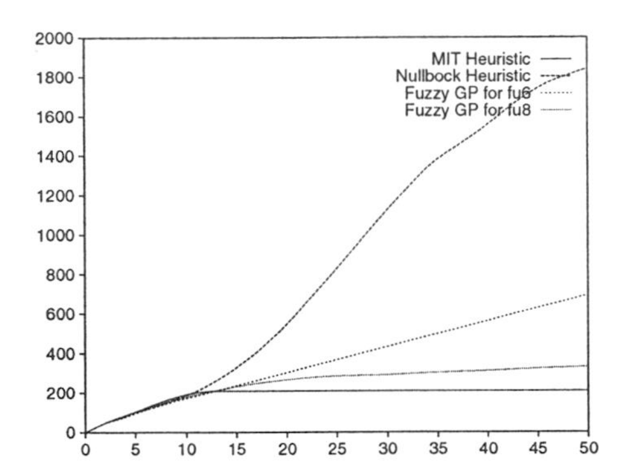

Figure 11.26 Comparing Inventory Cost for Fuzzy GP Strategies

11.4.3 Non-Deterministic Environments

So far we have paid lip service to the importance of selecting "good" sets of scenarios for training and testing, but we have provided almost nothing to help the reader with this task. In this section we explain the following approaches to this important task:

1. Scenarios as cluster means of empirical data.
2. Scenarios from qualitative analysis.
3. Sampling of environments.
4. Coevolution of agents and environments.

Scenarios as cluster means of empirical data. Frequently we encounter the following situation in process automation projects: Control of a complex, nonlinear process which is currently controlled by human operators with different efficiency should be improved. Due to automated data gathering systems we literally have thousands of process trajectories with system state, control actions, yield, and operator. A way to improve total process yield across all operators even without a model of the process is to search a control strategy which minimizes some distance-measure (e.g. mean-square error) from the trajectories of the most efficient human operators. To reduce the computational cost of using thousands of trajectories in the genetic search of the control strategy which best fits the trajectories of the control group, clustering process data and using only the cluster means as training set is recommended.

Scenarios from qualitative analysis.

In economic and management science complex decisions are often considerably simplified by qualitative analysis of possible scenarios which give a crude classification of future events. For an introduction to such portfolio approaches in marketing, see [98] and [112]. Such a portfolio of scenarios which combines different inventory levels in the MIT beer distribution game with typical customer order processes is used in the following to show that robust strategies can be bred by fuzzy genetic programming.

The game board in figure 11.27 shows four sets of initial conditions and in figure 11.30 four order streams are shown. The first three initial conditions differ only with regard to their inventory level, namely medium, low, and high, and in which the system is in equilibrium. For the fourth initial condition the system is out of equilibrium.

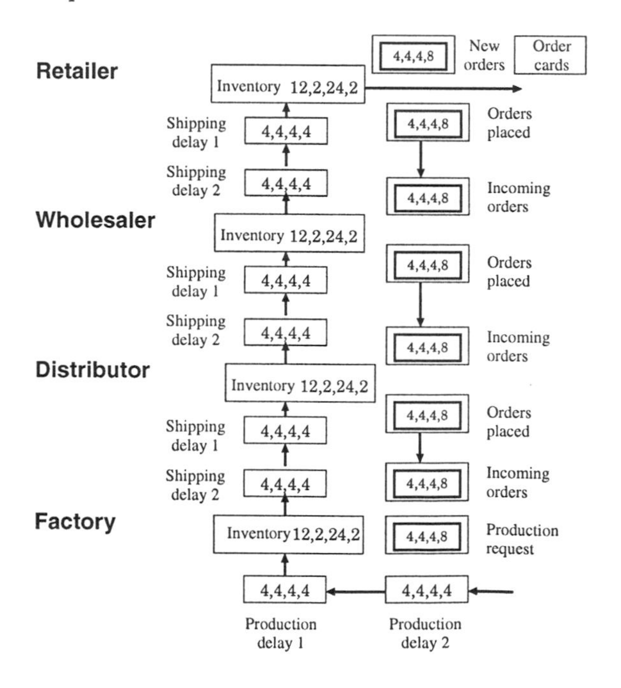

Figure 11.27 Game Board with Initial Conditions (Training)

The order streams in figure 11.28 characterize the following situations: Order stream 1 shows a demand jump from one level to another. Such demand

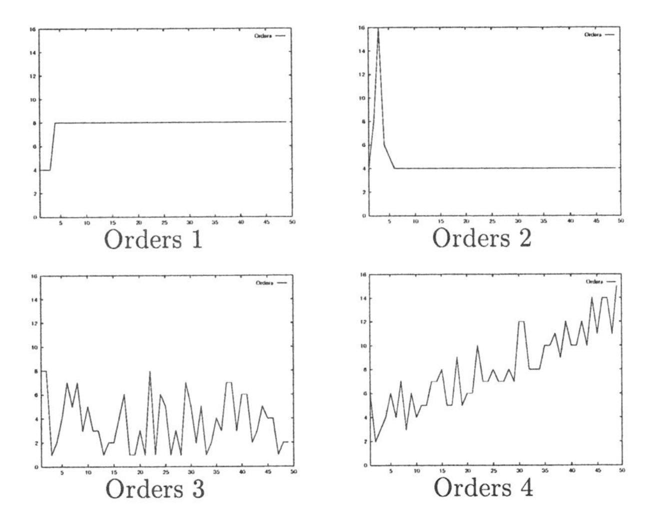

Figure 11.28 Customer Order Streams (Training)

jumps may be attributed e.g. to a competitor dropping out of the market. Order stream 2 is a demand peak levelling off to a constant demand for product replacement. This is typical for the introduction of new and innovative products. Order stream 3 is a constant demand with random fluctuations and order stream 4 is a linearly increasing demand with random fluctuations.

For initial condition 1 and order stream 1, table 11.6 summarizes the results of Sterman's experiments with human players (Mean), J. Sterman's benchmark solution which was computed by complete enumeration over the plausible parameter space of an adaptive ordering heuristic (MIT), a simple (binary) genetic algorithm's solution for the same adaptive ordering heuristic (GA), the currently best known solution bred by genetic programming with a context-free language for arithmetic expressions (GP), the best fuzzy genetic programming solution (FGP 1,1) and the fuzzy genetic programming solution which has been trained with all combinations of order streams and initial conditions shown in figure 11.28.

However, in order to test the robustness of the order heuristic we generated a test scenario with similar test cases (see figures 11.29 and 11.30). In table 11.7 we show the rule base of the best ordering heuristic of all combinations of the initial conditions and order streams, table 11.8 summarizes the performance of this ordering heuristic for the different combinations of intial conditions and order streams for the training and test scenario. We have compared the total inventory cost of the trainig scenario (24 554.0) with total inventory cost of the

Table 11.6 Comparing Inventory Cost

	Mean	MIT	GA	GP	FGP 1,1	FGPall 1,1
Team	2028	210	181.5	180	330.5	408.5
Retailer	383	44	36	36	61	148
Wholesaler	635	52	42	42	68	80.5
Distributor	630	60	48	48	101	83.5
Factory	380	54	55.5	54	100.5	96.5

Table 11.7 Rule Base for fu10

Retailer, Wholesaler, and Distributor:

```
Expected Loss Rule Base:
((LOWER (ABOVE LOW)) IF (EL IS MEDIUM))
Loss Rule Base:
((LOW IF ((EL IS MEDIUM) AND (L IS (LOWER (VERY LOW))))
OR ((BELOW MEDIUM) IF ((ST IS HIGH) AND (EL IS HIGH)))
```

Factory:

```
Expected Loss Rule Base:
((LOW IF ((EL IS MEDIUM) AND (L IS (LOWER(VERY LOW))))
OR ((BELOW MEDIUM) IF ((ST IS HIGH) AND (EL IS HIGH)))
Loss Rule Base:
((VERY HIGH) IF (EL1 IS MEDIUM))
```

test scenario (25 895.5). The cost difference of 5.46% between training and test scenario shows that the order heuristic of table 11.7 is quite robust and works well - even in previously unknown situations.

Sampling of Environments. In [44] Grefenstette and Fitzpatrick investigate the trade-off between sample size and sample error for non-deterministic environments and its effects on the performance of a genetic algorithm. Their recommendation is to prefer relatively small sample sizes with high sampling error over large samples with small sample error in the expected fitness. The genetic algorithm is quite robust with respect to noise in the fitness function.

Table 11.8 Inventory Cost

	Initial Conditions	Order Stream 1	2	3	4	Sum
Training	1	408.5	1393.5	1975.5	1344.0	
Training	2	1954.0	764.0	1264.5	1869.0	
Training	3	1168.0	2277.5	2798.0	1747.5	
Training	4	1452.0	989.0	1518.0	1631.0	24554.0
Test	1	1027.0	1513.0	1938.5	1637.0	
Test	2	621.5	762.5	1138.0	2282.0	
Test	3	1955.5	2410.5	2726.5	1920.0	
Test	4	491.0	905.5	1283.0	2693.0	25895.5

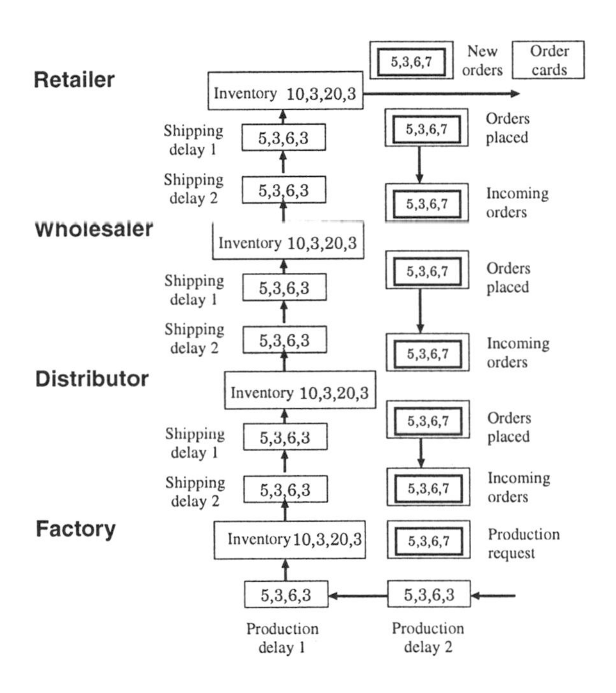

Figure 11.29 Game Board with Initial Conditions (Test)

In the case of no a priori information about the probability distribution and risk-neutrality of the decision-maker, it is recommended to sample environ-

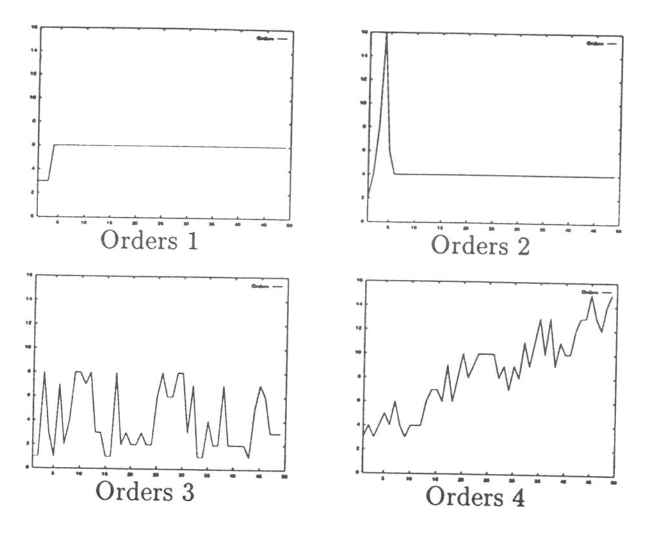

Figure 11.30 Customer Order Streams (Test)

ments uniformly from the region of interest. If, however, a probability distribution on the environment exists, environments should be sampled accordingly. The case of a risk-adverse decision-maker is treated below.

Coevolution of Agents and Environments. If the task is to develop very robust and reliable control strategies, none of the previous approaches makes sure that the environments get more challenging as the performance of the controller improves. To solve this problem Hillis applied predator-prey coevolution to improve genetic search (see [55] and [56]). In nature there are many examples of organisms that evolve defenses to parasites that attack them only to have parasites evolve ways to break the defense again. A threatening example of such a biological arms race is the growing immunity of many bacteria to antibiotics. In decision theory we can characterize predator-prey coevolution as a non-cooperative two person game.

The basic idea of coevolution of agents and environments is to introduce a second genetic algorithm on a population of environments which evolves environments which minimize the fitness of agents (see figure 11.31). For the MIT beer distribution game this implies coding the game state and the customer order process on a chromosome. For the cart centering problem, the chromosome contains the initial condition (position x and velocity v). The approach is not yet very popular, due to higher computational costs. However, it has successfully been applied for breeding a fuzzy controller for the collision avoidance

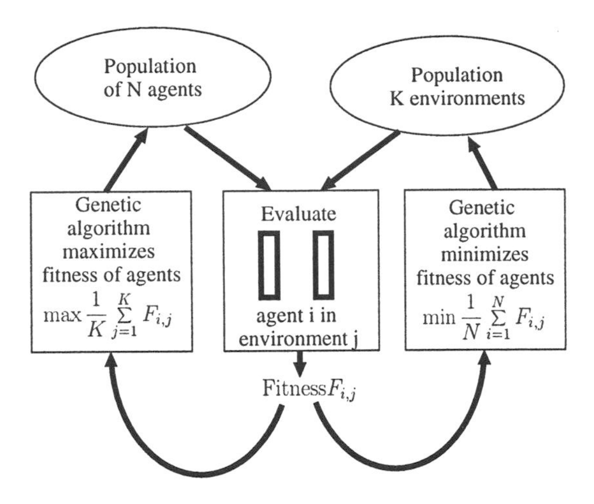

Figure 11.31 Coevolution of Agents and Environments

module of a mobile robot in [59] for breeding test-environments with obstacles of increasingly complex shape.

11.4.4 Results and Applications

For an impressive collection of genetic programming applications we refer the reader to [100], [101], [84], [3], [99], and [102]. The concept of grammar-based genetic algorithms for general languages (including unrestricted and context-sensitive languages) and the development of a syntactically closed crossover operator have been first proposed in [4]. The existence of genetic algorithms over context-free languages has been tacitly assumed for the comparison of crisp and fuzzy rule-languages in [25] (a full version of this appeared as [27]), although no algorithms of this kind existed at that time. The main theoretical result of this early work is the proof and quantification of the reduction of search space size by using context-free fuzzy rule languages. In 1994 explicit representations of grammars have been used in two applications, namely for evolving 3-D models of jet-planes with a hybrid GA/GP combination [118] and for genetic micro programming of neural networks [48]. The first extension of simple genetic algorithms to context-free languages and their analysis with the help of formal power series is in [28] which finally appeared as [29] and revised as [32]. The same class of algorithms has independently been developped by P. Whigham ([152], [153], and [154]), albeit without analysis by formal power series. In [49], Gruau adopts a similar approach for improving genetic programming performance. A uniform initialization procedure for this class is presented in [9], the proof of being an optimal minimax strategy in the case of no a priori information and the application to Koza's genetic programming variant is in [10].

In a logic programming context Wong and Leung have developped a genetic algorithm based on definite clause grammars [158] which evolved from [157]. One application of their algorithm is learning a program for a fuzzy relation in fuzzy Prolog. Variants of fuzzy control strategies for the MIT beer distribution game have been investigated in [33], [30], and [31].

Probably the first application of fuzzy genetic programming with Koza's genetic programming variant is reported in [18]. [2] derive a fuzzy control strategy for the cart-centering problem with Koza's genetic programming variant. However, they specify a fuzzy rule-language with a finite number of adjectives. This has the disadvantage that the membership functions defining the adjectives cannot be tuned by adverbs. [146] applied the same method to breed a fuzzy control strategy to properly steer a mobile robot along a prescribed path. They report that the fuzzy control strategy which has been trained with a small set of preselected environments generalized properly to other initial conditions. Improved results based on a modified fitness function can be found in [145].

References

[1] Aho, A. V. and Ullman, J. D. (1972). *The Theory of Parsing, Translation and Compiling, Volume I: Parsing*, Prentice–Hall, Inc., Englewood Cliffs, N.J.

[2] Alba, E., Cotta, C., and Troya. J. M. (1996). Type-constrained genetic programming for rule-base definition in fuzzy logic controllers. In Koza et al., 255–260.

[3] Angeline, P. J. and Kinnear, K. E. editors. (1996). *Advances in Genetic Programming 2*, Complex Adaptive Systems. MIT Press, Cambridge, MA.

[4] Antonisse, H. J. (1991). A grammar-based genetic algorithm. In Rawlins [123], pp. 193–204.

[5] Bäck, T. (1995). Generalized convergence models for tournament and (μ, λ)-selection. In Eshelman [19], pp. 2-8.

[6] Belew, R.K. and Booker,L.B. , editors. (1991). *Proceedings of the Fourth International Conference on Genetic Algorithms*, San Mateo, California, Morgan Kaufmann Publishers.

[7] Biethahn, J. , Höhnerloh, A. , Kuhl, J. and Nissen, V. , editors. *Fuzzy set-Theorie in betriebswirtschaftlichen Anwendungen*, Vahlen, München.

[8] Blickle, T. and Thiele, L. (1995). A mathematical analysis of tournament selection. In Eshelman [19], pp. 9–16.

[9] Böhm, W. and Geyer-Schulz, A. (1996). Exact uniform initialization for genetic programming. In R. Belew and M. Vose (Eds.) *Proc. FOGA'4*, pp. 255–269.

[10] Böhm, W. and Geyer-Schulz, A. (1997). Exact uniform initialization for genetic programming. In R. Belew and M. Vose (Eds.) *Foundations of Genetic Algorithms*, Morgan Kaufmann, San Francisco, pp. 379-407.

[11] Carse, B. Fogarty, T. C. and Munro, A. (1996). Evolving fuzzy rule based controllers using genetic algorithms. *Fuzzy Sets and Systems*, 80(3):273–293.

[12] Copeland, T. E. and Weston, J. F. (1988). *Financial Theory and Corporate Policy.* Addison-Wesley, Reading, 3rd edition.

[13] Cyert, R. and March, J. (1963) *A Behavioral Theory of the Firm.* Prentice-Hall, Englewood Cliffs.

[14] Davidor, Y. and Schwefel, H. P. , editors. (1994). *Parallel Problem Solving From Nature - PPSN III*, volume 866 of *Lecture Notes in Computer Science.* Springer- Verlag, Berlin.

[15] Dawid, H. (1995). Genetic learning in economic systems. Dissertation 2/107/95, Technische Universität Wien.

[16] Deb, K. (1996). Genetic algorithms for function optimization. In Herrera and Verdegay [53], pp. 3–29.

[17] DeJong, K. and Sarma, J. (1995). On decentralizing selection algorithms. In Eshelman [19], pp. 17–23.

[18] Edmonds, A.N. Burkhardt, D. and Adjei, O. Genetic programming of fuzzy logic production rules. *IEEE Conference on Evolutionary Computation*, 1995. Perth, Australia, IEEE Press.

[19] Eshelman, L. J. , editor. (1995). *Proceedings of the Sixth International Conference on Genetic Algorithms*, San Francisco, California. Morgan Kaufmann Publishers.

[20] Feferman, S. (1989). *The Number Systems.* Chelsea Publishing Company, New York, N.Y., 2nd edition.

[21] Filipic, B. and Juricic, D. (1996). A genetic algorithm to support learning fuzzy control rules from examples. In Herrera and Verdegay [53], pp. 403–418.

[22] Forrest, S. and Mitchell, M. (1993). What makes a problem hard for a genetic algorithm? some anomalous results and their explanation. *Machine Learning*, **13**(2/3), 285–319.

[23] Forrest, S. , editor. (1993). *Proceedings of the Fifth International Conference on Genetic Algorithms*, San Mateo, California. Morgan Kaufmann Publishers.

[24] Furuhashi, T. , editor. (1995). *Advances in Fuzzy Logic, Neural Networks and Genetic Algorithms*, volume 1011 of *Lecture Notes in Artificial Intelligence* . Springer Verlag, Berlin.

[25] Geyer-Schulz, A. (1991). Fuzzy classifier systems. In R. Lowen and M. Roubens, editors, *Computer, Management & Systems Science*, volume 1, 86–89, Brussel, Belgium, July 1991. IFSA.

[26] Geyer-Schulz, A. (1992). On learning in a fuzzy rule-based expert system. *Kybernetika*, 28:33–36.

[27] Geyer-Schulz, A. (1993). Fuzzy classifier systems. In R. Lowen and M. Roubens, editors, *Fuzzy Logic: State of the Art*, Series D: System Theory, Knowledge Engineering and Problem Solving, pages 345–354, Dordrecht, 1993. Kluwer Academic Publishers.

[28] Geyer-Schulz, A. (1994). Fuzzy rule-based expert systems and genetic machine learning. Habilitationsschrift, Wirtschaftsuniversität Wien, Augasse 2-6, A-1090 Vienna, Austria, June 9th, 1994.

[29] Geyer-Schulz, A. *Fuzzy Rule-Based Expert Systems and Genetic Machine Learning*, volume 3 of *Studies in Fuzziness*. Physica-Verlag, Heidelberg, 1995.

[30] Geyer-Schulz, A. (1996a). Das Lernen von Bestellregeln in Distributionsketten: Eine betriebswirtschaftliche Anwendung von Fuzzy Genetic Programming. In Biethahn et al. [7], pp. 92–106.

[31] Geyer-Schulz, A. (1996b). Fuzzy genetic programming and dynamic decision making. *Proc. ICSE'96*, 686–691.

[32] Geyer-Schulz, A. (1996c). *Fuzzy Rule-Based Expert Systems and Genetic Machine Learning*, volume 3 of *Studies in Fuzziness and Soft Computing*. Physica-Verlag, Heidelberg, 2nd revised edition, 1996.

[33] Geyer-Schulz, A. (1996d). The MIT beer distribution game revisited: Genetic machine learning and managerial behavior in a dynamic decision making experiment. In Herrera and Verdegay [53], pp. 658–682.

[34] Geyer-Schulz, A. (1997). The next 700 programming languages for genetic programming. *Proc. GP'97, (J. Koza (Ed.)*, page 9p.

[35] Goldberg, D. E. and Segrest, P. (1987). Finite markov chain analysis of genetic algorithms. In Grefenstette [47], pp. 1–8.

[36] Goldberg, D. E., Deb, K. and Korb, B. (1990a). Messy genetic algorithms revisited: Studies in mixed size and scale. *Complex Systems*, 4(4):415–444.

[37] Goldberg, D. E., Korb, B. and Deb, K. (1990b). Messy genetic algorithms: Motivation, analysis, and first results. *Complex Systems*, 3(5):493–530.

[38] Goldberg, D. E., Deb, K. and Korb, B. (1991). Don't worry, be messy. In Belew and Booker [6], pp. 24–30.

[39] Goldberg, D. E. (1989a). Genetic algorithms and Walsh functions: Part I, a gentle introduction. *Complex Systems*, 3(2):129–152.

[40] Goldberg, D. E. (1989b). Genetic algorithms and Walsh functions: Part II, deception and its analysis. *Complex Systems*, 3(2):153–171.

[41] Goldberg, D. E. (1990). A note on Boltzmann tournament selection for genetic algorithms and population-oriented simulated annealing. *Complex Systems*, 4(4):445–460.

[42] Gonzalez, A. and Perez, R. (1996). A learning system of fuzzy control rules based on genetic algorithms. In Herrera and Verdegay [53], pp. 202–225.

[43] Grefenstette, J. J. and Baker, J. E. (1989). How genetic algorithms work: A critical look at implicit parallelism. In Schaffer [131], pp. 20–27.

[44] Grefenstette, J. J. and Fitzpatrick, J. M. (1985). Genetic search with approximate function evaluation. In Grefenstette [45], pp. 112–120.

[45] Grefenstette, J. J., editor. (1985). *Proceedings of the First International Conference on Genetic Algorithms and their Applications.* Hillsdale, New Jersey.

[46] Grefenstette, J. J. (1986). Optimization of control parameters for genetic algorithms. *IEEE Transactions on Systems, Man, and Cybernetics*, SMC-16(1):122–128.

[47] Grefenstette,J. J., editor. (1987). *Genetic Algorithms and their Applications.* Hillsdale, New Jersey.

[48] Gruau, F. (1994). Genetic micro programming of neural networks. In Kenneth E. Kinnear [84], pp. 495–518.

[49] Gruau, F. (1996). On using syntactic constraints with genetic programming. In Angeline and Kinnear [3], pp. 377–394.

[50] Gupta, M. M. Saridis, G. N. and Gaines, B. R. editors. (1977). *Fuzzy Automata and Decision Processes*, New York. North–Holland.

[51] Hanebeck, U. D. and Schmidt, G. K. (1996). Genetic optimization of fuzzy networks. *Fuzzy Sets and Systems*, 79(8):59–68, 1996.

[52] Herrera, F. and Lozano, M. (1996). Adaptation of genetic algorithms parameters based on fuzzy logic. In Herrera and Verdegay [53], pp. 95–125.

[53] Herrera, F. and Verdegay, J. L. , editors. *Genetic Algorithms and Soft Computing*, volume 8 of *Studies in Fuzziness and Soft Computing*, Heidelberg, September 1996. Physica-Verlag.

[54] Herrera, F., Lozano, M. and Verdegay, J. L. (1995). Tuning fuzzy logic controllers by genetic algorithms. *International Journal of Approximate Reasoning*, 12(3/4):299–315.

[55] Hillis, D. W. (1990). Co-evolving parasites improve simulated evolution in an optimization procedure. *Physica D*, 42:228–234.

[56] Hillis, W. D. (1992). Co-evolving parasites improve simulated evolution as an optimization procedure. In Langton et al. [104], pp. 313–324.

[57] Hirsch, M. W. and Smale, S. (1974). *Differential Equations, Dynamical Systems, and Linear Algebra.* Pure and Applied Mathematics. Academic Press, New York.

[58] Hoffmann, F. and Pfister, G. (1996). Learning of a fuzzy control rule base using messy genetic algorithms. In Herrera and Verdegay [53], pp. 279–305.

[59] Hoffmann, F. (1997). *Entwurf von Fuzzy Reglern mit Genetischen Algorithmen.* DUV Informatik. DUV-Verlag, Leverkusen.

[60] Holland, J. H. (1973). Genetic algorithms and the optimal allocation of trials. *SIAM Journal of Computing*, 2(2):88–105.

[61] Holland, J. H. (1975). *Adaptation in Natural and Artificial Systems.* The University of Michigan Press, Ann Arbor, Michigan.

[62] Holmblad, L.P. and Ostergaard, J. J. (1995). The F, L, S application of fuzzy logic. *Fuzzy Sets and Systems*, 70(2/3):135–146.

[63] Homaifar, A. and McCormick, E. (1995). Simultaneous design of membership functions and rule sets for fuzzy controllers using genetic algorithms. *IEEE Transactions on Fuzzy Systems*, 3(2):129–139.

[64] Hopf, J. and Klawonn, F. (1994). Learning the rule base of a fuzzy controller by a genetic algorithm. In Kruse et al. [103], pp. 63–74.

[65] Joo, Y. H., Hwang, H.S. Kim, K.B. and Woo,K. B. (1997). Fuzzy system modeling by fuzzy partition and G, A hybrid schemes. *Fuzzy Sets and Systems*, 86(3):279–288, 1997.

[66] Karr, C.L. and Gentry, E. J. (1992). A genetics based adaptive p, H fuzzy logic controller. *Proc. International Fuzzy Systems and Intelligent Control Conference (IFSICC'92)*, 255–264, 1992. Louisville.

[67] Karr, C. L. and Gentry, E. J. (1993). Fuzzy control of p, H using genetic algorithms. *IEEE Transactions on Fuzzy Systems*, 1(1):46–53, 1993.

[68] Karr, C. L. and Sharma, S. K. (1994). An adaptive process control system based on fuzzy logic and genetic algorithms. *Proc. American Control Conference*, 3:2470–2474. Baltimore.

[69] Karr, C. L. and Stanley, D. A. (1991). Fuzzy logic and genetic algorithms in time-varying control problems. *Proc. NAFIPS'91*, 285–290.

[70] Karr, C. L., Freeman, L. M. and Meredith, D. L. (1989). Improved fuzzy process control of spacecraft autonomous rendezvous using a genetic algorithm. *Intelligent Control and Adaptive Systems*, 1196:274–288. G. Rodriguez (ed), The International Society of Photo-Optics Instrumentation Engineers (SPIE), Philadelphia.

[71] Karr, C. L., Freeman, L. M. and Meredith, D. L. (1990a). Genetic algorithms based fuzzy control of spacecraft autonomous rendezvous. *Proc. Fifth Conference on Artificial Intelligence for Space Applications*, 62:43–51. Huntsville.

[72] Karr, C. L., Meredith, D. L. and Stanley, D. A. (1990b). Fuzzy process control with a genetic algorithm. *Control'90 – Mineral and Metallurgical Processing*, 53–60. R. K. Rajamani and J. A. Herbst (ed), Society for Mining, Metallurgy, and Exploration Inc., Colorado.

[73] Karr, C. L., Sharma, S. K., Hatcher, W. J. and Harper, T. R. (1992). Control of an exothermic chemical reaction using fuzzy logic and genetic algorithms. *Proc. Int. Fuzzy Systems and Intelligent Control Conf. 1992*, 246–254. Louisville, KY.

[74] Karr, C. L., Sharma, S. K., Hatcher, W. J. and Harper, T. R. (1993a). Fuzzy control of an exothermic chemical reaction using genetic algorithms. *Engineering Applications of Artificial Intelligence 6*, 6:575–582.

[75] Karr, C. L., Sharma, S. K., Hatcher, W. J. and Harper, T. R. (1993b). Fuzzy logic and genetic algorithms for the control of an exothermic chemical reaction. *Modelling, Simulation, and Control of Hydrometallurgical Processes*, 227–236. V. G. Papangelakis and G. P. Demopoulos (eds), Canadian Institute of Mining, Metallurgy, and Petroleum, Montreal.

[76] Karr, C. L. (1991a). Design of a cart-pole balancing fuzzy logic controller using a genetic algorithm. *Proc. SPIE Conference on the Applications of Artificial Intelligence*, 26–36. Bellingham.

[77] [77] Karr, C. L. (1991b). Design of an adaptive fuzzy logic controller using a genetic algorithm. In Belew and Booker [6], pp. 450–457.

[78] Karr, C. L. (1991c). Applying genetics to fuzzy logic. *AI Expert*, 6(3):38–43.

[79] Karr, C. L. (1991d). Genetic algorithms for fuzzy controllers. *AI Expert*, 6(2):26–33.

[80] Karr, C. L. (1993a). Adaptive process control with fuzzy logic and genetic algorithms. *Sci. Comput. Autom.*, 9(10):23–30.

[81] Karr, C. L. (1993b). Real time process control with fuzzy logic and genetic algorithms. *Proc. Symposium on Emerging Computer Techniques for the Minerals Industry*, pages 31–37. Schneider B. J. and Stanley D. A. (eds), Society Mining Engineers, AIME, Littleton.

[82] Karr, C. L. (1994). Adaptive control with fuzzy logic and genetic algorithms. In Yager and Zadeh [159], pp. 345–367.

[83] Karr, C. L. (1996). Designing precise fuzzy systems with genetic algorithms. In Herrera and Verdegay [53], pp. 331–348.

[84] Kinnear, K. E. Jr., editor. (1994). *Advances in Genetic Programming*, Complex Adaptive Systems, MIT Press , Cambridge, MA.

[85] Kim, J. and Zeigler, B. (1996). Hierarchical distributed genetic algorithms: A fuzzy logic controller design application. *IEEE Expert*, 11(3):76–84.

[86] Kinzel, J., Klawonn, F. and Kruse, R. (1994a). Anpassung genetischer algorithmen zum erlernen und optimieren von fuzzy-reglern. In Reusch [124], pp. 103–110.

[87] Kinzel, J., Klawonn, F. and Kruse, R. (1994b). Modifications of genetic algorithms for designing and optimizing fuzzy controllers. *Proc. IEEE Conf. on Evolutionary Computation*, 1:28–33. IEEE, Orlando, Florida.

[88] Klement, E. P. and Slany, W., editors. (1993). *Fuzzy Logic in Artificial Intelligence*, volume 695 of *Lecture Notes in Artificial Intelligence*. Springer-Verlag, Berlin.

[98] Kotler., P. (1991). *Marketing Management – Analysis, Planning, Implementation, and Control*. Prentice–Hall, Englewood Cliffs, New Jersey, 7 edition.

[99] Koza, J. R. Goldberg, D. E. Fogel, D. B. and Riolo, R. L., Editors. (1996). *Genetic Programming 1996: Proceedings of the First Annual Conference*. MIT Press, Cambridge, MA.

[100] Koza, J. R. (1992). *Genetic Programming: On the Programming of Computers by Means of Natural Selection*. MIT Press, Cambridge, MA.

[101] Koza, J. R. (1994). *Genetic Programming II: Automatic Discovery of Reusable Programs*. MIT Press, Cambridge, MA.

[102] Koza, J. R. editor. (1996). *Late Breaking Papers at the Genetic Programming 1996 Conference*, Stanford. Stanford University.

[103] Kruse, R., Gebhardt, J. and Palm, R., editors. (1994). *Fuzzy Systems in Computer Science*, Braunschweig. Vieweg.

[104] Langton, C. G., Taylor, C. Farmer, J. D. and Rasmussen, S. editors. (1992). *Artificial Life II*, volume X of *Studies in the Sciences of Complexity*. Addison Wesley, Redwood City, Calofornia.

[105] Lee, M. A. and Esbensen, H. (1996). Evolutionary algorithms based multiobjective optimization techniques for intelligent system design. In Smith et al. [137], pp. 360–364. ISBN-0-7803-3225-3.

[106] Lee, M. A. and Takagi, H. (1993). Dynamic control of genetic algorithms using fuzzy logic techniques. In Forrest [23], pp. 76–83.

[107] Lee, M. A. and Takagi, H. (1995). A framework for studying the effects of dynamic crossover, mutation, and population sizing in genetic algorithms. In Furuhashi [24], pp. 111–126.

[108] Leitch, D. and Probert, P. (1995). Genetic algorithms for the development of fuzzy controllers for mobile robots. In Furuhashi [24], pp. 148–172.

[109] Leitch, D. (1995). A new genetic algorithm for the evolution of fuzzy systems. Thesis, Department of Engineering Science, University of Oxford.

[110] Leitch, D. (1996). Genetic algorithms for the evolution of behaviors in robotics. In Herrera and Verdegay [53], pp. 306–328.

[111] Li, R. and Zhang, Y. (1996). Fuzzy logic controller based on genetic algorithms. *Fuzzy Sets and Systems*, 83(1):1–10.

[112] Lilien, G. L., Kotler, P. and Moorthy, K. S. (1992). *Marketing Models*. Prentice–Hall, Inc., Englewood Cliffs, N.J.

[113] Lim, M. H., Rahardja, S. and Gwee, B. H. (1996). A G, A paradigm for learning fuzzy rules. *Fuzzy Sets and Systems*, 82(2):177–186.

[114] Magdalena, L. and Velasco, J. R. (1996). Fuzzy rule-based controllers that learn by evolving their knowledge base. In Herrera and Verdegay [53], pp. 172–201.

[115] Michalewicz, Z. (1992). *Genetic Algorithms + Data Structures = Evolution Programs.* Springer, Berlin.

[116] Mitchell, M. (1996). *An Introduction to Genetic Algorithms.* Complex Adaptive Systems. MIT Press, Cambridge, MA.

[117] Naur, P. (1963). Revised report on the algorithmic language A, L, G, O, L 60. *Communication of the ACM*, 6(1):1–17.

[118] Nguyen. T. and Huang, T. (1994). Evolvable 3D modeling for model-based object recognition systems. In Kenneth E. Kinnear [84], pp. 459–476.

[119] Nix, A. and Vose, M. D. (1992). Modeling genetic algorithms with Markov chains. *Annals of Mathematics and Artificial Intelligence*, 5:79–88.

[120] Nowak, M. and Sigmund, K. (1992). Tit for tat in heterogenous populations. *Nature*, 355.

[212] Ostergaard, J. J. (1977). Fuzzy logic control of a heat exchanger process. In Gupta et al. [50], pp. 285–320.

[122] Prügel-Bennett, A. and Shapiro, J. L. (1994). An analysis of genetic algorithms using statistical mechanics. *Physical Review Letters*, 72(9):1305–1309.

[123] Rawlins, G. J. E., editor. (1991). *Foundations of Genetic Algorithms.* Morgan Kaufmann, San Mateo.

[124] Reusch, B., editor. (1994). *Fuzzy Logik.* Springer, Berlin.

[125] Rohatgi, V. K. (1976). *An Introduction to Probability Theory and Mathematical Statistics.* John Wiley & Sons, New York.

[126] Rosca, J., editor. (1995). *Proceedings of the Workshop on Genetic Programming: From Theory to Real-World Applications.* Morgan Kaufmann, San Mateo.

[127] Rudolph, G. (1994). Convergence analysis of canonical genetic algorithms. *IEEE Transactions on Neural Networks*, 5(1):96–101.

Russel, S. and Norvig, P. (1995). *Artificial Intelligence: A Modern Approach.* Prentice-Hall, Upper Saddle River.

[129] Saffo, P. (1997). Sensors: The next wafe of innovation. *Communications of the ACM*, 40(2):93–97.

[130] Satyadas, A. and Krishnakumar, K. (1996). E, F, M-based controllers for space station attitude control: Application and analysis. In Herrera and Verdegay [53], pp. 152–171.

[131] Schaffer, J. D., editor. (1989). *Proceedings of the Third International Conference on Genetic Algorithms.* Morgan Kaufmann, San Mateo.

[132] Schröder, M., Klawonn, F. and Kruse, R. (1996). Sequential optimization of multidimensional controllers using genetic algorithms and fuzzy situations. In Herrera and Verdegay [53], pp. 419–444.

[133] Shapiro, J. L. and Prügel-Bennett, A. (1996). Genetic algorithm dynamics in two-well potentials with basins and barriers. *Proc. FOGA'4, (R. Belew and M. Vose (Eds.)*, 324–338.

[134] Shimojima, K., Kubota, N. and Fukuda, T. (1996). Virus-evolutionary genetic algorithm for fuzzy controller optimization. In Herrera and Verdegay [53], pp. 369–388.

[135] Simon, H. A. (1969). *The Sciences of the Artificial* MIT Press, Cambridge, MA.

[136] Simon, H. A. (1982). *Models of Bounded Rationality.* MIT Press, Cambridge, MA.

[137] Smith, M. H., Lee, M. A., Keller, J. and Yen, J., editors. (1996). *1996 Biennial Conference of the North American Fuzzy Information Processing Society - NAFIPS*, 445 Hoes Lane, Box 1331, Piscataway, NJ 08855-1331. IEEE Service Center.

[138] Sterman, J. D. (1989). Modeling managerial behavior: Misperceptions of feedback in a dynamic decision making experiment. *Management Science*, 35(3):321–339.

[139] Surmann, H., Kanstein, A. and Goser, K. (1993). Self-organizing and genetic algorithms for an automatic design of fuzzy control and decision systems. In Zimmermann [163], pp. 1097–1104.

[140] Surmann, H. (1996). Genetic optimization of fuzzy rule-based systems. In Herrera and Verdegay [53], pp. 389–402.

[141] Suzuki, L. (1995). A Markov chain analysis on simple genetic algorithms. *IEEE Transactions on Systems, Man, and Cybernetics*, 25(4):6–659.

[142] Takagi, H. and Lee, M. (1993). Neural networks and genetic algorithm – approaches to auto-design of fuzzy systems. In Klement and Slany [88], pp. 68–79.

[143] Tarng, Y. S., Yeh, Z. M. and Nian, C. Y. (1996). Genetic synthesis of fuzzy logic controllers in turning. *Fuzzy Sets and Systems*, 83(3):301–310.

[144] Thrift, P. (1991). Fuzzy logic synthesis with genetic algorithms. In Belew and Booker [6], pp. 509–513.

[145] Tunstel, E. and Jamshidi, M. (1996). On genetic programming of fuzzy rule-based systems for intelligent control. *International Journal of Intelligent Automation and Soft Computing*, 2(2). to appear.

[146] Tunstel, E., Akbahrzadeh-T, M. R., Kumbla, K. and Jamshidi, M. (1996). Soft computing paradigms for learning fuzzy controllers with applications to robotics. In Smith et al. [137], pp. 355–359.

[147] Tversky, A. and Kahneman, D. (1974). Judgement under uncertainty: Heuristics and bias. *Science*, 185(9):1124–1131.

[148] Vose, M. D. and Liepins, G. E. (1991). Punctuated equilibria in genetic search. *Complex Systems*, 5:31–44.

[149] Vose, M. D. (1993). Modeling simple genetic algorithms. In Whitley [155], pp. 63–73.

[150] Vose, M. D. (1996). Modeling simple genetic algorithms. *Evolutionary Computation*, 3(4):453–472.

[151] Wenstop, F. (1980). Quantitative analysis with linguistic variables. *Fuzzy Sets and Systems*, 4(2):99–115.

[152] Whigham, P. A. (1995a). Grammatically-based genetic programming. In Rosca [126], pp. 33–41.

[153] Whigham, P. A. (1995b). Inductive bias and genetic programming. In Zalzala [160], pp. 461–466.

[154] Whigham, P. A. (1996). Search bias, language bias, and genetic programming. In Koza et al. [99], pp. 230–237.

[155] Whitley, L. D., editor. (1993) *Foundations of Genetic Algorithms 2.* Morgan Kaufmann, San Mateo.

[156] Wong, C. C. and Feng, S. M. (1995). Switching-type controller design by genetic algorithms. *Fuzzy Sets and Systems*, 74(2):175–185.

[157] Wong, M. L. and Leung, K. S. (1995). Inducing logic programs with genetic algorithms: The genetic logic programming system. *IEEE Expert*, 9(5):68–76.

[158] Wong, M. L. and Leung, K. S. (1996). Learning recursive functions from noisy examples using generic genetic programming. In Koza et al. [99], 238–246.

[159] Yager, R. R. and Zadeh, L. A., editors. (1994). *Fuzzy Sets, Neural Networks, and Soft Computing.* Van Nostrand.

[160] Zalzala, A. M. S. editor. (1995). *First International Conference on Genetic Algorithms in Engineering Systems: Innovations and Applications, GALESIA*, volume 414, London, September 1995. IEE.

[161] Zimmermann, H. J., editor. (1993). *First European Congress on Fuzzy and Intelligent Technologies - EUFIT'93*, volume 2, Aachen, September 1993. Verlag der Augustinus Buchhandlung.

[162] Zimmermann, H. J., editor. (1994a). *Second European Congress on Intelligent Techniques and Soft Computing - EUFIT'94*, volume 3, Promenade 9, D-52076 Aachen, September 20-23 1994. Verlag der Augustinus Buchhandlung.

[163] Zimmermann, H.J., editor. (1994b). *Second European Congress on Intelligent Techniques and Soft Computing - EUFIT'94*, volume 2, Promenade 9, D-52076 Aachen, September 20-23 1994. Verlag der Augustinus Buchhandlung.

[164] Zurada, J. M., Marks, R. J. and Robinson, C. J., editors. (1994). *Computational Intelligence: Imitating Life.* IEEE Press.

12 FUZZY SYSTEMS, VIABILITY THEORY AND TOLL SETS

J. P. Aubin[1] and O. Dordan[2]

[1]CEREMADE, Université de Paris-Dauphine
F-75775 Paris cx(16), France
Jean-Pierre.Aubin@ens.fr, dordan@u-bordeaux2.fr

[2]Laboratoire de Mathématiques Appliquées de Bordeaux
Université de Bordeaux II
146 rue Léo Saignat, 33176 Bordeaux, France

12.1 INTRODUCTION

"Fuzzification" of dynamical and control systems, for which convexity plays a crucial mathematical role, led us to use indicators instead of characteristic functions for imbedding subsets into functions. Therefore, instead of obtaining the genuine fuzzy sets by a convexification procedure, we obtain the "toll sets", the membership function being a cost function taking values between 0 and ∞. With this tool at hands, we can define "fuzzy" — or actually, "toll" — differential inclusions and control systems and solve the viability problem or the toll viability problem by building fuzzy controllers which regulate viable solutions (satisfying at each instant viability and toll viability constraints).

Furthermore, it happens that the Cramer transform which was introduced to study large deviations maps probability measures to toll sets! It then reconciles in some way the heated debate between proponents of probabilities and the ones of fuzzy sets — or their clones, toll sets.

There are other reasons for introducing and studying fuzzy sets beyond the ones motivating the seminal and pioneering paper[1] by Zadeh [46], published in 1965.

The one we emphasis here is rooted in the need for convexification procedures which leads us "naturally" to fuzzy sets and toll sets which we shall introduce, study and apply to dynamical and control systems below.

"Fuzzification" of dynamical and control systems, for which convexity plays a crucial mathematical role, led us to use indicators instead of characteristic functions, and thus, the introduction of toll sets.

It happens that the Cramer transform which we introduce below maps probability measures to toll sets ! It then reconciles in some way the heated debate between proponents of probabilities and the ones of fuzzy sets — or their clones, toll sets (see Aubin [7], [8], and Aubin and Dordan [12]).

12.2 CONVEXIFICATION PROCEDURES

Indeed, the first general convexification procedure led to probability measures. Let us explain this on the simple case of a finite set M. We always can embed the discrete set M to $\mathbb{R}^m$ by the "Dirac measure" associating with any $j \in M$ the jth element of the canonical basis of $\mathbb{R}^m$. The convex hull of the image of M by this embedding is the **probability simplex** of $\mathbb{R}^m$. Let X be any set. If we denote by $\mathcal{S}(X)$ the set of functions from a set X to $\mathbb{R}$ supplied with the pointwise convergence topology, one can check that its topological dual $\mathcal{S}^\star(X)$ is the set of finite measures (finite linear combinations of Dirac measures $\delta(x) : f \in \mathcal{S}(X) \mapsto f(x) \in \mathbb{R}$). Then the Dirac map transforms points $x \in X$ to Dirac measures $\delta(x) \in \mathcal{S}^\star(X)$, and the convex hull of its image is the set of discrete probability measures. When X is compact, one can check that this set is dense in the closed (actually, weakly compact) convex set of Radon probability measures on X.

Even if one never uses the rich probabilistic interpretation of the convex hull of the set of Dirac measures, this universal embedding of a set without any structure into a convex subset of a vector space is still very useful.

Although is convexification procedure is universal, it is not unique. One can for instance use other embeddings taking advantages of added structures of a given set X.

The sets we consider from now on are power spaces. When E is a set, we shall denote by $\mathcal{P}(E)$ or $\mathbf{2}^E$ the **shape space** or the **power space** of E, i.e., the family of all subsets of E, including the empty set. For instance, when N is a discrete space with n elements, the shape space $\mathbf{2}^N$ has 2^n elements.

Embedding the power set 2^E into a vector space is quite advantageous, since it allows to exploit the many properties of vector-spaces.

However, we can define several specific embeddings from $\mathbf{2}^E$ to a vector space, each having its own list of advantages and inconveniences. We shall review some of them and use the two first ones of this list for convexification purposes.

1. *Characteristic Map and Fuzzy Sets*
   The characteristic function $\chi_K : E \mapsto \{0,1\}$ associates with $x \in E$ either 1 when $x \in K$ and 0 when $x \notin K$

$$\chi_K(x) := \begin{cases} 1 & \text{if } x \in K \\ 0 & \text{if not} \end{cases}$$

   We thus can embed the power set 2^E into the vector space $I\!R^E$ of real-valued functions defined on E through the map

$$\chi : K \in 2^E \mapsto \chi_K \in I\!R^E$$

   We observe that the image of the map $\chi : 2^E \mapsto I\!R^E$ associating with each subset $K \subset E$ its characteristic function χ_K is equal to

$$\chi\left(2^E\right) = \{0,1\}^E$$

   which is the set of the vertices of the hypercube $[0,1]^E$.
   It is then quite natural from a mathematical viewpoint to use the closed convex hull $[0,1]^E$ of the image $\{0,1\}^E$ of the power set 2^E.
   Since we can regard such a characteristic function χ_K as a membership function of an element x to K, by assigning the value 1 whenever x belongs to K and 0 whenever x does not belong to K, and since any element $c \in [0,1]^E$ is a function from E into $[0,1]$, we can interpret it as a membership function of a fuzzy set introduced in 1965 by L. Zadeh and which have been the object of a tremendous development.
   If a discrete set N contains n elements, the set of fuzzy sets being the convex hull of the power set $\{0,1\}^N$, we can write any fuzzy set in the form

$$c = \sum_{K \in 2^E} m_K \chi_K \text{ where } m_K \geq 0 \ \& \ \sum_{K \in 2^E} m_K = 1$$

   The memberships are then equal to

$$\forall\, x \in N, \ c(x) = \sum_{K \ni x} m_K$$

   Consequently, if m_K is regarded as the probability for the set K to be formed, the membership of the element x to the fuzzy set c is the sum of the probabilities of the sets containing the element x.

Let $(\Omega, \mathcal{A}, d\mu)$ be a measure space and $\chi : \mathcal{A} :\mapsto L^1(\Omega)$ the map associating with any $K \in \mathcal{A}$ its characteristic function χ_K, which imbeds the σ-algebra $\mathcal{A}$ to the set $L^1(\Omega, \{0,1\})$. We thus can interpret its closed convex hull $L^1(\Omega, [0,1])$ as the set of measurable fuzzy sets. *When the measure μ is nonatomic, C. Castaing derived from the Lyapunov Convexity Theorem that the set $L^1(\Omega, \{0,1\})$ of measurable sets is dense in the set $L^1(\Omega, [0,1])$ of measurable fuzzy sets* (when $L^1(\Omega, \mathbb{R})$ is supplied with the weak topology).

Remark. The choice of the scale $[0,1]$ is arbitrary. For instance, when N is a set of players in game theory and when subsets $A \subset N$ are regarded as coalitions of players, we can, for instance, introduce negative memberships when players enter a coalition with aggressive intents. This is mandatory if one wants to be realistic ! A positive membership is interpreted as a cooperative participation of the player i in the coalition, while a negative membership is interpreted as a non-cooperative participation of the ith player in the generalized coalition. Then, one can replace the cube $[0,1]^n$ by any product $\prod_{i=1}^{n}[\lambda_i, \mu_i]$ for describing the cooperative or noncooperative behavior of the players. The idea of using fuzzy coalitions has already been used in the framework of cooperative games with and without side-payments. For instance, it has been shown that in the framework of cooperative games with side payments involving fuzzy coalitions, the concepts of Shapley value and core did coincide with the (generalized) gradient of the "characteristic function" at the "grand coalition". The differences between these concepts for usual games is explained by the different ways one "fuzzyfy" a characteristic function defined on the set of usual coalitions. See Aubin [4], [5], Chapter 13 in [6].

Fuzzy coalitions are also used in dynamical economic theory to explain the evolution of coalitions of economic agents together with the evolution of their consumptions and the evolution of the market prices in order to satisfy the scarcity constraints of each coalitions (see Aubin [11]). □

2. *Indicator Map and Toll Sets*

Unfortunately, this embedding is not convenient when we want to use the numerous tools of functional analysis, where convexity plays a key role. Indeed, the convexity of a subset of a vector space cannot be characterized by some properties of its characteristic function. But another scaling, amounting to replace characteristic functions by indicators, allows us to characterize convex subsets.

Indeed, any subset $K \subset E$ can be characterized by its *"indicator"* ψ_K, which is the nonnegative extended function defined by :

$$\psi_K(x) := \begin{cases} 0 & \text{if } x \in K \\ +\infty & \text{if } x \notin K \end{cases}$$

It can be regarded as a "cost function" or a "penalty function", assigning to any element $x \in E$ an infinite cost when x is outside K, and no cost at all when x belongs to K.

We observe that K *is convex (respectively closed) if and only if its indicator is convex (respectively lower semicontinuous).*

The set of indicators being $\{0, \infty\}^E$, its closed convex hull is the set $[0, \infty]^E$ of nonnegative extended functions U from E to $\mathbb{R}_+ \cup \{+\infty\}$. They provide another implementation of the idea underlying "fuzzy sets", in which indicators replace characteristic functions. Instead of using membership functions taking values in the interval $[0, 1]$, we deal with *membership cost functions* taking their values anywhere between 0 and $+\infty$. Following a suggestion of Dubois and Prade, we shall call them "toll sets".

Observe also that in mathematical morphology, toll sets represent grayscale shapes, i.e., functions associating with each point of the shape its intensity in grey levels. The choice of the representation depends upon the scale. Actually, the real computer scale is $\{0, 256\}$!

3. *Gauge Map*

Let X be a Banach space. We shall say that the extended function $j_K : x \in X \to j_K(x)$ from X to $[0, \infty]$ defined by

$$j_K(x) = \inf\{\lambda \text{ such that } \lambda > 0 \text{ and } x \in \lambda K\} \subset [0, \infty]$$

is the gauge of K, because, when K is a closed convex subset containing the origin, we prove that

$$K = \{x \in X \mid j_K(x) \leq 1\}$$

The gauges satisfy the following properties

$$\begin{cases} i) & j_K(\lambda x) = \lambda j_K(x) \text{ for all } \lambda \geq 0 \\ ii) & j_K(x + y) \leq j_K(x) + j_K(y) \end{cases} \quad (12.1)$$

One observe easily that the gauge of an intersection is the supremum of the gauges.

4. *Support Map*

Let K be a nonempty subset of a Banach space X. We associate with any linear form $p \in X^\star$

$$\sigma_K(p) := \sigma(K,p) := \sup_{x \in K} \langle p, x \rangle = \sup_{x \in X} (\langle p, x \rangle - \psi_K(x)) \in I\!R \cup \{+\infty\}$$

The function $\sigma_K : X^\star \mapsto I\!R \cup \{+\infty\}$ is called the support function of K. We say that the subset of $X^\star$ defined by

$$K^- := \{p \in X^\star \mid \sigma_K(p) \le 0\}$$

is the (negative) polar cone, of K.
We observe that

$$\forall \lambda, \mu > 0, \ \sigma_{\lambda L + \mu M}(p) = \lambda \sigma_L(p) + \mu \sigma_M(p)$$

and in particular, that if P is a cone, then

$$\sigma_{M+P}(p) = \begin{cases} \sigma_M(p) & \text{if} \quad p \in P^- \\ +\infty & \text{if} \quad p \notin P^- \end{cases} \quad \square$$

The Separation Theorem can be stated in the following way : *Let K be a nonempty subset of a Banach space X. Its* closed convex hull *is characterized by linear constraint inequalities in the following way :*

$$\overline{co}(K) = \{x \in X \mid \forall p \in X^\star, \ \langle p, x \rangle \le \sigma_K(p)\}$$

Furthermore, there is a bijective correspondence between nonempty closed convex subsets of X and nontrivial lower semicontinuous positively homogeneous convex functions on $X^\star$. *Finally, a cone P is closed and convex if and only if $P = P^{--}$ is equal to its bipolar.* (See for instance Aubin [6] for proofs and more details).

We also observe that *the support function of K is the gauge of the polar set.* We also obtain a general Cauchy-Schwarz-Buniakowski inequality : *Let K be a closed convex containing the origin. Then*

$$\forall x \in X, \forall p \in X^\star, \ \langle p, x \rangle \le \sigma_K(p) \, j_K(x)$$

5. *Distance Map*

We denote by $\mathcal{K}(E)$ the family of nonempty compact subsets of a metric space E.

We can associate with any subset $K \subset E$ the distance function defined by

$$d_K(x) := d(x,K) := \inf_{y\in K} d(x,y)$$

We point out that

$$d_{K\cup L} = \min(d_K, d_L)$$

and that

$$d_K(x) = d_{\overline{K}}(x)$$

When K is nonempty a closed convex subset, the function d_K is convex. We introduce the Hausdorff distance

$$\boldsymbol{d}(K,L) := \max(h^\sharp(K,L), h^\sharp(L,K)) \tag{12.2 }$$

where

$$h^\sharp(K,L) = \sup_{x\in K} d(x,L) = \sup_{x\in K}\inf_{y\in L} d(x,y)$$

and we observe that

$$h^\sharp(K,L) := \sup_{x\in E}(d(x,L) - d(x,K))$$

If E is a compact set, one can check that the map $d : K \in \mathcal{K}(E) \mapsto d(\cdot,K) \subset \mathcal{C}(E)$ is an isometry when $\mathcal{K}(E)$ is supplied with the Hausdorff distance and when the space $\mathcal{C}(E)$ of continuous functions is supplied with the uniform convergence topology.

We just provided this nonexhaustive list of embeddings to show that we could incarnate and substantiate Zadeh's idea in many other way, according to the nature of the problems at hand.

12.3 TOLL SETS

Let X be a finite dimensional vector space. A function $V : X \mapsto I\!R \cup \{\pm\infty\}$ is called an extended (real-valued) function. Its domain is the set of points at which V is finite :

$$\text{Dom}(V) := \{x \in X \mid V(x) \neq \pm\infty\}$$

A function is said to be nontrivial if its domain is not empty. Such a function is said to be proper in convex and non smooth analysis. We chose this terminology for avoiding confusion with proper maps. Any function V defined on a subset $K \subset X$ can be regarded as the extended function V_K equal to V on K and to $+\infty$ outside of K, whose domain is K.

12.3.1 Toll Sets and Toll Maps

An extended function $V : E \mapsto I\!R_+ \cup \{+\infty\}$ is characterized by its *epigraph*

$$\mathcal{E}p(V) := \{(x, \lambda) \in E \times I\!R \mid V(x) \leq \lambda\}$$

An extended function V is convex (resp. positively homogeneous) if and only if its epigraph is convex (resp. a cone.)

We recall that any subset $K \subset X$ can be characterized by its *"indicator"* ψ_K, and thus, regarded as a "cost function" or a "penalty function", assigning to any element $x \in X$ an infinite cost when x is outside K, and no cost at all when x belongs to K.

We also recall that K *is convex (respectively closed, a cone) if and only if its indicator is convex (respectively lower semicontinuous, positively homogeneous).* For this we only have to observe that

$$\mathcal{E}p(\psi_K) = K \otimes I\!R_+$$

We are led to regard any nonnegative extended function U from X to $I\!R_+ \cup \{+\infty\}$ as another implementation of the idea underlying "fuzzy sets", in which indicators replace characteristic functions.

Definition 12.1 *We shall regard an extended nonnegative function $U : X \mapsto I\!R_+ \cup \{+\infty\}$ as a* toll set. *This terminology has been coined by Dubois and Prades (see Dubois and Prades [23]). Its* domain *is the domain of U, i.e., the set of elements x such that $U(x)$ is finite, and the* core *of U is the set of elements x such that $U(x) = 0$. The* complement *of the toll set U is the complement of its domain and the complement of its core is called the* toll boundary.

We shall say that the toll set U is convex *(respectively* closed, *a cone) if the extended function U is convex (respectively lower semicontinuous, positively homogeneous).*

We observe that the membership function of the empty set is the constant function equal to $+\infty$.

Definition 12.2 *We shall say that a set-valued map $\mathbf{U} : X \rightsquigarrow Y$ associating with any $x \in X$ a toll subset $U(x)$ of Y is a* toll set-valued map. *Its* graph *is the toll subset of $X \times Y$ associated with the extended nonnegative function $(x, y) \mapsto U(x, y) := U(x)(y)$ and its* domain *is*

$$\text{Dom}(\mathbf{U}) := \{x \in X \mid U(x, y) < +\infty \text{ for some } y\}$$

A toll set-valued map $\mathbf{U}$ is said to be closed *if and only if its graph is closed, i.e., if its membership function is lower semicontinuous. Its values are closed* *(respectively convex) if and only if the toll subset $U(x)$ are closed (respectively convex). It has linear growth if and only if, for some positive constant c,*

$$U(x,v) < +\infty \implies \|v\| \leq c(\|x\| + 1)$$

A nontrivial closed toll set-valued map $\mathbf{U}$ *with convex images and linear growth is called a* Marchaud toll set-valued map.

If $\mathbf{U}$ is a toll set-valued map from X to Y and $\mathbf{V}$ is a toll set-valued map from Y to Z, we define the (composition) product as the toll set-valued map $\mathbf{W} := \mathbf{V} \circ \mathbf{U}$ from X to Z by the cost function

$$W(x,z) := \inf_{y \in Y} (U(x,y) + V(y,z))$$

and the square product $\mathbf{W}_\square := \mathbf{V} \square \mathbf{U}$ by the cost function

$$W_\square(x,z) := \sup_{y \in U} (U(x,y) + V(y,z))$$

12.3.2 Operations on Toll Sets

One can define on toll sets the following operations :

Inclusion. We shall say that a tall set U_1 is "contained" in a toll set U_2 if and only if their cost functions satisfy $U_1 \geq U_2$.

Intersection of Toll Sets. We shall say that the cost function of an "intersection" of toll sets U_i is the sum $\sum_{i \in I} U_i$ of the cost functions.

Minkowski Addition and Difference of Toll Sets. Since toll sets are non-negative extended functions, we can use the epigraphical view point advocated in convex and nonsmooth analysis, which amounts to define operations on functions through operations on their epigraphs.

The closure of the epigraph of the cost function of a toll set is regarded as the epigraph of the cost function of a toll set, called the epiclosure or closure of a toll set.

Let K and B two subsets of a vector space. We set

$$\begin{cases} K + B := \{y+z\}_{y \in K, z \in B} = \bigcup_{y \in B} (K + y) \\ \text{(called Minkowski sum of } K \text{ and } B) \\ K - B := \{y-z\}_{y \in K, z \in B} = \bigcup_{y \in K} (y - B) \\ \\ \text{(is called the symmetric of } B) \\ K \ominus B := \{y \mid B + y \subset K\} = \bigcap_{y \in B} (K - y) \\ \text{(called the Minkowski difference of } K \text{ and } B) \\ \text{(also called the erosion of } K \text{ by } B) \end{cases}$$

The (Minkowski) sum and difference of epigraphs of two cost functions of toll sets V and W are regarded as epigraphs of two cost functions $V \oplus_\uparrow W$ and $V \ominus_\uparrow W$ built from V and W in the following way :

Definition 12.3 *If $V, W : X \mapsto \mathbb{R}_+ \cup \{+\infty\}$ are two nonempty toll sets, the function $V \oplus_\uparrow W$ defined by*

$$\mathcal{E}p(V \oplus_\uparrow W) = \mathcal{E}p(V) + \mathcal{E}p(W)$$

is the cost function of the episum *of the toll sets V and W and its epiclosure the* inf-convolution *of the toll sets V and W.*
The cost function $V \ominus_\uparrow W$ defined by

$$\mathcal{E}p(V \ominus_\uparrow W) = \mathcal{E}p(V) \ominus \mathcal{E}p(W)$$

is the cost function the epidifference *of the toll sets V and W.*

Proposition 12.1 *The epiclosure of the episum $V \oplus_\uparrow W$ is given by the formula*

$$\overline{(V \oplus_\uparrow W)}(x) := \inf_{y+z=x} (V(y) + W(z)) = \inf_{y \in X} (V(x-y) + W(y))$$

When the epigraph of V is closed, then

$$(V \ominus_\uparrow W)(x) := \sup_{y \in X} (V(x+y) - W(y))$$

The epiclosure of the episum is also called the inf-convolution for obvious reasons. This operation appears in several domains of convex analysis (since the *conjugate of the sum of two functions is the inf-convolution of the conjugates*) and statistics (law of large numbers).

Examples

1. When $W_i := \psi_{K_i}$ are the cost functions of usual sets K_i, the episum of the cost functions

$$(\psi_{K_1} \oplus_\uparrow \psi_{K_2})(x) := \psi_{K_1+K_2}(x)$$

is the cost function of the sum $K_1 + K_2$.
This is the reason why we regard the episum

$$(U_1 \oplus_\uparrow U_2)(\cdot)$$

of the cost functions of two toll sets U_1 and U_2 as the cost function of the "sum" of these two toll sets.

2. When $W := \psi_K$ is the cost function of a usual set, the inf-convolution can be written

$$\overline{(V \oplus_{\uparrow} hB)}(x) \ := \ \inf_{y \in B} V(x - hy)$$

Conjugation Let $V : X \to I\!R \cup \{+\infty\}$ be the cost function of a toll set. Its conjugate function $V^\star : X^\star \to I\!R \cup \{+\infty\}$ is defined on the dual of X by

$$\forall\, p \ \in \ X^\star, \ \ V^\star(p) \ := \ \sup_{x \in X}(< p, x > - V(x))$$

The map $V \mapsto V^\star$ is called the *Fenchel transform*.

For instance, the conjugate of a usual subset K is the support function of K.

We observe the following Fenchel inequality

$$\forall\, x \ \in \ X, \ \ p \ \in \ X^\star, \ \ < p, x > \leq \ V(x) + V^\star(p)$$

One of the two basic theorem of convex analysis states that *a nontrivial extended function $V : X \to I\!R \cup \{+\infty\}$ is convex and lower semicontinuous if and only if it coincides with its biconjugate.*

In particular, if f U is a toll cone, its conjugate is the cost function of the polar toll cone U^- defined by

$$U^- \ := \ \{p \in X^\star \mid \forall\, x \in X, \ \ \langle p, x \rangle \leq U(x)\}$$

because

$$U^\star(p) \ = \ \sup_{x \in X}(\langle p, x \rangle - U(x)) \ = \ \psi_{U^-}(p)$$

Furthermore, Moreau and Rockafellar did introduce in the early sixties the concept of subdifferential $\partial V(x)$ of V at x as the subset

$$\partial V(x) \ := \ \{p \in X^\star \mid \langle p, x \rangle \ = \ V(x) + V^\star(p)\}$$

Therefore, if $V : X \to I\!R \cup \{+\infty\}$ is convex and lower semicontinuous, *the inverse of the subdifferential $\partial V(\cdot)$ is the subdifferential $\partial V^\star(\cdot)$ of the conjugate function* :

$$p \ \in \ \partial V(x) \ \iff \ x \ \in \ \partial V^\star(p)$$

In this case, since $-V^\star(0) \ = \ \inf_{x \in X} V(x)$, then $\partial V^\star(0)$ is the set of minimizers of V, i.e., the set of the "cheapest" elements of the closed convex toll set V.

Remark. As in the case of fuzzy logic, one can construct a "toll logic" (or a fuzzy quantic logic) by representing the set of elements satisfying a given property by a closed convex cone of a finite dimensional vector space. We define the "toll conjunction" as the intersection of toll closed convex cone, the "toll disjunction" as the episum of toll closed convex cone, the "toll negation" through conjugation and the "toll implication" through inclusion of toll sets. □

12.3.3 The Cramer Transform

There is another (mathematical) reason for which toll sets provide a sensible mathematical representation of the concept of randomness, but different from the representation by probabilities (see Akian, Quadrat and Vio [1], [2], [3]).

The Cramer transform C associates with any nonnegative measure $d\mu$ on a finite dimensional vector space $\mathbb{R}^n$ the nonnegative extended function C_μ : $\mathbb{R}^n \mapsto \mathbb{R}_+ \cup \{+\infty\}$ defined on $\mathbb{R}^n$ (identified with its dual) by :

$$C_\mu(p) \ := \ \sup_{x \in \mathbb{R}^n} \left(\langle p, x \rangle - \log \left(\int_{\mathbb{R}^n} e^{\langle x,y \rangle} d\mu(y) \right) \right)$$

This Cramer transform plays an important role in statistics, and in particular, in the field of large deviations (see Azencott [15], for instance). It is the product of the *Laplace transform* $\mu \mapsto \int_{\mathbb{R}^n} e^{\langle x,y \rangle} d\mu(y)$, of the logarithm and of the *Fenchel transform* (conjugate functions) $V(\cdot) \mapsto V^\star(\cdot)$.

Since C_μ is the supremum of affine functions with respect to p, this is a lower semicontinuous convex function. It satisfies

$$C_\mu(p) \ \geq \ \langle p, 0 \rangle - \log \left(\int_{\mathbb{R}^n} e^{\langle 0,y \rangle} d\mu(y) \right) \ = \ -\log \left(\int_{\mathbb{R}^n} d\mu(y) \right)$$

so that when $d\mu$ is a probability measure, its Cramer transform C_μ is nonnegative and thus, a toll set.

The indicators $\psi_{\{a\}}$ of singleta a are images of *Dirac measures* δ_a : Indeed, if δ_a is the Dirac measure at the point $a \in \mathbb{R}^n$, then

$$\begin{cases} C_{\delta_a}(p) \ = \ \sup_{x \in \mathbb{R}^n} (\langle p, x \rangle - \langle a, x \rangle) \\ = \begin{cases} 0 & \text{if } \ p = a \\ +\infty & \text{if } \ p \neq a \end{cases} \\ = \ \psi_a(p) \end{cases}$$

The Cramer transform of the Gaussian with mean m and variance σ is the quadratic function $G_{\sigma,\, m}$ defined by

$$G_{\sigma,\, m}(x) \ := \ \frac{1}{2} \left\| \frac{x - m}{\sigma} \right\|^2$$

which we can regard as a Gaussian toll set with mean m and variance σ. Such toll sets play the role of Gaussians in probability theory.

The function $x \mapsto \log\left(\int_{I\!R^n} e^{\langle x,y\rangle} d\mu(y)\right)$ is

1. *convex*

   Indeed, applying Hölder inequality with exponents $\frac{1}{\alpha_i}$, we obtain

$$\begin{cases} \int_{I\!R^n} e^{\langle \alpha_1 x_1 + \alpha_2 x_2, y\rangle} d\mu(y) = \int_{I\!R^n} \left(e^{\langle x_1,y\rangle}\right)^{\alpha_1} \left(e^{\langle x_2,y\rangle}\right)^{\alpha_2} d\mu(y) \\ \leq \left(\int_{I\!R^n} e^{\langle x_1,y\rangle} d\mu y\right)^{\alpha_1} \left(\int_{I\!R^n} e^{\langle x_2,y\rangle} d\mu y\right)^{\alpha_2} \end{cases}$$

   By taking the logarithms, we get the convexity of this function with respect to x.

2. and *lower semicontinuous*

   Since the measure $d\mu$ is nonnegative, Fatou's Lemma implies that if x_p converges to x, then

$$\int_{I\!R^n} e^{\langle x,y\rangle} d\mu(y) \leq \liminf_{p\to\infty} \int_{I\!R^n} e^{\langle x^p,y\rangle} d\mu(y)$$

   Hence the lower semicontinuity of the Laplace transform of $d\mu$ is established. Since the logarithm is increasing and continuous, it is continuous and nondecreasing.

Therefore

$$C_\mu^\star(x) = \log\left(\int_{I\!R^n} e^{\langle x,y\rangle} d\mu(y)\right)$$

It is actually differentiable and its gradient is equal to

$$\nabla C_\mu^\star(x) = \frac{\int_{I\!R^n} y e^{\langle x,y\rangle} d\mu(y)}{\int_{I\!R^n} e^{\langle x,y\rangle} d\mu(y)}$$

When $d\mu$ is the probability law of a random variable, then its mean is equal to $\nabla C_\mu^\star(0)$, which is centered if and only if its Cramer transform vanishes at 0.

Inf-convolution plays the role of the usual convolution product of two integrable functions f and g defined by

$$(f \star g)(x) := \int_{I\!R^n} f(x-y)g(y)dy$$

We thus deduce that the Laplace transform of a convolution product is the product of the Laplace transforms because

$$\begin{cases} \int_{\mathbb{R}^n} e^{\langle x,y\rangle} \int_{\mathbb{R}^n} f(y-z)g(z)dydz = \int_{\mathbb{R}^n}\int_{\mathbb{R}^n} e^{\langle x,z\rangle} g(z) e^{\langle x,y-z\rangle} g(y-z)dydz \\ = \int_{\mathbb{R}^n} e^{\langle x,z\rangle} g(z)dz \int_{\mathbb{R}^n} e^{\langle x,u\rangle} g(u)du \end{cases}$$

Therefore, taking the logarithm, we obtain

$$\begin{cases} \log\left(\int_{\mathbb{R}^n} e^{\langle x,y\rangle}(f\star g)(y)dy\right) \\ = \log\left(\int_{\mathbb{R}^n} e^{\langle x,y\rangle} f(y)dy\right) + \log\left(\int_{\mathbb{R}^n} e^{\langle x,y\rangle} g(y)dy\right) \end{cases}$$

Finally, one can prove that if

$$0 \in \operatorname{Int}\left(\operatorname{Dom}\left(\log \int_{\mathbb{R}^n} e^{\langle\cdot,y\rangle} f(y)dy\right) - \operatorname{Dom}\left(\log \int_{\mathbb{R}^n} e^{\langle\cdot,y\rangle} g(y)dy\right)\right)$$

the Fenchel conjugate of this sum is the inf-convolution of the Fenchel conjugates. This happens for instance when the support of one of these functions is compact. This allows us to conclude that, under this assumption,

$$C_{f\star g} = C_f \oplus_\uparrow C_g$$

In particular, the regularization of a function is obtained by taking its convolution by a Gaussian. The Cramer transform implies that it is the inf-convolution of a function by a quadratic function, called *the Moreau or Moreau-Yosida transform* of a function $V : X \to \mathbb{R} \cup \{+\infty\}$. It is defined by

$$f_\sigma(x) := \inf_{y\in X}\left[f(y) + \frac{1}{2}\left\|\frac{x-y}{\sigma}\right\|^2\right] \tag{12.3 }$$

In the same way that the convolution product by a Gaussian maps a c function to an indefinitely differentiable function, the inf-convolution by a quadratic function maps a lower semicontinuous convex function to a continuously differentiable convex function :

Theorem 12.1 *Let $V : X \to \mathbb{R} \cup \{+\infty\}$ be a nontrivial lower semicontinuous convex function from X to $\mathbb{R} \cup \{+\infty\}$. Then there exists a* unique *solution (denoted by $J_\sigma(x)$) of the minimization problem $V_\sigma(x)$:*

$$V_\sigma(x) = V(J_\sigma x) + \frac{1}{2}\left\|\frac{x - J_\sigma x}{\sigma}\right\|^2.$$

Furthermore, V_σ is convex, continuously differentiable :

$$DV_\sigma(x) = \frac{x - J_\sigma x}{\sigma^2}$$

When σ converges to 0,

$$\forall x \in \mathrm{Dom}(V), \quad V_\sigma(x) \to V(x) \text{ and } J_\sigma x \to x \tag{12.4}$$

and when $\sigma \to \infty$,

$$V_\sigma(x) \text{ converges to } \quad -V^\star(0) = \inf_{x \in X} V(x) \tag{12.5}$$

The Moreau-Yosida transform of an indicator ψ_K is the function

$$x \mapsto \frac{1}{\sigma^2} d(x, K)$$

The quadratic functions

$$G_{\sigma,\, m}(x) := \frac{1}{2}\left\|\frac{x - m}{\sigma}\right\|^2$$

are regarded as Gaussian toll sets with mean m and variance σ. They form a class stable by inf-convolution :

Proposition 12.2 *The Gaussian toll sets are stable under inf-convolution :*

$$\left(G_{\sigma_1,\, m_1} \oplus G_{\sigma_2,\, m_2}\right)(x) = G_{\sqrt{\sigma_1^2 + \sigma_2^2},\, m_1 + m_2}$$

Proof. One must compute the solution to the minimization problem

$$\inf_y \left(\frac{1}{2}\left\|\frac{x - y - m_1}{\sigma_1}\right\|^2 + \frac{1}{2}\left\|\frac{y - m_2}{\sigma_2}\right\|^2\right)$$

¿From Fermat's Rule, this problem achieves its minimum at

$$\bar{y} := \frac{\sigma_2^2(x - m_1) + \sigma_1^2 m_2}{\sigma_1^2 + \sigma_2^2}$$

so that

$$\begin{aligned} \bar{y} - (x - m_1) &= \frac{\sigma_1^2(m_1 + m_2 - x)}{\sigma_1^2 + \sigma_2^2} \\ \bar{y} - m_2 &= -\frac{\sigma_2^2(m_1 + m_2 - x)}{\sigma_1^2 + \sigma_2^2} \end{aligned}$$

Consequently,

$$\left\{\begin{array}{l} (G_{\sigma_1, m_1} \oplus G_{\sigma_2, m_2})(x) = \frac{1}{2}\left\|\frac{x-\bar{y}-m_1}{\sigma_1}\right\|^2 + \frac{1}{2}\left\|\frac{\bar{y}-m_2}{\sigma_2}\right\|^2 \\ = \frac{1}{2}\left\|\frac{x-(m_1+m_2)}{\sqrt{\sigma_1^2+\sigma_2^2}}\right\|^2 = G_{\sqrt{\sigma_1^2+\sigma_2^2},\, m_1+m_2}. \end{array}\right. \quad \square$$

12.3.4 Analogy between Integration and Optimization

Beyond the Cramer transform which is only defined for functions or measures defined on vector spaces $\mathbb{R}^n$, there exists an striking formal analogy between optimization and probability theory that we shall only sketch without entering details which may lead us too far.

To the integral

$$\int_{\Omega} x(\omega) d\mu(\omega)$$

of a nonnegative measurable function defined on a measured space $(\Omega, \mathcal{A}, d\mu)$ corresponds the *infimum on a closed toll set* U *of a lower semicontinuous function* $V : E \mapsto \mathbb{R} \cup \{+\infty\}$ *on a metric space* E defined by

$$\inf_{x \in E} (V(x) + U(x))$$

To the Dirac measure $\delta_a : x(\cdot) \mapsto x(a)$ corresponds the indicator ψ_a because

$$\inf_{x \in V} (V(x) + \psi_a(x)) = V(a)$$

To the integral $\int_A d\mu(\omega)$ of the characteristic function of a measurable set $A \in \mathcal{A}$ providing the measure $d\mu$ of a subset A corresponds the minimization problem of a function $V(\cdot)$ on the closed subset A

$$\inf_{x \in E} (\psi_A(x) + V(x)) = \inf_{x \in A} V(x)$$

Consequently, to the measure $d\mu$, which is a function from the σ-algebra $\mathcal{A}$ to the half-line $\mathbb{R}_+$ supplied with the operations $+$ and $\times$, corresponds the *Maslov measure* M_V, function from the family of closed subsets of E defined by

$$K \mapsto \inf_{x \in K} V(x)$$

which enjoys the following properties :

$$\begin{cases} i) & M_V(E) = \inf_{x \in E} V(x) \\ ii) & M_V(\emptyset) = +\infty \\ iii) & M_V(A \cup B) = \inf(M_V(A), M_V(B)) \end{cases}$$

The analogy then becomes algebraic, because $(\mathbb{R}_+, +, \times)$ supplied with the usual addition and multiplication on one hand, and $(\mathbb{R}_+, \inf, +)$ supplied with the infimum and the usual addition on the other hand, are two instances of "dioids" (see Quadrat [26]), which are kind of rings supplied with two operations which do not have inverses : they are defined as follows :

Definition 12.4 *A* **commutative dioid** *is a set D supplied with*

1. *an associative, commutative and idempotent operation $\vee$, the neutral element of which is denoted by $\emptyset$*
2. *an associative commutative operation $\oplus$, the neutral element of which is denoted by $\{0\}$*

satisfying the distributivity condition

$$\begin{cases} i) & \forall\, x, y, z \in \mathcal{D}, \ (x \vee y) \oplus z = (x \oplus z) \vee (y \oplus z) \\ ii) & \forall\, x \in \mathcal{D}, \ x \oplus \emptyset = \emptyset \end{cases}$$

Besides the dioids $(\mathbb{R}_+, +, \times)$ and $(\mathbb{R}_+, \inf, +)$ (where the neutral element of the infimum is $\{+\infty\}$), another dioid is famous in mathematical morphology, a branch of image processing : it is the dioid of the family of subsets of a finite dimensional vector space supplied with the union of sets and the (Minkowski) addition of sets.

12.4 FUZZY OR TOLL DIFFERENTIAL INCLUSIONS

12.4.1 Set-Valued Maps

Let X and Y be two spaces. A set-valued map F from X to Y is characterized by its *graph* $Graph(F)$, the subset of the product space $X \times Y$ defined by

$$\text{Graph}(F) := \{(x, y) \in X \times Y \mid y \in F(x)\}$$

We shall say that $F(x)$ is the *image* or the *value* of F at x (see Aubin and Frankowska [13]).

A set-valued map is said to be *nontrivial* if its graph is not empty, i.e., if there exists at least an element $x \in X$ such that $F(x)$ is not empty.

12.4.2 The Viability Theorem

We shall use fuzzy analysis for uncertain controlled systems (see Aubin and Frankowska [14], Leitmann, Ryan, and Steinberg [25]).

Let us consider a *control system* (U, f) defined by

- a feedback set-valued map $U : X \rightsquigarrow Z$
- a map $f : \text{Graph}(U) \mapsto X$ describing the dynamics of the system

governing the evolution

$$\begin{cases} i) & \text{for almost all } t, \ x'(t) = f(x(t), u(t)) \\ ii) & \text{where } u(t) \in U(x(t)) \end{cases} \tag{12.6 }$$

Let us remark that when we take for controls the velocities, i.e., $U(x) := F(x)$ and $f(x, u) := u$, we find the usual differential inclusion $x' \in F(x)$. Conversely, the above system is the differential inclusion $x' \in F(x)$ in disguise where $F(x) := f(x, U(x))$.

We say that a closed subset $K \subset \text{Dom}(U)$ is viable under (U, f) if from any initial state $x_0 \in K$ starts at least one solution on $[0, \infty[$ to the control system (12.6) *viable in* K (in the sense that for all $t \geq 0$, $x(t) \in K$) (see Aubin [8]).

The contingent cone was introduced by G. Bouligand in the early thirties : when K is a subset of X and x belongs to K, we recall that the *contingent cone* $T_K(x)$ to K at x is the closed cone of elements v such that

$$\liminf_{h \to 0+} \frac{d(x + hv, K)}{h} = 0$$

It happens to remain the best way to describe mathematical the concept of tangent direction to any subset.

We associate with any subset $K \subset \text{Dom}(U)$ the regulation map $R_K : K \rightsquigarrow Z$ defined by

$$\forall\, x \in K, \ \ R_K(x) := \{u \in U(x) \mid f(x, u) \in T_K(x)\}$$

where $T_K(x)$ is the contingent cone to K at $x \in K$.

We say that K is a viability domain of (U, f) if and only if the regulation map R_K is strict (has nonempty values).

The Viability Theorem holds true for the class of Marchaud systems, which satisfy the following conditions :

$$\begin{cases} i) & \text{Graph}(U) \text{ is closed} \\ ii) & f \text{ is continuous} \\ iii) & \text{the velocity subsets } F(x) := f(x, U(x)) \text{ are convex} \\ iv) & f \text{ and } U \text{ have linear growth} \end{cases} \tag{12.7 }$$

Theorem 12.2 (Viability Theorem) *Let us consider a Marchaud control system* (U, f). *Then a closed subset* $K \subset \mathrm{Dom}(U)$ *is viable under* (U, f) *if and only if it is a* viability domain *of* (U, f).

Furthermore, any "open loop" *control* $u(\cdot)$ *regulating a viable solution* $x(\cdot)$ *in the sense that*

$$\text{for almost all } t, \;\; x'(t) = f(x(t), u(t))$$

obeys the regulation law

$$\text{for almost all } t, \;\; u(t) \in R_K(x(t)) \tag{12.8 }$$

Otherwise, if K *is not a viability domain of the control system* (U, f), there exists a largest closed viability domain *of* (U, f) *contained in* K *(possibly empty), denoted* $\mathrm{Viab}(K)$, *called the* viability kernel *of* K, *and equal to the set of states* $x_0 \in K$ *from which starts a solution of the control system viable in* K.

12.4.3 Toll Differential Inclusion

We shall use fuzzy analysis for uncertain controlled systems (see Aubin and Frankowska [14], Leitmann, Ryan, and Steinberg [25]).

By using indicators, we can reformulate the differential inclusion

$$\text{for almost all } t, \;\; x'(t) \in F(x(t))$$

as

$$\text{for almost all } t, \;\; \psi_{F(x(t))}(x'(t)) < +\infty$$

Then we are led to define "fuzzy or toll dynamics" of a system by a toll set-valued map $\mathbf{F}$ associating with any $x \in X$ a toll set $\mathbf{F}(x)$ of velocities $\{v \mid F(x, v) < +\infty\}$ (see Dubois and Prade [19]). In this case, we can write the associated *fuzzy or toll differential inclusion* in the form

$$\text{for almost all } t \geq 0, \;\; F(x(t), x'(t)) < +\infty \tag{12.9 }$$

or, equivalently, in the form

$$\text{for almost all } t \geq 0, \; x'(t) \in \mathbf{F}(x(t))$$

which is a toll subset instead of a usual subset (see Aubin[7], [12], Kaleva [24], Seikkala [27]). We associate with the toll dynamics its cheapest cost

$$\forall\, x \in \mathrm{Dom}(\mathbf{F}), \;\; \lambda_F(x) := \inf_{v \in X)} F(x, v)$$

When $\mu(x) \geq \lambda_F(x)$ is upper semicontinuous, we can associate the "level set-valued map" Λ_F^μ defined by

$$\Lambda_F^\mu(x) := \{v \in X \mid F(x,v) \leq \mu(x)\}$$

We observe that *the "level set-valued map" Λ_F^μ is a Marchaud map whenever* **F** *is a toll Marchaud map and μ is upper semicontinuous.*

The "cheapest set-valued map" $\Lambda_F := \Lambda_F^{\lambda_F}$ corresponds to the choice of $\mu = \lambda_F$ (see Dubois and Prade [20]). It is Marchaud whenever λ_F is upper semicontinuous. It is easy to check that *if the cost functions $F(\cdot, v)$ of the toll map* **F** *are continuous, then Λ_F is upper semicontinuous.*

Therefore, for any upper semicontinuous function $\mu \geq \lambda_F$, we say that the solutions of the differential inclusion

$$\text{for almost all } t \geq 0, \quad x'(t) \in \Lambda_F^\mu(x(t))$$

are μ-selections of the toll differential inclusion. When λ_F is upper semicontinuous, the solutions to

$$\text{for almost all } t \geq 0, \quad x'(t) \in \Lambda_F(x(t))$$

are called *cheapest solutions to the toll differential inclusion.*

12.4.4 Example : Random Differential Equations

Consider for instance random variables associating with each state of the system a random velocity. We consider the probability law by $d\mu(x)$ of the velocities, with which associate their average mean

$$C^\star_{\mu(x)}(0) = \int_{I\!R^n} y d\mu(x)(y)$$

where $C_{\mu(x)}$ denotes the Cramer transform of the state dependent probability law $d\mu(x)$ and $C_{\mu(x)}$ its Fenchel transform, equal to the logarithm of its Laplace transform.

It is then natural to regard the differential equation

$$x'(t) = \int_{I\!R^n} y d\mu(x(t))(y)$$

as a random differential equation, associating with the state $x(t)$ at time t the state dependent average velocity.

By the Cramer transform, we know that this average velocity being the gradient at 0 of the conjugate function of the Cramer transform of $d\mu(x)$, it achieves the minimum of the cost function of the Cramer transform :

$$C_{\mu(x)}\left(\int_{\mathbb{R}^n} y d\mu(x)(y)\right) = \inf_{v\in X} C_{\mu(x)}(v)$$

Therefore, *the solution to the random differential equation is the cheapest solution of the toll differential inclusion*

$$\text{for almost all } t \geq 0, \ \ C_{\mu(x(t))}(x'(t)) < +\infty$$

12.4.5 Viability Theorem for Toll Differential Inclusions

We begin by characterizing usual subsets K enjoying the viability property for toll differential inclusion : from any initial state $x_0 \in K$ starts at least one solution $x(\cdot)$ to the toll differential inclusion (12.9) which is viable in K in the sense that

$$\forall\, t \geq 0, \ \ x(t) \in K$$

We shall denote by $\mathbf{R_K}$ the toll map associating with any $x \in K$ the toll intersection $\mathbf{R_K}(x) := \mathbf{F}(x) \cap T_K(x)$ of the toll set $\mathbf{F}(x)$ and the contingent cone $T_K(x)$. Its cost function is equal to

$$R_K(x,v) := F(x,v) + \psi_{T_K(x)}(v)$$

Definition 12.5 *We shall say that* $\mathbf{R_K}$ *is the* toll regulation map (or the fuzzy controller) *associated with the viability constraints* K *and that a subset* $K \subset Dom(\mathbf{F})$ *is a viability domain of the toll set-valued map* $\mathbf{F}$ *if and only if*

$$\forall\, x \in K, \ \ \mathbf{R_K}(x) \neq \emptyset$$

i.e., if and only if

$$\forall\, x \in K, \ \ \exists\, v \text{ such that } R_K(x,v) < +\infty$$

We begin by proving an extension of the Viability Theorem (see Theorem 3.3.5 , Aubin [8]) to toll differential inclusions.

Theorem 12.3 (Toll Viability) *Let us consider a Marchaud toll set-valued map* $\mathbf{F}$ *from a finite dimensional vector-space* X *to itself. Any closed subset* $K \subset Dom(\mathbf{U})$ *enjoying the viability property with respect to* U *is a viability domain.*

The converse holds true if

$$\forall\, x \in \mathrm{Dom}(\mathbf{R_K}), \ \ \lambda_{R_K}(x) := \inf_{v\in T_K(x)} F(x,v) \leq \mu(x) < +\infty$$

where μ is upper semicontinuous. In this case, from any initial state $x_0 \in K$, starts at least one viable μ-selection of the toll differential inclusion.

When the toll set-valued map **F** is continuous, we can select a viable solution to the toll differential inclusion (12.9) which is *cheapest*, in the sense that the cost of its velocity's membership is minimal :

$$\text{for almost all } t, \quad R_K(x(t), x'(t)) \;=\; \inf_{v \in T_K(x(t))} F(x(t), v) \tag{12.10 }$$

In order to obtain cheapest solutions, we have to check that the function λ_{R_K} is upper semicontinuous.

We need naturally more regularity on the toll dynamics, but also on the closed subset K. We recall (see Chapter 5 of Aubin [13] for instance) that K is *sleek* if the set-valued map $x \rightsquigarrow T_K(x)$ is lower semicontinuous. Convex subsets are sleek, and for sleek subsets, the contingent cones are closed convex cones.

Theorem 12.4 *We posit the assumptions of Theorem 12.3. We assume moreover that the restriction of the membership function U to its domain (the graph of* **U***) is continuous and that the viability domain K is sleek. Then there exists a cheapest viable solution to the differential inclusion (12.9) (i.e., which satisfies condition (12.10)).*

Proof. We introduce the function λ_{R_K} defined by

$$\lambda_{R_K}(x) \;:=\; \inf_{v \in T_K(x)} U(x, v) \;=\; \inf_{v \in X} R_K(x, v)$$

Since saying that K is sleek amounts to saying that the set-valued map $x \rightsquigarrow T_K(x)$ is lower semicontinuous, the Maximum Theorem (see for instance Theorem 2.1.6 of Aubin [13]) implies that the function λ is upper semicontinuous, because we have assumed that U is upper semicontinuous.

We then introduce the set-valued map G defined by

$$\Lambda_{R_K}(x) \;:=\; \{\, v \in T_K(x) \mid F(x, v) \le \lambda_{R_K}(x) \,\}$$

Then Λ_{R_K} has a closed graph, and the other assumptions of the Viability Theorem are satisfied. There exists a viable solution to differential inclusion $x'(t) \in \Lambda_{R_K}(x(t))$, which is a cheapest viable solution to the toll differential inclusion (12.9). □

12.4.6 Example : Moreau Transform of a Differential Inclusion

Starting with a viability problem for a differential inclusion $x' \in F(x)$ when the subset K is not viable under F, one can try to regularize by the Moreau transform the subsets $F(x)$ to obtain toll subsets whose cost functions are

$$F_\sigma(x,v) \;=\; \frac{1}{\sigma^2} d(v, F(x))$$

The associated toll differential inclusion is defined by

$$\text{for almost all } \geq 0, \;\; d(x'(t), F(x(t))) \;<\; +\infty$$

and the cost function of the regulation map is equal to

$$R_K(x,v) \;:=\; d(v, F(x)) + \psi_{T_K(x)}(v)$$

Therefore,

$$\lambda_{R_K}(x) \;=\; d(F(x), T_K(x))$$

Therefore, the cheapest solutions to this toll differential inclusion are the solutions to the projected differential inclusion

$$x'(t) \;\in\; \Pi_{T_K(x(t))}(F(x(t)))$$

It has solutions whenever the Marchaud map F is continuous and the subset K is sleek. This projected differential inclusion has been studied in Chapter 10 of Aubin [13]. Their solutions are actually the same than the solutions to the variational inequalities

$$x'(t) \;\in\; F(x(t)) - N_K(x(t))$$

12.4.7 Toll Viability Domains

Is it possible to speak of toll subsets having the viability property?

A way to capture this idea is to introduce a continuous function which are $w(t) = (w(0) - \frac{b}{a})e^{-at} + \frac{b}{a}$. $\varphi : I\!R_+ \to I\!R$ with linear growth (which is used as a parameter in what follows) and the associated differential equation

$$w'(t) = -\varphi(w(t)), \;\; w(0) \;=\; V(x_0) \tag{12.11 }$$

whose solutions $w(\cdot)$ set an upper bound to the membership of a toll subset when time elapses. The main instance of such a function φ is the affine function $\varphi(w) := aw - b$, the solutions of which are $w(t) = (w(0) - \frac{b}{a})e^{-at} + \frac{b}{a}$.

We shall say that a toll set $V \subset \mathrm{Dom}(\mathbf{F})$ enjoys the "toll viability property" (with respect to φ) if and only if from every initial state $x_0 \in \mathrm{Dom}(V)$ starts at least one solution to the toll differential inclusion (12.9) and to the differential equation (12.11) which are viable in the sense that

$$\forall\, t \geq 0, \;\; V(x(t)) \;\leq\; w(t), \;\; w(0) = V(x_0)$$

We have to adapt the concept of contingent cone to toll sets. This is possible by introducing the contingent cone to the epigraph of the cost function V of the toll set. We observe that the contingent cone to the epigraph of V at $(x, V(x))$ is the epigraph of a function denoted by $D_\uparrow V$:

$$\mathcal{E}p(D_\uparrow V(x)) = T_{\mathcal{E}p(V)}(x, V(x))$$

and which is equal to

$$\forall\, u \in X, \ D_\uparrow V(x)(u) = \liminf_{h \to 0+, u' \to u} (V(x + hu') - V(x))/h$$

We say that $D_\uparrow V(x)$ is the contingent epiderivative of V at $x \in \mathrm{Dom}(V)$.

We introduce now the "contingent set" $T_V^\varphi(x)$ (also denoted $T_V(x)$), the closed subset defined by :

$$T_V^\varphi(x) := \{\, v \in X \mid D_\uparrow V(x)(v) + \varphi(V(x)) \leq 0 \,\}$$

Definition 12.6 (Toll Viability Domain) *Let the continuous function φ with linear growth be given. We shall say that a toll subset V is a toll viability domain of a toll set-valued map $\mathbf{F}$ (with respect to φ) if and only if*

$$\forall\, x \in \mathrm{Dom}(V), \ \mathbf{R_V(x)} := \mathbf{F}(x) \cap T_V^\varphi(x) \neq \emptyset$$

i.e., if and only if

$$\forall\, x \in \mathrm{Dom}(V), \ \exists\, v \ \ \textit{such that} \ R_K(x,v) := F(x,v) + T_V^\varphi(x) < +\infty$$

Let us set

$$\lambda_{R_V}(x) := \inf_v R_K(x,v) = \inf_{v \in T_V^\varphi(x)} F(x,v)$$

Theorem 12.3 can be extended to toll viability domains :

Theorem 12.5 *The toll set-valued map $\mathbf{F}$ satisfies the assumptions of Theorem 12.3. We assume that $V \subset \mathrm{Dom}(\mathbf{F})$ is a closed toll subset which is contingently epidifferentiable. This means that for all $x \in \mathrm{Dom}(V)$, $\forall\, v \in X$, $D_\uparrow V(x)(v) > -\infty$ and that $D_\uparrow V(x)(v) < \infty$ for at least a $v \in X$. If a closed toll subset V enjoys the viability property, then it is a closed toll viability domain of $\mathbf{F}$ and the converse holds true if*

$$\forall\, x \in \mathrm{Dom}(\mathbf{F}), \ \lambda_{R_V}(x) := \inf_{v \in T_V^\varphi(x)} F(x,v) \leq \mu(x) < +\infty$$

where μ is upper semicontinuous.

We proceed by extending Theorem 12.4 on selection of toll viable solutions to toll differential inclusions which are cheapest, in the sense that

$$\text{for almost all } t, \quad F(x(t), x'(t)) \;=\; \inf_{v \in T_V^{\varphi}(x(t))} F(x(t), v) \qquad (12.12\ )$$

Theorem 12.6 *We posit the assumptions of Theorem 12.3. We assume moreover that the restriction of the membership function F to its domain (the graph of* **F***) is continuous and that the toll viability domain V satisfies*

$$x \rightsquigarrow T_V^{\varphi}(x) \quad \text{is lower semicontinuous}$$

Then there exists a cheapest viable solution to the differential inclusion (12.9) (which satisfies condition (12.12)).

Proof. The proof is the same as the one of Theorem 12.4, where the function μ is now defined by

$$\lambda_{R_V}(x) \;:=\; \inf_{v \in T_V^{\varphi}(x)} F(x, v) \;\square$$

12.4.8 Toll Viability Kernels

Let us consider now any closed toll subset of the domain of **F**, which is not necessarily a toll viability domain. The functions φ being given, we shall construct the largest closed toll viability domain V_φ contained in V.

Theorem 12.7 *The toll set-valued map satisfies the assumptions of Theorem 12.3. We assume that $V \subset$ Dom(***F***) is a closed toll subset which is contingently epidifferentiable.*

Then for any upper semicontinuous function $\mu > 0$, there exists a largest closed toll viability domain V_φ^μ contained in V (for the toll differential inclusion), which enjoys furthermore the property :

$$\text{for almost all } t \geq 0, \quad F(x(t), x'(t)) \;\leq\; \mu(x(t))$$

12.4.9 Fuzzy Control Problems

If X, Z denote the state and control spaces of a control system

$$x'(t) \;=\; f(x(t), u(t)) \qquad (12.13\ )$$

we consider a "fuzzy or toll" control as a toll set-valued map **U** defined by a cost function

$$(x, u) \in X \times Z \mapsto U(x, u) \in [0, +\infty] \qquad (12.14\ )$$

We look for solutions to the toll control problem

$$\forall\, t \geq 0,\ \ u(t) \text{ belongs to the toll set } \mathbf{U}(x(t)) \tag{12.15 }$$

in the sense that

$$\forall\, t \geq 0,\ \ U(x(t), u(t)) < +\infty \tag{12.16 }$$

We observe that solutions to the toll control system are solutions to the toll differential inclusion $x' \in \mathbf{F}(x)$ where the toll set-valued map $\mathbf{F}$ is the product of $\mathbf{U}$ with the map $f(x, \cdot)$, the cost function of which is equal to

$$F(x, v) \ := \ \inf_{\{u \mid f(x,u)=v\}} U(x, u)$$

Indeed, $F(x, v)$ is finite if and only if there exists a control $u \in \mathbf{U}(x)$ such that $f(x, u) = v$.

If we consider the toll viability problem of finding at least one viable solution

$$\forall\, t \geq 0,\ \ V(x(t)) \ \leq \ w(t) \tag{12.17 }$$

of the toll control problem starting from every initial state $x_0 \in K$, we introduce the "toll regulation map" or "fuzzy controller" $\mathbf{R_V}$ defined by its cost function

$$R_V(x, u) \ = \ U(x, u) + \psi_{T_V^{\varphi}(x)}(f(x, u)) \tag{12.18 }$$

In other words, $v \in \mathbf{R_V}(x)$ if and only if there exists a control $u \in \mathbf{U}(x)$ such that

$$D_{\uparrow}V(x)(f(x, u)) + \psi(V(x)) \ \leq \ 0$$

We set

$$\lambda_{R_V}(x) \ := \ \inf_{v \in X} \inf_{\{u \mid f(x,u)=v\}} (U(x, u) + \psi_{T_V^{\varphi}(x)}(f(x, u)))$$

Hence we observe that the toll set V is a viability domain of the toll control system if and only if

$$\forall\, x \in \mathrm{Dom}(V),\ \ \mathbf{R_V}(x) \ \neq \ \emptyset$$

In this case, for any upper semicontinuous function $\mu \geq \lambda_{R_V}$, from any initial state $x_0 \in \mathrm{Dom}(V)$ starts at least one viable μ-selection of the toll control system, regulated by controls satisfying the toll regulation law

$$\text{for almost all } t \geq 0, \begin{cases} u(t) \in \mathbf{R_U}(x(t)) \\ U(x(t), u(t)) \ \leq \ \mu(x(t)). \end{cases}$$

Notes

1. Since then, it has been wildly successful, above all in many areas outside mathematics. Lately, we found the following quotation by A. Bercoff : *"Aujourd'hui, nous nageons dans la poésie pure des sous ensembles flous" ...*

François Mitterand

in "LA LUTTE FINALE", Michel Lafon (1994), p.69. Before, the French novelist Jacques Laurent published a novel untitled *Les sous-ensembles flous.*

References

[1] Akian, M., Quadrat, J.-P., and Viot, M. (1994) Bellman Processes, **199**, *Springer Verlag, Lect. Notes Control Inf. Sci.*, **199** pp. 302-311.

[2] Akian, M., Quadrat, J.-P., and Viot, M. (1996) Duality between probability and optimization, *Idempotency* , Gunarwadernar J. Ed, Cambridge University Press.

[3] Akian, M., Quadrat, J.-P., and Viot, M. (1995) Bellman Processes with independent increments, *Proceedings of the 17th IFIP Conference* pp. 207-210

[4] Aubin, J.-P. (1981) Cooperative fuzzy games, *Math. Op. Res.*, **6**, 1-13.

[5] Aubin, J.-P. (1981) Locally lipchitz cooperative games, *J. Math. Economics*, **8**, 241-262.

[6] Aubin, J.-P. (1983) Analyse Non Linéaire et ses Motivations Economiques, *Masson, Paris.*

English translation: (1994) *Optima and Equilibria*, Springer-Verlag..

[7] Aubin, J.-P. (1990) Fuzzy differential inclusions, *Problems of Control and Information Theory*, **19**, 55-67.

[8] Aubin, J.-P. (1991) Viability Theory, *Birkhäuser.*

[9] Aubin, J.-P. (1994) Initiation à l'Analalyse Appliquée, *Masson, Paris.*

[10] Aubin, J.-P. (1997a) (to appear) Dynamical Economic Theory: a viability approach, *Springer-Verlag.*

[11] Aubin, J.-P. (1997b) (to appear) Mutational and morphological analysis: tools for shape regulation and optimization, *Birkhäuser.*

[12] Aubin, J.-P. and Dordan, O. (1996) Rétroactions floues de systèmes contrôlés *JESA* **vol 30, 5** .

[13] Aubin, J.-P. and Frankowska, H. (1990a) Set-Valued Analysis, *Birkhäuser.*

[14] Aubin, J.-P. and Frankowska, H. (1990b) Controllability and Observability of Control Systems Under Uncertainty, *Annales Polonici Mathematici*, Volume dedicated to Opial, 37-76.

[15] Azencott, R. (1980) Grandes déviations et applications, *Lecture Notes in Mathematics*, **774**, Springer-Verlag.

[16] Bacelli F., Cohen G., Olsder G., and Quadrat J.-P. (1992) Synchronisation and linearity, *J. Wiley, New York.*

[17] Dordan, O. (1995) Analyse qualitative, *Masson, Paris.*

[18] Dubois, D. and Prade, H. (1984) Fuzzy numbers an overview, *The analysis of Fuzzy Information (J.C. Bezdek, Ed), CRC Press, Boca Raton, Fl.*, pp. 3-40.

[19] Dubois, D. and Prade, H. (1987c) On several definitions of the differential of a fuzzy mapping, *Fuzzy sets and systems*, **24**, 117-120.

[20] Dubois, D. and Prade, H. (1987d) The mean value of a fuzzy number, *Fuzzy sets and systems*, **24**, 279-300.

[21] Dubois, D. and Prade, H. (1992) Upper and lower images of a fuzzy set induced by a fuzzy relation, *Information Sciences*, **64**, 203-232.

[22] Dubois, D. and Prade, H. (1988) Théorie des possibilités, *Masson, Paris.*

[23] Dubois, D. and Prade, H. (1993) Toll sets and toll logic, *Fuzzy Logic*, Lowen R. and Roubens, M. Eds, Kluwer, pp. 169-177.

[24] Kaleva, O. (1987) Fuzzy differential equations, *Fuzzy sets and systems journal*, **24**, 301-317.

[25] Leitmann, G., Ryan, E.P., and Steinberg, A. (1986) Feedback control of uncertain systems: robustness with respect to neglected actuator and sensor dynamics, *Int. J. Control*, **43**, 1243-1256.

[26] Quadrat, J.P. (1995) Max-plus algebra and applications to system theory and optimal control, *Chatterji, S. D. (ed.), Proceedings of the international congress of mathematicians, ICM '94, August 3-11, 1994, Zuerich, Switzerland. Vol. II. Basel: Birkhäuser*, 1511-1522.

[27] Seikkala, S. (1987) On the fuzzy initial value problem, *Fuzzy sets and systems journal*, **24**, 319-330.

[28] Zadeh, L.A. (1965) Fuzzy sets, *Information and Control*, **8**, 338-353.

13 CHAOS AND FUZZY SYSTEMS

Phil Diamond

Mathematics Department
University of Queensland
St Lucia, Qld 4072 Australia

13.1 INTRODUCTION.

13.1.1 What is Chaos?

Chaos is a concept and a paradigm which has captured both the public imag ination and the attention of workers in science and engineering. Despite this intense interest, there is still no universally accepted mathematical definition of chaos. Nevertheless, there are characteristics which are widely held to be incorporated in any notion of a chaotic situation.

- Chaotic systems have *aperiodic longterm behaviour*: as time evolves, typical trajectories do not settle down to equilibrium states such as fixed points or periodic cycles, nor to quasiperiodic orbits.

- Chaos is a property of *deterministic systems* and is induced by certain types of nonlinearities rather than random disturbances in the environment.

- In chaos there is *sensitive dependence upon initial conditions*. By this is meant that if two runs of the system begin from very similar configurations, their subsequent behaviours rapidly become very different. This is the phenomenon evocatively called "The Butterfly Effect".

The net result of these properties is very irregular behaviour, as in Figure 8.1, which shows the evolution in time of the $x-$coordinate in the Hénon map [17]

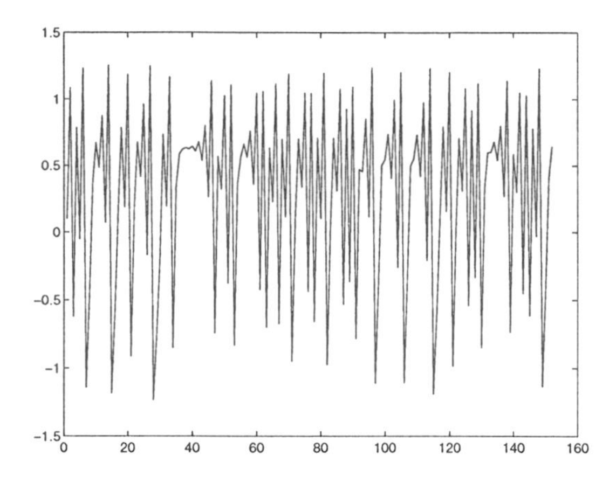

Figure 13.1 Irregular behaviour.

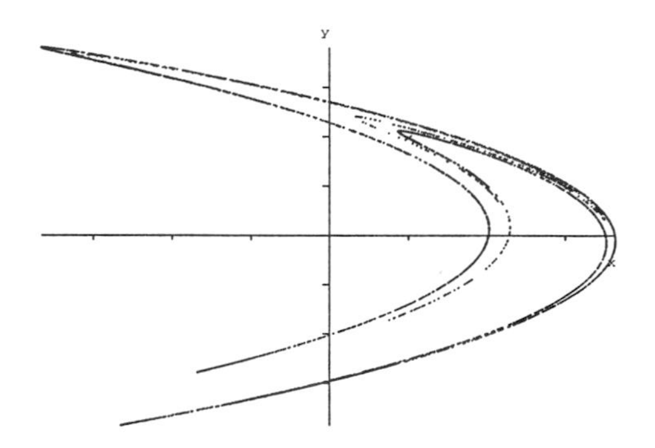

Figure 13.2 The Hénon attractor.

defined by the recurrence

$$x_{n+1} = y_n + 1 - 1.4x_n^2, \quad y_{n+1} = 0.3x_n\,. \tag{13.1 }$$

Although a chaotic system is not attracted to a simple state, it is drawn into a very complicated set called a *strange attractor*. As the system wanders through the strange attractor, it exhibits irregular and erratic behaviour which relect the complicated structure of the attractor. Figure 8.2 shows the Hénon attractor for the system induced by the recurrence (13.1)

If one takes ever magnified snapshots of parts of the attractor, it has fractal like properties. Cross sections are so complicated that they are Cantor sets.

13.1.2 Chaos and modeling.

It is now widely recognised in systems modeling and controller design that chaos can exist in the dynamical systems with which we deal and can produce chaotic and erratic behaviour in practical situations. This behaviour is not necessarily something to be avoided at all costs, but it is an aspect which should be recognised and taken into account when dealing with a control system. Indeed, aspects of chaotic behaviour can be exploited to give periodic control of chaotic systems [26,29].

Any mathematical definition of chaos must be more precise than simply offering the signatures highlighted above. Common descriptions are in terms of positive topological entropy [3], Li–Yorke scrambled sets [23], sensitive dependence of initial conditions [7,31] and positive Liapunov exponents [15]. The precise relationships between these competing signatures is unclear and some notions are not even independent [1,5].

Even granted that there is no unequivocal concept of chaos, it is still inadequate to describe complicated, erratic behaviour as chaotic without further criteria. Numerous authors have given sufficient conditions for types of chaos to appear. Among many we mention Li and Yorke [23], Kloeden [21], Marotto [24], Wiggins [31]. A very simple criterion was given in [8] but it only produces periodic sets of arbitrary periods rather than periodic points. ... The situation can be still more complicated with some fuzzy controllers. With Sugeno and other similar controllers, although a fuzzy decision base is used, the system is nonlinear but essentially crisp and the usual definitions are still applicable. However, there are systems which, while still dynamical systems, operate directly with functions (fuzzy sets) rather than with points in a Euclidean space. Nevertheless, mathematical descriptions of chaos can be formulated in much the same way and there has been some progress in giving sufficient conditions for the behaviour [11,14,22]. Some specific examples can be found in [9,30]. It has been shown that if $f : \Omega \to \Omega$ is chaotic, so also is its fuzzification

$$\begin{aligned} &\tilde{f} : \mathfrak{D}(\Omega) \to \mathfrak{D}(\Omega) \, , \\ &(\tilde{f}A)(y) = \sup\{A(x) : x \in f^{-1}(y)\} \, , \end{aligned} \tag{13.2}$$

defined on the space $\mathfrak{D}(\Omega)$ of fuzzy sets on a set Ω. However, this chaos is in a sense degenerate, because the iterates are asymptotically crisp and, ultimately, one sees chaos of a mapping of ordinary sets rather than of fuzzy output [9]. Apparently, the usual fuzzification and sup–min composition are inadequate to describe all of the complexities which may arise in fuzzy control. This is not completely unexpected since other t–norm/conorm operations are being used [18] and more general extension principles have been developed [16].

In this paper we draw some of these threads together. After some preliminaries on spaces of fuzzy sets, a brief overview of discrete dynamical systems is given and several mathematical definitions of chaos are described. Then chaos is studied in two distinct metric spaces of fuzzy sets. We examine the way in which a measure of information content changes, during the evolution of a dynamical system. More general fuzzifications are also considered. These do not necessarily degenerate under chaotic iteration. It is shown that if the fuzzification operation is reasonably well–behaved then chaotic maps on Ω induce chaos in the corresponding fuzzy system. A relationship between Li–Yorke chaos, positive topological entropy and the simple criterion of [8] is demonstrated. We give a theorem which generalises and improves the chaotic theorem for set valued mappings of Benhabib and Day [2].

Below, Section 3.2 gives definitions and some background. Section 3.3 gives an overview of dynamical systems and chaos, while Section 3.4 introduces a measure of information content of fuzzy sets. Section 3.5 discusses how information can degrade under iteration of fuzzified chaotic mappings when the usual fuzzification is used. More general classes of Γ–fuzzifications are considered in Section 3.6 and a general theorem linking various types of chaos to a simple sufficient condition is proved in Section 3.7. An example of a chaotic fuzzified mapping with nondegenerate periodicities is given in Section 3.8, and a number of examples of fuzzy chaos are studied in the last section. A brief appendix defining topological entropy concludes the paper.

13.2 PRELIMINARIES.

Let Ω be a compact metric space with metric ρ. Denote by $\mathfrak{D}(\Omega)$ the totality of fuzzy sets

$$A : \Omega \to [0,1] = I$$

which are *upper semicontinuous*. That is, for each x_0 in the domain of A,

$$A(x_0) \geq \overline{\lim}_{x \to x_0} A(x) \ .$$

Note that the *characteristic function* χ_Y of a set $Y \subset \Omega$, defined by

$$\chi_Y(x) = \begin{cases} 1 \text{ if } \ x \in Y \ , \\ 0 \text{ otherwise} \ , \end{cases}$$

is a fuzzy set.

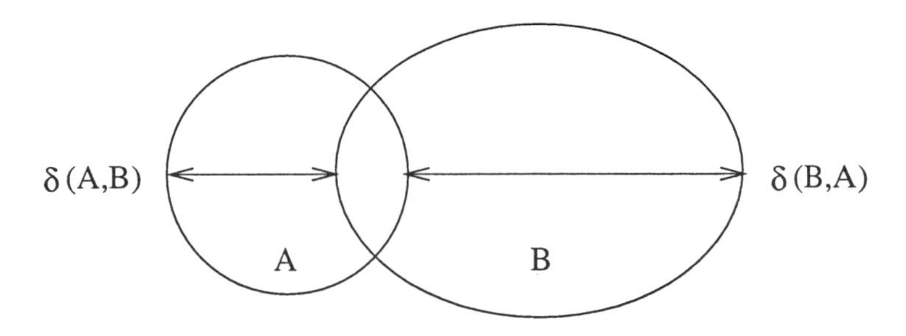

Figure 13.3 The Hausdorff separations of A and B.

Let X, Y be any two nonempty closed subsets of a metric space $(\mathfrak{P}, \bar{\rho})$. The *Hausdorff separation* of X from Y is defined to be

$$\bar{\rho}_H(X, Y) = \sup_{x \in X} \bar{\rho}(x, Y),$$

where $\bar{\rho}(x, Y) = \inf_{y \in Y} \bar{\rho}(x, y)$. The separation $\bar{\rho}_H$ is not a metric because it is not symmetric (See Figure 3.3). It may also be shown that $\bar{\rho}_H(X, Y) = 0$ if and only if $X \subseteq Y$. The quantity

$$d_H(X, Y) = \max \{\bar{\rho}_H(X, Y)\, , \, \bar{\rho}_H(Y, X)\}$$

is said to be the *Hausdorff distance* between X and Y. It is a metric on the space $\mathcal{S}(\mathfrak{P})$ of closed nonempty subsets of Ω. The metric space $(\mathcal{S}(\mathfrak{P}), d_H)$ is compact provided $\mathfrak{P}$ is compact [6].

For any $A \in \mathfrak{D}(\Omega)$ define

$$\text{send}\,(A) = \{(x, \alpha) \in \Omega \times I : A(x) \geq \alpha\} \ .$$

Since A is upper semicontinuous, send (A) is closed in $\mathfrak{P} = \Omega \times I$ with respect to the natural product metric $\bar{\rho}((x, \alpha), (y, \beta)) = \rho(x, y) + |\alpha - \beta|$. Consequently, we can define a distance function on $\mathfrak{D}(\Omega)$ by

$$H(A, B) = d_H\,(\text{send}\,(A), \text{send}\,(B))\, , A, B \in \mathfrak{D}(\Omega).$$

This is indeed a metric on $\mathfrak{D}(\Omega)$ and it is straightforward but lengthy to show

Theorem 13.1 *([20]) The metric space* $(\mathfrak{D}(\Omega), H)$ *is compact.*

For any $\delta \in (0, 1]$, define $\mathfrak{D}_\delta(\Omega)$ to be the class of all $A \in \mathfrak{D}(\Omega)$ satisfying

$$\max_{x \in \Omega} A(x) \ \geq \ \delta \ .$$

Observe that the class of all normal fuzzy sets on Ω is just $\mathfrak{D}_1(\Omega)$ and so the $\mathfrak{D}_\delta(\Omega)$ can be thought of as generalising the idea of normality.

Yet another metric may be defined on the class of normal upper semicontinuous fuzzy sets $\mathfrak{D}_1(\Omega)$. This metric is especially important when levels of possibility are part of the context, and is defined in terms of level sets $[A]^\alpha = \{x \in \Omega : A(x) \geq \alpha\}$, $0 < \alpha \leq 1$. We put

$$d_\infty(A, B) = \sup_{0<\alpha\leq 1} d_H([A]^\alpha, [B]^\alpha) \ .$$

Theorem 13.2 *([27]). The metric space $(\mathfrak{D}(\Omega), d_\infty)$ is compact.*

It may be shown that convergence in the d_∞ metric implies convergence in the sendograph metric H, but not conversely. Hence the topology of H is weaker than d_∞.

If the fuzzy sets are defined on a space which is not compact, for example $\Re^n$ where only bounded and closed sets are compact spaces, it is more natural to define the space $\mathfrak{D}^n \stackrel{\text{def}}{=} \mathfrak{D}(\Re^n)$ of upper semicontinuous fuzzy sets with *compact support.* The metrics H, d_∞ are defined as above and the space is *complete* with respect to each.

Theorem 13.3 *The metric spaces $(\mathfrak{D}^n, H)$ and $(\mathfrak{D}^n, d_\infty)$ are complete.*

For a more complete description of metric properties, the reader is directed to [12].

Let $C(\Omega)$ be the set of all continuous maps $f : \Omega \to \Omega$. Suppose that there is a correspondence between each $f \in C(\Omega)$ and a mapping

$$\mathfrak{F}(f) : \mathfrak{D}(\Omega) \to \mathfrak{D}(\Omega).$$

Such a correspondence is called a *fuzzification.* The most common fuzzification is $\tilde{f}$ defined by

$$\left(\tilde{f}A\right)(y) = \sup\{A(x) : x \in f^{-1}(y)\} \qquad (13.3)$$

for all $y \in f(\text{supp}(A))$, and $\left(\tilde{f}A\right)(y) = 0$ for all other $y \in \Omega$. Further fuzzifications are discussed in Section 3.6.

13.3 DYNAMICAL SYSTEMS AND CHAOS

A dynamical system in a space $\mathfrak{X}$ describes the way quantities in the space change with the passage of time. Usually, the space $\mathfrak{X}$ is a space of states of a physical system. A differential equation generates dynamics for such a system, as does a difference equation. But so also does the interaction of a practitioner with an expert system, or a fuzzy controller, or a decision maker working with vague and imperfect information. However, it is all a matter of what space (or universe) $\mathfrak{X}$ that one is working in, and all these systems have common features. Broadly speaking, each is a smooth mapping $\phi_t : \mathfrak{X} \to \mathfrak{X}$. More precisely, a *dynamical system* is a continuously differentiable map $\phi : \mathbf{P} \times \mathfrak{X} \to \mathfrak{X}$, where $\mathbf{P} = \Re$ or $\mathbb{Z}$, written as $\phi_t : \mathfrak{X} \to \mathfrak{X}$, $t \in \mathbf{P}$, and where ϕ_t satisfies

- ϕ_0 is the identity on $\mathfrak{X}$, $\phi_0(x) = x, \quad$ for all $\quad x \in \mathfrak{X}$;
- The composition $\phi_s \circ \phi_t = \phi_{s+t}$ for all $s, t \in \mathbf{P}$, that is $\phi_s(\phi_t(x)) = \phi_{s+t}(x)\ \forall\ x \in \mathfrak{X}$.

Typically, a differential equation $x' = f(t,x), x(t_0) = x_0$ generates a mapping $\phi_t : \Re^n \to \Re^n$, with $t \in \Re$ (a *continuous* dynamical system), defined by $\phi_t(x_0) = X(t, t_0; x_0)$, where this denotes the solution of the initial value problem at time t. Similarly, a difference equation $x_{k+1} = f_k(x_k)$ generates $\phi_k : \mathfrak{X} \to \mathfrak{X}$, $k \in \mathbb{N}$ (a *discrete* dynamical system or *iterated function system*), defined by

$$\phi_k(x) = f_k \circ f_{k-1} \circ \cdots \circ f_1(x) \ .$$

The discrete systems are described by iterations of one or more functions. We shall restrict our discussion to systems consisting of iterates $f^k(x)$ of a single continuous function f, where x is an element of a metric space $\mathfrak{X}$. These can arise as solutions of the difference equation $x_{k+1} = f(x_k)$.

The object of study in such systems are the *orbits* $\{x, f(x), f^2(x), \ldots\}$ of points $x \in \mathfrak{X}$. *Periodic points* of period k correspond to k–cycles, that is to orbits of finite length k, for then $f^k(x) = x$.

Discrete dynamical systems are capable of enormously complicated behaviour, and certain aspects are said to be "chaotic". One definition of chaos is due to Devaney [7] and uses the criteria of

1. f is *transitive*; for all disjoint open sets $U, V \subset \mathfrak{X}$, $f^k(U)$ meets V eventually for some k.
2. Periodic points of f are dense in $\mathfrak{X}$.
3. *Sensitive dependence on initial conditions*; there is a sensitivity constant δ such that for any unequal x, y, eventually (for some k) $f^k(x)$ and $f^k(y)$ are more than a distance δ apart.

It has recently been shown that **1&2** $\Rightarrow$ **3** [1] and, even more, transitivity seems the key element [5].

A closely related definition, that is easier to handle in many respects, involves the notion of a *scrambled set* [20,23]. The criteria here are that

(LY1) There exist k-cycles, for all $k \geq K$.

(LY2) There exists a set $S \subset \mathfrak{X}$, containing no cycles, such that

(a) $f(S) \subset S$;

(b) for every $x, y \in S$, $x \neq y$,

$$\limsup_{k \to \infty} ||f^k(x) - f^k(y)||0 \ ;$$

(c) for every $x \in S$ and cyclic point p,

$$\limsup_{k \to \infty} ||f^k(x) - f^k(p)||0 \ .$$

(LY3) There exists an uncountable subset $S_0 \subset S$ such that for all $x, y \in S_0$,

$$\liminf_{k \to \infty} ||f^k(x) - f^k(y)|| = 0 \ .$$

Definition 13.1 *A mapping ϕ of a metric space $\mathfrak{X}$ to itself is said to be* **LY–chaotic** *if it satisfies (LY1) and (LY2) above.*

Definition 13.2 *The map $\phi : \mathfrak{X} \to \mathfrak{X}$ is said to be* **E–chaotic** *if it has positive topological entropy $E(\phi)$. See the appendix for a brief description of topological entropy.*

Now suppose that we have an iterative system on a space $\mathfrak{X}$,

$$x_{k+1} = f(x_k) \ , \ k = 0, 1, 2, \ldots \ ,$$

where f is a continuous mapping from $\mathfrak{X}$ into itself. Numerous sufficient conditions for chaos have been found: see [8,10,20,23,24], to list just a few. When $\mathfrak{X} = \Re$, the "period three is chaos" result of Li and Yorke [23] is especially evocative.

Definition 13.3 *A map $f \in C(\Omega)$ is said to be* **erratic** *if there exists a nonempty compact set $Y \subset \Re^n$ satisfying*

(C1) $Y \cap f(Y) = \emptyset$;

(C2) $Y \cup f(Y) \subset f^2(Y)$.

It should be noted that if the set Y is also specified to be convex, erratic maps have much the same behaviour as the Kloeden conditions [20] and are considerably simpler to verify. When $\Omega = \Re$ then Y is a compact interval and these erratic conditions include the period three criterion. However, all these conditions have similar structure. There is first an expansion in one direction, often accompanied by shrinkage in others. This is followed by a wraparound step, which not only keeps the iterates in a bounded region but also jumbles them up.

In Section 3.7 we will show that erratic mappings induce both E–chaos and LY–like chaos on $(\mathfrak{D}(\Omega), H)$, provided that the fuzzification of the erratic map is well-behaved in the sense that it preserves possibility levels.

13.4 INFORMATION CONTENT OF FUZZY SETS

If one thinks of greyscale images as fuzzy sets, it is clear that some fuzzy sets convey more information than others. A crisp set Y, given by the characteristic function χ_Y (also a fuzzy set), is totally white on the greyscale and we would want to say that it had less information content than an image which included various shades and gradations of grey.

Information has most commonly been defined in a probabilistic context. Recently, the theory has been extended to uncertainty based on discrete *possibility distributions* [19]. Continuous information measures were introduced for elements of $\mathfrak{D}^1$ by Ramer [28] and involve the concept of a *rearrangement* compatible with the possibility levels of fuzzy sets. The rearrangement of functions of a single variable is a nonincreasing function which generalises the idea of a permutation in decreasing order. Rearrangements of functions defined on $\Re^n$ do not have such a natural interpretation, because this domain is only partially ordered for $n > 1$, nor is the rearrangement a function on $\Re^n$.

Denote by $m(X)$ the measure or content of a set $X \subset \Re^n$, and let $A \in \mathfrak{D}^1$. Write $P_A(\beta) = m\left(\{x : A(x) \geq \beta\}\right) = m\left([A]^\beta\right)$. Then $P_A(\cdot)$ is a nondecreasing function and has a generalised inverse P_A^{-1}, and so

$$\widetilde{A}(\xi) = P_A^{-1}(\xi) \ , \ m([A]^1) \leq \xi \leq m([A]^0) \ ,$$

is well-defined and $0 \leq \widetilde{A}(\xi) \leq 1$. In $\mathfrak{D}^1$, Ramer proposes the measure

$$I(A) = \int_{m([A]^1)}^{m([A]^0)} \frac{1 - \widetilde{A}(\xi)}{\xi} \, d\xi \ , \tag{13.4 }$$

This measure is additive on joint distributions of independent possibility assignments, subadditive on marginal assignments, symmetric and expansible (see [28] for an explanation of these concepts).

Note that the definition of $P_A(\beta)$ depends only on the measure of the β–level sets of A. This suggests that the same construction for a rearrangement $\widetilde{A}$ of

$A \in \mathfrak{D}^n$ can be carried out, even though the function $\widetilde{u}$ is a function of a single variable. In particular, it follows that the measure of the set in which $\widetilde{A}(\xi) \geq \beta$ is $P_A(\beta)$, that is

$$m([\widetilde{A}]^\beta) = m([A]^\beta) ,$$

even though $[\widetilde{A}]^\beta \subset \Re$, while $[A]^\beta \subset \Re^n$. Here, the measure $m(\cdot)$ denotes length in $\Re$ and n–dimensional measure in $\Re^n$ respectively, according to context.

Unfortunately, the integral (1) may not exist, as can be seen for

$$A(x) = 1 + (\ln x)^{-1} , \ 0 < x \leq 1/e .$$

Here, $[A]^\beta = (0, e^{-1/(1-\beta)}]$, $P_A(\beta) = e^{-1/(1-\beta)}$ and $\tilde{A}(\xi) = 1 + 1/\ln \xi$. Consequently the integral

$$I(A) = \int_0^{1/e} \frac{-1}{\xi \ln \xi} d\xi$$

does not exist because it diverges at the origin. However, there is a further reason for seeking another measure for studying the behaviour of information in chaotic systems. The measure I is somewhat unintuitive in so far as the information content of *every* triangular fuzzy number in $\mathfrak{D}^1$ is 1. To see this, note that $P_A(\beta) = m([A]^\beta) = (1-\beta)m([A]^0)$ and so $\tilde{A}(\xi) = 1 - \xi/m([A]^0)$, giving $I(A) = 1$.

For discussing triangular functions, and keeping in mind the difficulty in defining $\widetilde{A}$ for functions A of more than one variable, it is more suitable to define a measure of information directly in terms of the level sets, using the function P_A, by

$$\mathfrak{I}(A) = \int_0^1 \frac{P_A(0) - P_A(\beta)}{\beta} d\beta , \tag{13.5 }$$

Both definitions imply that $I(\chi_Y) = \mathfrak{I}(\chi_Y) = 0$ for crisp connected subsets Y of $\Re^n$. For triangular fuzzy numbers A,

$$\mathfrak{I}(A) = \int_0^1 \frac{m([A]^0) - (1-\beta)m([A]^0)}{\beta} d\beta = m([A]^0).$$

The integral in (4) also need not exist, for example if

$$A(x) = \exp(1 + \frac{1}{x-1}) , \ 0 \leq x < 1 .$$

The measure $\mathfrak{I}$ has virtually all of the desirable properties (additivity is replaced by a form of weighted subadditivity) and will suffice for our purposes here.

13.5 CHAOTIC MAPPINGS ON $(\mathfrak{D}^N, D_\infty)$

Let f be a continuous mapping of $(\mathfrak{D}^n, d_\infty)$ into itself. Such mappings can be induced from mappings on the base space $\Re^n$, using the classical extension principle. For
$T : \Re^n \to \Re^n$, and $A \in \mathfrak{D}^n$, define $TA \in \mathfrak{D}^n$ by

$$(TA)(x) = \begin{cases} \sup_{x \in T^{-1}(y)} A(x) & \text{if } T^{-1}(y) \neq \emptyset \ , \\ 0 & \text{otherwise} \ , \end{cases}$$

where $x \in T^{-1}(y)$ iff $y = T(x)$. In terms of level sets, this means that

$$[TA]^\beta = T\left([A]^\beta\right) \text{ and so } [T^k A]^\beta = T^k\left([A]^\beta\right) \ , \ k = 1, 2, \ldots \ . \qquad (13.6\ )$$

However, mappings from $\mathfrak{D}^n$ to itself can be constructed which are not induced in this way, as is the case with the first and last examples of Section 3.9.

Consequently, if the mapping T satisfies some conditions such as (C1), (C2) of Section 3.3, chaotic iterates will be generated by the difference scheme. This chaos can lead to loss of information. To see how this might occur, suppose that the iteration has the effect of increasing the relative measure of $[A]^\beta$ in $[A]^0$, so that

$$m([TA]^0) - m([TA]^\beta) \leq m([A]^0) - m([A]^\beta) \ , \ 0 \leq \beta \leq 1 \ ,$$

Then $\mathfrak{I}(TA) \leq \mathfrak{I}(A)$.

Another way that information might be lost is when the iterative systems have an attractor Λ which is a set (for example, a strange attractor). So, under iteration, the level sets $[T^k A]^\beta = T^k([A]^\beta)$ of $T^k A$ will approach the same limit set Λ for each $0 \leq \beta \leq 1$. Hence,

$$\mathfrak{I}(T^k A) = \int_0^1 \frac{P_{T^k A}(0) - P_{T^k A}(\beta)}{\beta} \, d\xi \longrightarrow \mathfrak{I}(\chi_\Lambda) = 0$$

as $k \to \infty$. Thus, there are a number of mechanisms whereby the information content can decay along the orbit of a fuzzy dynamical system – see examples of the next section.

13.6 Γ–FUZZIFICATION.

A *t-norm* is an associative binary operation $T(\cdot, \cdot)$ on I that satisfies

(i) T is nondecreasing in each place,

$$T(x, y) \leq T(x', y')$$

for all $x, x', y, y' \in I$ such that $x \leq x'$ and $y \leq y'$.

(ii) 1 is an identity of T,

$$T(x, 1) = T(1, x) = x \ , \ \forall x \in I.$$

(iii) T is commutative,

$$T(x, y) = T(y, x).$$

As a simple consequence of (i) and (ii) we have

$$T(x, x) \leq x \ , \ x \in I \ . \tag{13.7 }$$

Some common examples are $M(x, y) = \min(x, y)$, $\Pi(x, y) = xy$ and $W(x, y) = \max(x + y - 1, 0)$. Define the corresponding *t–conorm* by

$$T^*(x, y) = 1 - T(1 - x, 1 - y) \ , \ \forall x, y \in I.$$

A function Γ is a t–conorm if there is a t–norm T such that $\Gamma = T^*$. The $t-$conorms corresponding to those above are $M^*(x, y) = \max(x, y)$, $\Pi^*(x, y) = x + y - xy$, and $W^*(x, y) = \min(x + y, 1)$. It follows that

$$W < \Pi < M < M^* < \Pi^* < W^*.$$

If Γ is any $t-$norm or any t–conorm, define the *diagonal of* Γ, $\Delta_\Gamma : I \to I$ by

$$\Delta_\Gamma(x) = \Gamma(x, x).$$

Let Γ be any t-norm or t-conorm. A mapping $f : \Omega \to \Omega$ has a Γ-*fuzzification* $\tilde{f}_\Gamma : \mathfrak{D}(\Omega) \to \mathfrak{D}(\Omega)$ defined by

$$\left(\tilde{f}_\Gamma A\right)(y) = \sup_{x, x' \in f^{-1}(y)} \{\Gamma(A(x), A(x'))\} \ ,$$

for $y \in \Omega$ and $A \in \mathfrak{D}(\Omega)$. By virtue of property (i), this can be rewritten in terms of the diagonal as

$$\left(\tilde{f}_\Gamma A\right)(y) = \sup_{x \in f^{-1}(y)} \{\Delta_\Gamma(A(x))\} \, . \tag{13.8 }$$

An *α-level set with respect to* Γ is defined as

$$[A]^\alpha_\Gamma = \{x \in \operatorname{supp}(A) : \Delta_\Gamma(A(x)) \geq \alpha\}$$

for all $A \in \mathfrak{D}(\Omega)$ and $0 \leq \alpha \leq 1$.

Example For $\Gamma = \Pi$, $\Delta_\Pi(A) = A^2$ and so

$$[A]^\alpha_\Pi = \{x : A(x)^2 \geq \alpha\} = [A]^{\sqrt{\alpha}},$$

and similarly, for Π^*, $\Delta_{\Pi^*}(A) = 2A - A^2$ and so

$$[A]^\alpha_{\Pi^*} = [A]^{1-\sqrt{1-\alpha}} .$$

Lemma 13.1 *Let Γ be any t-norm or t-conorm and let $f \in C(\Omega)$. Then for any $A \in \mathfrak{D}(\Omega)$ and all $\alpha \in I$*

$$[A]^\alpha_\Gamma \subseteq [A]^\alpha , \tag{13.9 }$$

$$f\left([A]^\alpha_\Gamma\right) \subseteq \left[\tilde{f}_\Gamma A\right]^\alpha . \tag{13.10 }$$

Proof. Let $x \in [A]^\alpha_\Gamma$. That is, $\Delta_\Gamma(A(x)) \geq \alpha$. But then (8) gives

$$A(x) \geq \Delta_\Gamma(x) \geq \alpha$$

and the first inclusion follows. For $x \in [A]^\alpha_\Gamma$, if $y = f(x) \in f\left([A]^\alpha_\Gamma\right)$, then

$$\begin{aligned} \left(\tilde{f}_\Gamma A\right)(y) &= \sup\{\Delta_\Gamma(A(\xi)) : y = f(\xi)\} \\ &\geq \Delta_\Gamma(A(x)) \geq \alpha \end{aligned}$$

and the second inclusion is proved. □

Theorem 13.4 *Let Γ be any t−norm or t−conorm. If $f \in C(\Omega)$ and $A \in \mathfrak{D}(\Omega)$, then a necessary and sufficient condition for*

$$f\left([A]^\alpha_\Gamma\right) = \left[\tilde{f}_\Gamma A\right]^\alpha , \quad \alpha \in (0,1] , \tag{13.11 }$$

is that for every $y \in f(\mathrm{supp}\,(A))$,

$$\sup_{\xi \in f^{-1}(y)} \{\Delta_\Gamma(A(\xi))\}$$

is attained.

Proof. Necessity: let $f\left([A]^\alpha_\Gamma\right) = \left[\tilde{f}^\alpha_\Gamma\right]$. Choose y such that

$$\sup\{\Delta_\Gamma(A(\xi)) : y = f(\xi)\} = \alpha .$$

Then $y \in [f_\Gamma A]^\alpha$ and so, by hypothesis, $y \in f\left([A]_\Gamma^\alpha\right)$. But then $y = f(x')$ for some $x' \in [A]_\Gamma^\alpha$ which means that $\Gamma(A(x'), A(x')) \geq \alpha$. But

$$\begin{aligned} \alpha &= \sup_{\xi \in f^{-1}(y)} \{\Delta_\Gamma(A(\xi))\} \\ &\geq \Gamma(A(x'), A(x')) \\ &\Rightarrow \Gamma(A(x'), A(x')) = \alpha \end{aligned}$$

and the supremum is thus attained. Conversely, suppose that the supremum is attained. Take $y \in [f_\Gamma A]^\alpha$. That is,

$$(f_\Gamma A)(y) = \sup_{\xi \in f^{-1}(y)} \{\Delta_\Gamma(A(\xi))\} \geq \alpha .$$

Since this is attained, there is x' such that

$$y = f(x') \text{ and } \Delta_\Gamma(A(x')) \geq \alpha .$$

Hence $x' \in [A]_\Gamma^\alpha$ which means that $y \in f\left([A]_\Gamma^\alpha\right)$, that is $\left[\tilde{f}_\Gamma A\right]^\alpha \subseteq f\left([A]_\Gamma^\alpha\right)$, and (11) now follows from (10). □

As with (6), the relation (11) provides a procedure for computing the iterates of the fuzzified mapping, in this case of $\tilde{f}_\Gamma$ through the level sets induced by Γ. Since Δ_Γ is nondecreasing, there is a generalized inverse Δ_Γ^{-1} and

$$[A]_\Gamma^\alpha = [A]^{\Delta_\Gamma^{-1}(\alpha)} .$$

From this it follows that

Proposition 13.1 *Let Γ be any t-norm or t-conorm. If $f \in C(\Omega)$ and $A \in \mathfrak{D}(\Omega)$ satisfy the condition of Theorem 6.2, then for each positive integer k*

$$\left[\tilde{f}_\Gamma^k A\right]^\alpha = f^k\left([A]^{\Delta_\Gamma^{-k}(\alpha)}\right) .$$

Proof. Since

$$\begin{aligned} \left[\tilde{f}_\Gamma^{k+1} A\right]^\alpha = \left[\tilde{f}_\Gamma\left(\tilde{f}_\Gamma^k A\right)\right]^\alpha &= f\left(\left[\tilde{f}_\Gamma^k A\right]^{\Delta_\Gamma^{-1}(\alpha)}\right) \\ &= f\left(f^k\left([A]^{\Delta_\Gamma^{-k}(\Delta_\Gamma^{-1}(\alpha))}\right)\right) \\ &= f^{k+1}\left([A]^{\Delta_\Gamma^{k+1}(\alpha)}\right) , \end{aligned}$$

the result follows by induction. □

13.7 CHAOS AND FUZZIFICATION

A fuzzification $\mathfrak{F}$ is said to be *H–continuous* if every mapping $\mathfrak{F}(f)$, $f \in \Omega$, is continuous in $\mathfrak{D}(\Omega)$ with respect to the sendograph metric H. The H–continuous fuzzification $\mathfrak{F}$ is said to be *δ–normal* if the following properties are satisfied: For any $A \in \mathfrak{D}(\Omega)$ and any $f \in C(\Omega)$

$$\text{supp}\,(\mathfrak{F}(f)A) \subseteq f\,(\text{supp}\,(A)) \ ,$$

and any function of the form

$$\mathfrak{F}(f)A \ , \ A \in \mathfrak{D}_\delta(\Omega) \ , \ f \in C(\Omega)$$

belongs to the set $\mathfrak{D}_\delta(\Omega)$.

Denote by $\mathfrak{B}(N)$ the set of all binary strings

$$w = \{w_1, w_2, \ldots, w_N\} \ , \ w_i \in \{0,1\} \ , \ 1 \leq i \leq N$$

of length N that do not contain two adjacent zeros and begin with $w_1 = 1$. Write the cardinality of $\mathfrak{B}(N)$ as $\#\mathfrak{B}(N)$. Clearly

$$2^{(N-1)/2} \leq \#\mathfrak{B}(N) \leq 2^{N-1} \ .$$

Consequently,

$$\gamma = \lim_{N\to\infty} \left(N^{-1}\ln\left(\#\mathfrak{B}(N)\right)\right) \ .$$

exists and

$$\gamma \geq \frac{\ln 2}{2} \ .$$

Our principal result is

Theorem 13.5 *Let f be erratic on the set $X \subset \Omega$. Let $\mathfrak{F}$ be a δ–normal fuzzification. Then the following statements hold:*

(I) *The map $\mathfrak{F}(f)$ is E–chaotic and its topological entropy satisfies*

$$E\left(\mathfrak{F}(f)\right) \geq \gamma \ . \qquad (13.12)$$

(II) *The map $\mathfrak{F}(f)$ is LY–chaotic and, for each natural number k, it has at least $\#\mathfrak{B}(N)$ distinct periodic points in $\mathfrak{D}_\delta(\Omega)$.*

Remark. This theorem shows that the situation is somewhat different from that for erratic maps on $\Re^n$ as described in [8]. In $\Re^n$, rather than periodic points it is only possible to show that *periodic sets* occur in general. The difference arises because compact sets of fuzzy sets have stronger fixed point properties than compact sets in $\Re^n$ [13]. Theorem 3.5 also strengthens the theorem of [2] in the same manner, as well as generalizing it from the special case of mappings of crisp compact sets, whose characteristic functions are specific types of fuzzy sets.

The proof proceeds through several lemmas. All statements below assume that f is erratic. Denote by $\mathfrak{B} = \lim_{N\to\infty} \mathfrak{B}(N)$. Write

$$X_0 = X \quad , \quad X_1 = f(X) \; .$$

For each $w \in \mathfrak{B}$, denote by $\mathfrak{N}(w)$ the set of all $x \in \Omega$ satisfying

$$x \in X \; , \; f^i(x) \in X_{w_i} \; , \; i = 1, 2, \ldots \; .$$

Lemma 13.2 **(a)** *For each $w \in \mathfrak{B}$, the set $\mathfrak{N}(w)$ is nonempty and closed in Ω.*

(b) *Let the word $w \in \mathfrak{B}$ be $k-$periodic. Then the sequence of sets*

$$f^i(\mathfrak{N}(w)) \; , \; i = 1, 2, \ldots$$

is also $k-$periodic. If w and w^* are two distinct words in $\mathfrak{B}$, then the sets $\mathfrak{N}(w)$ and $\mathfrak{N}(w^*)$ are disjoint.

Lemma 3.2 follows directly from the definitions.

Lemma 13.3 **(a)** *Let w, w^* be distinct elements of $\mathfrak{B}$. Let $A \in \mathfrak{D}_\delta(\mathfrak{N}(w))$ and $B \in \mathfrak{D}_\delta(\mathfrak{N}(w^*))$. Then if $w_i \neq w_i^*$ for an infinite number of i,*

$$\overline{\lim}_{k\to\infty} H\left(\mathfrak{F}(f)^k A, \mathfrak{F}(f)^k B\right) \geq \min\{\sigma, \delta\} \; ,$$

where $\sigma = d_H(X_0, X_1)$.

(b) *Let the word $w \in \mathfrak{B}$ be $k-$periodic. Then*

$$\mathfrak{F}(f)\left(\mathfrak{D}_\delta(\mathfrak{N}(w))\right) \subseteq \mathfrak{D}_\delta(\mathfrak{N}(w)) \; .$$

If w, w^* are two distinct $k-$periodic words, then the corresponding sets $\mathfrak{D}_\delta(\mathfrak{N}(w))$ and $\mathfrak{D}_\delta(\mathfrak{N}(w^*))$ are disjoint.

Proof. Statement (a) follows immediately from the defined properties of a $\delta-$normal fuzzification, since supp $(\mathfrak{F} A)$ and supp $(\mathfrak{F} B)$ are disjoint and compact. Now, (b) will follow from the same properties and (b) of Lemma 7.2. □

Let us return to the proof of the theorem. To prove (I) it is sufficient to prove (12), that is $E(\mathfrak{F}(f)) \geq \gamma$. Let U_0 and U_1 respectively be neighbourhoods in $\mathfrak{D}(\Omega)$ of the sets $\mathfrak{D}_\delta(X_0)$ and $\mathfrak{D}_\delta(X_1)$. Since $X_0 \bigcap X_1 = \emptyset$ and are compact, the sets U_0, U_1 can be chosen to be disjoint. Introduce the open cover $\mathfrak{A}$ with elements U_0, U_1 and

$$U_2 = \mathfrak{D}(\Omega) \setminus \left(\mathfrak{D}_\delta(X_0) \bigcup \mathfrak{D}_\delta(X_1)\right)$$

Let w be a fixed word in $\mathfrak{B}(N)$ and let A_w be a member of $\mathfrak{D}(\mathfrak{N}(w))$. From the $\delta-$normal properties it follows that

$$\mathfrak{F}(f)^i A_w \in \mathfrak{D}_\delta(X_{w_i}) \ , \ i = 1, 2, \ldots, N \ .$$

Hence the unique element of the cover $\mathfrak{A}(\mathfrak{F}(f), N)$ which contains A_w is the element

$$U_0 \bigcap \mathfrak{F}(f)^{-1} U_{w_1} \bigcap \cdots \bigcap \mathfrak{F}(f)^{N-1} U_{w_n}.$$

So $\#\mathfrak{A}(\mathfrak{F}(f), N) \leq \#\mathfrak{B}(N)$ and (12) follows.

To prove (II), let us first construct a set $\mathcal{S}$ that satisfies (i) and (ii) of (LY2). Let $\mathcal{Q}$ be an uncountable set of words from $\mathfrak{B}$ such that no elements of $\mathcal{Q}$ are periodic, and each pair of distinct words from $\mathcal{Q}$ differ on an infinite number of letters. Let $\mathcal{Q}^*$ be the set of elements $A_w \in \mathfrak{D}(\Omega)$ each of which belongs to the disjoint sets $\mathfrak{D}_\delta(\mathfrak{N}(w))$:

$$\mathcal{Q}^* = \{A_w \in \mathfrak{D}_\delta(\mathfrak{N}(w)) : w \in \mathcal{Q}\}.$$

Then $\mathcal{Q}^*$ satisfies (i) and (ii) of (LY2) because of the property (a) of Lemma 7.3. The property (iii) can be shown to follow by a similar construction.

It remains to show the existence of $k-$periodic points of the map $\mathfrak{F}(f)$. Let k be fixed. From (b) of Lemma 7.3 there exist at least $\#\mathfrak{B}(k)$ disjoint invariant sets of the form $\mathfrak{D}_\delta(N(w))$. But sets of this type have the fixed point property [13] (see also the appendix) and consequently have fixed points of $\mathfrak{F}(f)^k$. The theorem is proved. □

13.8 NONDEGENERATE PERIODICITIES AND CHAOS

Consider the $t-$norm $W(x, y) = \max(x + y - 1, 0)$. The diagonal map is $\Delta_W(A) = \max(2A - 1, 0)$. Hence $\left[\tilde{f}_W A\right]^\alpha = f\left([A]^{(1+\alpha)/2}\right)$, and for all positive integers n

$$\left[\tilde{f}_W^n A\right]^\alpha = f\left([A]^{\alpha_n}\right),$$

where $\alpha_n = \Delta_W^{-n}(\alpha) = 1 - (1 - \alpha)/2^n$, since

$$\alpha_{n+1} = \frac{1 + \alpha_n}{2} = 1 - \frac{(1 - \alpha)}{2^n} \ .$$

Observe that the fuzzification associated with W is not $\delta-$normal because, if $A \in \mathfrak{D}_\delta(\Omega)$ we can in general only infer that $\tilde{f}_W A \in \mathfrak{D}_{\delta'}(\Omega)$, where $\delta' = \Delta_W(\delta)$. Nevertheless, we will use this fuzzification to construct a type of nondegenerate chaos on $\mathfrak{D}_1(\Omega)$. That is, not all orbits asymptotically become crisp sets.

Let Ω be the unit interval I and let

$$f(x) = \begin{cases} 2x, & \text{if } 0 \leq x \leq 1/2\ , \\ 2 - 2x, & \text{if } 1/2 < x \leq 1\ . \end{cases}$$

If we take $X = \mathfrak{D}_1([\frac{5}{18}, \frac{4}{9}])$, then $\tilde{f}_W(X) = \mathfrak{D}_1([\frac{5}{9}, \frac{8}{9}])$, $\tilde{f}_W^2(X) = \mathfrak{D}_1([\frac{2}{9}, \frac{8}{9}])$ and each of these sets is compact in $\mathfrak{D}_1(I)$. So, arguing as in [8], $\tilde{f}_W$ satisfies (LY2), and has *periodic sets* of any order. We show that in fact $\tilde{f}_W$ has periodic points, that is, periodic fuzzy sets, of any order and is consequently nondegenerate. Let k be any positive integer and let

$$p_k = \frac{2}{1+2^k}\ ,$$

so that $f^k(p_k) = p_k$. Consider the isosceles triangular fuzzy number A_k with "feet" at $x = p_k \pm \lambda$ where λ is chosen so that $[p_k - \lambda, p_k + \lambda] \subset [0, 1/2]$. In addition, choose λ so small that

$$f^{k-2}(p_k + \lambda/2^k) = \frac{2^{k-1}}{1+2^k} + \frac{\lambda}{2^2} < \frac{1}{2}\ .$$

Note that the α_k-level sets of A are

$$[A]^{\alpha_k} = [p_k - \lambda + \lambda\alpha_k, p_k + \lambda - \lambda\alpha_k]\ .$$

Write $\beta = \lambda(1-\alpha)$. Then

$$\begin{aligned} \left[\tilde{f}_W^k A_k\right]^\alpha &= f^k\left([p_k - \beta/2^k, p_k + \beta/2^k]\right) \\ &= \left[f^k(p_k - \beta/2^k), f^k(p_k + \beta/2^k)\right] \\ &= [p_k - \beta, p_k + \beta] \\ &= [A_k]_W^\alpha\ , \end{aligned}$$

and $f^k(A_k) = u_k$. This situation is in marked contrast to the case when the $\Gamma = \min$ fuzzification is used, where the only $k-$periodic points are singleton sets of the form $\{p_k\}$.

In this example there are just a countable number of periodic fuzzy sets. A lower bound of the actual number is given by (II) of Theorem 3.5.

13.9 EXAMPLES OF FUZZY CHAOS

13.9.1 Example 1.

This is due to Kloeden [22]. Consider the space $\mathfrak{D}^1([0,1])$, the set of all $A \in \mathfrak{D}^1$ with support $[A]^0 \subseteq [0,1]$. For each such A, there exist functions $a, b : [0,1] \to \Re$ such that the β–level sets of A are the intervals $[A]^\beta = [a(\beta), b(\beta)]$. Denote by $\mathfrak{T}$ the subset

$$\mathfrak{T} = \{A \in \mathfrak{D}^1 : a(\beta) = \beta b(0)/2, b(\beta) = (1 - \beta/2)b(0)\}$$

of $\mathfrak{D}^1$. Define the mapping $T : \mathfrak{T} \to \mathfrak{T}$ by

$$[TA]^\beta = \left[g\left(b(0)\right) a(\beta), g\left(b(0)\right) b(\beta)\right] ,$$

where $xg(x) : [0,1] \to [0,1]$ is the tent function θ,

$$\theta(x) = xg(x) = \begin{cases} 2x & \text{if } 0 \leq x \leq 1/2 \\ 2 - 2x & \text{if } 1/2 \leq x \leq 1. \end{cases}$$

The map T is obviously a H–continuous and δ–normal fuzzification of θ. Clearly, any $A \in \mathfrak{T}$ is completely determined by the value of $b(0)$, written b, and will be denoted by A_b. Then $T(A_b) = A_{\theta(b)}$.

Let $Y = \{b : 1/4 \leq b \leq 4/9\}$ and note that

$$\theta(Y) = \{b : 1/2 \leq b \leq 8/9\} ,$$

$$\theta^2(Y) = \{b : 2/9 \leq b \leq 1\} ,$$

so that $Y \cap T(Y) = \emptyset$ and $T^2(Y) \supseteq Y \cup T(Y)$, and the mapping T is erratic.

Now, for $A_b \in \mathfrak{T}$, $P_{A_b}(\beta) = (1 - \beta)b$ and so

$$\mathfrak{I}(A_b) = \int_0^1 \frac{b - b(1 - \beta)}{\beta} \, d\beta = b .$$

Hence, the information of $T^k A_b$,

$$\mathfrak{I}(T^k A_b) = \mathfrak{I}(A_{\theta^k(b)}) = \theta^k(b)$$

and the information in the fuzzy iterates $T^k(A)$ does not degrade. The information content varies in a chaotic fashion as $k \to \infty$, because the trajectory $\{\theta^k(b) : k = 1, 2, \ldots\}$ is chaotic for every $0 < b < 1$, but it does not continually decrease. Observe that, if the measure $I(\cdot)$ were used here, then $I(T^k A_b) = 1$ for $k = 1, 2, \ldots$.

We note finally that if $\mathcal{Y} = \{A_b \in \mathfrak{T} : 1/4 \leq b \leq 4/9\}$, then also $Y \cap T(Y) = \emptyset$ and $T^2(Y) \supseteq Y \cup T(Y)$.

13.9.2 Example 2.

Let $f : [0,1] \to [0,1]$ be given by $f(x) = 3.9x(1-x)$. Define $F : \mathfrak{D}^1([0,1]) \to \mathfrak{D}^1([0,1])$ by

$$[TA]^\beta = f([A]^\beta)\ .$$

It is not difficult to show that the whole of the interval $\Lambda = [.0950625, .975]$ is an attractor for the iterative system f^k on $[0,1]$. It follows that for any subinterval $J \subset [0,1]$, no matter how small, the images $f^k(J)$ expand to cover Λ as $k \to \infty$. Now consider a fuzzy set $A \in \mathfrak{D}^1$, and observe that for $0 \le \alpha \le \beta \le 1$ we have $[A]^0 \supseteq [A]^\alpha \supseteq [A]^\beta$. From the continuity of T , $[TA]^0 \supseteq [TA]^\alpha \supseteq [TA]^\beta$.

Let

$$Y = \mathfrak{D}^1([0.15, 0.45]) = \{A \in \mathfrak{D}^1 : [A]^0 \subseteq [0.15, 0.45]\}.$$

Then

$$\begin{aligned} T\left(\mathfrak{D}^1([.15, .45])\right) &= \mathfrak{D}^1([.497, .965])\,, \\ T^2\left(\mathfrak{D}^1([.15, .45])\right) &= \mathfrak{D}^1([.131, .975])\,. \end{aligned}$$

Consequently, the sufficient conditions (C1), (C2) are satisfied and a chaotic regime exists.

However, for this system information is lost as the system evolves with increasing k. To see this, for any $A \in \mathfrak{D}^1$ choose any β–level set, $[A]^\beta$, $0 \le \beta \le 1$. Then for k sufficiently large, $T^k([A]^\beta) = \Lambda$, and $\mathfrak{I}(T^k A) = 0$. Note how, in Figure 1, the iterates of the fuzzy set are driven towards a crisp set on Λ.

13.9.3 Example 3.

Let $h : \Re^2 \to \Re^2$ be the Hénon map

$$\begin{aligned} x_{k+1} &= 1.4 + 0.3\, y_k - x_k^2 \\ y_{k+1} &= x_k\ . \end{aligned}$$

Define $H : \mathfrak{D}^2(K) \to \mathfrak{D}^2(K)$ by $[HA]^\beta = h([A]^\beta)$, where

$$K = \{(x, y) \in \Re^2 : |x|, |y| \le 1.904\}\ .$$

It is well known that h maps K into itself, so the map H is well–defined. It seems certain (but is still an open question) that the Hénon mapping has a strange attractor Λ, and that each level set of $A \in \mathfrak{D}^2(K)$ is attracted to Λ. Consequently, information degrades as the chaotic iteration proceeds, since $h^k([A]^\beta)$ becomes close to Λ as k increases, that is the fuzzy set $H^k u$ becomes very like the characteristic function χ_Λ, whose information content is zero.

13.9.4 Example 4.

We take an example from [9], where chaos was defined on the Banach space of functions in which fuzzy sets were embedded. We restrict ourselves to the subspace $\mathcal{E}_L^2 \subset \mathcal{E}^2$ consisting of fuzzy convex members of $\mathfrak{D}^2$ such that the map $\beta \mapsto [A]^\beta$ is Lipschitz with respect to the Hausdorff distance, $d_H([A]^\alpha, [A]^\beta) \leq M|\alpha - \beta|$ for all $0 < \alpha, \beta \leq 1$ and some fixed M depending only on A. Then there exists an isometric embedding $j : \mathcal{E}_L^2 \to C([0,1] \times S^1)$, as a cone. Now, if θ is a chaotic mapping on the embedded cone in that function space, then T defined by $j \circ T = \theta \circ j$ is a chaotic mapping on $\mathcal{E}_L^2$. We construct such a mapping θ on the positive cone of $C(\{\beta\} \times S^1)$ for each $0 \leq \beta \leq 1$. For a fixed $\beta \in [0,1]$, the domain of each support function $s_{[A]^\beta}$ can be identified with the interval $[0, 2\pi)$, and we are essentially looking at the 2π–periodic functions. The support function s_A of A is given explicitly by, for each $p = (\cos t, \sin t) \in S^1$,

$$\widetilde{s}_A(\beta, t) \triangleq s_A(\beta, p) \triangleq s_{[A]^\beta}(p) = \max_{x \in [A]^\beta} (x \cos t + y \sin t)$$

for $0 \leq t < 2\pi$. For a given kernel $V(t, \tau)$, periodic in t, define θ by

$$(\theta \widetilde{s}_A)(\beta, t) = 3.9(1 - \widetilde{s}_A(\beta, t)) \int_0^{2\pi} V(t, \tau) \widetilde{s}_A(\beta, \tau)\, d\tau.$$

It is possible to show that, for suitable kernels V, the mapping θ^2 is chaotic on the unit ball of the function space. Hence, $T : \mathcal{E}_L^2 \to \mathcal{E}_L^2$, defined by

$$[j \circ T(A)]_\beta = j \circ T([A]^\beta) = \theta^2(s_A(\beta, \cdot))$$

is chaotic on $\mathcal{E}_L^2$.

13.10 CONCLUSION

- Fuzzy dynamical systems can be generated by either fuzzification of functions on the underlying base space or by intrinsic mappings on the various spaces of fuzzy sets.
- Quite different fuzzifications are produced by different t–norms and t–conorms and their general properties in terms of level sets were investigated.
- Chaos can arise from all these types of maps, provided certain conditions are met. Fuzzy chaotic mappings are most richly described in a *metric space* (like $(\mathfrak{D}^n, d_\infty)$).
- Several definitions of fuzzy chaos were shown to be linked and could each be produced from simple sufficient conditions.

- A measure of information content of fuzzy dynamical states was introduced and discussed.
- Chaos may or may not result in degradation of information along orbits, and examples were given of both circumstance.

13.11 APPENDIX.

Topological Entropy

Roughly speaking, a system is chaotic if the trajectories are mixing. Essentially, the Li–Yorke definition says that such mixing occurs, while topological entropy measures the speed of such mixing. So the system is chaotic if the topological entropy is positive.

Let p be a fixed positive integer and let Ω be a compact metric space. Suppose that

$$\mathfrak{A} = \{U_0, U_1, \ldots, U_N\}$$

be an open cover of Ω. Denote by $\mathfrak{A}_{\min}$ the subcover of $\mathfrak{A}$ which contains a minimum number of elements of $\mathfrak{A}$ covering Ω. Write $\#\mathfrak{A}_{\min}$ for the cardinality of the minimal subcover.

Definition A1. *The quantity*

$$E(\mathfrak{A}) = \ln\,(\#\mathfrak{A})$$

is called the **topological entropy** *of the cover* $\mathfrak{A}$.

Let $f : \Omega \to \Omega$ generate a dynamical system and suppose that $\mathfrak{A}$ as above is a fixed open cover of Ω. Then for any positive integer k, the class

$$\mathfrak{A}_{f,k} = \{f^{-k}(U_0), \cdots, f^{-k}(U_N)\}$$

is also an open cover of Ω. Define the open cover

$$\mathfrak{A}(f, K) = \left\{\bigcap_{k=1}^{K} V_k : V_k \in \mathfrak{A}_{f,k}, 1 \le k \le K\right\}.$$

Lemma A.1. *Let K, L be positive integers. Then*

$$\#\mathfrak{A}(f, K)\,\#\mathfrak{A}(f, L) \;\ge\; \#\mathfrak{A}(f, K + L)\;.$$

From this it follows that

Theorem A.2. *The following limit exists:*

$$E(f;\mathfrak{A}) = \lim_{K\to\infty} K^{-1}E(\mathfrak{A}(f,K)) .$$

Definition A.2. *The topological entropy of f is defined to be*

$$E(f) = \sup_{\mathfrak{A}} E(f;\mathfrak{A}) .$$

It can easily be shown that if $\mathfrak{A}_i$ is a sequence of open covers of Ω with $\operatorname{diam}\mathfrak{A}_i \to 0$, then $E(f;\mathfrak{A}_i) \to E(f)$.

There are relationships between the number of k−periodic points of a mapping and its entropy, especially if additional conditions are placed on f. One typical example is

Theorem A.3 . *If f is an expansive homeomorphism of a compact metric space, then*

$$E(f) \geq \overline{\lim}_k \frac{1}{k} \ln P_k(f)$$

where $P_k(f)$ is the number of fixed points of f^k. and the limit is finite.

Further details can be found in [4].

The Fixed Point Property

Definition B.1. *A metric space (Π,ρ) is said to have the* **fixed point property** *(f.p.p.) if any continuous mapping*

$$f : \Pi \to \Pi$$

has at least one fixed point.

It was shown in [13] that quite general classes of fuzzy sets have the f.p.p., provided that the topology of the sendograph metric is used. The conditions of fuzzy convexity and normality can be dropped and the totality $\mathfrak{D}(\Omega)$ of upper semicontinuous fuzzy sets on a compact metric space (Ω,ρ) can be shown to have the f.p.p. This is so even when Ω is not connected, in which case Ω itself does not even have the f.p.p.

A subset $\mathfrak{X}$ of Π is said to be a *retract* of Π if there exists a continuous mapping, called a *retraction*, $f : \Pi \to \mathfrak{X}$ such that $f(x) = x$ for all $x \in \mathfrak{X}$. The set $\mathfrak{X}$ is called an *absolute retract* if for any pair $\Pi_0 \subseteq \Pi$ of metric spaces and each continuous map $\phi : \Pi_0 \to \mathfrak{X}$, there exists a continuous map $\phi_* : \Pi \to \mathfrak{X}$ such that $\phi = \phi_*|_{\Pi_0}$.

The connection between retracts and the f.p.p. is given by the following two well–known theorems.

Theorem B.2. *If* Π *has the f.p.p. and* $\mathfrak{X}$ *is a retract of* Π, *then* $\mathfrak{X}$ *also has the f.p.p.*

Theorem B.3. *Every compact absolute retract has the f.p.p.*

The principal result of [13] is

Theorem B.4. *The metric spaces* $(\mathfrak{D}(\Omega), H)$ *and* $(\mathfrak{D}_\delta(\Omega), H)$, *for* $0 < \delta \leq 1$, *each have the f.p.p.*

This may be proved directly, or by showing that $(\mathfrak{D}(\Omega), H)$, $(\mathfrak{D}_\delta(\Omega), H)$ are each an absolute retract. By virtue of Theorem B.3, this also gives a proof that each has the f.p.p.

Proofs, references and further details may be found in [13].

Figure 13.4 Iterates of Example 8.9.2.

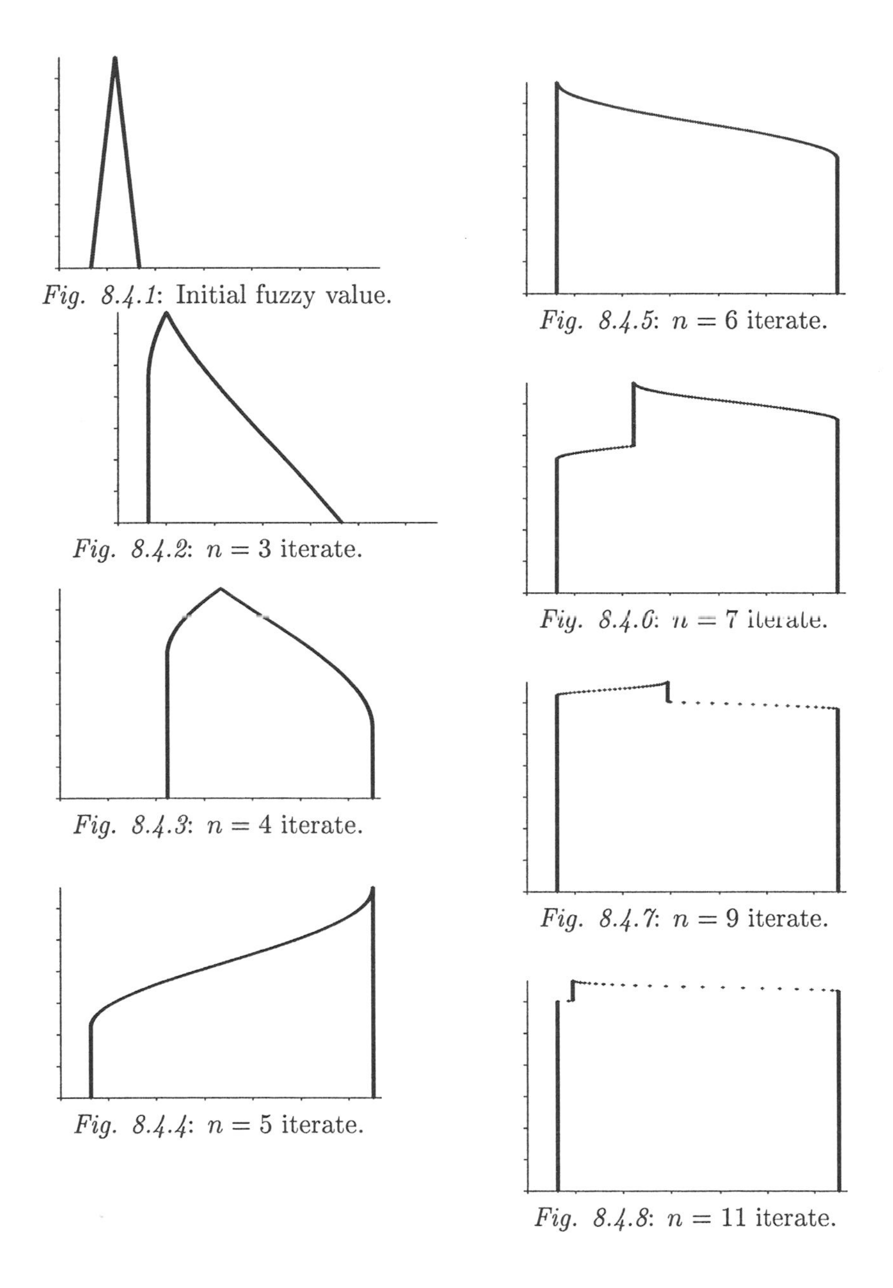

Fig. 8.4.1: Initial fuzzy value.

Fig. 8.4.2: $n = 3$ iterate.

Fig. 8.4.3: $n = 4$ iterate.

Fig. 8.4.4: $n = 5$ iterate.

Fig. 8.4.5: $n = 6$ iterate.

Fig. 8.4.6: $n = 7$ iterate.

Fig. 8.4.7: $n = 9$ iterate.

Fig. 8.4.8: $n = 11$ iterate.

References

[1] Banks, J., Brooks, J, Cairns, G., Davis G. and Stacey, P (1992) On Devaney's definition of chaos, *Amer. Math. Monthly* **99**, 332–334.

[2] Benhabib, J. and Day, R.H. (1981) Rational choice and erratic behaviour, *Rev. Econ. Stud.* **48**, 459–471.

[3] Blank, M.L. (1989) *Small perturbations of chaotic dynamical systems, Russian Math. Surveys* **44**:6, 1–33.

[4] Bowen, R. (1970) Topological entropy and Axiom A, *Proceedings of Symposia in Pure Mathematics,* Volume XIV, American Mathematical Society, Providence, Rhode Island.

[5] Crannell, A. (1995) The role of transitivity in Devaney's definition of chaos, *Am. Math. Month.*, **102**, 788–793.

[6] Debreu, G. (1967) Integration of correspondences, *Proc. Fifth Berkeley Symp. Math. Statist. Probability,* Vol. II, 351–372.

[7] Devaney, R.L. (1989) *An Introduction to Chaotic Dynamical Systems*, Addison - Wesley, New York.

[8] Diamond, P. (1976) Chaotic behaviour of systems of difference equations, *Intern. J. System Sci.* **7**, 953–956.

[9] Diamond, P. (1992) Chaos and fuzzy representations of dynamical systems, *Proc. 2nd International Conference on Fuzzy Logic & Neural Networks* (Iizuka, Japan, July 17–22, 1992), 51–58, FLSI, Iizuka, Japan.

[10] Diamond, P. (1987) Nonlinear integral operators and chaos in Banach spaces, *Bull. Austral. Math. Soc.* **35**, 275–290.

[11] Diamond, P. (1994) Chaos in iterated fuzzy systems, *J. Math. Anal. Applns.* **184**, 472–484.

[12] Diamond, P. and Kloeden, P. (1994) *Metric Spaces of Fuzzy Sets: Theory and Applications*, World Scientific, Singapore.

[13] Diamond, P., Kloeden, P. and A. Pokrovskii, A. (1996) Absolute retracts and a general fixed point theorem for fuzzy sets, *Fuzzy sets and Systems* to appear.

[14] Diamond, P. and Pokrovskii, A. (1994) Chaos, entropy and a generalized extension principle, *Fuzzy Sets and Systems* **61**, 277–283.

[15] Eckmann J.-P and Ruelle, D. (1985) Ergodic theory of chaos and strange attractors, *Rev. Mod. Phys.* **57**, 617–656.

[16] Fullér, R. and Keresztfalvi, T. (1991) On generalization of Nguyen's theorem, *Fuzzy Sets and Systems* **41**, 371–374.

[17] Hénon, M. (1976) A two-dimensional mapping with a strange attractor, *Commun. Math. Phys.* **50**, 69–77.

[18] Gupta, M.M. and J. Qi, J. (1991) Design of fuzzy logic controllers based on generalized T–operators *Fuzzy Sets and Systems* **40**, 473–489.

[19] Higashi, M. and Klir, G. (1983) Measures of uncertainty and information based on possibility distributions, *Int. J. Gen. Systems* **9**, 43–58.

[20] Kloeden, P. (1980) Compact supported endographs and fuzzy sets, *Fuzzy Sets and Systems* **4**, 193–201.

[21] Kloeden, P. (1981) Chaotic difference equations in $\Re^n$, *J. Australian Math. Soc. A* **31**, 217–225.

[22] P. Kloeden, P. (1991) Chaotic iterations of fuzzy sets, *Fuzzy Sets and Systems* **42**, 37–42.

[23] Li, T.Y. and Yorke, J.A. (1975) Period three implies chaos, *Amer. Math. Monthly* **82**, 985–992.

[24] Marotto, F.R. (1978) Snap–back repellers imply chaos in $\Re^n$, *J. Math. Anal. Appl.* **63**, 199–223.

[25] Nguyen, H.T. (1978) A note on the extension principle for fuzzy sets, *J. Math. Anal. Appl.* **64**, 369–380.

[26] Ott, E., Grebogi, C. and Yorke J.A. (1990) Controlling chaos, *Phys. Rev. Lett.* **64**, 1196–1199.

[27] Puri, M.L. and Ralescu, D.A. (1985) The concept of normality for fuzzy random variables, *Ann. Probab.* **13**, 1373–1379.

[28] Ramer, A. (1990) Concepts of fuzzy information measures on continuous domains, *Int. J. Gen. Systems*, **17**, 241–248.

[29] Shinbrot, T., Ott, E., Grebogi, C. and Yorke, J.A. (1990) Using chaos to direct trajectories to targets, *Phys. Rev. Lett.* **65**, 3215–3218.

[30] Teodorescu, H.N. (1992) Chaos in fuzzy systems and signals, *Proc. 2nd International Conference on Fuzzy Logic and Neural Networks* (Iizuka, Japan, July 17–22, 1992), 21–50, FLSI, Iizuka, Japan.

[31] Wiggins, S. (1988) *Global Bifurcations and Chaos*, Springer–Verlag, Berlin.

Index